MANUAL CS

Commercial Design Manual

Applications Systems and Equipment

Air Conditioning Contractors of America
1330 Braddock Place, Suite 350
Alexandria, VA 22314
PHONE: 703 • 575 • 4477
FAX: 703 • 575 • 4449
www.acca.org

Acknowledgments

Manual CS is a comprehensive guide for designing commercial, industrial and institutional HVAC systems.

The development and publication of this manual was commissioned by the **Air Conditioning Contractors of America** (ACCA). Much of the information in this manual was secured from, or based on information published in, the American **Society of Heating, Refrigerating and Air-Conditioning Engineers (ASHRAE)** handbooks, the **Sheet Metal and Air Conditioning Contractors National Association** (SMACNA) manuals and from various design manuals and engineering data manuals published by equipment manufacturers.

Grateful acknowledgment is made to those members of **ACCA** who contributed to the preparation of this manual.

This manual was written by **Hank Rutkowski,** P.E., ACCA Technical Director.

A special note of appreciation is extended to **Marion Wright** who produced the final art for the tables and drawings.

ACCA Manual CS Task Committee:

Clint Van Der Kolk
Metzger's Inc.
Kalamazoo, MI

Glenn Friedmann
Engineered Air Systems
Richmond, CA

James B. Isaac
Isaac Heating and Air Conditioning Inc.
Rochester, NY

Warren Lupson
Beltway Heating and Air Conditioning Inc.
Forestville, MD

Richard J. Pasterkamp
Pasterkamp Heating and Air Conditioning Co.
Denver, CO

Alfred A. Piff
A-C Contracting Co. Inc.
Mobile, AL

Randy Seaman
Seaman's Air Conditioning and Refrigeration Inc.
Grand Rapids, MI

Reprint 2017

Reader Response

ACCA is dedicated to providing its members and users of all ACCA Manuals with accurate, up-to-date and useful information. If you believe that any of the information contained in this manual is incomplete or inaccurate, if you have suggestions for improving the manual, or if you have general comments, we'd like to hear from you. Please write your comments below and return this form to:

Air Conditioning Contractors of America
1330 Braddock Place, Suite 350
Alexandria, VA 22314
PHONE: 703 • 575 • 4477
FAX: 703 • 575 • 4449
www.acca.org

Comments on ACCA's Manual CS:

____

____

____

____

____

____

Name ______________________________

Company ______________________________

Address

____

____

Phone ______________________________

Introduction

HVAC systems that do not provide adequate comfort and efficiency, and systems which are not cost effective, are common. Poorly conceived HVAC systems are found in older buildings and unfortunately, they are all too common in new buildings. In almost every case, chronically poor performance — in terms of comfort and efficiency — can be traced to system selection errors, inadequate design calculations and inappropriate design procedures. It is important to remember that after the system has been installed, any problems that were created by an inappropriate design concept or an incorrect design procedure will be difficult and expensive to correct. Therefore, a comprehensive engineering effort (on the original design) is always a good investment.

Building owners and system designers should appreciate that there are no "quick fixes" for upgrading older (obsolete or inadequately designed) HVAC systems. Retrofit designs normally take more engineering talent, effort, time and money than original designs. This added commitment occurs because retrofit work requires the same calculation effort that a new design would require; plus the additional effort that is associated with inspecting (and documenting) the control hardware, the control cycles and the various components of the existing system. Retrofit work also is more demanding and time consuming because the designer has to work out a series of compromises in order to salvage as much of the existing system as possible.

This manual focuses on the methods and procedures that should be used to design small commercial HVAC systems. The subject material includes information about system selection, system performance characteristics, equipment sizing and selection procedures and equipment performance characteristics. This manual also includes information about the HVAC system performance requirements that are associated with various types of buildings (applications).

Other ACCA manuals that pertain to small commercial HVAC system design include **Manuals N** (loads), **P** (psychrometrics), **Q** (ducts and fans), **T** (basic air distribution) and **H** (heat pumps). There will be additional small commercial design manuals published in the future.

Do not Use Residential Manuals for Small Commercial Applications

It is important to emphasize that the residential design procedures that are documented in ACCA **Manuals J**, **S** and **D** should not be used for small commercial design work. The following paragraphs list some obvious differences between the small commercial and residential design procedures and the equipment performance characteristics.

Zoning

It is common for commercial equipment to serve more than one zone. Most residential equipment is designed to serve a single zone.

Load Calculations

In residential work, the internal loads are relatively small and fairly constant. In commercial work, the internal loads due to lighting, people, appliances and equipment make up a much larger percentage of the total load. In addition, these loads usually vary in accordance with some type of operating schedule.

In residential work, the **Manual J** glass, wall and roof loads are based on average values that are assumed to be constant throughout the day. In small commercial work, the **Manual N** glass, wall and roof loads depend on the time of day. In **Manual N**, the solar loads are adjusted for latitude, but a latitude adjustment is not used in **Manual J**.

In commercial work, the exhaust systems are much larger and more complex than the exhaust systems that are found in residential structures. Commercial exhaust systems can have a large effect on infiltration loads and ventilation loads.

In **Manual N**, the designer must determine which hour of the day corresponds to the most demanding set of load conditions. **Manual J** does not require time of day calculations.

Psychrometrics

A psychrometric analysis is usually included in the commercial design procedure. The psychrometric chart is seldom used in residential work.

Entering Conditions

The condition (in terms of dry-bulb and wet-bulb temperature) of the air that enters commercial equipment is noticeably different than the condition of the air that enters a residential unit. This occurs because there are differences in the ventilation loads. Commercial buildings always require outdoor air for ventilation, but ventilation is not (commonly) used in residential work. Other factors that affect the entering condition are associated with appliances or process equipment that produce large latent loads and with the practice of using ceiling cavities as return air plenums.

Leaving Conditions
The condition of the air that is leaving a piece of commercial equipment is often quite different than the condition of the air that is discharged from a residential unit. This difference is usually associated with the latent load and with humidity control. (The larger range of room sensible heat ratios that are encountered in commercial applications translate into a wider range of leaving air temperatures.) In some commercial applications, reheat may be required.

Analysis of Part-Load Conditions
Commercial equipment selection requires an analysis of the part-load conditions. Heating and cooling equipment control strategies are based on knowledge of the minimum load condition. Heating and cooling equipment capacity may be staged or modulated. Compressors may be equipped with unloaders. Economizer cycles are often used because commercial buildings normally require cooling during cold weather. Economizer cycles have a significant effect on the minimum cooling load that is imposed on the compressor. Occasionally, hot gas bypass is required to prevent compressor short-cycling.

Limitations of Residential Equipment
Most residential equipment is designed for single-zone applications. It is not well suited for applications that have large ventilation loads and it is not available with economizer accessories. Low ambient mechanical cooling is available, but there are no capacity control options for cooling or heating (variable speed equipment is an exception). The supply fans cannot develop much pressure, usually less than 0.70 IN.WG. and return fans or exhaust fans are not available as an accessory option. The cooling capacity is limited to 3 or 4 TONS and the power supply is predominantly single phase.

Different Duct Sizing Procedures
Residential duct systems are designed to work with the fan that is supplied with the heating and cooling equipment. In this case, the fan pressure is relatively small and the designer can only make minor performance adjustments (by changing the fan speed). In commercial work, the duct system is designed without any specific knowledge of the fan performance. After the system is designed, the fan performance requirements are calculated and the performance parameters are specified. This insures that the fan performance will be compatible with the pressure and air flow requirements of the duct system. (Fan performance is controlled by specifying pulley sizes or RPM, and fan motor horsepower.)

Different Air Distribution Devices
Floor, ceiling and sidewall outlets are found in residential and commercial systems. Other types of devices and supply outlets such as reheat boxes, mixing boxes, VAV boxes, fan boxes, VAV outlets and slot diffusers are associated with commercial systems and they are not commonly installed in residential systems.

Different Control Strategies
Commercial systems have more operating modes than residential systems. This occurs because commercial systems require features like zone control, simultaneous heating and cooling capability, economizer cycles, day- and night-cycles and a wide range of heating-cooling changeover points. Commercial systems also require staged or modulating control instead of the simple on/off control that is used with residential systems.

Be careful About Using Small Commercial Calculation Procedures

It is also important to emphasize that there are some procedures that are used for large commercial, industrial and institutional work that go beyond the procedures that are documented in ACCA **Manuals N**, **P**, **Q**, **T**, and **H**. The following paragraphs list some of the obvious differences between small commercial applications and large commercial, industrial and institutional applications.

Zoning
Small commercial buildings are usually divided into a few zones that can be conditioned with packaged equipment. Multiple single- zone systems, packaged multizone systems, simple VAV systems, packaged variable volume variable-temperature-changeover systems, simple hydronic heating systems and radiant heating systems provide examples of the types of systems that are installed in small commercial buildings. Large, complex, multi-use buildings may require many zones. In some cases, two or more packaged systems can be used, but customized built-up systems are common. Dual duct systems, water-loop heat pump systems, complex VAV systems, large terminal reheat systems and complex hydronic heating systems fall into this category.

Equipment
Packaged reciprocating water chillers, reciprocating DX cooling units, small boilers, fossil fuel furnaces, electric furnaces, air- and water-source heat pumps, packaged terminal air conditioners and packaged terminal heat pumps provide examples of the type of equipment that can be found in small commercial buildings. The same type of equipment

that is used in small commercial buildings is also used in large buildings, but large central water chillers (usually centrifugal or absorption units) and large central boilers also are common.

Load Calculations
In some cases, the load calculation procedures that are associated with very large HVAC systems need to be more detailed than the **Manual N**, small commercial, load calculation procedures. For example, the ASHRAE load calculation procedures allow the designer to calculate the loads for any hour of the day and for any month of the year. The ASHRAE procedure also provides latitude corrections for solar loads, wall loads and roof loads. There are other differences between the ASHRAE procedure and the **Manual N** procedure, but they are mostly associated with details and secondary factors.

Different Duct Sizing Procedures
Small commercial duct systems are predominately low-velocity, low-pressure systems that are designed by the equal friction method or the extended plenum method. These systems also are used for large commercial applications. In addition, some large commercial designs require high-velocity, and medium or high-pressure, duct systems. These types of systems are usually designed by the static regain method or the constant velocity method.

Different Controls
The same basic control problems apply to both types of applications (small and large commercial). Most of the differences are related to the size and the number of systems that must be controlled; the number of zones, and the complexity of the occupancy and operating schedules. Small systems usually have electric and electronic controls and they are often equipped with microprocessor-based controls. Large systems are commonly equipped with pneumatic controls and/or electronic controls (with the associated microprocessors) and possibly, a central computer.

How to Use this Manual

This manual can be read from cover to cover, if the reader wants to become familiar with all of the information that pertains to small commercial system design. This manual also can be used as a reference manual. For example, a reader that is interested in all-air systems and all-electric equipment would not read the sections that cover water systems and fossil fuel equipment. However, the first five sections contain information that applies to any type of design. These sections, which are considered to be required reading, are:

Section 1 — Overview of the Design Procedure
Section 2 — System Selection
Section 3 — Applications by Building Type and Use
Section 4 — Applied Load Analysis
Section 5 — Fluid Temperature and Flow Rates

The information that is contained in the remaining sections can be used on an ad-hoc basis. These sections include:

Appendix 2 — Applications (information sheets about specific applications)
Section 6 — Cooling Equipment (air- and water-source)
Section 7 — Heating Equipment (except radiant heaters)
Section 8 — Heat Pump equipment (air- and water-source)
Section 9 — Radiant Heating Equipment
Section 10 — Humidification Equipment
Section 11 — All-Air Systems
Section 12 — All-Water Systems
Section 13 — Unitary Refrigerant-Based Systems

The scope of this manual is limited to providing information about applications, systems and small commercial equipment. But, the appendix material does contain supplemental information and it provides an overview of some of the other work that is associated with the design process. This material, which can be used on an ad-hoc basis, includes:

Appendix 3 — Air-Side Design (air outlets, returns, air-side devices, ducts and fans)
Appendix 4 — Piping Design (refrigerant, water, gas, oil and steam)
Appendix 5 — Refrigerant-Side Capacity Control
Appendix 6 — Venting (gas- and oil-fired equipment)

Table of Contents

Section 4

Applied Load Analysis

Section 5

Applied Load Analysis

Section 6

Selecting Cooling Equipment

Section 7

Selecting Heating Equipment

Section 8

Selecting Heat Pump Equipment

Section 9

Radiant Heat

Section 10

Selecting Humidification Equipment

Section 11

All-Air Systems

Section 12

Water Systems

Section 13

Unitary Refrigerant-Based Systems

Appendix 1

Charts, Tables and Equations

Appendix 2

Applications

Appendix 3

Air-Side Design

Appendix 3 — Continued

Air-Side Design

Appendix 4

Pipe Sizing

Appendix 5

Refrigerant-Side Capacity Control

Appendix 6

Venting Gas and Oil Heating Equipment

Section 1
Overview of the Design Procedure

Good design work is one of the most cost effective investments a building owner can make. The HVAC system designer should make sure that the building owner or manager understands this point. This section provides a brief overview of the small commercial design procedure. These procedures will be discussed in detail in the following sections.

1-1 General Comments

The design procedure consists of selecting an HVAC system and performing the calculations and procedures that are required to size and locate the HVAC equipment and associated components. Specification of controls and control cycles also is part of the design process.

One subtle point about HVAC design that is often overlooked or unappreciated, is that the design calculations are not a series of independent, uncoupled procedures. Actually, the design calculations are interrelated; therefore, they must be performed (somewhat) simultaneously. Good designers instinctively keep an overview of the entire design problem in the back of their minds, even when working on small details. This cross referencing process is continuously applied throughout the calculation procedure as the designer juggles the requirements and parameters that are associated with the particular application. This "systems approach" results in a fully inte-

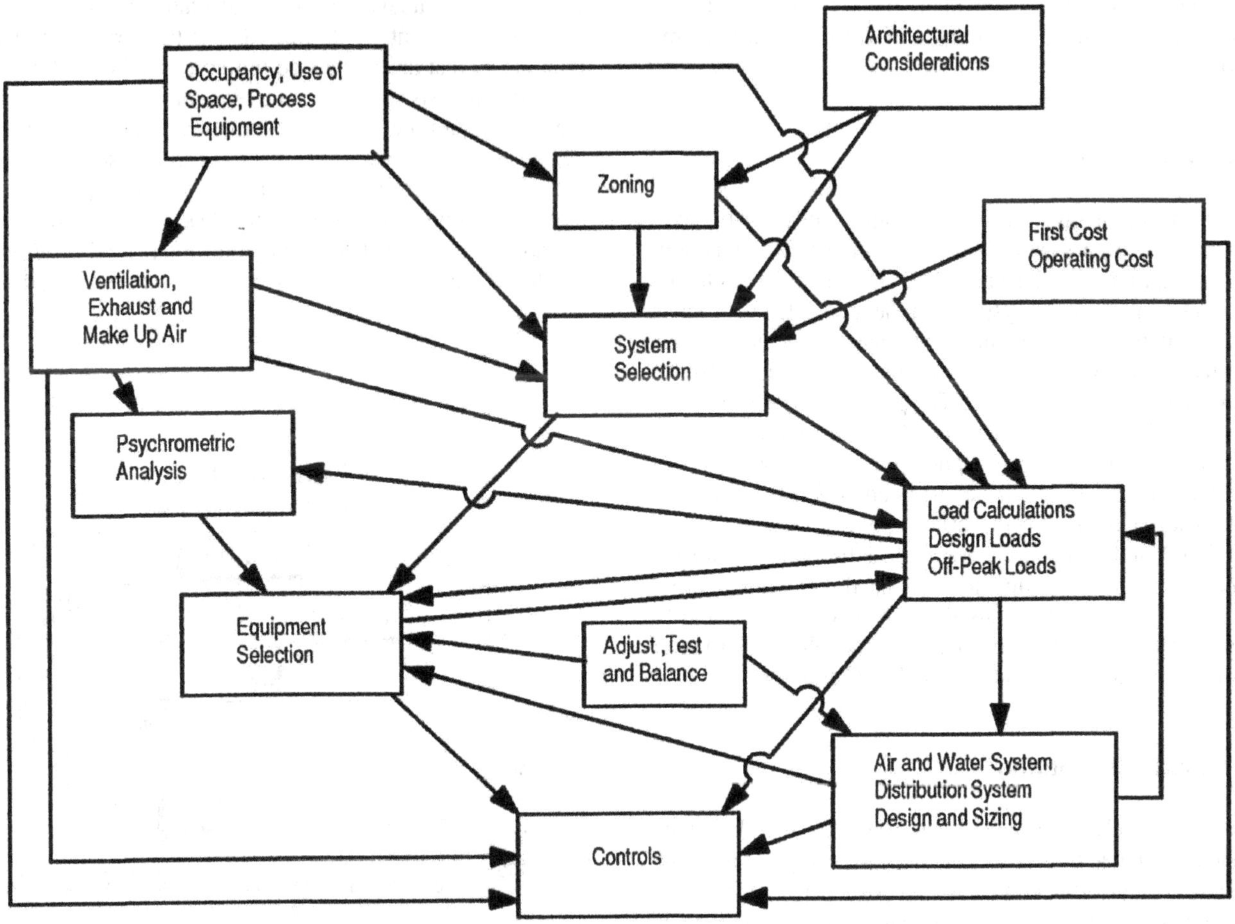

System Design Calculations are Interrelated

grated design that will be compatible with the constraints and requirements that are associated with the building that is being conditioned.

For example, a load calculation is affected by zoning requirements, system type, outdoor air requirements, duct and piping losses, fan heat, water pumping losses, and in some cases, the influence of a return air plenum ceiling. In principle, the load calculation cannot be made until the system is completely designed, but the system cannot be designed until the load calculation is made. This dilemma requires that some assumptions be made before proceeding with the load calculation. If these assumptions are correct, the load calculation will be accurate. If the assumptions are not correct, new assumptions must be made and the loads must be recalculated.

Note that the "systems" concept is especially important in retrofit work because the alterations must meet operating cost goals and conform to budget constraints. In addition, the retrofit work must not severely disrupt the owners operating schedules. If the retrofit design is done correctly, the redesigned system should provide both comfort and economy, with no surprise maintenance or operating problems. Lack of attention to the possible effects of the retrofit work, as it pertains to economy and performance, can produce disappointing results. This is why many quick fix, bolt-on, black-box solutions result in inadequate comfort, marginal economy and damaged equipment.

1-2 System Selection

The first step in any design is to decide exactly what type of HVAC system or systems will be required and what type of HVAC equipment will be installed. The performance characteristics of the HVAC system and the associated equipment must match a set of requirements that are dictated by the thermal characteristics of the building envelope and by the way the building will be used. The physical characteristics of the HVAC system (or systems) and equipment must be compatible with the architecture and the physical constraints of the building envelope. Refer to Section 2 for a complete discussion of the factors that affect system selection. Refer to Section 3 and Appendix 2 for more information about the specific requirements that are associated with various types of buildings and applications. Refer to Sections 11, 12 and 13 and to page 2-10 for information about the performance characteristics of various types of HVAC systems.

1-3 Load Calculations

The load calculation is the single most important task that the system designer must perform, because every facet of the design is somehow related to the load calculation. Load information is required to select heating and cooling equipment, chillers, boilers and water towers. Load information is also required to select fans and pumps, or to design air distribution systems and hydronic systems. The load calculations also provide important information that is required to design the control system.

The load analysis usually begins with calculations that determine the design loads. There are two general types of design loads to consider—space loads and equipment loads. Equipment loads are further classified as zone loads and block loads. Depending on the type of HVAC system, the equipment capacity may be based on a peak zone load, the sum of a group of peak zone loads or a peak block load. Refer to Section 4 for a discussion of applied load analysis. Refer to ACCA **Manual** N for details about the load calculation procedure.

1-4 Part-Load Conditions and Load-Line Analysis

Design loads are only experienced for a few hours per year. Part-load conditions are in effect for the vast majority of the equipment operating hours. This means that the HVAC equipment is subjected to continuously changing loads, and the equipment must be able to track these load changes. Therefore, an analysis of the heating and cooling loads, at part-load conditions, is an important design requirement.

Classical load calculations — the calculations that are used to size the equipment — do not address the performance requirements that are associated with part-load conditions. A load-line analysis is required to analyze the "off-design" loads that will be imposed on the heating and cooling equipment. The results of this load-line analysis can be used to specify the devices and control cycles that protect the equipment or allow the equipment to track the load changes. Refer to Section 4 for a complete discussion of part-load conditions and for a complete explanation of how to perform a load-line analysis.

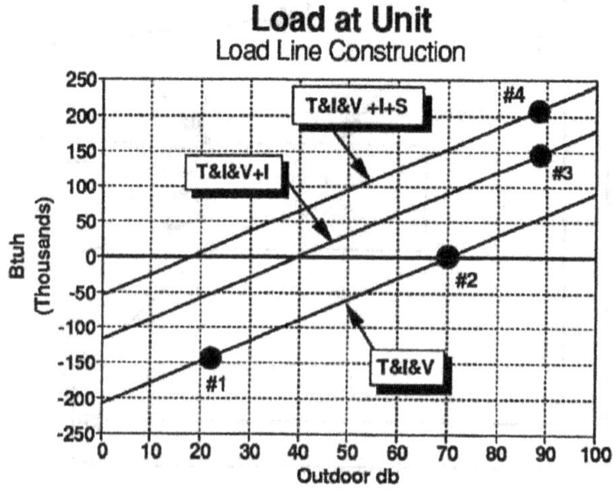

Load-Line Analysis

1-5 Psychrometric Analysis

The psychrometric chart should be used to determine the room sensible heat ratio (RSHR) and the coil sensible heat ratio (CSHR). The psychrometric chart also should be used to determine the condition of the air — in terms of dry-bulb (DB) and wet-bulb temperature (WB) — entering the cooling coil and the design condition (WB and DB) of the air leaving the coil. Refer to Section 4, for a brief review of psychrometric principles. Refer to ACCA **Manual P** for a complete discussion of psychrometrics.

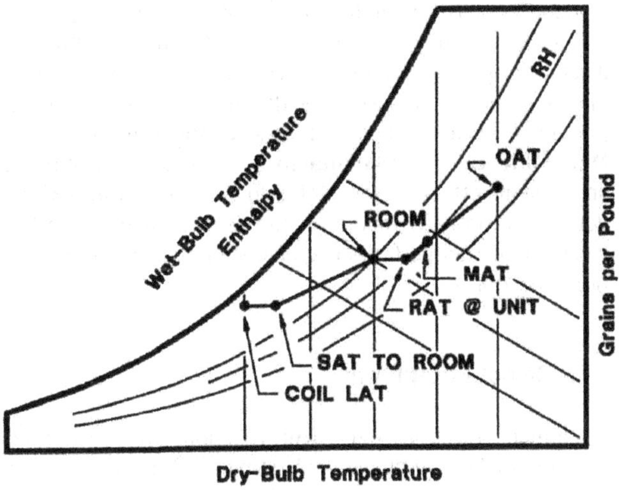

Psychrometric Analysis

1-6 Temperature of Heating and Cooling Fluids

Comfort is affected by fluid temperatures. The supply air discharged from any type of cooling coil must be cool enough and dry enough to offset the sensible and latent loads that are associated with the conditioned space. The supply air off any type of heating coil or heat exchanger must be hot enough to offset the heating loads that are associated with the conditioned space. Note that the transport capacity of the heating and cooling fluids is related to flow rate (CFM or GPM).

Equipment performance is affected by fluid temperatures. The entering and leaving conditions of the air flowing through a cooling coil, heating coil or heat exchanger affect the capacity of the device. Hot water and chilled water coil capacity is also affected by entering and leaving water temperatures. Chiller and boiler performance is affected by entering and leaving water temperatures. Air temperatures and water temperatures may be limited by the operating range of the equipment. For example, temperatures that are too cold can cause frost and icing problems and temperatures that are too hot can damage equipment.

Refer to Section 5 for a more detailed discussion of the design procedures, as they pertain to fluid temperatures. Also refer to manufacturer's engineering data for information about the performance characteristics and operating range limitations that are associated with any particular device or piece of equipment.

1-7 Flow Rates for Heating and Cooling Fluids

Comfort is affected by flow rates. The supply CFM discharged from any type of cooling coil must be sufficient to offset the sensible and latent loads that are associated with the conditioned space. The supply CFM discharged from any type of heating coil or heat exchanger must be sufficient to offset the heating loads that are associated with the conditioned space. Also note that the transport capacity of the heating and cooling fluids is related to the temperature of the fluid.

Equipment performance is affected by fluid flow rates. The CFM of air flowing through a cooling coil, heating coil or heat exchanger has an effect on the capacity of the device. Hot water and chilled water coil capacity also is affected by the GPM of water flowing through the device, as is chiller and boiler performance. Note that flow rates may be limited by the operating range of the equipment. For example, direct expansion (DX) coils usually have a minimum CFM requirement. Flow rates also may be limited by the air-side or water-side pressure losses that are associated with a device.

Refer to Section 5 for a more detailed discussion of the design procedures as they pertain to air-side and water-side flow rates. Also refer to manufacturer's engineering data for information about the performance characteristics and operating range limitations that are associated with any particular piece of equipment.

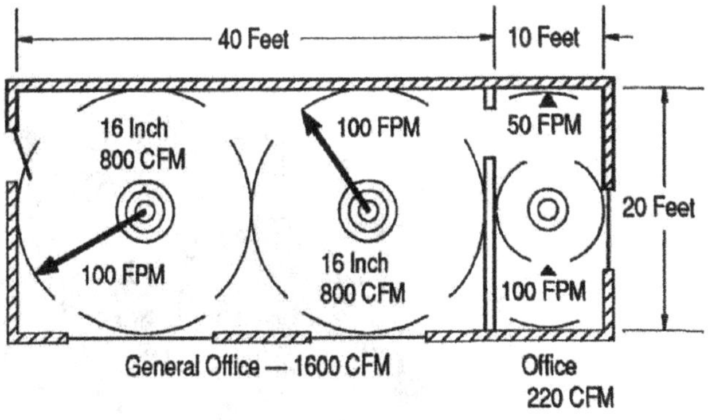

1-8 Air Distribution

Proper air distribution is required to insure comfort. Problems with drafts, stagnate air and noise are closely related to the style, size and location of the supply air outlets. Returns also are important. Some method of return is required for every isolated room and all returns must have ample face area. However, the location — in terms of whether they are near the ceiling or near the floor — of the returns is not critical. Refer to Appendix 3 for a brief summary of the design work that is related to air distribution. Refer to ACCA **Manual T** for a complete discussion of air distribution theory and applications. Refer to manufacturer's engineering data for information about the performance of specific devices.

1-9 Ventilation and Exhaust System Requirements

In commercial work, outdoor air is always required to provide fresh air for the occupants. Outdoor air also may be required to replace air that is exhausted from the conditioned space or to pressurize the conditioned space. When outdoor air is introduced at the return side of the air-handling equipment, it is commonly referred to as "ventilation." The ventilation and exhaust requirements always have a direct effect on the air balance in the conditioned space and the air balance at the air-handling equipment. Refer to Appendix 3 for a brief summary of the design work that is related to the exhaust, ventilation and air balance requirements. Refer to ACCA **Manual Q**, Section 17 for a complete discussion of air-side design as it pertains to building pressurization and air balance. Refer to **Manual Q**, Section 18 for information about exhaust system design.

1-10 Supply, Return and Exhaust Ducts

Duct system sizes are usually determined by applying the equal friction method. In some cases, a small duct system may take the form of an extended plenum. Use the ACCA Duct Sizing Slide Rule for this work. Refer to Appendix 3 for a brief summary of the design work that is related to duct sizing. Refer to ACCA **Duct Slide Rule User Manual** for a quick summary of the equal friction design procedure. Refer to ACCA **Manual Q** for detailed information about all aspects of small commercial duct sizing procedures.

1-11 Fan Performance Requirements

In commercial work, the fan performance — in terms of CFM, static pressure, total pressure, RPM and horse power — is specified after the duct system has been designed. Most of this work is associated with calculating the pressure losses that are associated with the straight runs, fittings, air-side components and air-side devices. The system effect at the fan inlet and discharge opening also must be considered. Refer to Appendix 3 for a brief summary of the design work that is related to pressure loss calculations and fan selection. Refer to **ACCA Duct Slide Rule User Manual** for a brief discussion of fitting and system effect losses. Refer to ACCA **Manual Q** for detailed information about all the aspects of pressure loss calculations, fan performance characteristics and fan selection procedures.

1-12 Select Equipment

Equipment selection begins with function. The basic functions of HVAC equipment are heating, cooling, and ventilating. Equipment may be designed to provide one, two or all three of these functions.

The equipment configuration also must be considered. The designer must evaluate the advantages and disadvantages

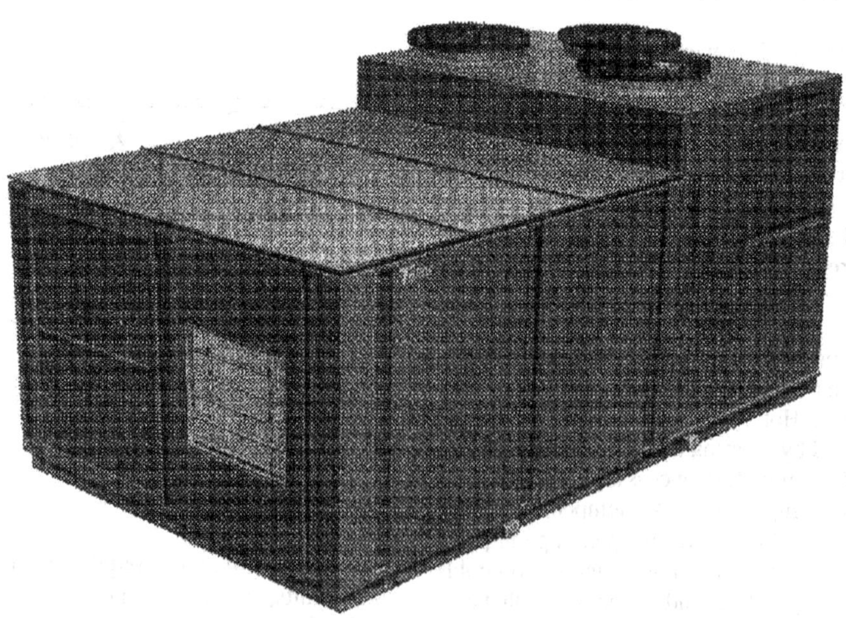

Packaged Rooftop Heating and Cooling Equipment

associated with central built-up equipment, central packaged equipment or distributed unitary equipment. The size, weight, serviceability and appearance of the equipment must be compatible with the spatial constraints and architectural features of the structure.

Operating characteristics also are important. Obviously, the equipment must use the type of fuel that is available at the site. If heat sinks and/or heat sources are required, they must be compatible with the codes, regulations and the physical conditions that prevail at the site. The efficiency of the equipment must be compatible with prevailing codes, standards and operating cost goals. It also is necessary to specify all the accessory items that will be required for installation, operation and safety.

Zoning capability and the distribution fluid are additional considerations. Obviously, the zoning capability of the equipment must match the zoning requirements of the structure. Air, water, steam and hydrocarbon refrigerants are commonly used as distribution fluids.

Also consider the method that is used to control capacity of the distribution fluid. Some equipment is designed to maintain a constant flow rate, while the fluid temperature varies. Other equipment operates at a constant temperature, with a variable flow rate. And, some equipment is designed to adjust both the flow rate and the fluid temperature.

The capacity of the cooling or heating equipment must be equal to or greater than the loads that are imposed on the equipment. For cooling equipment, the sensible and latent capacity must be greater than the sensible and latent loads. The capacity of heating equipment must be greater than the heating load. Whenever manufacturer's capacity data is used, it is necessary to make sure that the data corresponds to the same set of operating conditions that will be encountered on the site.

Refer to Section 2 for a general discussion of the factors that are listed above. Also refer to Sections 6, 7 and 8 for information about selecting cooling equipment, heating equipment and heat pumps.

1-13 Water-Side Design

Water piping is required for chilled water systems, hydronic heating systems and for water-loop heat pump systems. This design work is related to selecting a piping arrangement, pipe sizing, pressure loss calculations, pump selection, specifying hydronic accessories and specifying operating and safety controls. Refer to Appendix 4 for a short discussion of this subject. Refer to manufacturer's engineering data for information about the performance of specific pieces of equipment.

1-14 Refrigerant Piping

Refrigerant piping is required for split DX systems. This work consists of sizing suction lines, hot gas lines and liquid lines. In some cases it may be necessary to design risers. It also is necessary to provide traps and refrigerant-side accessories. Refer to Appendix 4 for a short discussion of this subject. Refer to manufacturer's engineering data for information about designing refrigerant lines for a particular piece of equipment.

1-15 Gas, Propane and Oil Piping and Venting

Piping is required to connect fossil fuel equipment with storage tanks or utility distribution systems. This design work is related to selecting a piping arrangement, pipe sizing and specifying operating and safety controls. If fuel oil is required, the work includes piping pressure loss calculations and pump selection. If oil or propane is required, a storage facility must be designed. All fossil fuel equipment must be properly vented. Refer to Appendix 4 for information about pipe system design. Refer to Appendix 6 for information about venting gas- and oil-fired equipment. Also refer to manufacturer's engineering data for information about piping and venting fossil fuel equipment.

1-16 Specify Control Cycles

Controls may be electric, electronic or pneumatic but the control cycles are generic. The designer must specify how each operating device should perform during every possible operating condition. The designer also must specify how safety controls should perform. (Comments about control cycles are made throughout this manual.) Refer to Appendix 5 for information about refrigerant-side capacity controls. Refer to manufacturer's engineering data for information about packaged control options.

Section 2
System Selection

Each building and each zone within a building has a set of unique design requirements. Some of these requirements are functional and others are financial. The HVAC system or systems that are installed in a building must satisfy these requirements. Usually, some of these requirements are in conflict with each other.

Conflicting requirements make system selection somewhat of an art. The designer must consider all the design requirements and the relationships between those requirements, then specify a system that makes the best compromise between the competing requirements. For example, a requirement for precise control of temperature and humidity usually conflicts with a requirement for low operating cost. Zoning decisions may be compromised by limitations on first cost. However, there are limits to performance-related compromises. Targeted efficiency ratings and economic goals become irrelevant if system performance fails to satisfy the minimum functional requirements. Fuel source, owner preference, esthetics and space restrictions also constrain the system selection options.

2-1 The Design Process

The design process can be represented by an inverted tree. The tree begins with a general system concept that branches to secondary system concepts that branch to the required design procedures. The design procedures are then divided into ever smaller and more detailed tasks. Subjective conceptualization decreases and quantitative evaluation increases as the branching process continues. As the work progresses, the

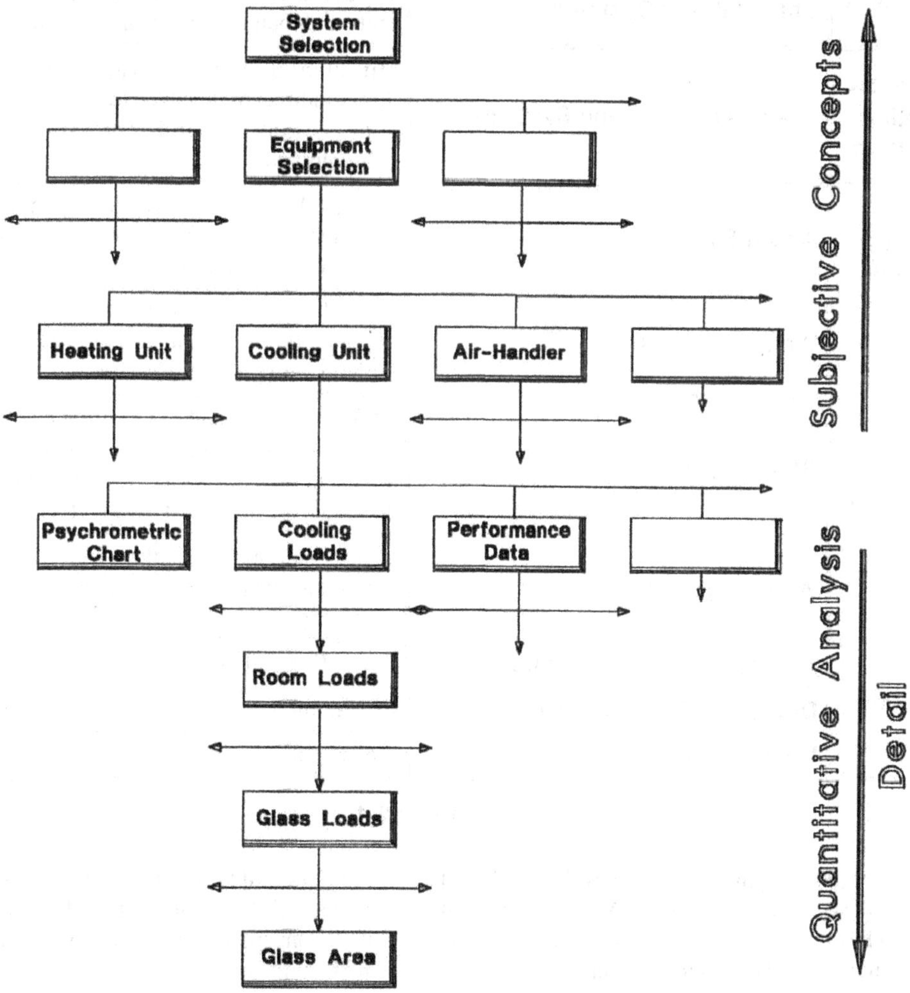

System Functions and Equipment		
Ventilation Only	**Power Exhaust Systems**	Propeller fans, exhaust fans, exhaust hoods
	Gravity Systems	Relief dampers, gravity vents
Heating Only	**Forced Air Systems**	Electric or fossil fuel furnace with no outdoor air capability; electric, hot water or fossil fuel unit heaters, duct furnaces
	Convective Heating Systems	Electric or hot water baseboard and cabinet convectors
	Radiant Heating Systems	Electric radiant heating panels, hot water or electric floor radiation, high intensity electric or fossil fuel heating fixtures, low intensity gas-fired heating pipes
Heating and Ventilation	**Forced Air Systems**	Electric or fossil fuel make up air units, electric or fossil fuel furnace with outdoor air capability, central station air handlers with outdoor air capability
	Hot Water Systems	Distributed fan-coil units with outdoor air capability, unit ventilators
Cooling and Dehumidification, no Ventilation	**All-Air Systems**	Unitary or split DX equipment with no outdoor air capability
	Chilled Water Systems	Central station air handlers with no outdoor air capability, fan-coil units with no outdoor air capability
Cooling, Dehumidification and Ventilation	**All-Air Systems**	Unitary or split DX equipment with outdoor air capability, packaged terminal air conditioners (PTAC)
	Chilled Water Systems	Central station air handlers with outdoor air capability, fan-coil units with outdoor air capability, unit ventilators
Cooling, no Dehumidification and Ventilation	**Evaporative Cooling Systems**	Direct and indirect evaporative cooling equipment; direct/indirect multistaged evaporative cooling equipment. (In some situations indirect units may provide partial dehumidification.)
Heating, Cooling, Dehumidification; no Ventilation	**All-Air Systems**	Package heating and DX cooling equipment with no outdoor air capability, unitary or split heat pump equipment with no outdoor air capability
	Water Systems	Central station air handlers with no outdoor air capability, distributed fan-coil units, water-loop heat pump equipment with no outdoor air capability
Heating, Cooling, Dehumidification and Ventilation	**All-Air Systems**	Package heating and DX cooling equipment with outdoor air capability, unitary or split heat pump equipment with outdoor air capability, packaged terminal air conditioners (PTAC), packaged terminal heat pumps (PTHP).
	Water Systems	Central station air handlers with outdoor air capability, unit ventilators, water-loop heat pump equipment with outdoor air capability
Humidification	**Ducted — Steam Systems**	Steam Humidifiers
	Ducted — Water Systems	Pan, atomizing and wetted element humidifiers
	Nonducted Systems	Steam, pan, atomizing and wetted humidifiers

Table 2-1

designer gains greater insight about how well the design concept satisfies the design requirements. All system-related decisions should be continuously evaluated and revised during the design process. If the design concept fails to satisfy even one important design requirement, the system performance will be unsatisfactory. Errors associated with the design concept or the design procedure can be costly and difficult to correct.

2-2 Classification of Systems

Classification is an important part of the system selection process because classification summarizes system performance characteristics and attributes. HVAC systems can be classified by function, distribution fluid, the method that is used to control the capacity of the distribution fluid, configuration, zoning capability, type of fuel, heat sink and/or heat source.

Function

The basic functions provided by HVAC systems are heating, cooling, dehumidification, humidification and ventilation. Some systems provide just one of these functions — usually heating — while others may provide two, three, four or all five of these functions. In some cases, two or more separate systems may be required to provide the necessary functions. Table 2-1, on the previous page, summarizes the functions that are provided by various types of systems.

Distribution Fluid

Distribution fluids are used to transfer heat between central equipment and the conditioned spaces. Air, water, refrigerants and steam can be used as distribution fluids. Most systems use one distribution fluid, but some systems use two distribution fluids. In some cases, two separate systems may be required to provide the necessary functions. Some typical "distribution fluid" classifications are listed below.

All-air systems rely on fans and ducts to distribute hot and cold air to the conditioned spaces. The terminal equipment in the conditioned space may consist of a simple constant volume supply outlet or some type of modulating supply outlet. Outdoor air is introduced at the central equipment. All-air systems, which are very common, are well suited for small commercial buildings.

All-water systems rely on pumps and pipes to distribute hot and cold water to the conditioned spaces. The terminal equipment in the conditioned space may consist of a simple convector, unit heater, fan-coil unit or unit ventilator. Outdoor air can be introduced at the terminal equipment if that equipment consists of a unit ventilator or a fan-coil unit that has outdoor air capability. A separate ventilation system is required if the terminal equipment does not have outdoor air capability. All-water cooling systems, or combination heating and cooling systems, are more suited for large buildings. All-water heating systems are suited for both large and small commercial buildings.

Water-loop heat pump systems are a variation of the classical all-water system. These systems rely on a closed water loop to distribute heat to, or to remove heat from, the conditioned spaces. The terminal equipment, which is located in the conditioned space consists of a water-source heat pump that uses the water loop as a heat source and as a heat sink. The loop water also is routed through a boiler and an evaporative cooler to maintain a heat balance during any operating condition. Outdoor air may or may not be introduced at the terminal heat pump. A separate ventilation system is required if the terminal heat pump equipment does not have outdoor air capability. Water-loop systems are suited for medium and large buildings that require simultaneous heating and cooling for a large number of hours per year. This system can be used in small commercial buildings if an economic evaluation indicates that the system is more cost effective than one of the more traditional systems.

Air-water systems rely on fans, pumps, ducts and pipes to distribute (hot and cold) air and water to the conditioned spaces. The terminal equipment in the conditioned space may consist of an especially designed induction unit or a fan-coil unit. Both air and water are supplied to the terminal unit. Outdoor air is introduced at the central equipment. Air-water heating and cooling systems are suited for large buildings, but they are seldom, if ever, used for new construction. Air-water systems are not a good choice for small commercial buildings.

Distributed DX systems rely on refrigerant lines to supply refrigerant to remote terminal equipment located in or near the conditioned spaces. The terminal equipment may consist of a DX fan-coil cooling unit or a heat pump fan-coil unit. Outdoor air may or may not be introduced at the terminal equipment. A separate ventilation system is required if the terminal equipment does not have outdoor air capability. Distributed DX systems are not usually a good choice for large buildings, but they are often used to condition small commercial buildings.

Steam systems rely on pressure differences and pipes to distribute steam to the heated spaces. The terminal equipment in the conditioned space may consist of a simple convector, unit heater, fan-coil unit or unit ventilator. Outdoor air can be introduced at the terminal equipment if that equipment consists of a unit ventilator or a fan-coil unit that has outdoor air capability. A separate ventilation system is required if the terminal equipment does not have outdoor air capability. Steam heating systems are more suited for large buildings. All-air systems or hot water heating systems would be a much better choice for small commercial buildings.

Capacity of Distribution Fluid

The heating or cooling capacity of distribution fluid can be controlled by varying the supply temperature of the fluid or by varying the flow rate. Some systems control capacity by varying both the temperature and the flow rate. These performance characteristics are commonly used to classify all-air systems as constant volume, variable temperature systems; constant temperature, variable volume systems or variable volume, variable temperature systems. Similar classifications apply to all-water systems, which can be classified as constant flow, variable temperature systems; constant temperature, variable flow systems or variable flow, variable temperature systems.

Configuration
Central all-air systems can be built up or packaged. Built-up systems are typically found in large buildings and they are usually associated with central chillers and boilers. Packaged systems are available in a wide range of sizes and they are used in all types of buildings. Packaged systems usually have an integral heating plant and/or cooling plant. Central systems can provide some, or all, the functions and distribution fluid options that are described by Table 2-1. Central systems can satisfy any of the zoning requirements described below in Section 2-3.

Distributed systems consist of a combination of central equipment and satellite equipment. For example, an all-water system may have a central chiller or boiler that serves a collection of terminal devices. (In large buildings the terminal equipment could consist of one or more large central air handlers.) A central condensing unit that serves two or more remote DX fan-coil units is another example of a distributed system.

Unitary systems are simply a collection of packaged, single-zone units. Each unit has an integral heating plant and/or cooling plant, fan or fans, dampers, filters and controls. Unitary equipment can be further classified as outdoor equipment and indoor equipment. Indoor cooling equipment must have a water-cooled condenser or a remote air-cooled condenser. Unitary equipment can be large — as demonstrated by some single-zone rooftop units — or small, as demonstrated by a a simple window air conditioner. Air- and water-source heat pumps and PTAC equipment are also considered to be unitary equipment.

Zoning
HVAC systems can either be classified as single-zone systems or multizone systems. A multizone system may accommodate as few as two zones or it could be required to provide service to dozens of zones. Depending on the way the equipment is designed and the size of the zone loads; a single, packaged, multizone unit can accommodate 6 to 14 zones. Built-up central multizone systems can accommodate 20 or more zones. Dual-duct systems and VAV systems can accommodate an unlimited number of zones. Systems that serve perimeter zones are called perimeter systems. Systems that serve interior zones are called interior systems.

Fuel
The fuel options for heating equipment are normally limited to natural gas, oil, propane and electricity. Occasionally live steam is available from a district heating system. Small commercial cooling systems are usually electric. Gas-fired cooling equipment is also available, but it is not nearly as common as electric cooling equipment. Absorption cooling equipment is occasionally installed in large buildings.

Heat Source and Heat Sink
Cooling equipment is usually classified as air-cooled or water-cooled. Air-cooled equipment expels heat directly to the outdoor air. Water-cooled equipment expels heat to the water that flows through the condenser. Water-cooled equipment usually requires a cooling tower. A "once-through" water-cooled system can be installed, providing that water is plentiful, inexpensive and unprotected by codes and regulations.

Heat pump equipment is classified by its heat source and by its heat sink. By convention, the fluid that flows through the outdoor unit is named first and the fluid that flows through the indoor unit is named second. The most common types of heat pump equipment are classified as air-to-air and water-to-air. Water-to-water and air-to-water heat pumps also are available.

2-3 Factors that Affect Zoning Decisions

Zoning is one of the most important factors that affect system selection. The zoning requirements must always be analyzed before any type of system can be specified. There are many reasons for creating a zone. The factors that affect zoning decisions are discussed below.

Different Heating and Cooling Functions
Interior rooms, and perimeter rooms that have large internal loads usually require cooling, but they may never require heat. Perimeter rooms, or some interior rooms that have an exposed roof or floor, usually require cooling and heating. Some spaces, a storage room for example, may just require heat. Rooms with similar heating and cooling requirements can be grouped into a zone. Rooms with different heating and cooling requirements should not be grouped into a zone.

Different Exposures
If a conditioned space is subject to solar gains, the heating and cooling loads will vary according to the month of the year and the hour of the day. The HVAC system must be designed to track these monthly and hourly load variations. Interior rooms, which are not affected by solar gains, can be grouped together. Perimeter rooms, and interior rooms that have an exposed surface, must be grouped by exposure.

Perimeter rooms should be grouped according to the compass direction of the exterior walls. Corner rooms that have a large amount of glass area on both exterior walls — especially rooms with southeast and southwest exposures — may need to be treated as separate zones. Corner rooms that have a large amount of glass area on just one exterior wall can be combined with other rooms that have the same exposure as the glass. Interior rooms that have an exposed ceiling can be grouped with north rooms that have an exposed ceiling and a moderate amount of glass area.

Load Patterns
The heating or cooling loads in some spaces — usually an internal zone — can be practically constant. However, it is more common for room loads to vary by the month of the year

and by the hour of the day. Differences in load patterns can be attributed to exposure (as described above), scheduled or random variations in internal load patterns and scheduled or random variations in ventilation requirements. Rooms with similar load patterns can be grouped together within a zone. Rooms with incompatible load patterns should not be combined to form a zone.

Latent Loads and Room Sensible Heat Ratios
During cooling, the maximum allowable supply air temperature and the corresponding cooling coil temperature (apparatus dew point) is related to the room sensible heat ratio (RSH). Rooms that have low sensible heat ratios will require supply air that is colder than rooms that have high sensible heat ratios. Rooms with similar sensible heat ratios can be combined into a zone. Rooms with different sensible heat ratios can be collected into one zone, providing that humidity control is not a critical and providing the spread in the RSH values is not too large. Rooms with incompatible sensible heat ratios should not be combined into a zone — especially if humidity control is a critical requirement or if the spread in the RSH values is relatively large.

Occupancy Schedules
Zoning by occupancy schedule increases system efficiency and reduces operating cost. Equipment can be shut off or thermostat setpoints can be set up or set back when a zone is unoccupied. Zoning by occupancy should always be investigated when there are two or more occupancy and/or production schedules. Consider which hours of the day and which days of the week the various spaces will be occupied. Also determine if the schedules are regular, irregular or random.

Temperature and Humidity
Zoning by temperature and humidity requirements is essential whenever the design conditions of two or more spaces are significantly different. For example, general offices, which only require comfort conditioning, are not compatible with laboratories or process environments, which require close control of temperature and humidity.

Fresh Air Requirements
Zoning decisions are affected by ventilation requirements. The minimum amount of outdoor air that is required for any room depends on what the room is used for. Rooms that are associated with general comfort conditioning only require a small percentage of outdoor air. Rooms that are associated with chemical or biological contaminants may require a large percentage of outdoor air (as much as 100 percent) to dilute contaminants. Rooms with similar fresh air requirements can be grouped together to form a zone. Rooms with incompatible fresh air requirements should not be combined into a single zone.

Make Up Air Requirements
Zoning decisions are affected by local exhaust systems requirements. The minimum amount of outdoor air that is required for any room may depend on how much air is exhausted from that room. The exhaust systems that are associated with general comfort conditioning — typically toilet exhaust systems — only require a small percentage of outdoor air. Rooms that are associated with laboratory or process exhaust systems may require a large percentage (up to 100 percent) of outdoor air. Rooms with similar make up air requirements can be grouped together within a zone. Rooms with incompatible make up air requirements should not be combined to form a zone.

Air Filtration Requirements
Rooms that are associated with general comfort conditioning do not require special filters. Clean rooms and rooms that are associated with laboratory or process work often have specific filtration requirements. Rooms with similar filtration requirements can be grouped together within a zone. Rooms with incompatible filtration requirements should not be combined into a single zone.

Pressure in the Conditioned Space
Rooms that are associated with general comfort conditioning are usually maintained at neutral or at a slightly positive pressure. Clean rooms and rooms that are associated with laboratory or process work often have specific — positive or negative — pressurization requirements. In some cases, the designer may find it necessary to group rooms according to the pressure requirements.

Heating-Cooling Changeover
Some rooms — depending on construction, exposure and internal loads — require cooling during every hour of the year. Most rooms switch from a cooling requirement to a heating requirement (also depending on construction, exposure and internal loads) when the outdoor temperature drops below a certain value. Rooms with similar changeover requirements can be grouped together within a zone. Rooms with incompatible changeover requirements should not be combined to form a zone.

Tenant Billings
Sometimes buildings are zoned so that the tenants can be held responsible for their utility bills. In this case, separate systems and meters will be required for every rental space.

2-4 Zoning Evaluation Table

The Zoning Evaluation Table, which is located in Apendix 1, can be used to evaluate the zoning requirements that are associated with each room of a building. The following example shows how to use this table.

Example
All of the rooms in a six-room building have the same temperature, humidity, filter and pressure requirements. Room 5 has an exhaust system that requires 100 percent make up air. Rooms 1, 2, 3 and 4 are occupied from 7 AM to 5 PM. Rooms 5 and 6 are occupied on a 24-hour basis. Room 1 is an interior

Zoning Evaluation Table

Zoning Factors	Selection Schedule	Room Numbers①											
		1	2	3	4	5	6	7	8	9	10	11	12
Functions	Heat	✗	✗	✗	✗	✗	✗						
	Cool	✗	✗	✗	✗	✗	✗						
	Dehumidification	✗	✗	✗	✗	✗	✗						
	Ventilation	✗	✗	✗	✗	✗	✗						
	Humidification												
Space Conditions	Comfort Conditions	✗	✗	✗	✗	✗	✗						
	Special Temperature												
	Special Humidity												
Filter Requirements	Standard Filters	✗	✗	✗	✗	✗	✗						
	Special Filters												
Space Pressure	Neutral	✗	✗	✗	✗	✗	✗						
	Negative												
	Positive												
Exhaust From Room	Less Than 10%	✗	✗	✗	✗		✗						
	More Than 10%												
	100 Percent					✗							
Outdoor CFM	0% to 10%	✗	✗	✗	✗		✗						
	11% to 20%												
	Above 20%												
	100%					✗							
Occupancy Pattern	Schedule A	✗	✗	✗	✗								
	Schedule B					✗	✗						
	Schedule C												
Exposure②	Interior Zone	✗											
	N, E, S, W		W	W	S	N	N						
	SE or SW (Corner)												
Peak Load③	Constant	✗				✗	✗						
	9 AM, Noon, 3, 6 PM		6	6	3								
Heating-Cooling Change-over	Cool Only	✗											
	OAT Above 50 F												
	OAT 35 F to 50 F												
	OAT 20 F to 35 F		✗	✗	✗	✗	✗						
	OAT Below 20 F												
Room SHR	Below 0.75												
	0.76 to 0.85												
	Above 0.85	✗	✗	✗	✗	✗	✗						
Utility Bills	Separate Billing												
	Central Billing	✗	✗	✗	✗	✗	✗						

Notes:
① Enter an ✗ or a ✔ in the room columns to indicate choice from selection schedule.
② Enter compass point in room columns to indicate exposure.
③ Enter a 9, N, 3 or 6 in room columns to indicate time of day peak.

zone. Rooms 2 and 3 face west. Room 4 faces south. Rooms 5 and 6 face north. Rooms 2 and 3 have a peak cooling load at 6 PM. Room 4 has a peak cooling load at 3 PM. Rooms 5 and 6 have a relatively steady cooling load. All the rooms have a sensible heat ratio that is above 0.85. Room 1 is a cooling-only zone with a constant cooling load. The other rooms change to the heating mode when the outdoor temperature is between 20 and 35 degrees. The zoning evaluation table (see the previous page) allows the designer to make the following observations:

- Room 1 must be a separate zone because it is the only interior, cooling-only zone.
- Rooms 2 and 3 can be combined because they have identical evaluations.
- Room 4 may need to be a separate zone because the south exposure causes the cooling load to peak at 3 PM. However, this room can be combined with rooms 2 and 3 if the 6 PM cooling load is within 10 percent of the 3 PM cooling load.
- The room 5 occupancy pattern, exposure and peak load factors are compatible with room 6. But, room 5 is a unique zone because of the 100 percent exhaust and outdoor air requirements.
- The room 6 occupancy pattern, exposure and peak load factors are compatible with room 5. But, room 6 will be a separate zone because it has exhaust and outdoor air requirements that are incompatible with room 5.

2-5 Simultaneous Heating and Cooling

Simultaneous heating and cooling is not required for single-zone applications because the operating mode of the equipment is controlled by the conditions in one zone. Simultaneous heating and cooling capability is required when central equipment is used to heat and cool two or more zones that have incompatible load patterns. For example, simultaneous heating and cooling is required if cooling-only interior zones are combined with perimeter zones. Also note that the load patterns associated with perimeter zones may be incompatible, depending on the exposure, the month of the year and the time of the day.

2-6 Architectural Considerations

The HVAC system must be physically, functionally and esthetically compatible with the structure. This requirement always presents a challenge because the building design and the HVAC system design are seldom executed as one simultaneous, interactive process. Most of the time the HVAC system designer is not consulted until after the spatial, structural and aesthetic details of the building have been finalized. In any case, the HVAC system designer must accommodate the performance requirements and work within the constraints that are created by the structural design and the architectural intent.

Building Codes and Regulations
The HVAC system designer has an unconditional responsibility to produce a design that conforms to all codes and regulations. Codes and regulations are typically published by branches of state and local governments and regulations may be issued by utilities. Codes and regulations are usually concerned with health, fire, safety and energy efficiency. They also may specify methods, materials and procedures.

Space Available for Ducts and Pipes
Supply, return, exhaust and outdoor air ducts are commonly installed within the building envelope, usually in cavities and chases. Obviously, the size and shape of the ducts must be compatible with the size and shape of the corresponding cavities and chases. Ducts that are attached to the exterior of the building have less constraint on size and shape, but they must not detract from the appearance of the structure. It also is necessary to route ducts and pipes so that they do not interfere with the structural components of the building or with the components of any other building system. However, it is desirable to maintain reasonable duct aspect ratios. In some cases a duct will be "jogged" or "eased" to avoid a beam or a pipe. When this situation occurs, it is important to use aerodynamic fittings and gradual transitions.

Space Available for Equipment
Major pieces of equipment are normally installed on the roof, in an equipment room, on a ground pad or in a basement. Sometimes, air handlers and packaged equipment are installed in a dead space or ceiling cavity. Irrespective of location, each piece of equipment and each individual component must be accessible so that it can be tested, adjusted and serviced. Also make sure that there is enough room to remove and replace components such as filters, coils, fans and compressors. If equipment is located outdoors, it should be housed in a weather proof enclosure. Outdoor air intakes and outdoor heat exchangers should not be located close to hot roofs and walls or in unventilated spaces. Air intakes should be shielded from the wind, rain and snow; and they should not be located near an exhaust opening or any other source of thermal, chemical, gaseous or particulate contamination. All outdoor air-side equipment should be protected from contamination by natural and man-made agents and debris. All equipment should be located so that there is no objectionable noise inside or outside the structure.

Optimize the Mechanical Systems Layout
Equipment location also affects first cost and operating costs. Ideally, the equipment should be located so that the use of duct and piping materials are minimized. Short duct and pipe runs also reduce fan and pump energy requirements.

Avoid Noise Problems
The noise level in every occupied space must be below some recommended level. If equipment is installed within the occupied space, it must not cause objectionable noise. If equipment is located near an occupied space, the machinery noise should not be propagated through the duct system or transmitted through the structural components of the building. Duct systems should be designed to attenuate mechanical noise and to prevent acoustic coupling between rooms.

Effect on Structure and Envelope
The size and location of roof openings, wall openings and any other penetrations must be compatible with the visual appearance and structural integrity of the building. The structural components of the building must support the weight of every piece of HVAC equipment. Mechanical vibration must not be transmitted to, or amplified by, the structure.

Visual Appearance
Mechanical equipment, ducts and pipes usually need to be completely hidden, or at least camouflaged. If equipment cannot be hidden it should be located where it would be least noticed. Sometimes, ducts and pipes can be integrated with the interior decor. The appearance of devices that cannot be hidden, such as supply outlets, return inlets, exhaust hoods, air intake and air exhaust louvers must be visually compatible with the interior or exterior finish.

Flexibility and Adaptability
Buildings are seldom used in the same way year after year. Always consider how important it is for the system to accommodate changes in floor plans, occupancy schedules and the way the conditioned space is used. Also consider the possibility that the central equipment may have to accommodate the loads associated with future expansions.

2-7 System Efficiency

Overall system efficiency is important because it affects operating costs and because codes and regulations may dictate minimum efficiency levels. Note that the overall system efficiency is related to, but is not the same as, equipment efficiency.

Equipment efficiency is usually described by a "single-value" descriptor such as EER, SEER, COP, AFUE or HSPF. Overall system efficiency is harder to evaluate; therefore, systems are usually rated on the basis of annual or seasonal energy consumption. Energy consumption estimates may be summarized by a value that represents the total energy requirement for the whole building, or as a "per square foot" value. Some factors that affect overall system efficiency, which are not reflected in the equipment efficiently rating include: duct and piping losses; damper leakage; economizer free cooling; reheat loads; hot and cold fluid mixing loads; set-up or set-back schedules; and daily and weekly operating schedules.

2-8 Safety

The system must be compatible with all fire and smoke control regulations. Special provisions must be made for systems that operate in explosive environments. Special provisions must be made for systems that are associated with chemical or biological contaminants. Equipment must be located and guarded so that it does not present a safety hazard.

2-9 Site Services

Obviously, the type of equipment will depend on the type of fuel that is available at the site. Electricity is normally available at every site, but the designer must make sure that the electrical service provides the required voltage, amperage and phase. Natural gas, oil and propane may or may not be available on site. (Of course, oil or propane service could be installed as part of the project.) Even if a particular type of fuel is available at the site, the designer must make sure that the capacity and dependability of the source is adequate. If natural gas is available, the designer should ask the gas company about the pressure at the meter. The designer also should consider the unit costs, rate schedules and demand charges that are associated with the various fuels. It also is important to consider the first cost that is associated with installing a given service. In addition, all fuel storage and distribution systems must comply with codes and regulations.

Earth-coupled heat sources and heat sinks also fall into the site services category. Wells, surface water or the soil can be used as a heat source or as a heat sink. Therefore, the designer must evaluate the ability of the source or sink to meet the demands of the HVAC system. The designer also must evaluate the first cost and operating costs that are associated with developing and using a source or sink. It also is necessary to make sure that the design complies with codes and regulations.

Sewers and drains must be adequate. Usually the load on the waste water disposal system is limited to the condensate that drains off of the cooling coil(s). Water discharged from water-source heat pumps or from "once-through" condensers should be disposed of according to local codes and regulations.

2-10 Economics

Economics always has a large influence on the design concept. The installed cost, energy costs, maintenance costs and replacement costs must be compatible with the owner's (or the developer's) budget. Of course, the emphasis will vary, depending on the owner's goals and objectives. In some cases, a low first cost may be the primary (or only) objective. In other cases, a low operating cost may be the overriding concern. Some owners or developers will want to maximize their return on investment or minimize the payback period. Occasionally, the economic evaluation will be based on a life cycle analysis. Sometimes, the tax laws, a utility rebate or some other type of incentive program will make a decisive difference in the economic analysis. However, the designer must never forget that the economic goals become irrelevant when system performance characteristics fail to satisfy the minimum functional requirements.

2-11 Intangibles

Some of the factors that affect system selection cannot be quantized. For example, air conditioning is expensive, but it could be justified by increased productivity due to improved comfort. In some cases, an exceptionally efficient or innovative HVAC system has a subjective, but intangible value that is related to public relations and company image. Occasionally, the owner or developer may prefer a certain type of system or piece of equipment because of a reciprocal business arrangement with a client or supplier. Of course, both the owner and the designer should evaluate the availability of maintenance personnel that will be required to operate the system and the ability of the local companies to service the system. In all cases, the system must be reliable. And, in some cases, the system might require a certain amount of backup equipment (redundancy). Finally, owner preferences and prejudices must be considered, even when they are not completely rational.

2-12 Summary of System Performance

The System Performance Characteristics table provided on the next page, can be used to compare the performance of various types of systems. The following comments provide additional information about how the various performance characteristics affect system selection.

Function
Some systems only provide one function while others provide two or three functions. A fourth function (humidification) is usually provided by equipment that is added to the basic system.

Zones
Buildings that have more than one zone can be equipped with multiple single-zone units or with a central multizone system. The number of zones that a central system can accommodate varies. Packaged multizone systems may provide 6 to 14 zones. Variable volume, variable temperature changeover systems, terminal reheat systems, variable volume systems and water systems can accommodate more than 14 zones.

Mode
Some central multizone systems cannot provide simultaneous heating and cooling. These systems should not be used unless the designer is absolutely sure that all the zones have the same load patterns.

Comfort
Some systems — a reheat system, for example — provide close control over temperature and humidity. Other systems — a variable volume system, for example — control temperature accurately, but lose control over humidity. And, some systems — a two-pipe water system, for example — can lose control over both temperature and humidity.

Air Quality
Ventilation (outdoor air) and filtration are required for health and comfort. In some cases, pressurization is required to prevent contaminated air from entering the occupied space. Some systems provide one or more of these requirements, other systems cannot satisfy any of these requirements.

Efficiency Factors
Reheat or mixing hot and cold fluids decreases system efficiency. An economizer cycle increases system efficiency. Certain reclaim strategies improve system efficiency. It is the designer's job to evaluate all energy saving options.

Spatial Requirements
Equipment located inside the structure uses valuable floor space. If this space is not available, the equipment must be located outdoors. An indoor location is preferred for the duct runs, but in some cases they are installed outdoors.

Condenser
Air-cooled equipment is preferred for small commercial applications. Water-cooled equipment is less desirable because this type of equipment usually requires a water tower.

Cooling Coil
DX coils are typically found in packaged equipment. Chilled water coils are usually associated with remote air-handling units. Water coils are easier to control at part-load conditions.

Heating Coil
Packaged equipment usually has electric or fossil fuel heat. Hot water coils are usually associated with remote air-handling units. Water or electric coils are simple to control at part-load conditions. Indirect-fired heating equipment with on/off control should not be used for make up air applications. (Direct-fired equipment is specifically designed for make up air applications.) Indirect-fired equipment can be used for make up air applications if the burner is designed to provide an adequate turn-down ratio.

Flexibility
Some systems can accommodate a new set of load conditions — a new floor plan or a new internal load schedule, for instance — without modification. Other systems must be modified or rebalanced when the load conditions vary from the original design. In some cases, a system may be designed with excess capacity for a future expansion.

System Performance Characteristics

General Characteristics and Performance Details		Single-Zone, Unitary & Self-Contained DX											All-Air Systems						Water Systems							
		Multiple 1-Zone Units	Rooftop H&C	Rooftop Heat Pump	Split System — Cooling	Split-System Heat Pump	Indoor H&C Unit	PTAC Cooling Unit	PTAC Heat Pump	Furnaces and MUA Units	Unit /Duct Heaters	Radiant Heating	Terminal Reheat	Dumping VAV ②	Modulating VAV ②	Blow-Through Multizone	Draw-Through Multizone	VVT-Changeover	2-Pipe System	3-Pipe System	4-Pipe System	Water-Loop Heat Pumps	Fan-Coil Terminal Units	Unit Ventilators	Unit Heaters	Baseboard Radiation
Function	Heat	✗	✗	✗		✗	✗		✗	✗	✗	✗	✗	✗	✗	✗	✗	✗	✗	✗	✗	✗	✗	✗	✗	✗
	Cool / Dehumid	✗	✗	✗	✗	✗	✗	✗	✗				✗	✗	✗	✗	✗	✗	✗	✗	✗	✗	✗	✗		
	Ventilation	✗	✗	✗	①	①	①	✗	✗	✗			✗	✗	✗	✗	✗	✗	✗	✗	✗			✗		
Zones	One		✗	✗	✗	✗	✗	✗	✗	✗	✗	✗											✗	✗	✗	✗
	Many	F											B	G	B	F	F	F	P	B	B	B				
Mode (H&C)	Heat or cool		✗	✗		✗	✗		✗					✗	✗			✗	✗				✗	✗		
	Simultaneous	✗											✗			✗	✗			✗	✗	✗				
Comfort	Temp control	G	G	G	G	G	G	G	G	G	F	F	B	G	G	G	G	F	P	B	B	G	G	G	G	G
	RH control	G	G	G	G	G	G	G	G				B	F	F	G	B	F	P	B	B	G	G	F		
Air Quality	Circulation	B	B	B	B	B	B	F	F	G	F		B	G	P	G	G		G	F	F	G	F	F	F	
	Fresh Air	G	G	G	①	①	①	F	F	G			B	F	P	G	G	F	G	G	G			G		
	Filtration	G	G	G	G	G	G	P	P	G			B	G	G	G	G	G	P	P	P	F	P	P		
	Pressurization	✗	✗	✗	①	①	①			✗			B	P	P	F	F	P								
Efficiency Factors	Reheat												✗													
	Mixing															✗	✗			✗	✗					
	Economizer	✗	✗	✗	①	①	①						✗	✗	✗	✗	✗	✗						✗		
	Reclaim																					✗				
Spatial Requirement	Equip outdoors	✗	✗	✗	✗	✗		✗	✗	✗			✗	✗	✗	✗	✗	✗				✗				
	Equip inside				✗	✗	✗			✗	✗	✗	✗	✗	✗	✗	✗		✗	✗	✗	✗	✗	✗	✗	✗
	Ducts inside	✗	✗	✗	✗	✗	✗			✗	✗		✗	✗	✗	✗	✗	✗				✗				
Condenser	Air Cooled	✗	✗	✗	✗	✗		✗	✗				✗	✗	✗	✗	✗	✗	✗	✗	✗					
	Water Cooled						✗						✗	✗	✗	✗	✗		✗	✗	✗	✗				
Cooling Coil	DX	✗	✗	✗	✗	✗	✗	✗	✗				✗	✗	✗	✗	✗	✗				✗				
	Water												✗	✗	✗	✗	✗		✗	✗	✗		✗	✗		
Heating Coil	Electric	✗	✗	✗		✗	✗		✗	✗	✗	✗	✗	✗	✗	✗	✗	✗								
	Water												✗	✗	✗	✗	✗		✗	✗	✗		✗	✗	✗	✗
	Fossil	✗	✗				✗			✗	✗	✗	✗	✗	✗	✗	✗	✗								
Flexibility	Floor plan revs.												✗	✗	✗			✗								
Op-Cost	Energy Use	B	G	B	G	B	G	F	G	G	F	G	P	G	B	F	P	G	G	P	G	B	G	F	F	G

Note: ① Option not standard — Provide OA/EA ducts; field fabricated economizer section. ② No reheat P = poor, F = fair, G = good, B = best, Blank = not available

Operating Cost

Single-zone systems are inherently efficient because the equipment capacity is directly related to the load in the conditioned space. Reheat and mixing hot and cold fluids is not usually required for single-zone applications. (Sometimes, single-zone reheat is required if the application requires close control over humidity.) Multizone systems — even variable volume systems — may be penalized by reheat loads or mixing loads. However, variable volume systems do tend to have lower operating costs than the other types of central systems. An economizer cycle will increase the cooling efficiency of all-air systems. A water-loop heat pump system can be very efficient if the building has the correct combination of interior cooling zones and perimeter heating zones.

Section 3
Applications by Building Type and Use

Every application has some characteristic design requirements. Some of these requirements are related to the architectural features of the building and others are related to how the building is used. It is the HVAC system designer's responsibility to match system performance characteristics to the applications performance requirements. The proposed system also must be compatible with the owner's budgetary and investment goals.

3-1 Single Purpose Occupancy

A single purpose occupancy is characterized by spaces that are used for a single activity. In this case, the indoor design conditions, air quality requirements, ventilation requirements and internal load schedules will be the same for each occupied space. Normally, a single purpose occupancy will only require one system, but not necessarily a single-zone system. Zoning will be required when multiple spaces have different load patterns or operating schedules.

3-2 Multiple Purpose Occupancy

A multiple purpose occupancy is characterized by two or more spaces that are used for different activities. In this case, the indoor design conditions, air quality requirements, ventilation requirements and internal load schedules may not be the same for each occupied space. Sometimes, a multiple purpose occupancy can be conditioned by a single, central, zoned system and sometimes multiple systems will be required.

3-3 Application Design Requirements

In order to select an HVAC system for a given application, it is necessary to identify the load characteristics and the design requirements that are associated with the application. This subsection provides a general discussion of these items. Refer to Appendix 2 for information about the load characteristics and the design requirements that are associated with specific applications.

Function

Some applications only require heat or only require cooling. Other applications may require heating and cooling. Dehumidification is usually, but not always, required when cooling is required. Almost every commercial application requires outdoor air for ventilation or makeup air. When ventilation is required, the outdoor air CFM can vary from some minimum amount, say 15 CFM per occupant, to as much as 100 percent of the supply CFM.

Zoning Requirements

Zoning may be required for a single purpose occupancy when the architectural plan creates spaces that have different exposures, envelope construction details or infiltration rates. (Differences in envelope load characteristics translate into incompatible seasonal load patterns, time of day load patterns and cooling/heating changeover temperatures.) Zoning also will be required if two or more spaces have different indoor design conditions, air quality requirements, internal load schedules or operating schedules. Multiple systems may have to be installed if renting and leasing operations require separate utility bills for each tenant. Refer to Section 2-3 for a more detailed discussion of zoning factors.

A multiple purpose occupancy application can be treated as a collection of single purpose occupancy applications. As mentioned above, one or more zones will be required for each occupancy type. However, some types of occupancies are not compatible. In this case, separate systems may be required.

Lighting Loads

Lighting loads vary widely from application to application. Applications that have high lighting loads generally require more sensible cooling and less heating. Sometimes, the lighting loads can be large enough to offset the heating load, but if heating is required, the cooling/heating changeover usually occurs at a low outdoor temperature. Large lighting loads make it easier to control humidity because the sensible gains tend to increase the room sensible heat ratio.

Occupancy Loads

Applications with large population densities tend to require less heat, more sensible cooling and more dehumidification. This occurs because the occupants produce substantial sensible and latent loads. If the occupants are not too active, the size of the sensible and latent components are about equal. As activity levels increase, the latent component begins to exceed the sensible component. Relatively large, occupant related, latent loads make it more difficult to control humidity because they reduce the room sensible heat ratio and increase the dehumidification requirement.

Some occupancy loads can be large enough to offset the design heating load. Even if heating is required, the cooling/heating changeover point may occur at a low outdoor temperature. Under these circumstances, humidity control can be a problem because the latent load stays fairly constant as the sensible cooling load gets smaller.

Appliance and Equipment Loads

Appliance and equipment loads may be sensible, latent or both sensible and latent. In some situations — food display cases, for example — the sensible and latent loads may be negative. Large sensible loads decrease the heating requirement, increase the cooling requirement, reduce the cooling/heating changeover point, increase the room sensible heat ratio and make it easier to control humidity. Large latent loads increase the cooling requirement, reduce the room sensible heat ratio and make it harder to control humidity.

Cooling Load

The peak cooling load depends on the combined effects of the envelope, ventilation and internal loads. Note that these individual load components may not peak simultaneously. Internal loads and ventilation loads depend on the occupancy or operating schedule. Envelope loads depend on the month of the year and the time of the day. Some applications peak because of the occupancy and equipment operating schedules, others peak because of the envelope loads.

The peak load is not the only design consideration. Part-load performance is just as important — possibly more important — than peak load performance. Some applications have cooling loads that are fairly constant. Most applications experience load swings, which depend on the outdoor temperature, time of day, occupancy schedule and internal load schedule. It is easier to design a system for applications that have a steady load. Special equipment and controls are required for applications that are characterized by wide load swings.

Operating Mode

Some applications — internal zones, for example — never require heating. Perimeter zone applications that have relatively high internal loads may not require heat until it is quite cold outside. The cooling/heating changeover point determines the operating mode for the equipment. Single-zone applications will either be in the cooling mode or the heating mode. Simultaneous heating and cooling will be required for multiple-zone applications when two or more zones have noticeably different heating changeover points.

Humidity Control

For most cooling applications the indoor humidity must be controlled during full-load and part-load conditions. Note that it is more difficult to control humidity at part-load conditions, especially if the application has a low room sensible heat ratio. (Low sensible heat ratio applications require a lower cooling coil temperature and possibly a reheat coil.) Winter humidification is used to control static electricity, to provide an environment that is compatible with health standards or to satisfy (production) equipment and process requirements.

The desired humidity level and the tolerance for changes in the humidity level depend on the application. Comfort conditioning humidity levels are defined by the ASHRAE comfort zone, which is defined by the figure below. A few applications, laboratories for example, require extremely close control over humidity during every hour of the year. In some cases, the humidity levels that are required for experiments, processes or equipment may not be comfortable for humans.

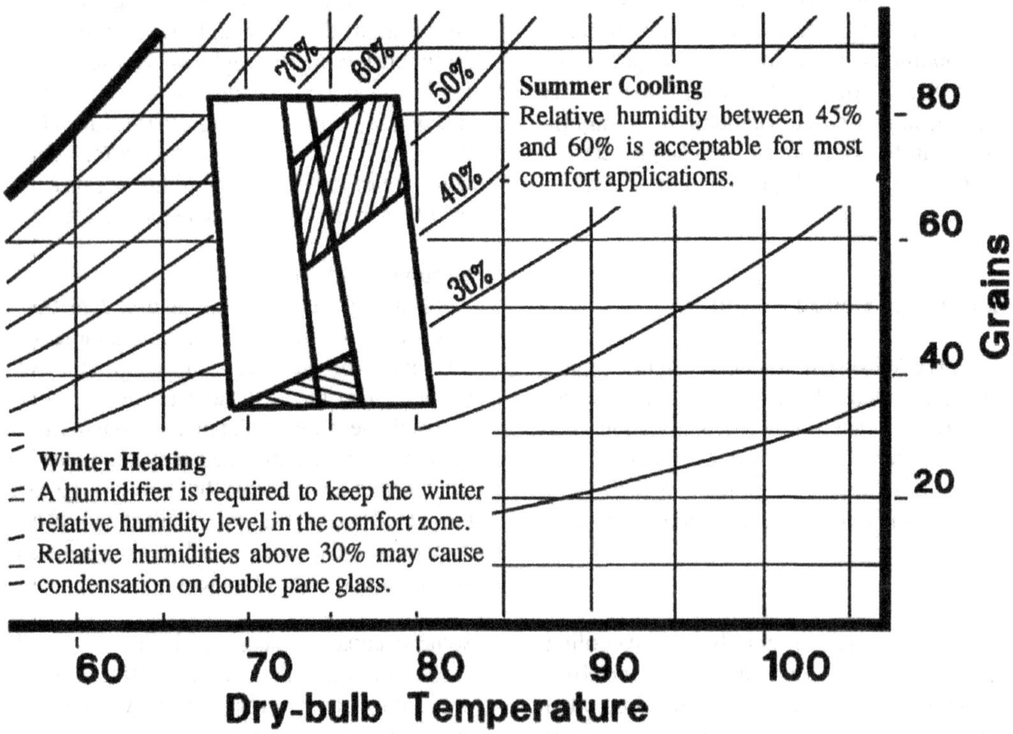

ASHRAE Comfort Chart

Suggested ASHRAE Dust Spot Efficiencies for NonSmoking[1] Applications

Apartments	20% to 30%
Arcades	20% to 30%
Arenas	
Public areas	30% to 35%
Floor	30% to 35%
Ticket Booth	30% to 35%
Stages & Work Rooms	30% to 35%
Bakeries	20% to 30%
Ball / Dance Rooms	30% to 35%
Bank	35% to 40%
Barber Shops	20% to 30%
Bars and Discos	20% to 30%
Beauty	20% to 30%
Bowling Centers	
Bowling & Spectator	20% to 30%
Cafeterias	35% to 40%
Church	
Assembly	30% to 35%
Classroom	30% to 35%
Meeting / Activity	30% to 35%
Coin Laundries	20% to 30%
Convenience Centers	
Stores	20% to 30%
Court Club/Fitness Center	
Tennis	20% to 30%
Racquetball / Squash	20% to 30%
Exercise Room	20% to 30%
Department Stores	
Shopping Areas	20% to 30%
Diners or Fast Food	30% to 35%
Doctor/Dentist Office	
Waiting	30% to 35%
Examination	30% to 35%
Medical Procedure	90% +
Dormitories	
Bedrooms	20% to 30%
Suites	20% to 30%
Lobbies	20% to 30%
Meeting / Game	20% to 30%
Dry Cleaners	
Commercial	20% to 30%
Coin	20% to 30%
Print / Duplicating	20% to 30%
Exhibition Hall	
Public areas	30% to 35%
Floor	30% to 35%
Ticket Booth	30% to 35%
Stages & Work Rooms	30% to 35%
Florist	20% to 30%
Funeral Parlor	30% to 35%
Garages	Below 20%

Government Buildings	
Court Room	30% to 35%
Legislative Chambers	30% to 35%
Committee-Conference	30% to 35%
Offices	30% to 35%
Press Room lounge	20% to 30%
Toilets	Below 20%
Jail Cells	20% to 30%
Guard Stations	20% to 30%
Police Station	20% to 30%
Fire House	20% to 30%
Gymnasiums	
Floor	20% to 30%
Spectator	20% to 30%
Locker	Below 20%
Hospital	
Patient Room	90% & PF
Medical Procedures	90% & PF
Operate Delivery	90% & PF
Recovery / Intens Care	90% & PF
Autopsy	90% & PF
Physical Therapy	90% & PF
Cafeteria	90% & PF
Amphitheater	30% to 35%
Industrial	10% to 99.%
Kitchens	Below 20%
Laboratories	80% to 99%
Lab Animal Rooms	Below 20%
Library or Museum	
General Areas	35% to 60%
Work Rooms	35% to 60%
Rare books & Exhibits	80% to 99%
Shopping Mall	
Mall	20% to 30%
Store	20% to 30%
Motels	
Bedrooms	20% to 30%
Suites	20% to 30%
Lobbies	20% to 30%
Conference	20% to 30%
Assembly Hall	20% to 30%
Nursing Homes	
Patient Room	80%
Medical Procedures	80%
Physical Therapy	80%
Office Building	
Office	35% to 60%
Bull Pen	35% to 60%
Conference	35% to 60%
Reception	35% to 60%
Pet Shops	20% to 30%

Pharmacy	
Work Room	30% to 35%
Sales Area	30% to 35%
Photo Studio	
Studio	30% to 35%
Darkroom	30% to 35%
Pool Hall	20% to 30%
Restaurant / Nite-club	35% to 40%
School	
Classroom	30% to 35%
Laboratories	30% to 60%
Shops	30% to 35%
Music Rooms	30% to 35%
Libraries	30% to 35%
Auditoriums	30% to 35%
Gyms	20% to 30%
Locker Rooms	Below 20%
Common Rooms	30% to 35%
Offices	30% to 35%
Lunch Room	30% to 35%
Shoe Repair	20% to 30%
Skating Rinks	20% to 30%
Small Stores	20% to 30%
Smoking Room	80%
Supermarket	35% to 40%
Swimming Pools	
Pool & Deck	20% to 30%
Spectator	20% to 30%
Theater	
Public areas	30% to 35%
Floor	30% to 35%
Ticket Booth	30% to 35%
Stages & Work Rooms	30% to 35%
Public Toilet	Below 20%
Transportation Terminal	
Public Areas	30% to 35%
TV-Radio	
Booths	35% to 40%
Studios	35% to 40%
Electronic Equipment	80% & PF
Audience	35% to 40%
Stages	35% to 40%
Variety Stores	20% to 30%
Veterinarian	
Kennel	10% to 15%
Operating Room	30% to 35%
Waiting Room	30% to 35%
Warehouse	10% to 35%

① Filter efficiency must be 80% or higher to remove smoke.

② PF = Prefilter

ASHRAE Standard 62-89 Ventilation Rates (CFM per Person)③

Apartments ① ②	30 / Room
Arcades	0.3 / SQ.FT.
Arenas	
Public areas	15
Floor	20
Ticket Booth	20
Stages & Work Rooms	15
Bakeries	15
Ball / Dance Rooms	25
Bank	20
Barber Shops	15
Bars and Discos	30
Beauty	25
Bowling Centers	
Bowling & Spectator	25
Cafeterias	20
Church	
Assembly	15
Classroom	15
Meeting / Activity	15
Coin Laundries	15
Convenience Centers	
Stores	0.30 / SQ.FT.
Court Club/Fitness Center	
Tennis	20
Racquetball / Squash	20
Exercise Room	20
Department Stores	
Shopping Areas	0.30 / SQ.FT.
Diners or Fast Food	20
Doctor/Dentist Office	
Waiting	15
Examination	15
Medical Procedure	30
Dormitories	
Sleeping Area ②	15
Suites	30 / Room
Lobbies	15
Meeting / Game	20
Dry Cleaners	
Commercial	30
Coin	15
Print / Duplicating	0.50 / SQ.FT.
Exhibition Hall	
Public Areas	15
Floor	20
Ticket Booth	20
Stages & Work Rooms	15
Florist	15
Funeral Parlor	15
Garages	1.5 / SQ.FT.

Government Buildings	
Court Room	15
Legislative Chambers	15
Committee-Conference	20
Offices ②	20
Press Room lounge	60
Toilets	50 / Fixture
Jail Cells	20
Guard Stations	15
Police Station	15
Fire House	15
Gymnasiums	
Floor	20
Spectator	15
Locker	0.5 /SQ.FT.
Hospital	
Patient Room	25
Medical Procedures	15
Operate Delivery	30
Recovery / Intens Care	15
Autopsy	100
Physical Therapy	15
Cafeteria	20
Amphitheater	15
Industrial	Survey
Kitchens	15
Laboratories	20
Lab Animal Rooms	0.50 / SQ.FT.
Library or Museum	
General Areas	15
Work Rooms	15
Rare books & Exhibits	15
Shopping Mall	
Mall	0.20 / SQ.FT.
Store	0.30 / SQ.FT.
Motels	
Bedrooms ②	30 / Room
Suites	30 / Room
Lobbies	15
Conference	20
Assembly Hall	15
Nursing Homes	
Patient Room ②	25
Medical Procedures	15
Physical Therapy	15
Office Building	
Office ②	20
Bull Pen	20
Conference	20
Reception	15
Pet Shops	1.0 / SQ.FT.

Pharmacy	
Work Room	15
Sales Area	0.3 / SQ.FT.
Photo Studio	
Studio	15
Darkroom	0.5 / SQ.FT.
Pool Hall	20
Restaurant / Nite-club	20
School	
Classroom	15
Laboratories	20
Shops	20
Music Rooms	15
Libraries	15
Auditoriums	15
Gyms	20
Locker Rooms	0.5 / SQ.Ft.
Smoking Lounges	60
Offices ②	20
Lunch Room	20
Shoe Repair	20
Skating Rinks	0.5 / SQ.FT.
Small Stores	0.3 / SQ.FT.
Smoking Room	60
Supermarket	15
Swimming Pools	
Pool & Deck	0.5 / SQ.FT.
Spectator	15
Theater	
Public areas	20
Floor	15
Ticket Booth	20
Stages & Work Rooms	15
Public Toilet	50 / Fixture
Transportation Terminal	
Public Areas	15
TV-Radio	
Booths	20
Studios	20
Electronic Equipment	20
Audience	15
Stages	15
Variety Stores	0.30 / SQ.FT.
Veterinarian	
Kennel	1.0 / SQ.FT.
Operating Room	30
Waiting Room	15
Warehouse	0.15 / SQ.FT.

① 100 CFM kitchen Exhaust
② 50 CFM Private Bath Exhaust
③ Unless otherwise noted

Air Quality
Every application has a unique set of air quality requirements. These requirements are dictated by the surroundings and the activities that are associated with the application. In some cases, comfort conditioning for example, the goal may just be to protect the HVAC equipment and to provide a dust free, odor free environment. In other cases, the health and safety of the occupants or the integrity of a process may be involved.

Filters, outdoor air and good circulation are required for controlling the contaminants that are associated with comfort conditioning applications. The table on page 3-4 lists the ASHRAE 62-1989 ventilation requirements for various applications and the table on page 3-3 recommends ASHRAE dust spot (filter) efficiencies for various applications.

Most filters just trap lint, dust and some mists. They protect fans, coils and heat exchangers and keep the larger particles out of the room. High efficiencies are required to remove small particles and smoke. Filter efficiencies of 80 percent or higher are required to remove ordinary cigarette smoke! Most filters are not effective at removing odors, fumes, chemical, biological or radioactive contaminants. Serious air quality problems require measures that reduce the source of contamination. Institutional or industrial applications that produce chemical, biological and particulate contamination require special filters, air washers, large amounts of outdoor air, sophisticated exhaust systems and carefully controlled space pressures. Refer to . . . **TLVs—Threshold Limit Values and Biological Exposure Indices for 1986-87, American Conference of Governmental Industrial Hygienists** for information about institutional or industrial applications not listed in the ASHRAE 62-1989 Standard.

The filter efficiency recommendations provided in this manual do not take priority over codes and regulations that call for higher filter efficiency levels. However, it is permissible to use the efficiencies that are suggested in this manual when codes and regulations do not provide guidance or when they call for a lower efficiency level.

Exhaust CFM
The designer must consider the exhaust requirements that are associated with a given application. An air balance is required in every room that is associated with a given application. Sometimes this translates into maintaining a neutral pressure in the occupied space and sometimes the space pressure must be positive or negative. In most cases, air that is exhausted from a zone must be replaced with outdoor air. This makeup air requirement increases the load on the heating and cooling equipment. Dedicated makeup air units may be required if an application has a large exhaust system.

Outdoor Air
The outdoor air CFM requirements will be dictated by the amount of fresh air that is required for the occupants or by the amount of air that is exhausted from the space, which ever is greater. Some applications require a large amount of outdoor CFM for each person and others require less. As noted above, the exhaust requirements can vary widely from application to application. The table on page 3-4 lists the ASHRAE 62-1989 ventilation requirements for various applications.

The ASHRAE ventilation recommendations do not take priority over codes and regulations that call for higher ventilation rates. However, if the codified ventilation rates are lower or if the code does not provide guidance, it is permissible to use the ASHRAE 62-1989 ventilation rates.

The outdoor air in some cities may not be acceptable for ventilation. Outdoor air quality is considered acceptable if the contaminants do not exceed the levels listed in the following table. Refer to data provided by government pollution control agencies to verify compliance with this table. If the quality of the outdoor air is not acceptable, it must be filtered, washed or treated to control the offending contaminants.

Outdoor Air Quality Standard

Pollutant	Long Term		Short Term	
	ug/m^3	Average	ug/m^3	Average
Sulfur dioxide	80	1 Year	365	24 Hours
Total particulate	75	1 Year	260	24 Hours
Carbon monoxide			40000	1 Hour
Carbon monoxide			10000	8 Hours
Oxidants (ozone)			235	1 Hour
Nitrogen dioxide	100	1 Year		
Lead	1.5	3 Months		

ug/m^3 = Microgram Per Cubic Meter

Space Pressure
There are no special space pressure requirements for simple comfort applications. In these situations the space is usually maintained at a neutral pressure, but to offset some of the infiltration, the designer may elect to keep the space at positive pressure. However, positive pressure is not recommended if the CFM exhausted from the space exceeds 10 percent of the supply air CFM.

Industrial and institutional applications may require carefully controlled space pressures to maintain the desired air quality.

Recommended Supply Outlet Noise Levels for Various Applications

Application	NC Level	Maximum Velocity
Private Residence		
Rural/Suburban	20 to 25	700
Urban	25 to 30	700
Apartments	30 to 35	700
Hotels/Motels		
Private Room or Suite	30 to 35	700
Meeting/Banquet Room	30 to 35	700
Halls, Corridors, Lobbies	35 to 40	800
Service and Support Areas	40 to 45	900
Offices		
Executive	25 to 30	700
Conference Rooms	25 to 30	700
Private	30 to 35	700
Open Bay Areas	35 to 40	800
Computer Room	35 to 40	900
Business Machine Room	40 to 45	900
Waiting Rooms	40 to 45	900
Retail Stores		
Clothing Department	35 to 40	800
General Sales Areas	40 to 50	900
Supermarkets	40 to 50	900
Hospitals/Nursing Homes/Clinics		
Private/Semi-private Rooms	25 to 30	700
Operating Rooms	25 to 30	700
Wards	30 to 35	700
Churches and Schools		
Sanctuaries	20 to 30	700
Libraries	30 to 35	700
Classrooms	30 to 35	700
Laboratories and Shops	35 to 40	800
Gyms and Rec. Areas	40 to 50	900
Kitchens	40 to 50	900

Application	NC Level	Maximum Velocity
Public Assembly		
Concert Halls	20 to 25	700
Legitimate Theaters	20 to 25	700
Auditoriums	25 to 30	700
Motion Picture Theaters	30 to 35	700
Libraries / Museums / Courts	30 to 35	700
Coliseums	30 to 35	700
Convention Halls	30 to 35	700
Exhibition Halls	35 to 40	800
Bowling Centers	40 to 50	1000
Bank / Post Office	35 to 40	800
Lobbies	35 to 40	800
Swimming Pools	40 to 55	1000
Transportation Terminals		
Ticketing Areas	30 to 35	700
General Waiting	35 to 45	800
Studios		
Recording Studios	20 to 25	700
Broadcasting Studios	20 to 25	700
TV audience Studios	30 to 35	700
Restaurants/Cafeterias/Lounges		
Restaurant Eating Area	35 to 40	800
Lounges and Bars	35 to 45	800
Night Clubs	35 to 45	800
Cafeterias	40 to 50	900
Kitchens	40 to 50	900
Industrial Production Areas	Background noise sets noise limits	

Note: *Maximum Velocity* refers to the face or neck velocity of the supply outlet (FPM).

For example, clean rooms are normally pressurized and hospitals require carefully controlled space pressures — which may be positive or negative — throughout the facility.

Schedules

Some applications have very specific occupancy schedules and production schedules, but others have rather unpredictable schedules. Knowledge of occupancy schedules and production schedules is important because they affect the loads on the heating and cooling equipment and the ventilation requirement. HVAC equipment operating hours also are related to these schedules (subject to a manual override.) Some applications only require conditioning for a few hours per day or week; others require continuous, uninterrupted service.

Noise

Noise is not critical for some applications, but most applications have specific noise level requirements. Refer to the table above for a list of the supply outlet noise criterion (NC) requirements that are associated with various applications.

Special Problems
Some applications have one or more special problems. These problems will always be related to one of the items that are discussed above. Problems that are peculiar to a specific application are discussed in Appendix 2.

3-4 Indoor Design Conditions

Refer to local codes and regulations to determine the indoor dry-bulb temperature and relative humidity requirements for a given application. If there are no codes or regulations, the indoor design conditions for "comfort conditioning" applications should be based on the ASHRAE Comfort Chart, which is provided on page 3-2. This chart shows that a 75°F/50% RH condition is a good choice for cooling and that a 70°F dry-bulb temperature is a good choice for heating. The chart also shows that the cooling condition will remain in the comfort zone even if the relative humidity is allowed to drift — say between the 45 and 60 percent. In cold climates, winter humidification is desirable, but not always mandatory.

Some applications do not fall into the "comfort conditioning" category — a laboratory, manufacturing process, health facility or warehouse for example. Some of these applications are covered by codes, regulations and standards — a health care facility for example. Other applications depend on the requirements dictated by the owner, manager or administrator — an electronics laboratory, for example.

3-5 Qualitative Analysis

The System Evaluation Form that is provided in Appendix 1 can be used to match the system performance characteristics to the application performance requirements. First, apply the concepts that are outlined in Section 3-3 and the information provided in Appendix 2 to establish the application performance requirements. Then, evaluate the suitability of candidate systems by comparing these requirements with the information that is provided by the System Performance Characteristics table, which also is provided in Appendix 1. (Refer to Section 2-12 for more information about this table.)

Example
An example that shows how to use the System Evaluation Form is provided below. This example compares the performance characteristics of a rooftop DX cooling, gas heating unit; a VVT-changeover heat pump system and a rooftop DX cooling, gas heating, multizone unit.

The example application consists of an office building that has six zones. One of these zones is an interior conference room that is a cooling-only application. There is also a reception area that has a large amount of west glass. This zone has wide load swings and it will require cooling when other areas require heat and heat when other areas require cooling. The remaining zones have cooling/heating changeover points that are somewhat similar to each other. Simultaneous heating and cooling will be required to maintain the design conditions in each zone. Temperature, humidity control and air quality should be good, but not necessarily perfect. An economizer is needed because of the low ambient cooling requirement. The roof is preferred location for the equipment. Ducts can be installed in the ceiling cavity. Gas is available for heating. Noise levels should not exceed NC-35. Capital and operating cost are always important.

System Selection Form

General Characteristics and Performance Details		Application	1-Zone H/C	VVT-CO	Multizone
Function	Heat	✗	✗	✗	✗
	Cool / Dehumid	✗	✗	✗	✗
	Ventilation	✗	✗	✗	✗
Zones	One		1		
	Many	6		6	6
Mode (H&C)	Heat or Cool		✗	✗	
	Simultaneous	✗			✗
Comfort	Temp. Control	G	P	P	G
	RH Control	G	F	F	G
Air Quality	Circulation	G	G	F	G
	Fresh Air	G	G	F	G
	Filtration	G	G	G	G
	Pressurization				
Efficiency Factors	Reheat				
	Mixing				✗
	Economizer	✗	✗	✗	✗
	Reclaim				
Spatial Requirement	Equip. Outside	✗	✗	✗	✗
	Equip. Inside				
	Ducts Inside	✗	✗	✗	✗
Condenser	Air Cooled	✗	✗	✗	✗
	Water Cooled				
Cooling Coil	DX	✗	✗	✗	✗
	Water				
Heating Coil	Electric			✗	
	Water				
	Fossil	✗	✗		✗
Noise	NC Rating	35	✗	✗	✗
Flexibility	Floor Plan Revs.				
Cost Analysis	Installed Cost				
	Energy Use		F	F	F
	Maint. Costs				

P = Poor, F = Fair, G = Good, B = Best

✗ = Available or acceptable

Blank = Not available or not applicable

Single-Zone Heating and Cooling Unit
The evaluation form on the previous page shows that the performance of the rooftop heating and cooling unit is not adequate because it can only serve one zone and because it cannot provide simultaneous heating and cooling. (Note that the interior cooling-only zone will over heat if the rooftop unit is equipped with a central heating plant.)

However, a cooling-only rooftop unit could be used if devices such as variable-volume bypass supply outlets, reheat or baseboard radiation are added to the system. (Humidity control, air circulation and the percentage of ventilation air will be somewhat compromised if variable volume features are added. If VAV devices are used, the supply CFM should not be allowed to modulate to less than 40 percent of the design CFM. Reheat is required to maintain a minimum air flow after a supply outlet modulates to the 40 percent position.) Of course, the operating costs will increase if some form of local heat is used, but some zones will be out of control if the local heat is not used.

VVT-Changeover System
The VVT-changeover system provides the required number of zones, but it does not provide true simultaneous heating and cooling. This system can be expected to lose control over zones that have wide variations in the cooling/heating changeover points. This system cannot be expected to control the temperature in the interior cooling-only zone. Temperature control, humidity control, air circulation and the percentage of ventilation air are somewhat compromised by the variable-volume changeover features of this system. The temperature in 5 of 6 zones can be controlled if reheat is added to the zones that have high cooling/heating changeover points. Operating costs will increase if reheat is used. Some zones will be out of control if reheat is not used. A second system might be appropriate for the cooling-only zone.

Blow-Through Multizone System
The rooftop multizone unit meets all of the application design requirements. The multizone unit is penalized by a loss in efficiency that is associated with mixing cold and hot fluids, but the other two systems are burdened with reheat requirements.

3-6 Industrial Buildings

Industrial buildings can present unusual challenges to the HVAC system designer. Production equipment and manufacturing process produce unique design requirements regarding space conditions and air quality. The sensible and latent loads and load patterns associated with production equipment and processes may not be documented. Exhaust, ventilation and cooling may be required on a very local basis. Space pressures may be hard to control.

Refer to the **1987 ASHRAE handbook — HVAC Systems and Applications** and to the American Conference of Governmental Industrial Hygienist manual — **Industrial Ventilation (20th Edition)** for information about industrial applications. Unfortunately, some of the design problems that are related to process equipment are not discussed in any of the standard HVAC reference manuals. When this situation occurs, the HVAC system designer may have to resort to a text book application of basic engineering principles.

Section 4
Applied Load Analysis

A successful system design is closely related to the accuracy of the load analysis. A partial list of reasons for making accurate load calculations would include:

- Equipment sizing — refrigeration machine, fans, coils, ducts, room terminal units, furnace or boiler
- Determine fluid flow rates — CFM associated with supply air outlets, fans and coils, GPM associated with chilled water equipment, hydronic equipment and pumps
- Control system design — select sensor locations and setpoints, set changeover temperatures, start and stop heating and cooling equipment, setpoint optimization and scheduling, control refrigeration machine at part load, control heating equipment at part load
- Estimating energy requirements — operating hours, heating and cooling equipment energy inputs, fan and pump energy inputs
- Estimating cost — installation cost, operating cost

4-1 Terminology

Usually the term "load" is assumed to refer to the maximum heating load on a boiler, furnace or heating device or the maximum cooling load on a coil or compressor. Obviously, these loads are very important, but there are other types of loads of equal importance. A review of some load calculation terminology is provided below.

Space Load
For heating, the space load includes the transmission loads and the infiltration load. For cooling, the space load includes the transmission loads, the infiltration load, the solar loads, and the internal loads. The ventilation load is not added to the space unless raw unconditioned air is introduced directly into the space by ventilating equipment. The **Manual N**, Form N-1, can be used to calculate the space load — refer to lines 1 through 6 for cooling and lines 13 through 16 for heating.

Room Load
The room load is the space load (heating or cooling) that is associated with a single room. It also could be a zone load.

Zone Load
The zone load is the space load (heating or cooling) that is associated with a single zone. A zone may be a single room or a combination of rooms. When a group of rooms is combined into a single zone, they should all have similar load profiles. That is, they should have similar load vs. time-of-day profiles and load vs month of year profiles. (Some examples of incompatible time of day load profiles are provided on the following page.)

Peak Zone Load
The peak zone load is the maximum cooling load that a zone will ever experience during the entire cooling season. Zone cooling loads may peak during any hour of the day or during any month of the year. Typical factors that contribute to zone peaks include solar gains, occupancy loads, equipment and process loads and lighting loads. It is the designer's responsibility to determine exactly when peak zone load occurs. This may require making load calculations for more than one time of day or for more than one month of the year. The following table indicates when zone loads are likely to peak.

Space Load Peaks			
Zone	**Month**	**Hour**	**Notes**
East	July / Aug	9 - 11 AM	Glass forces morning peak
West	July / Aug	5 - 7 PM	Glass, walls and roof peak late in day
North	July / Aug	5 - 7 PM	Roof peaks in the late afternoon
South	Sept to Nov	1 - 3 PM	Glass can force peak in early afternoon
Internal	Any	Any	Occupancy, lights and equipment can peak at any time
Block	July / Aug	5 - 7 PM	Assuming there is not a large amount of south glass
Glass not shaded by overhangs, external devices etc.			

Table 4-1

Equipment Load
The load at the HVAC equipment includes the space loads (as described above) plus the ventilation load and the distribution loads (duct losses and piping losses, for example). The **Manual N**, Form N-1, can be used to calculate the equipment sizing loads — refer to lines 1 through 12 for cooling and lines 13 through 21 for heating.

Block Load
The block load is the combined load that is imposed on central equipment that serves more than one zone. The block load usually refers to a cooling load, but it also could refer to a heating load.

Peak Block Load
The peak block load is the maximum combined cooling load that is imposed on central equipment during the entire cooling season. The peak block load may occur during any hour of the day or during any month of the year. The peak block load is usually less than the sum of the corresponding peak zone loads. This difference is the result of the diversity that is created when the load profiles that are associated with the zones are not synchronized (see below). It is the designer's responsibility to determine exactly when the peak block load occurs. This may require making load calculations for more than one time of day or more than one month of the year.

Sizing Load
The equipment sizing load is the maximum load that will be imposed on the heating equipment during the heating season or the maximum load that will be imposed on the cooling equipment during the cooling season. The sizing load associated with cooling coils is divided into sensible and latent components.

Off-Peak Loads
The off-peak loads consist of all the possible heating and cooling load conditions that may be experienced during part-

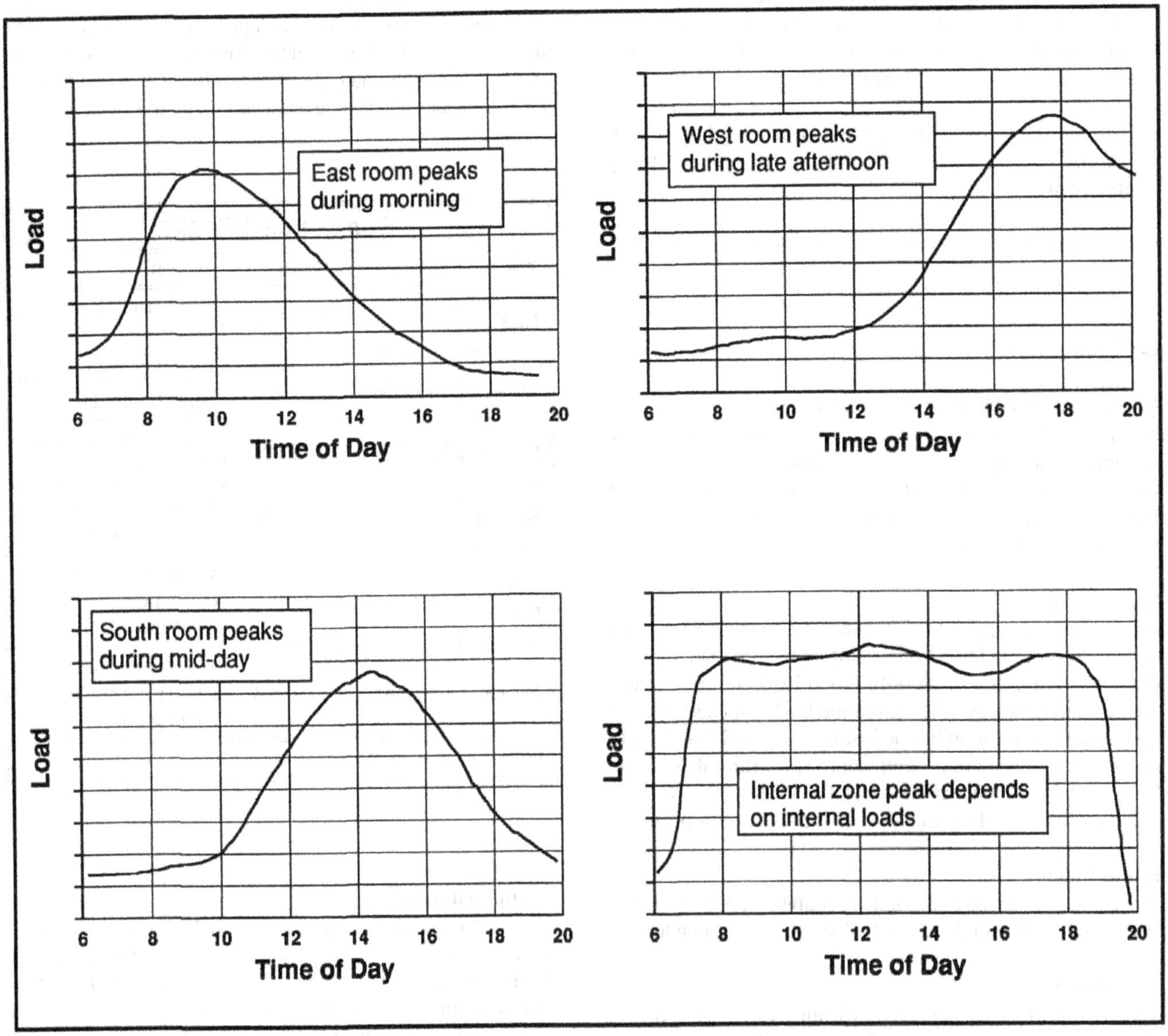

Examples of Time of Day Load Profiles

load operation. The off-peak concept applies to space loads and equipment loads.

Minimum Cooling Load
The minimum cooling load is the smallest load that will be experienced during the entire cooling season. This concept applies to space loads and to equipment loads.

4-2 Manual N Load Calculation Procedure

Manual N provides a procedure for calculating the cooling loads at 9 AM, Noon, 3 PM and 6 PM. This should be adequate for most small commercial applications. The **Manual N** procedure is valid for the months of June, July and August. Peak cooling loads normally occur in these months — unless a zone has an unusual amount of south facing glass. (Refer to the ASHRAE load calculation procedure for tables that pertain to cooling load calculations for other months of the year or hours of the day.)

4-3 Equipment Sizing Loads

Load analysis begins with calculations that determine the loads that are used to size central equipment, unitary equipment and distribution devices. Different types of loads are used to size different types of equipment. In some cases the sizing load is equal to the space load. In other cases the sizing load is equal to the equipment load.

Supply Air Outlets
The supply CFM that is required for any room or zone is based on the space load (as defined on page 4-1). When cooling is required, the supply CFM is based on the sensible component of the peak space load. When heating is required, the supply CFM is based on the peak space heating load. When both heating and cooling are required, the design CFM is based on the larger of the two CFM values.

$$\textit{Cooling CFM} = \frac{\textit{Room Sensible Cooling Load (BTUH)}}{1.1 \times (\textit{Room DB} - \textit{Supply Air DB})}$$

$$\textit{Heating CFM} = \frac{\textit{Room Heating Load (BTUH)}}{1.1 \times (\textit{Supply Air DB} - \textit{Room DB})}$$

Heating and Cooling Equipment and Devices
When single-zone heating and cooling equipment (normally installed within the space) has no outdoor air capability, it is sized to satisfy the space load (as defined on page 4-1). Since there is only one zone, this load will be equal to the peak room load or the peak zone load. Space loads also are used to determine the flow rate (CFM or GPM) of the fluids that are associated with heating and cooling a room or zone. Supply air outlets, unit heaters, convectors, and fan-coil units (that have no outdoor air capability) provide examples of the type of equipment that is sized to satisfy the space load. (Refer to the following comments about single-zone HVAC equipment if outdoor air is introduced through the unit.)

Single-Zone HVAC Equipment
The sizing load associated with single-zone heating, cooling and ventilating equipment is equal to the equipment load (as defined on page 4-1). Since there is only one zone, this load will be based on a peak room load or a peak zone load. Single-zone rooftop units, fan-coil units with outdoor air capability and unit ventilators provide examples the type of equipment that is sized to satisfy the equipment load.

Central Multizone HVAC Equipment
The sizing load associated with central equipment that serves two or more zones is equal to the equipment load (as defined on page 4-1) at the central unit. Obviously, this load must include the space loads that are associated with the various zones, but the procedure for combining these zone loads depends on the system type. In some cases, the equipment load is based on a peak block load and in other cases, it is based on the total of the peak zone loads. Table 4-2 shows the relationship between the system type and the method that is used to combine the zone loads.

Combining Zone Loads

System Type	Cooling Capacity	Coil CFM	Fan CFM	Duct Diversity
TRH	Sum Zones	Sum Zones	Sum Zones	No
M-VAV	Block	Block	Block	Yes
BP-VAV	Block	Sum Zones	Sum Zones	No
DT-DD	Sum Zones	Sum Zones	Sum Zones	Yes
BT-DD	Block+XS	Block+XS	Sum Zones	Yes
DT-Mult	Sum Zones	Sum Zones	Sum Zones	No
BT-Mult	Block+XS	Block+XS	Sum Zones	No

XS = An allowance to compensate for air that is diverted through the hot-deck — approximately 10 percent.

TRH — Terminal reheat
M-VAV — Full modulating variable air volume
BP-VAV — Bypass variable air volume
DT-DD — Draw-through dual-duct
BT-DD — Blow-through dual-duct
DT-Mult — Draw-through multizone
BT-Mult — Blow-through multizone

Table 4-2

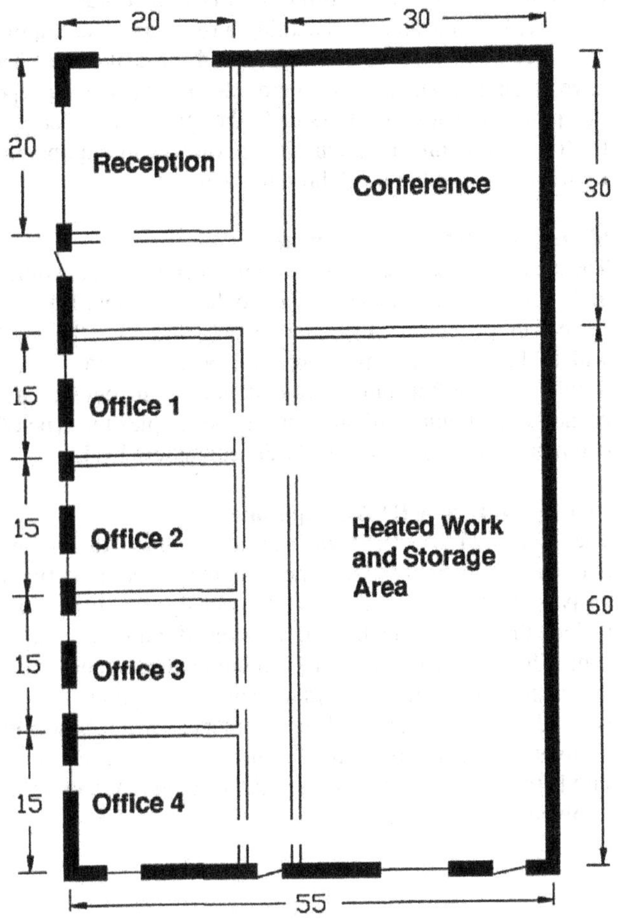

Figure 4-1

Equipment Loads — Sum of Peak Zone Loads					
	Sensible	Latent	CFM	Time	Occ
Conference	27,000	9,400	1,100	6 PM	15
Reception	26,200	4,850	1,160	6 PM	6
Office-1	10,250	1,850	440	3 PM	2
Office-2	10,250	1,850	440	3 PM	2
Office-3	10,250	1,850	440	3 PM	2
Office-4	12,450	2,100	550	Noon	2
Totals	96,400	21,900	4,130		
Equipment Loads Based on Block Load					
	Sensible	Latent	CFM	Time	Occ
Block	85,750	14,700	3,690	3 PM	15

Table 4-3

4-4 Example — Equipment Load vs System Type

The calculated equipment loads associated with the offices in a HVAC contractor's building (Figure 4-1) are summarized in Table 4-3. This table shows that approximately 10 TONS of cooling capacity will be required for the systems that are sized by totaling the peak zone loads — 5 PTAC units, a central reheat system or a draw-through multizone system, for example. The table also shows that a little more than 8 TONS of capacity are required for the systems that are sized by using the block load — a dumping VAV system or a blow-through multizone system, for example.

4-5 Ventilation Load Calculation Procedure

Outdoor air is intentionally supplied through the equipment for ventilation, exhaust system make up air and for economizer cycle cooling. Sometimes, outdoor air is used to pressurize a building. Outdoor air also may be unintentionally introduced through leaky dampers. The following steps should be taken to determine the outdoor air CFM.

Step 1) Determine the fresh air required for occupants.
State and local building codes set minimum fresh air requirements, which are mandatory. Also refer to the ASHRAE Standard — 62-1989 for outdoor air CFM values (Section 3, page 3-4). Due to the potential liability that is associated with a "poor air quality" law suit, the designer should use the ASHRAE values whenever they exceed the code values.

Step 2) Determine the make up air requirements.
Toilet and locker room exhaust requirements are specified by codes. Kitchen exhaust also may be specified by codes. (The code requirements usually represent the minimum acceptable values.) In any case, kitchen and toilet exhaust systems should control odors by maintaining the room at a slightly negative pressure with respect to other areas. (Make up air hoods can be used to reduce kitchen exhaust requirements.) Also investigate the equipment and process exhaust requirements associated with each space. Refer to ACCA **Manual Q**, Section 18, for more information about exhaust systems.

Step 3) Determine the pressurization CFM.
Outdoor air can be used to pressurize a space. In this case, the amount of outside air supplied to the space should be equal to the exhaust CFM plus 5 to 10 percent of the supply CFM. *(Buildings or rooms that are loosely constructed or any space that has an exhaust system that expels more than 10% of the supply CFM should not be pressurized.)* Refer to ACCA **Manual Q**, Section 17, for more information about building pressure. Refer to ACCA **Manual N**, Table 13, for information about how building pressure affects infiltration.

Pressurization CFM = Exhaust CFM + % of Supply CFM

Step 4) Determine the damper leakage.
The leakage associated with a closed damper is usually 5 to 10 percent of the "open damper" CFM. This means that the leakage associated with a closed economizer damper will be 5 to 10 percent of the supply CFM.

Leakage CFM = Open damper CFM x Percent leakage

Step 5) Determine the ventilation requirement.
The outdoor air CFM will be determined by the fresh air requirement, exhaust system make up air requirement, damper leakage or the CFM that is required to pressurize the space — which ever is greater.

Example
The building on the previous page (Figure 4-1) requires 3,690 CFM of supply air — based on the block load. Determine the outdoor air CFM for the following situations (cases). Use 15 CFM of outdoor air per person to meet the code requirement.

1) 15 people, 200 CFM toilet exhaust, 5% economizer leakage, no pressurization

2) 15 people, 300 CFM exhaust, 5% economizer leakage, no pressurization

3) 15 people, 300 CFM exhaust, 10% economizer leakage, no pressurization

4) 15 people, 200 CFM exhaust, 5% economizer leakage, 5% space pressure = 3690 x 0.05 + 200 = 385 CFM

5) 15 people, 400 CFM exhaust, 5% economizer leakage, 5% space pressure = 3690 x 0.05 + 200 = 385 CFM

Solutions to Example Problem

Case	Fresh Air CFM	Exhaust CFM	Leakage CFM	Pressure CFM
1	225	200	184	na
2	225	300	184	na
3	225	300	369	na
4	225	200	184	385
5	225	400	184	Void[1]

(1) Exhaust exceeds 10% of supply, do not pressurize

4-6 System Loads

System loads include duct losses, fan heat, ventilation loads, damper leakage, piping losses, water pump heat, return air ceiling plenum loads, reheat loads and mixing loads. The ACCA **Manual N** — Form N-1, provides a method for accounting for duct losses and return air ceiling loads (refer to lines 7, 11 and 10). Form N-1 also accounts for ventilation loads and outdoor air damper leakage — providing that the outdoor air CFM is calculated in accordance with procedure outlined in the previous section. The following comments apply to those items that are not included on Form N-1.

Fan Heat
Fan heat is usually (or should be) accounted for in the manufacturer's performance data. That is, the sensible capacity of the cooling equipment should be derated by an amount equal to the fan heat. Refer to Section 6, an example that illustrates this point. If the manufacturer's performance data does not accounted for fan heat, refer to ACCA **Manual N**, Table 11 (column B or C) and add the fan heat to the line 12 sensible load.

Piping Losses and Water Pump Heat
Piping losses can be added to the line 12 cooling load or the line 21 heating load (Form N-1). Water pump heat can be added to the line 12 cooling load. Refer to Table 11 (column B) to estimate the heat that is added to the water.

Reheat Loads
The reheat loads associated with properly designed reheat systems do not increase the equipment sizing load. Reheat only increases the "part-load" cooling load.

Mixing Loads
Mixing hot and cold fluids may increase the equipment sizing loads. Mixing loads do not increase the equipment load if mixing devices shut tight when throttled to the closed position. Mixing loads will increase the equipment load if a device cannot provide "tight shut-off." Mixing boxes and mixing dampers provide examples of devices that increase the equipment load. Refer to the manufacturer's data for information about leakage losses and add these losses to the line 12 cooling load or the line 21 heating load on Form N-1.

4-7 Operating Loads

During normal operation, the HVAC equipment and devices must track loads that are continuously changing. Cooling loads vary in accordance with changes in solar gains and internal loads. Cooling and heating loads are affected by changes in outdoor temperature, wind velocity, exhaust air requirements and outdoor air requirements. Day-cycle and night-cycle operating modes and setpoint schedules vary in accordance with occupancy schedules and production schedules. Economizer cycles and reclaim devices reduce the loads on the primary equipment.

Internal Zone Space Load
The space load associated with internal zones can vary from the no-load condition to the design cooling load condition. Heat is never required. When the space is occupied the cooling load will vary with changes in lighting levels, occupancy and the internal loads that are associated with equipment and appliances. Usually the minimum load occurs when the space is unoccupied (no lights, people or internal loads).

Perimeter Zone Space Load
The space load associated with perimeter zones can vary from the design heating load condition to the design cooling load condition. When the space is occupied, the cooling load will

vary with changes in the envelope loads, lighting levels, occupant density and the internal loads that are associated with equipment and appliances. The cooling load decreases as the outdoor air temperature drops. Eventually, the load will changeover from cooling to heating. The outdoor temperature that corresponds to this cooling/heating changeover point depends on the relative size of the internal loads and envelope loads. It is possible that a perimeter zone may never require heat when it is occupied. Usually, the maximum heating load occurs when the space is unoccupied (no lights, people or internal loads).

Equipment Load — Any Type of Zone
The equipment load associated with any type of zone can vary from the design heating load condition to the design cooling load condition. When the space is occupied, the cooling load will vary with changes in the envelope loads, internal loads, ventilation loads and other system loads. The cooling load decreases as the outdoor air temperature drops. Eventually, the load may changeover from cooling to heating. The outdoor temperature that corresponds to this cooling/heating changeover point depends on the relative size of the internal loads, envelope loads and system loads. It is possible that equipment that serves internal zones — or perimeter zones that have large internal loads — may never switch to the heating mode when the space is occupied.

4-8 Part-Load Analysis

Part-load analysis provides the information that is required for system selection, equipment selection and for developing control strategies. Some of the information that can be determined from a part-load analysis is listed below.

- Estimates of the minimum load on cooling equipment, which is important for specifying how the cooling equipment should unload or for deciding if hot gas bypass is required (refrigerant-side capacity control).
- An estimate of minimum load on heating equipment, which is important for specifying how the heating equipment should be controlled. (Minimum heating loads establish turn down ratios and burner staging requirements.)
- An estimate of cooling/heating changeover temperature that is associated with the equipment, which is used for deciding if an economizer is required and for developing a control strategy for the cooling/heating changeover.
- An estimate of the minimum cooling load that is associated with the conditioned space, which is used for making zoning decisions and for choosing between constant volume, variable volume and reheat systems. If a variable volume system is installed, this information is used to decide if reheat is required to maintain adequate air circulation rates and ventilation rates at part-load conditions.
- Estimates of the heating/cooling changeover temperatures associated with the conditioned spaces, which is required for making zoning decisions and for developing a control strategy for the cooling/heating changeover.

Problems Associated with Part-Load Operation

Temperature drifts out of comfort zone
Poor humidity control
Improper ventilation, poor control of fresh air
Improper air motion, poor circulation
No air balance
Heat required but not available
Cooling required but not available
Incorrect changeover from cooling to heating
Loads fall below the throttling range of the controls
Improper control of the compressor
Improper control of DX coil capacity
Improper control of the heating plant
Operating or safety controls cycle equipment
Unstable valve or damper performance
Inefficient operation

4-9 Load Line Basics

Load lines are simple graphs that show how space loads or equipment loads vary with outdoor temperature. These diagrams are the basis for making the part-load analysis.

Figure 4-2 (see next page) shows the load lines that are associated with the various types of load components. Note that the transmission, infiltration and ventilation loads vary linearly with outdoor temperature and that internal loads are not affected by outdoor temperature. Solar loads are also shown as constant because they are not directly coupled to the outdoor temperature. For example, the solar gain on a hot day in August is approximately the same as the solar gain on a cool day in August. (Technically, the solar load line does vary with outdoor temperature because the solar azimuth angle in the colder months is not the same as the azimuth angle in the warmer months. This technicality does not affect the load-line analysis because solar load is ignored when the cooling/heating changeover point is evaluated.)

Load-line diagrams associated with a conditioned space include the transmission loads, infiltration loads, solar loads and internal loads. Load-line diagrams associated with equipment include the loads that are associated with the space plus the ventilation and other system loads. The construction and

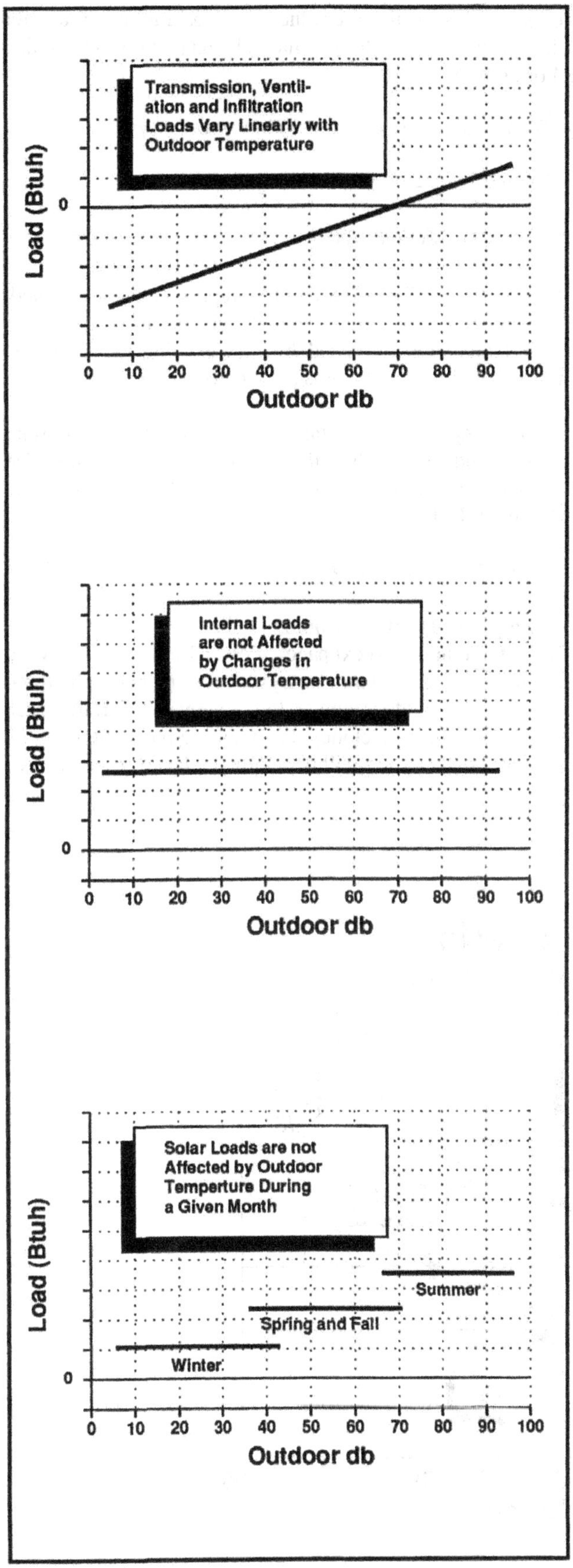

Figure 4-2

evaluation of these diagrams is explained by the following paragraphs. These explanations will be based on load data for a single-zone office area (see the floor plan below). The load information for this office area is summarized in the following table.

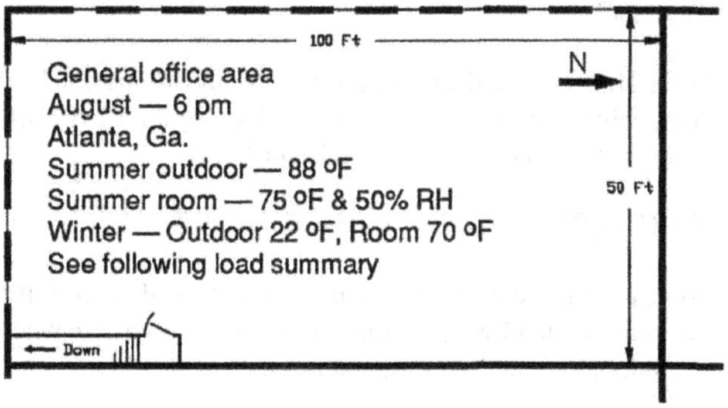

Form N-1 Load Summary (BTUH) — Office Area		
Component	**N-1 Line**	**Sensible Cooling**
Solar	2	14,899
Transmission	3	43,343
Internal	4	90,590
Infiltration	5	3,790
Cooling subtotal	6	152,622
Supply duct	7	4,579
Space load	8	157,201
Ventilation	9	12,870
RA Ceiling	10	37,780
Return duct	11	0
Equipment load	12	207,851
Component	**N-1 Line**	**Heating Load**
Transmission	14	70,753
Infiltration	15	21,912
Heating subtotal	16	92,665
Supply duct	17	2,780
Ventilation	18	47,520
Return duct	20	0
Equipment load	21	142,965

4-10 Load Lines Associated with the Space

The information on the ACCA **Manual** N, Form N-1 and some graph paper can be used to construct a load-line diagram for the conditioned space. (Refer to the floor plan and load summary above.) This load-line diagram will provide a graphical summary of the loads that will exist in the conditioned space during any operating condition. An example of a completed load-line diagram is provided by Figure 4-3. This diagram was constructed as follows:

Step 1) Draw a T&I line that represents the combined, heating season, transmission and infiltration load at any outdoor temperature. Proceed by drawing the T&I line through points #1 and #2.

Point #1 represents the total transmission and infiltration load at the winter design condition. The value of point #1 is equal to the sum of lines 16 and 17 on Form N-1.

Pt #1 = -(92,665 + 2,780) = -95,445 BTUH @ 22 °F

Point #2 represents the no-load condition. In this case, the transmission load and the infiltration load are both zero when the indoor and outdoor temperatures are equal.

Pt #2 = 0 BTUH @ 70 °F

Step 2) Draw a T&I+I line that represents the combined, transmission, infiltration and internal loads at any outdoor temperature. Proceed by drawing the T&I+I line through point #3 so that it is parallel to the T&I line.

Point #3 represents the total of the transmission, infiltration and internal loads at the summer design condition. The value of point #3 is equal to the value of the T&I line at the summer design temperature plus the internal load that is found on line 4 of Form N-1.

T&I load @ 88 °F = 36,000 BTUH {from graph}

Internal loads = 90,590 BTUH {N-1, line 4}

Point #3 = 126,590 BTUH

Step 3) Draw a T&I+I+S line that represents the combined, transmission, infiltration, internal and solar load at any outdoor temperature. Proceed by drawing the T&I+I+S line through point #4 so that it is parallel to the T&I line.

Point #4 represents the total of the transmission, infiltration, internal and solar loads at the summer design condition. The value of point #4 is equal to the summer design load on line 8 of Form N-1.

Point #4 = 157,201 BTUH

Changeover Points — Room

Refer to Figure 4-4 (next page). Point #1 indicates that when the space is occupied on a sunny day, cooling will be required whenever the outdoor temperature is above 10°F. If the space is occupied during a cloudy day or during the night, point #2 indicates that cooling will be required whenever the outdoor

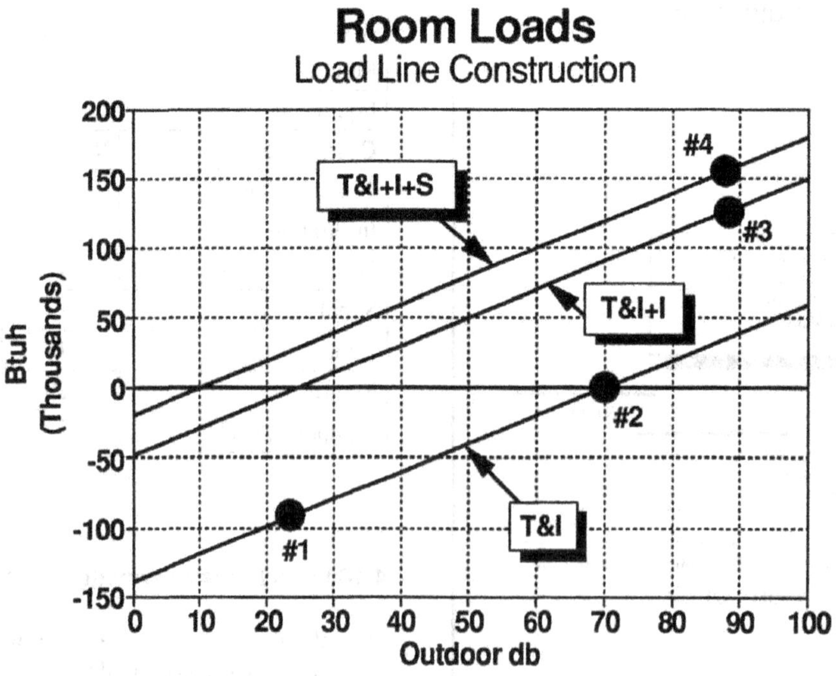

Figure 4-3

temperature is above 24°F. Point #3 indicates that when the space is not occupied, during a cloudy day or during the night, heating will be required when the outdoor temperature is below 70°F.

Control Strategy — Room

During occupancy (even though there are two exposed walls and an exposed roof), the internal loads make this room a cooling-only application. But in this case, the minimum cooling load is less than 40 percent of the design cooling load. (This occurs when the outdoor temperature is below 55°F and there is no solar load.) This means that if a VAV system is installed, reheat will have to be used to create a load that will require (at least) 40 percent of the design air flow.

To minimize operating costs, the cooling should be shut off when the space is not occupied. Heat will be required when the space is not occupied and the outdoor temperature is below 70°F. If night set back is used, heat will be required whenever the space is not occupied and the outdoor temperature is below the set-back temperature.

4-11 Load Lines Associated with the Unit

The cooling/heating changeover points at the unit will not occur at the same outdoor temperature as the room or zone changeover points. The changeover points that are associated with the unit are different because the outdoor air that is required for ventilation is introduced at the return side of the unit. The duct losses at the return side of the duct system also affect the heating/cooling changeover points at the unit. The unit must operate in the cooling mode when the outdoor temperature is above the cooling/heating changeover point. The heating mode must be activated when the outdoor temperature falls below this changeover point.

The Form N-1 information and some graph paper can be used to construct a load-line diagram for the unit. (Refer to the floor plan and load summary on page 4-7.) This load-line diagram summarizes (graphically) the loads that will be imposed on the unit during any operating condition. An example of a completed load-line diagram is provided on the next page by Figure 4-5. This diagram was constructed as follows:

Step 1) Draw a T&I&V line that represents the combined, heating season, transmission, infiltration and ventilation load at any outdoor temperature. Proceed by drawing the T&I&V line through points #1 and #2.

Point #1 represents the total transmission, infiltration and ventilation load at the winter design condition. The value of point #1 is equal to line 21 on Form N-1.

Pt #1 = -142,965 BTUH

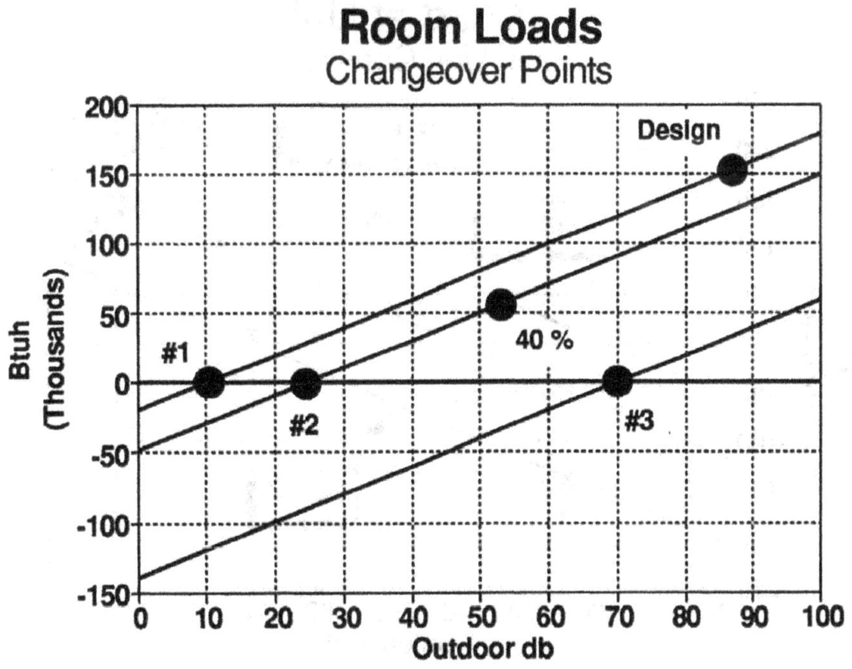

Figure 4-4

Point #2 represents the no-load condition. In this case, the transmission load and the infiltration load are both zero when the indoor and outdoor temperatures are equal.

Pt #2 = 0 BTUH @ 70 °F

Step 2) Draw a T&I&V+I line that represents the combined, transmission, infiltration, ventilation and internal load at any outdoor temperature. Proceed by drawing the T&I&V+I line through point #3 so that it is parallel to the T&I&V line.

Point #3 represents the total of the transmission, infiltration, ventilation and internal loads at the summer design condition. The value of point #3 is equal to the value of the T&I&V line at the summer design temperature plus the internal loads that are found on line 4 of Form N-1.

T&I&V load @ 88 °F = 54,000 BTUH {from graph}

Internal loads = 90,590 BTUH {N-1, line 4}

Point #3 = 144,590 BTUH

Step 3) Draw a T&I&V+I+S line that represents the combined, transmission, infiltration, ventilation, internal load and solar load at any outdoor temperature. Proceed by drawing the T&I&V+I+S line through point #4 so that it is parallel to the T&I&V line.

Point #4 represents the total of the transmission, infiltration, ventilation, internal and solar load at the summer design condition. The value of point #4 is equal to the summer design load on line 12 of Form N-1.

Point #4 = 207,851 BTUH

Design Loads at the Unit
Refer to Figure 4-6 on the next page. Point #1 corresponds to the 207,851 BTUH sensible design load on the unit. The design latent load on the unit is 42,835 BTUH, as indicated on line 12 of Form N-1.

Minimum Cooling Loads at the Unit
If the unit is equipped with an economizer cycle, the compressor and the DX coil will be deactivated when the outdoor air can provide the necessary cooling. This condition is usually satisfied when the outdoor temperature is between 55°F and 60°F — depending on the local climate and on the type of economizer control. Use the cooling loads that correspond to the 55°F outdoor temperature to estimate the minimum cooling loads on the unit.

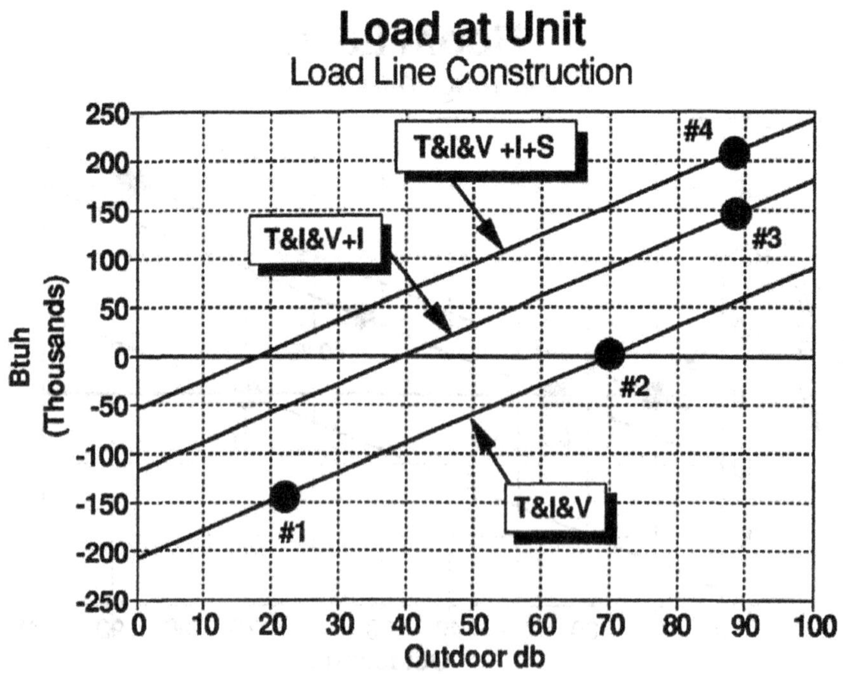

Figure 4-5

Refer to Figure 4-6. Point #2 indicates that when the space is occupied on a sunny day, 109,000 BTUH of sensible cooling will be required just before the mechanical refrigeration is de-energized. If the space is occupied during a cloudy day or during the night, point #3 indicates that 46,000 BTUH of sensible cooling will be required just before the mechanical refrigeration is de-energized.

Point #3 establishes the minimum sensible cooling load on the refrigeration equipment. The latent load that corresponds to point #3 will be less than the design latent load because the latent loads that are associated with infiltration and ventilation will be zero or negative at the 55°F (outdoor air) changeover temperature. Therefore, a conservative estimate of the minimum "total load" on the refrigeration equipment is represented by the point #3 sensible load.

Changeover Points — Unit

Refer to Figure 4-6. Point #4 indicates that when the space is occupied on a sunny day, a net cooling load will exist at the unit whenever the outdoor temperature is above 18°F. If the space is occupied during a cloudy day or during the night, point #5 indicates that cooling will be required whenever the outdoor temperature is above 40°F. Point #6 indicates that when the space is not occupied during a cloudy day or during the night, heating will be required when the outdoor temperature is below 70°F.

Heating Loads at the Unit

Refer to Figure 4-6. Point #7 indicates that when the space is occupied on a cloudy day or at night, the net winter design heating requirement will be equal to 52,000 BTUH. Point #8 corresponds to the 143,000 BTUH total heating capacity that must be available at the unit when there are no internal loads. This amount of heat would be required if the space is not occupied during a cloudy day or during the night and the outdoor temperature is equal to the winter design temperature.

Day-Cycle Control Strategy — Unit

The minimum cooling load on the compressor occurs just before the system switches to the economizer cycle. At this point, the compressor capacity control hardware must be capable of unloading the machine down to 18 percent of the design capacity.

46,000/(207,851+42,835) = 0.18

During occupancy, the internal loads make this a "cooling-only" application whenever the outdoor temperature is above 40°F. During the day cycle, heat must be available when the outdoor temperature is at or below 40°F. Note that only 52,000 BTUH of heat will be required on a winter design day. The day-cycle heating capacity is relatively small because a

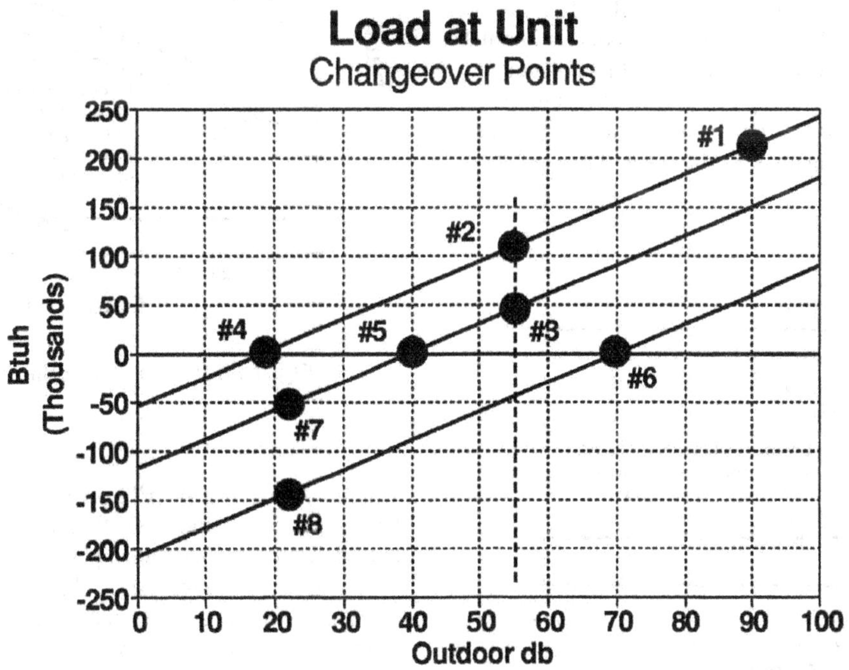

Figure 4-6

considerable amount of heat is provided by the internal loads. To prevent short cycling, the controls should be able to stage or modulate the output capacity of the heating equipment.

Night Cycle Control Strategy — Unit
To minimize operating costs, the cooling should be shut off and the outdoor air dampers should be closed when the space is not occupied. When the space is not occupied and the outdoor temperature is below 70°F, 143,000 BTUH of heating capacity will be required. If night set back is used, heat will be required whenever the space is not occupied and the outdoor temperature is below the set-back temperature.

4-12 Control Strategies for Conditioned Spaces

The concept of using load lines to establish a control strategy for the conditioned space has already been introduced in Section 4-10. In general, the load lines associated with the conditioned space can be used to evaluate the low ambient cooling requirements, to predict the minimum cooling load and to determine the cooling/heating changeover point.

Internal Zones
Figure 4-7 shows the load lines that are associated with internal zone space loads. These lines are horizontal because the cooling load is not affected by outdoor air temperature. The lines show that internal zones require year-round cooling. When the space is occupied, the maximum and minimum cooling loads are directly related to variations in the internal loads. Cooling is not usually required when the space is unoccupied. If "unoccupied cooling" is required, the load may be substantially smaller when the lights are off and the people are gone. In any case, the equipment that cools an internal zone must either be equipped with a low ambient cooling kit or an economizer cycle. Unitary equipment with outdoor air capability is appropriate for most single-zone applications. Constant volume, constant temperature systems are appropriate for cooling two or more internal zones that have steady cooling loads. Variable volume or variable temperature systems should be considered when two or more internal zones experience load swings that exceed 10 percent of the design cooling load.

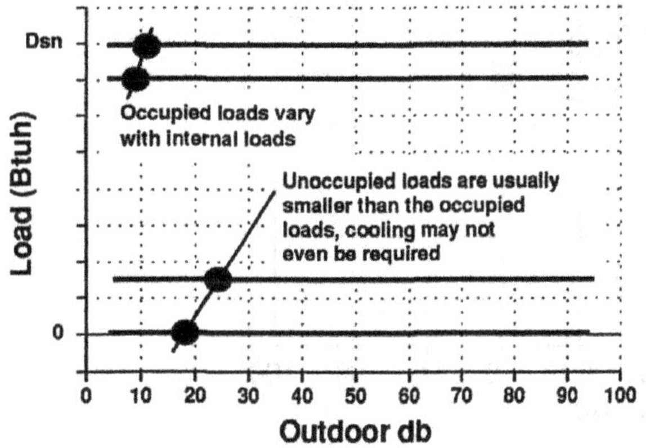

Figure 4-7 - Internal Zone

Perimeter "Cooling-Only" Zones
Figure 4-8 shows the load lines associated with the space load in a perimeter zone that has a relatively large internal load (compared to the envelope loads). These lines slope because the cooling load is affected by outdoor air temperature. The lines show that this type of perimeter zone requires year-round cooling. When the space is occupied, the minimum cooling load is related to variations in the internal loads and outdoor temperature. Cooling is not usually required when the space is unoccupied. If "unoccupied cooling" is required, the load may be substantially smaller when the lights are off and the people are gone. If the internal loads are very small or negligible when the space is unoccupied, the space load will changeover to heating. In any case, the equipment that cools this type of zone must either be equipped with a low ambient cooling kit or an economizer cycle. Unitary equipment with outdoor air capability is appropriate for most single-zone applications. Variable volume or variable temperature systems should be considered when central equipment serves two or more zones. When the space is occupied, a variable volume system should not be allowed to throttle the supply CFM to a value that is less than 40 percent of the design CFM.

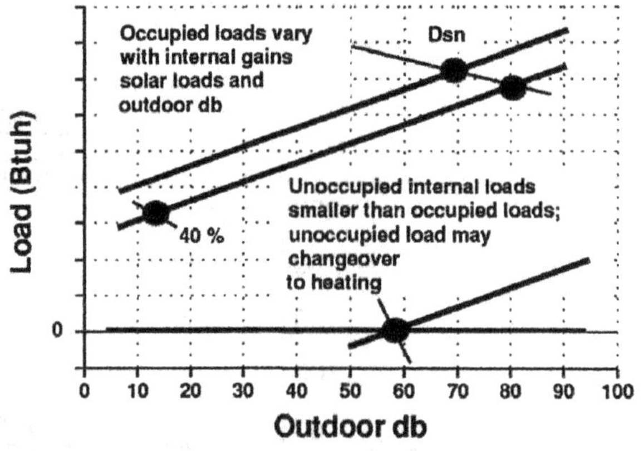

Figure 4-8 - Cooling-Only / Perimeter

Perimeter "Changeover" Zones
On the next page, Figures 4-9 and 4-10 show the load lines that are associated with the a perimeter zone space load that experiences a cooling/heating changeover when the zone is occupied. These figures show that the cooling/heating changeover point depends on the relative size of the internal loads and envelope loads. This type of perimeter zone requires heating and cooling during the occupied operating cycle. If the heating changeover point occurs at a low outdoor temper-

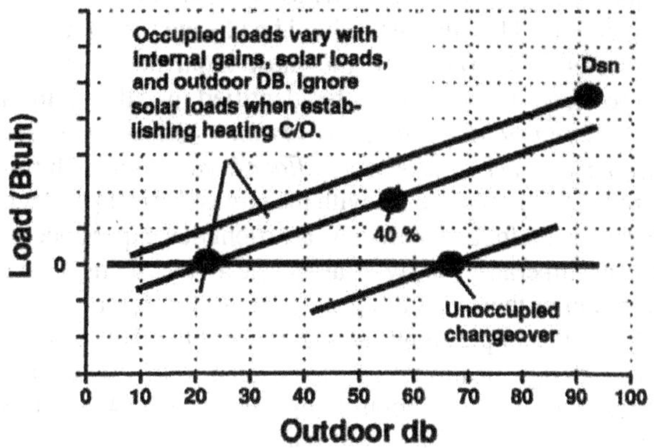

Figure 4-9

ature, the cooling equipment must have a low ambient cooling kit or an economizer cycle. Cooling is not usually required when the space is unoccupied. If "unoccupied cooling" is required, the load may be substantially smaller when the lights are off and the people are gone. If the internal loads are very small or negligible when the space is unoccupied, the space load will changeover to heating at a relatively mild outdoor temperature. Unitary equipment with outdoor air capability is appropriate for most single-zone applications. Variable volume or variable temperature systems should be considered when central equipment serves two or more zones. When the space is occupied, a variable volume system should not be allowed to throttle the supply CFM to a value that is less than 40 percent of the design CFM.

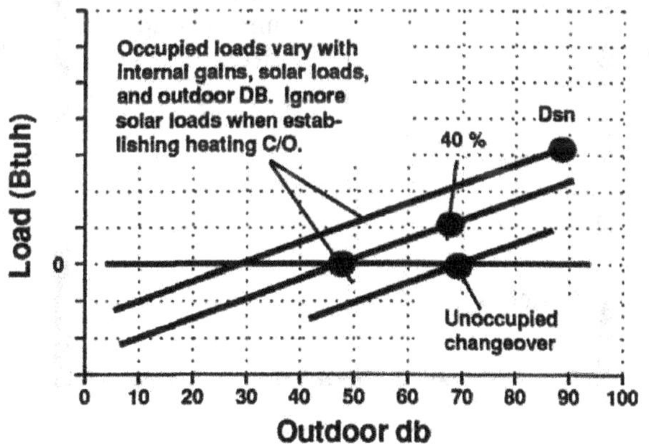

Figure 4-10

Simultaneous Heating and Cooling

Simultaneous heating and cooling will be required when a building has two or more rooms that have incompatible cooling/heating changeover points. For example, interior "cooling-only" zones are not compatible with exterior "changeover" zones. Exterior zones that changeover at a mild outdoor air temperature are not compatible with exterior zones that changeover at a low outdoor air temperature. Therefore, when a building is zoned, the rooms that make up a zone should have similar cooling/heating changeover points.

A simultaneous heating and cooling requirement can be satisfied by installing multiple single-zone units — one for each zone. This strategy is energy efficient because simultaneous heating and cooling is accomplished without mixing penalties or reheat penalties. Water-source, water-loop heat pump systems provide simultaneous heating and cooling and energy efficient operation. Central all-air systems must mix hot and cold fluids or reheat supply air to satisfy a simultaneous heating and cooling requirement. Variable volume, variable temperature, changeover (VVT) systems, which "flip-flop" between the heating and cooling modes, do not provide simultaneous heating and cooling. Four-pipe water systems provide simultaneous heating and cooling without mixing hot and cold fluids, but three-pipe systems are characterized by a mixing penalty. Two-pipe water systems cannot provide simultaneous heating and cooling.

Heating Changeover Point

The heating changeover point is defined by the outdoor air temperature that corresponds to a "no-load" condition in the occupied space. This condition occurs when the transmission and infiltration losses just balance the solar and internal gains. Because internal rooms do not have transmission or infiltration losses, they are "cooling-only" zones. Since internal rooms never require heat, they do not have a cooling/heating changeover point. However, exterior zones will lose heat during cold weather and they will have a cooling/heating changeover point.

Design Heating Changeover Temperature

The cooling/heating changeover temperature that is associated with an occupied room is affected by the size of the internal and solar loads. The highest changeover temperature occurs when the internal loads and solar loads are at their smallest combined value. The lowest heating changeover temperature occurs when the internal loads and solar loads are at their largest combined value. The design changeover temperature is equal to the highest changeover temperature. The simplest way to determine the changeover temperature is to draw the load-line diagram.

4-13 Humidity Control

Another neglected aspect of part-load analysis is related to the room sensible heat ratio (RSHR) fluctuations that occur as the room sensible and latent loads change. In some cases, equipment that is selected to provide adequate dehumidification at the summer design condition may be short of latent capacity at some part-load condition. This occurs because the sensible loads decrease faster than the latent loads as the outdoor temperature drops or the solar gains and internal loads diminish.

There is no easy way to predict how room humidity levels will vary throughout the cooling season. Fortunately, precise humidity control is not required for comfort cooling applications. Therefore, the room humidity can be allowed to drift— providing that it remains within the ASHRAE comfort zone. (Refer to Section 3, page 3-2 for more information about the ASHRAE comfort zone.) In general, rooms that have a high room sensible heat ratio (RSHR = 0.80 or higher) at the summer design condition will remain comfortable if the system is designed to maintain 50 percent relative humidity on a design day. Rooms with a low RSHR (0.70 or lower) may require reheat to maintain comfort at part-load conditions. Reheat is usually used for applications that require precise control over room humidity — a laboratory, for example.

The humidity control problem is more complicated when one air handler is used to condition more than one zone. In this case, the coil sensible heat ratio (CSHR) at the air handler may not be compatible with each of the RSHR values (refer to Figure 4-17 on page 4-17). Ideally, the building should be zoned so that the various rooms have similar RSHR values (see the Zoning Evaluation Table, Section 2, page 2-6). If the building cannot be zoned according to the RSHR values, some compromises will be necessary when the coil performance is specified. The following options are available to the system designer — assuming that none of the zones requires precise humidity control.

- Select the coil to satisfy the zone with the lowest RSHR. In this scenario, the humidity in the controlling zone will be maintained at the desired level on a design day. Since the coil is cold enough to take care of the most humid zone, the humidity levels in the other zones will be lower than the desired level.

- Select the coil to satisfy the zone that has a RSHR value that falls in the mid-range of the various RSHR values. In this scenario, the humidity in the controlling zone will be maintained at the desired level on a design day. However, the coil will be too cold or too warm for the other zones and the humidity levels in those zones will be lower or higher than the desired level.

4-14 Night Cycle Thermostat Location

There are many applications that allow the designer to reduce operating costs by activating set-up and set-back controls when the building is unoccupied. (In many cases, the cooling can be eliminated during the unoccupied hours.) One popular control strategy is to use a single "night-stat" to cycle central heating equipment (on and off) to maintain a night set-back temperature. This night-stat should be located in the room or zone that will experience the most rapid decay in space temperature. This room can be identified by calculating the *space heating load—per degree of winter design temperature difference — per square foot of floor area* — for each room or zone. The room or zone with the largest BTUH per degree °F per square foot value will be the room that experiences the fastest temperature decay — assuming all the rooms or zones have similar thermal mass per square foot of floor area. The following example shows how to make this calculation. The winter design heating loads associated with this example correspond to a 70°F room temperature and a 4°F outdoor temperature. The night-stat should go in the reception area.

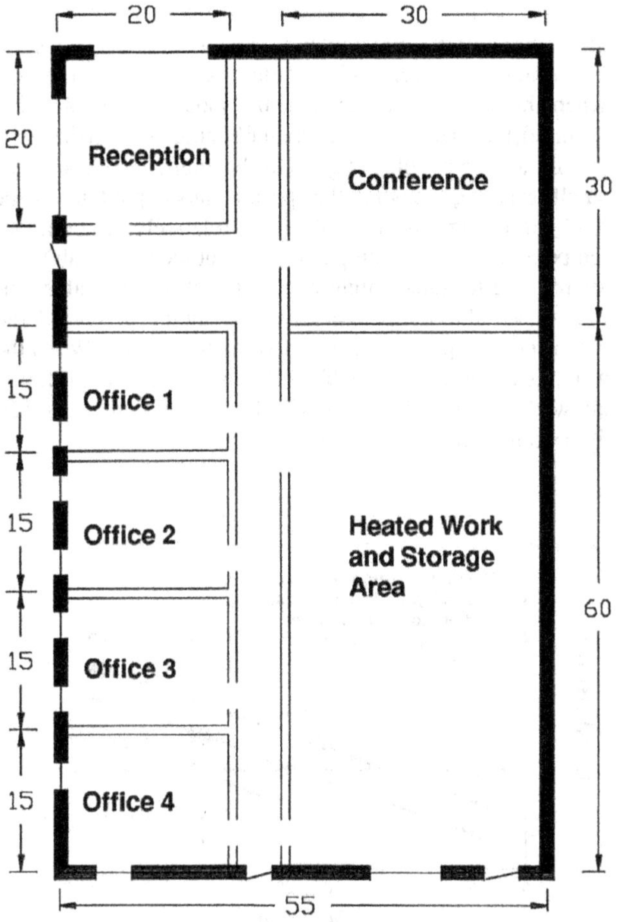

Zone	Load①	Floor SF	BTUH /°F /SF
Office 1,2,3	6000	300	0.30
Office 4	7100	300	0.36
Confer.	12200	900	0.21
Recept.	10100	400	0.38
① Form N-1, Line 16 . . . (TD = 70 - 4 = 66 °F)			

4-15 Equipment Loads vs Space loads

The load lines associated with the equipment are different from the load lines related to the space because they include the effect of the system loads — especially the ventilation load. Figure 4-11 shows the relationship between the equipment load and the space load. Note that the ventilation load and any other system loads cause the design heating and cooling loads at the equipment to be larger than the corresponding space loads. Also, note that the equipment load changes to the heating mode at a higher outdoor temperature than the space load.

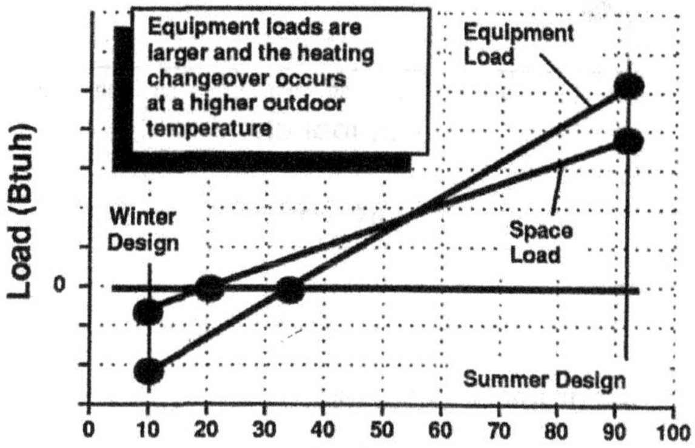

Figure 4-11

4-16 Control Strategies for Equipment

The concept of using load lines to establish the control strategy for HVAC equipment has already been introduced in Section 4-11. In general, the load lines that are associated with the equipment can be used to evaluate the minimum cooling load on the refrigeration machine, to find the cooling/heating changeover point and to evaluate the range of operating loads on the heating plant.

Minimum Cooling Load at the Unit

The minimum cooling load on the refrigeration equipment depends on the method that is used to provide low ambient cooling. If no air-side economizer is used, the refrigeration equipment must operate when the outdoor temperature is below 55 degrees. In this case, the minimum load would be equal to zero at the changeover point. If an economizer is used, the minimum refrigeration load will occur just before the system switches from mechanical cooling to economizer cooling. In this case, the minimum refrigeration load will most likely be greater than zero. Figure 4-12 illustrates this point.

In all cases, the refrigeration equipment — specifically the compressor and the cooling coil — must track the changes in the cooling load. This is accomplished by staging multiple compressors, equipping compressors with unloaders, installing variable speed compressors, equipping compressors with hot gas bypass accessories, installing row-split and face-split DX coils and using modulating water valves on chilled water coils. (Low ambient kits will be required for low ambient operation if the system has no air-side economizer.)

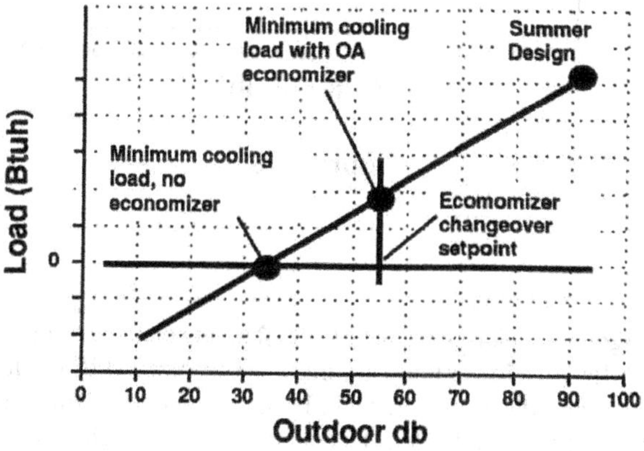

Figure 4-12

The minimum cooling load will dictate the amount of capacity control that is required. Compressor staging typically provides 50 percent, 33 percent and 25 percent capacity reduction increments — depending on how many compressors are staged. Unloaders can be installed if smaller increments of capacity reduction are required. Variable speed compressors provide modulating capacity control. Hot gas bypass accessories can be installed if the minimum cooling load is less than the smallest step of cooling capacity. Appendix 5 provides more information about capacity control.

Cooling/Heating Changeover Point at the Unit

The cooling/heating changeover point at the unit depends on whether the space is occupied or unoccupied and on the night-cycle control strategy. When the space is occupied, the internal loads tend to reduce the changeover temperature and the ventilation load tends to increase the changeover temperature. In any case, the changeover temperature is determined by the equipment load line that corresponds to the occupied condition when there are no solar gains. Figure 4-13, on the next page, illustrates this condition.

When the space is unoccupied, the internal loads are usually small (or zero) and the outdoor air damper will be closed. In this case, the cooling/heating changeover point corresponds to a load line that is based on the unoccupied internal loads, with an allowance for the outdoor air load that is associated with damper leakage. This changeover condition also is illustrated by Figure 4-13.

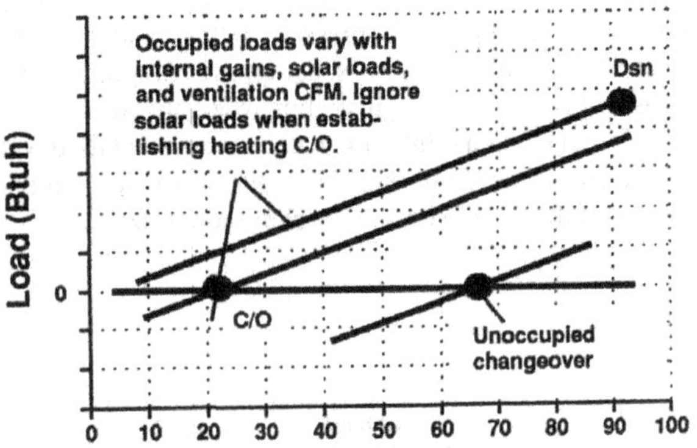

Figure 4-13

If night set back is used, the cooling/heating changeover point is shifted to a lower temperature. The size of the shift is equal to the amount of the set back. Figure 4-14 below illustrates how a 15°F set back affects the unoccupied changeover temperature.

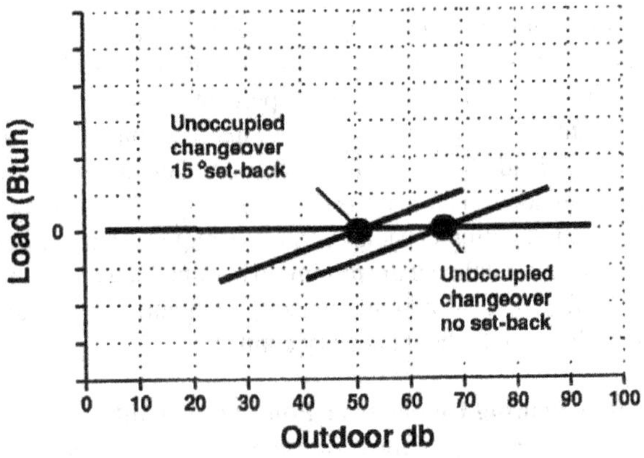

Figure 4-14

Controlling Heating Capacity

The design heating load during the occupied condition can be substantially smaller than the design heating load during the unoccupied condition. This situation is illustrated by Figure 4-15. In this case, the heating equipment should be able to adjust its capacity in proportion to the heating demand. This can be accomplished by equipping boilers and indirect-fired furnaces with on-off, staged or modulating fueltrain controls. (Direct-fired equipment features modulating capacity control and it can be used for heating applications. See Section 7-10.)

Electric heat can be staged by using simple controls or pulsed by SCR controls. Hot water coils can be equipped with modulating water valves or modulating face and bypass dampers.

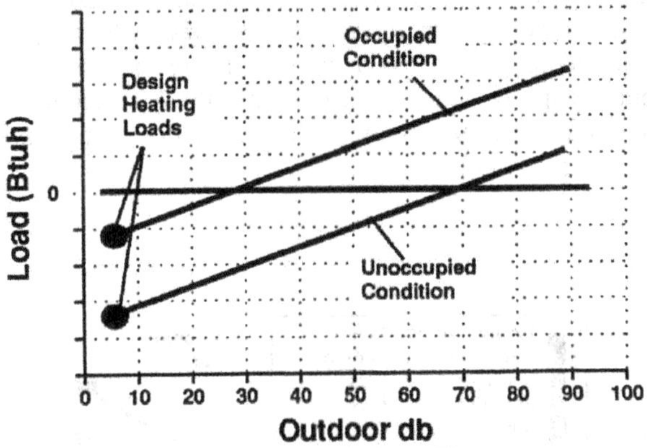

Figure 4-15

4-17 Psychrometric Analysis

Some basic psychrometric principles, as they pertain to space loads and equipment loads, are reviewed in this section. Refer to ACCA **Manual P** for a comprehensive discussion of psychrometric principles. In particular, refer to **Manual P**, Section 3 and Section 4, for information about sensible heat ratio lines.

Humidity Control in the Conditioned Space

The room sensible heat ratio (RSHR) line can be used to evaluate the ability of the supply air to absorb the sensible and latent loads that are associated with the conditioned space. The sensible capacity of the supply air will depend on its dry-bulb temperature. The latent capacity of the supply air will depend on its absolute humidity — measured in *grains per pound*. For a single-zone system, any supply air condition that falls on or below the RSHR line is an acceptable supply condition (from the standpoint of humidity control). This idea is illustrated by the shaded area on Figure 4-16 (see the next page); also refer to ACCA **Manual P**, page 29.

Multiple RSHR lines are associated with a central air handler that serves two or more zones. These lines may have similar slopes or they may "fan out" — depending on the RSHR values that are associated with each zone. Figure 4-17, on the next page, provides an example that shows three zones that have different RSHR values.

When the RSHR lines fan out, a supply air condition that is ideal for one or some of the zones may be less than ideal for the other zones. If the designer selects a supply condition that

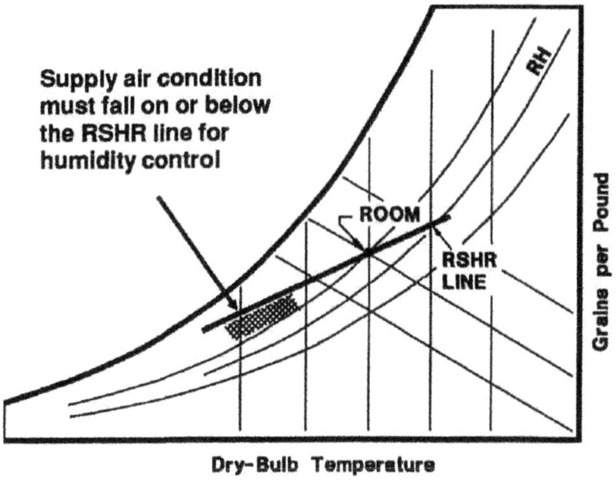

Figure 4-16

falls below the lowest RSHR line, the relative humidity in each zone will be maintained at or below the desired level. If the designer selects a supply condition that falls in the middle of the RSHR lines, the relative humidity in some zones will be higher than the desired level and the relative humidity in other zones will be lower than the desired level. Refer to Section 4-13 of this manual for more information about humidity control, also refer to ACCA **Manual P**, page 29.

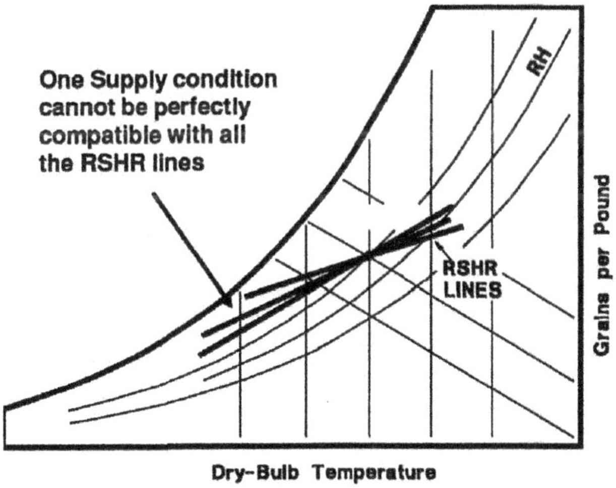

Figure 4-17

Psychrometric Analysis at the Unit

The coil sensible heat ratio (CSHR) line can be used to determine the ability of the supply air to absorb the sensible and latent loads that are associated with the equipment load. (The equipment load includes the space loads, the ventilation load and any other system load.) *To insure humidity control in the conditioned space, the supply air condition must fall on or below the CSHR line and it also must fall on or below the lowest RSHR line.* This idea is illustrated by the shaded area in Figure 4-18. Refer to ACCA **Manual P**, page 31 for a more comprehensive discussion of this point. Also refer to **Manual P**, pages 23 and 24 for an explanation of how to determine the return-side mixed air condition (MAT).

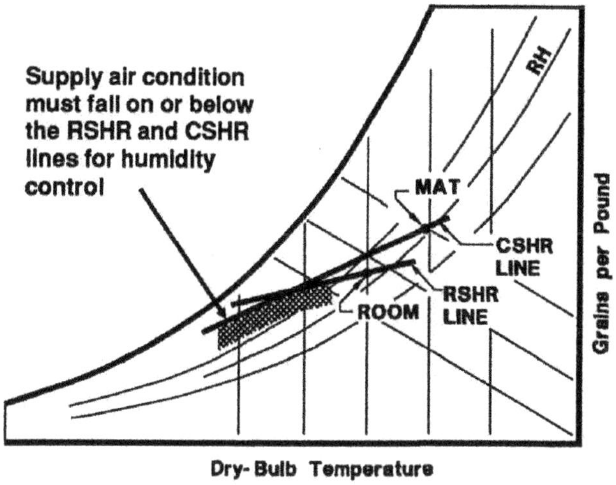

Figure 4-18

Equipment Selection

It is the designer's responsibility to make sure that the cooling coil will provide the desired supply air condition. Some equipment selection procedures allow the designer to specify the supply air condition explicitly — when specifying chilled water cooling coil performance, for example. In this case, the designer specifies the CFM, the condition of the entering air and the condition of the leaving air. Other equipment selection procedures do not specifically address the question of the supply air condition — packaged cooling equipment for example. In this case, the designer uses performance data to verify that the equipment has the required sensible and latent capacity at the design conditions. Equipment selection procedures are explained in more detail in the following sections of this manual.

Section 5
Fluid Temperatures and Flow Rates

The designer must either specify or determine the fluid operating temperatures and flow rates at various points in the HVAC system. This section discusses the temperatures and flow rates that are associated with the air side of cooling and heating systems, the water side of chilled water systems and the water side of hydronic heating systems.

5-1 Air-Side Temperatures — Cooling

The designer should be concerned with the dry-bulb and wet-bulb temperatures of the air entering and leaving the cooling coil and the condition of the air that is supplied to the occupied space. The entering and leaving conditions that are associated with the cooling coil are important because they affect the equipment selection procedure. The supply air dry-bulb temperature is important because it is directly related to the CFM that is supplied to the conditioned space.

Entering Conditions

Cooling coil performance, in terms of sensible and latent capacity, is affected by the dry-bulb and wet-bulb temperatures of the air entering the coil. Therefore, knowledge of the entering condition is a prerequisite to the equipment sizing

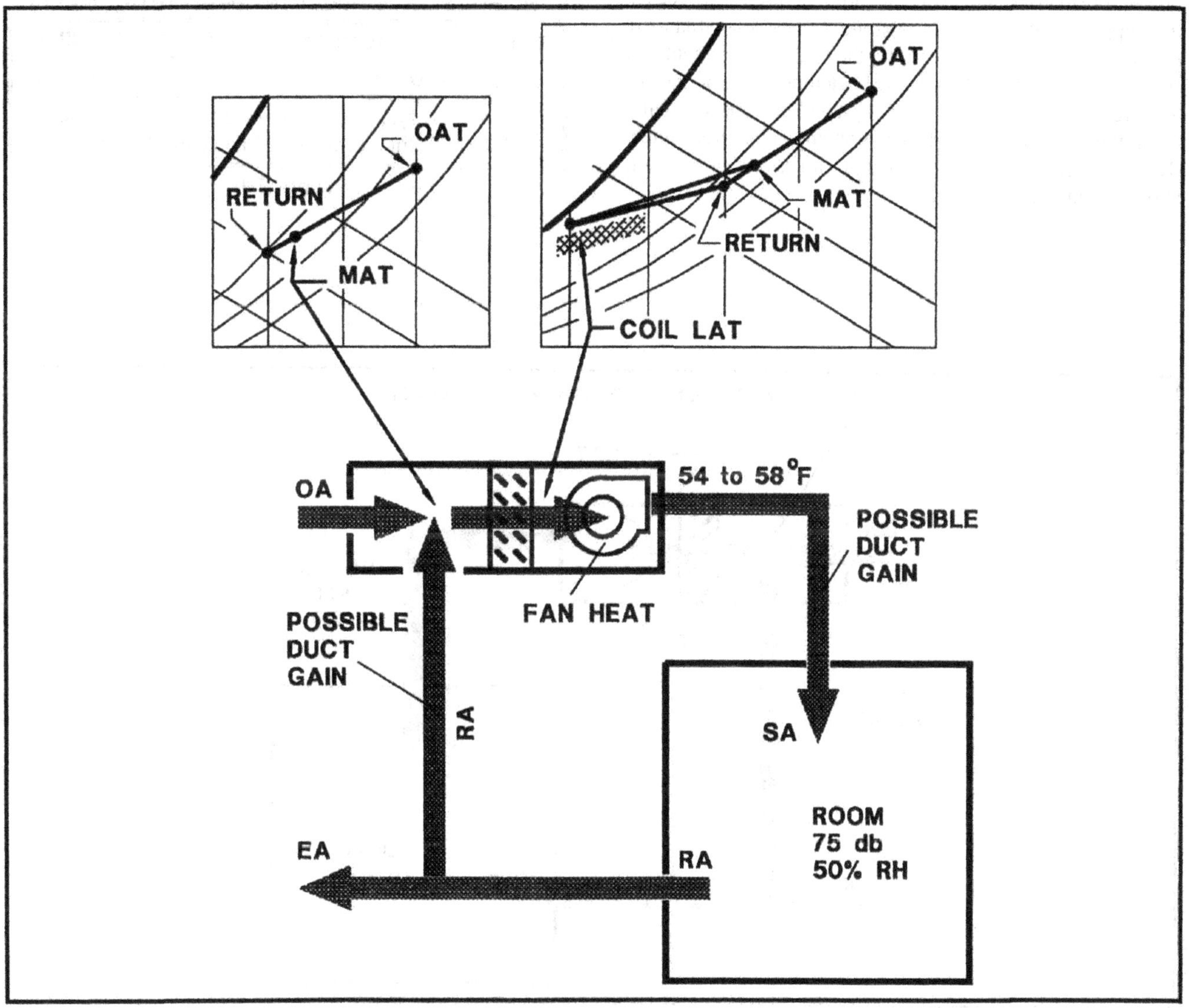

Figure 5-1

procedure. In commercial work, the condition of the air entering the cooling coil is usually the result of mixing return air (RA) with outdoor air (OA). This condition, as illustrated on the previous page by Figure 5-1, is determined by drawing the mixing process on the psychrometric chart. (Refer to ACCA **Manual P**, page 23, for a discussion of this procedure.)

Leaving Conditions
During the cooling season, the air off the coil (LAT) must be cold enough and dry enough to absorb the sensible and latent loads that are associated with the conditioned space. A psychrometric analysis, as illustrated by Figure 5-1, can be used to distinguish between acceptable and unacceptable leaving conditions. This analysis is discussed in Section 4-17 of this manual.

Supply Condition
The condition of the supply air (SA) entering the room may or may not be the same as the condition of the air off the cooling coil. Heat gains associated with a draw-through fan or a supply duct located in an unconditioned space will cause the supply air dry-bulb temperature to be higher than the temperature at the coil. However, the CFM requirement at a supply outlet can be based on the temperature of the air leaving the coil if the supply-side system loads are accounted for in the load calculation and equipment sizing procedures. (Refer to Section 4-6 of this manual for more information about system loads.)

Return Condition at the Equipment
The condition of the return air (RA) at the equipment may not be the same as the condition of the room air. Heat gains associated with a return fan, return air ceiling or a return duct located in an unconditioned space will cause the return air temperature to be warmer than the room temperature. This increase in return air temperature will cause the mixed air and the entering air temperatures to be a little warmer than indicated by the psychrometric analysis that is shown by Figure 5-1 on the previous page.

An estimate of the actual temperature of the air entering the cooling coil is desirable because manufacturer's performance data is based on the temperature of this air. Table 5-1, below, can be used to correct the mixed air temperature for the effects of the return-side system gains. (Refer to Section 4-6 of this manual for more information about system loads.)

Return-Side Mixed-Air Temperature Correction			
Return Load Percentage ①	**10 %**	**20 %**	**30 %**
Add to Mixed Air Dry-bulb	2 °F	3 °F	4 °F
① Return-side system loads divided by the space load — see Form N-1; lines 6, 10 and 11 and consider fan heat.			

Table 5-1

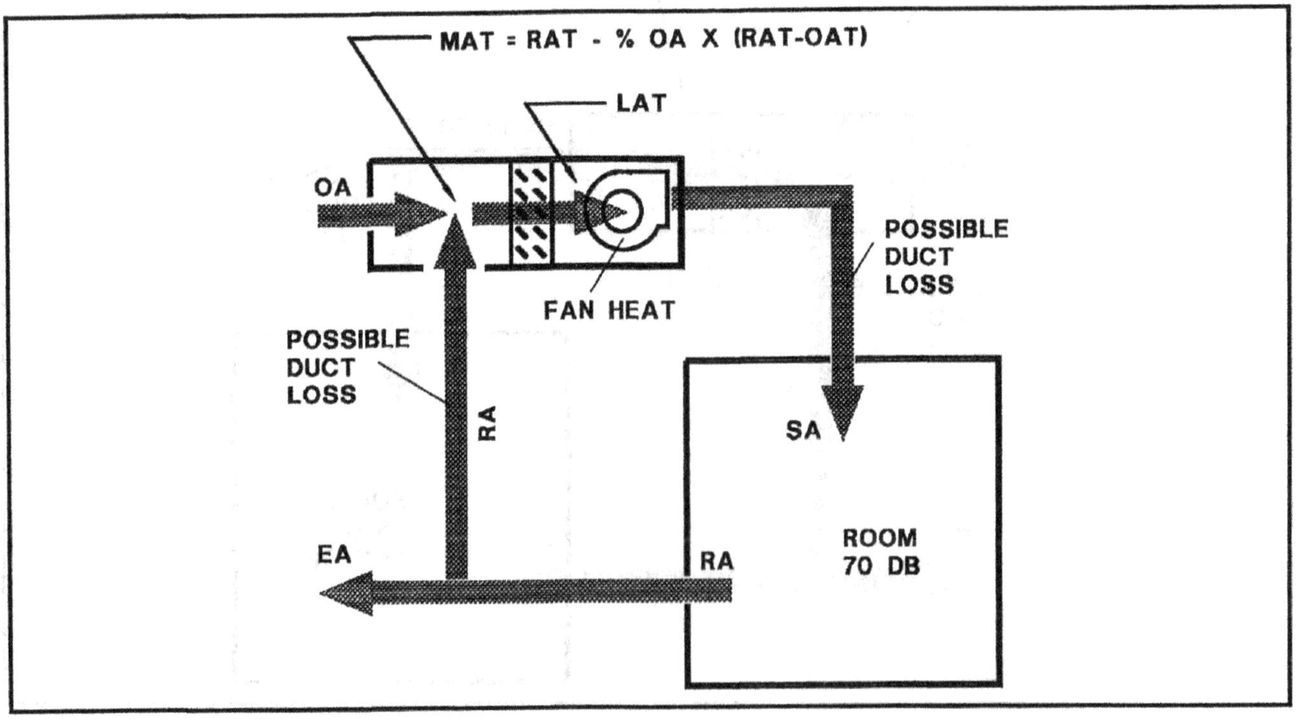

Figure 5-2

5-2 Air-Side Temperatures — Heating

The designer should be concerned with the dry-bulb temperature of the air entering and leaving the heating equipment and the temperature of the air that is supplied to the occupied space. The entering and leaving conditions that are associated with the heating equipment are important because they affect the equipment selection procedure. The supply air dry-bulb temperature is important because it is directly related to the CFM that is supplied to the conditioned space.

Entering Conditions
Hot water and steam coil performance is affected by the temperature of the air entering the coil (EAT). Electric coils and indirect-fired heat exchangers are subject to temperature rise limitations. Therefore, knowledge of the entering temperature is a prerequisite to the equipment sizing procedure. In commercial work, the condition of the air entering the heating equipment is usually the result of mixing return air (RA) with outdoor air (OA), as illustrated on the previous page by Figure 5-2. Note that the mixed air temperature (MAT) is the same as the entering air temperature. The following equation can be used to estimate the mixed air temperature (or EAT) associated with the heating cycle — where *RAT* is the return air temperature, *OAT* is the outdoor temperature and *Percent OA* is the percentage of outdoor air used for ventilation.

$$MAT = RAT - Percent\ OA \times (RAT - OAT)$$

Example
Find the mixed air temperature that corresponds to 15 percent outdoor air if the outdoor temperature is equal to 10 degrees and the return temperature is equal to 70 degrees.

$$MAT = 70 - 0.15 \times (70\text{-}10) = 61\ °F$$

Leaving Conditions
Air-source heat pump equipment is characterized by leaving air temperatures (LAT) that range from about 80°F to 100°F when the second stage heat is off and to 100°F or more when the second stage heat is energized. High efficiency furnaces are characterized by a moderate temperature rise that produces leaving air temperatures in the neighborhood of 100°F. Mid-efficiency furnaces usually have a larger temperature rise and hotter leaving air temperatures than high efficiency furnaces. Water, steam and electric heating coils provide a wide range of temperature rise and leaving air temperature values.

The temperature of the air discharged from a heating device (LAT) can be computed by adding the temperature rise to the entering air temperature (EAT). The following equation can be used to make this calculation. In this equation, *BTUH Added* is the heat added to the air that is flowing through the device and *CFM* is the rate at which the air flows through the device.

$$LAT = EAT + \frac{BTUH\ Added}{1.1 \times CFM}$$

Supply Temperature
The temperature of the supply air entering the room may or may not be the same as the temperature of the air leaving the heating device. Heat losses associated with a supply duct located in an unconditioned space will cause the supply air dry-bulb temperature to be lower than the temperature at the equipment. However, the CFM requirement at a supply outlet can be based on the temperature of the air leaving the heating device if the supply-side system loads are accounted for in the load calculation procedures. (Refer to Section 4-6 of this manual for more information about duct losses.)

Return Temperature at the Equipment
The temperature of the return air at the equipment may not be the same as the temperature of the room air. Heat loss associated with a return air ceiling plenum or a return duct located in an unconditioned space will cause the return air temperature to be colder than the room temperature. The temperature of the return air at the equipment can be estimated by using the following equation.

$$RAT_{equip} = Room\ DB - \frac{Return\text{-}Side\ Losses\ (BTUH)}{1.1 \times Return\ CFM}$$

Example
5000 CFM flows through a return duct that is in an unconditioned space. Form N-1 indicates that the duct loss is equal to 22,000 BTUH. If the outdoor temperature is equal to 10°F, and the room temperature is equal to 70°F; find the mixed air temperature that corresponds to 15 percent outdoor air.

$$RATequip = 70 - \frac{22000}{1.1 \times 5000} = 66\ °F$$

$$MAT = 66 - 0.15 \times (66\text{-}10) = 58\ °F$$

5-3 Temperatures — Chilled Water Systems

The designer should be concerned with the temperature of the water entering and leaving a cooling coil or a water chiller. These entering and leaving water temperatures are important because they are used in the equipment selection procedures. In general, water-side operating temperatures are subject to limitations associated with humidity control, freeze-up prevention, water-side GPM requirements and air-side CFM requirements.

Chilled Water Coils

The ARI standard for rating chilled water coils specifies that the entering water temperature (or glycol solution for temperatures below 40°F) can range from 35°F to 65°F. This range is much wider than the range that is required for simple comfort cooling applications. In most cases, chilled water coils are selected for an entering water temperature that falls between 42°F and 48°F. Colder water temperatures (42°F or lower) would only be required when latent loads are unusually large. Warmer temperatures (48°F or higher) may be desirable for dry-coil applications, but they may be incompatible with the operating limits of the chiller

Chilled water coil performance data may provide ratings for water temperature rise values that range between 6°F and 18°F. In theory, the coil could be selected for any temperature rise value that is supported by the coil manufacturer's data. In practice, the temperature of the water returning from the cooling coils must be compatible with the operating characteristics of the chiller — specifically the temperature drop through the chiller. Usually the water temperature rise across the coil is limited to a range of 8°F to 12°F with 10°F being the most popular choice.

Water Chillers

The ARI standard for rating water chillers specifies a 54°F entering water temperature and a 44°F leaving water temperature, with a 10°F drop across the chiller. Manufacturer's performance data usually provides a wider range of leaving water temperatures, say from 40°F to 50°F. Leaving water temperatures below 40°F require a glycol solution and temperatures above 50°F may cause hermetic motors to over heat.

The allowable water temperature drop across a chiller typically ranges from 8°F to 12°F. Usually, the manufacturer's performance data is based on a 10°F drop (performance with an 8°F or 12° F drop will be slightly different). Temperature drops beyond the 12°F range may require special chiller controls and temperature drops less than 8°F may result in inadequate refrigerant superheat.

Piping and Pumping Heat Gains

Piping and pumping heat gains (system loads) must be accounted for in the load calculation. (Refer to Section 4-6 for information about system loads.) Gains on the supply side of the chiller cause the temperature of the water entering the coil to be a little warmer than the temperature of the water leaving the chiller. Gains on the return side of the chiller cause the temperature of the water entering the chiller to be a little warmer than the temperature of the water leaving the coil. The following equation can be used to estimate the temperature *rise* associated with the piping or pumping *System Load* — where *GPM* is the water-side flow rate.

$$\text{Rise } (°F) = \frac{\text{System Load (Btuh)}}{500 \times GPM}$$

5-4 Temperatures — Hydronic Heating Systems

The designer should be concerned with the temperature of the water entering and leaving a heating coil or a boiler. These temperatures are important because they are used in the equipment selection procedures. Water-side operating temperatures are subject to limitations associated with thermal shock, system GPM requirements and air-side CFM requirements.

Hot Water Coils

The ARI standard for rating low-pressure hot water coils specifies that the entering water temperature can range from 120°F to 250°F. In actual practice, designers utilize this entire range of entering water temperatures.

In theory, a hot water coil could be selected for any temperature drop value that is supported by the manufacturer's performance data, say 10°F to 40°F. In practice, the temperature of the water returning from the heating coil must be compatible with the operating characteristics of the boiler — specifically the temperature rise through the boiler. Usually, a 20°F water temperature drop across the coil is specified to avoid thermal shock in the boiler (see below).

Low-Pressure Hot Water Boilers

Low-pressure water boilers can supply water at temperatures that range from 120°F to 250°F. Temperatures below 120°F are not recommended because of problems associated with condensation in the boiler vent and reabsorption of air into the boiler water. Temperatures above 250°F require boiler operation at pressures that exceed 30 PSI.

The temperature rise across a boiler is limited by the potential for causing thermal shock. Cast iron boilers can tolerate a 25°F to 30°F temperature rise. For steel boilers, the temperature rise should be limited to 20°F. The following equation can be used to estimate the water flow rate (*GPM*) that is required for a given temperature rise. In this calculation, the *Heating BTUH* is the boiler sizing load.

$$GPM = \frac{\text{Heating BTUH}}{500 \times \text{Boiler Rise } (°F)}$$

Piping Losses

Piping losses are system loads (refer to Section 4-6 of this manual). Supply-side losses cause the temperature of the water entering the coil to be a little cooler than the temperature of the water leaving the boiler. Return-side losses cause the temperature of the water entering the boiler to be a little cooler than the temperature of the water leaving the coil. Use the following equation to calculate the temperature drop associated with a *piping loss*, where *GPM* is the water-side flow rate.

$$\text{Drop } (°F) = \frac{\text{Piping Loss (Btuh)}}{500 \times GPM}$$

5-5 Temperatures — Series and Parallel Piping

The temperature rise across the boiler and the temperature drop across the heating coil (or coils) may not be independent of each other. If the temperature rise across the boiler is used as the design criteria, the coil temperature drop (Coil TD) will be determined by the following (Water-Side) equation. In this equation, the *Coil BTUH* is the heating load at the coil and the *Coil GPM* is the water flow rate through the coil. The relationship between the boiler temperature rise and the coil temperature drops also is affected by the piping arrangement, as explained below.

$$\textit{Coil TD}\ (^\circ F) = \frac{\textit{Coil BTUH}}{500 \times \textit{Coil GPM}}$$

Series Piping

Figure 5-3 shows a series piping arrangement. In this case, the boiler rise is equal to 20°F so the sum of the coil temperature drops also must be equal to 20°F. The Water-Side equation can be used to show that the boiler flow is 20 GPM. Because of the series piping arrangement, the flow through each coil also must be equal to 20 GPM. Since the flow through the heating coils is relatively large, the temperature drops across the coils will be relatively small and the water-side pressure drops across the coils will be relatively large. A parallel piping arrangement will be required if the designer wishes to reduce the flow rate through the coils. This, in turn, will increase the temperature drops across the coils and reduce the pressure drops across the coils.

Parallel Piping

Figure 5-4 shows a parallel piping arrangement. In this case, the boiler rise is equal to 20°F, so the temperature drop across each coil also must be equal to 20°F. The Water-Side equation can be used to show that the boiler flow is 20 GPM. Because of the parallel piping arrangement the flow through each coil also is determined by the Water-Side equation. These values are shown on Figure 5-4. In this case, the sum of the coil GPM values must be equal to the 20 GPM flowing through the boiler.

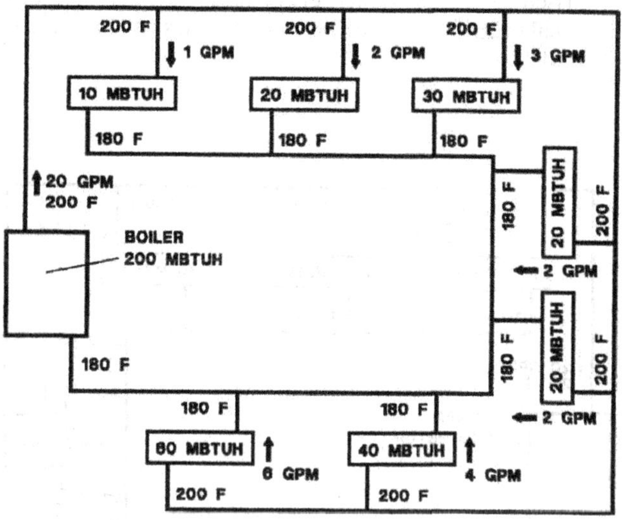

Figure 5-4

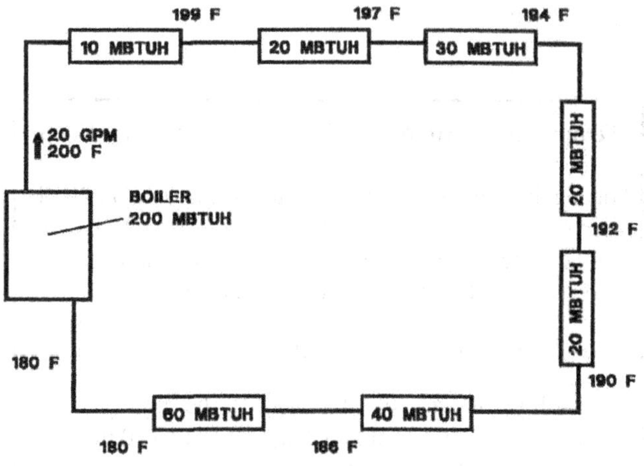

Figure 5-3

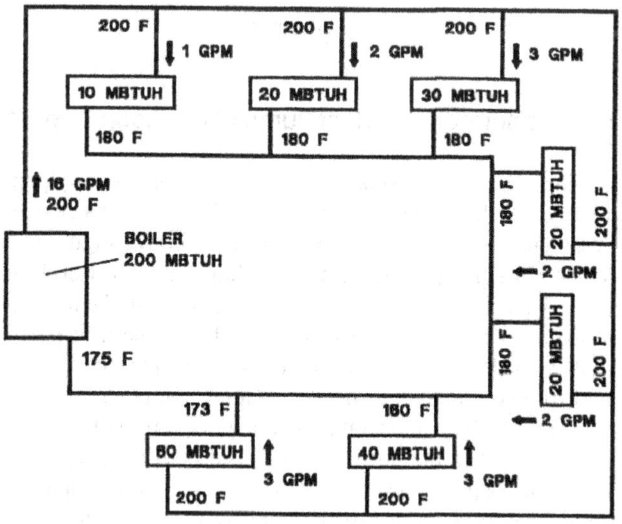

Figure 5-5

Note that the designer does not have the option to set the coil temperature drops or flow rates to any value that may be desired. Figure 5-5, on the previous page, shows what would happen if the designer decided to increase the temperature drops across the 40 MBTUH and 60 MBTUH coils in an attempt to limit the coil flow rates to a maximum of 3 GPM. (The designer may wish to do this in order to reduce the water-side pressure drops.) Note that the flow through the boiler is reduced to 16 GPM, causing the temperature rise across the boiler to exceed the 20°F limit.

Figure 5-6 shows how a bypass can be used to obtain the desired conditions (see paragraph above) at the boiler and at the coils. The figure shows that if 4 GPM of 200°F water is bypassed back to the boiler, the boiler temperature rise will be equal to 20°F and the coil flow rate values will not exceed 3 GPM.

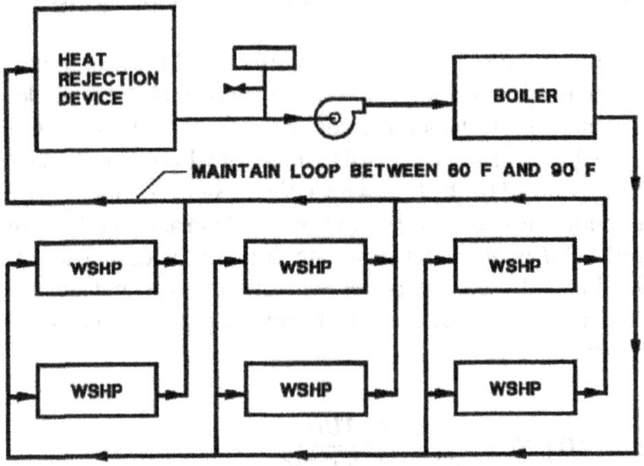

Figure 5-7

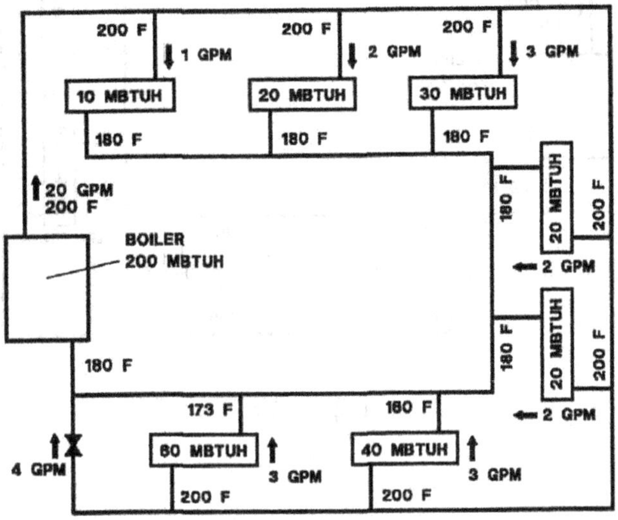

Figure 5-6

5-6 Temperatures — Heat Pump Water-Loop System

A closed-loop heat pump system consists of a water loop, water-to-air heat pumps, a boiler and a heat rejection device. A schematic diagram of this system is provided by Figure 5-7. When a heat pump operates in the cooling mode, it rejects heat to the water loop. When a heat pump operates in the heating mode, it extracts heat from the water loop. The heat pump units provide zone control. The loop temperature is allowed to drift when some units provide heating and others provide cooling. The loop temperature will begin to drop if too many units operate in the heating mode. The boiler will be fired when the loop temperature falls below 60°F. The loop temperature will begin to rise if too many units operate in the cooling mode. The heat rejection device will be activated when the loop temperature exceeds 90°F.

5-7 Temperatures — Cooling Tower Water

Cooling towers are usually sized so that, at design conditions, the leaving water temperature is about 85°F. (Nominal tower tonnage ratings are based on rejecting 15,000 BTUH/ton of heat, when operating at a flow rate of 3 GPM/ton and a 78°F outdoor wet-bulb temperature. Under these conditions, a tower is usually designed to produce a water temperature drop of 10°F — say from 95°F to 85°F.)

5-8 Temperatures — Earth-Coupled Water Loop

The temperature of the water in a earth-coupled water loop varies over a wide range, depending on the month of the year, the climate, the soil conditions and the loop geometry. In late winter or early spring, the water temperature may be below 30°F. In the late summer or fall, the water temperature may exceed 90°F. Guidelines for estimating these temperatures are provided below.

Temperature Extremes — Earth-Coupled Water Loop

For heating, assume that the water in the loop is 40 °F warmer than the coldest outdoor temperature (ATL) that will be experienced during the most severe winter conditions. (Use 25°F if this value is less than 25°F.)

Low Temp = ATL + 40 °F ... But not less than 25 °F

For cooling, assume that the water in the loop is 10 °F colder than the hottest outdoor temperature (ATH) that will be experienced during the most severe summer conditions.

High Temp = ATH - 10 °F

5-9 Temperatures — Well Water or City Water

The temperature of well water or city water depends on local conditions. A thermometer can be used to determine the temperature of the city water. Well drillers will know the temperature of the ground water.

5-10 Air-Side Flow Rates

The maximum CFM that is required to condition a room or zone depends on the design load that is associated with the space, the dry-bulb temperature in the space and the dry-bulb temperature of the supply air. The design load may be a heating load or a cooling load. The maximum CFM that is associated with a piece of central equipment depends on the system type. The minimum CFM that is required to condition a space depends on factors that are associated with maintaining air quality — for example, the air circulation rate and the percentage of fresh air.

Cooling CFM
The cooling CFM is computed by using the Sensible Heat Equation that is provided below. In this equation the *Space Sensible Load* is the sensible load associated with the conditioned space (not the sensible load at the coil) and the *Cooling TD* is the temperature difference between the room air and the supply air.

$$\text{Cooling CFM} = \frac{\text{Space Sensible Load}}{1.1 \times \text{Cooling TD}}$$

> The *Cooling TD* will depend on the relative size of the latent load. Since colder coil temperatures provide more dehumidifying capacity, a larger TD would apply when the latent load is relatively high, and a smaller TD would apply when the latent load is relatively small. Refer to Section 5-1 above for more information about the air-side temperatures associated with the cooling cycle.

Heating CFM — Heated Space
The heating CFM is also computed by using the Sensible Heat Equation. In this equation the *Space Heating Load* is the heating load associated with the heated space and the *Heating TD* is the temperature difference between the supply air and the room air. Refer to Section 5-2 for more information about the air-side temperatures associated with the heating cycle.

$$\text{Heating CFM} = \frac{\text{Space Heating Load}}{1.1 \times \text{Heating TD}}$$

Design CFM — Central Equipment
The design CFM that is associated with equipment that serves a single room or zone depends on the peak load that is associated with that room or zone. In this case, the CFM at the unit will be equal to the CFM that is required for the room or zone. (Refer to Form N-1, line 8 for and lines 16 and 17 for an estimate of the design loads.)

The CFM at central equipment that serves two or more zones may be based on the sum of the zone loads or on the peak block load, depending on the type of system. Refer to Section 4-3 in this manual for more information about central equipment sizing loads.

Minimum CFM
When the space is occupied, a variable volume system should not be allowed to throttle the supply CFM beyond the point where the supply air outlets become ineffective. Some designers assume that this point is equal to 40 percent of the design CFM, but manufacturer's air outlet performance data should be consulted for a more accurate value.

When the space is occupied, a variable volume system should not be allowed throttle the supply CFM beyond the point where the outdoor CFM becomes inadequate. Special controls may be required to insure that the outdoor air CFM is adequate during any operating condition.

5-11 Water-Side Flow Rates

The maximum GPM that is required to cool or heat a room or zone depends on the design load that is associated with the space. The maximum GPM that is associated with a piece of central equipment depends on the system type and the piping configuration.

Cooling or Heating GPM
The cooling or heating GPM associated with hydronic equipment is computed by using the Water-Side equation (below). In this equation, the *Water-Side Load* is the design cooling load (sensible plus latent) or the design heating load imposed on the equipment and the *Design TD* is the water-side temperature rise (or drop) across the equipment. Refer to Sections 5-3 and 5-4 in this manual for more information about the water-side temperatures associated with hydronic equipment.

$$\text{GPM} = \frac{\text{Water-Side Load}}{1.1 \times \text{Design TD}}$$

Design GPM — Terminal Equipment
The design GPM that is associated with hydronic equipment that serves a single room or zone depends on the peak load that is associated with that room or zone. In this case, the GPM

at the equipment will be equal to the GPM that is required to offset the room or zone load. (Refer to Form N-1, line 8 and lines 16 and 17 for an estimate of the design loads.)

Design GPM — Central Equipment
The design GPM at central equipment that serves two or more zones may be based on the sum of the zone loads or on the peak block load, depending on the type of system. (Refer to Section 4-3 in this manual for more information about equipment sizing loads.)

> The GPM at the central equipment is also related to the flow rates associated with the terminal devices and the piping arrangement. A parallel piping arrangement with a bypass circuit can be used to decouple the flow rate at the central equipment from the flow rates that are associated with the terminal devices. Refer to Section 5-5 above for additional discussion concerning this point.

5-12 Equipment Performance

The temperatures and flow rates that are determined from the equations and procedures presented above must be compatible with the operating range of the heating and cooling equipment. Always refer to manufacturer's performance data to verify that the equipment is capable of operating at the desired conditions. Manufacturer's data should provide the following information.

- Sensible capacity and latent capacity of DX cooling coils and condensing units — over a range of CFM or face velocity values, entering wet-bulb temperatures, entering dry-bulb temperatures and condensing temperatures.
- Sensible capacity and latent capacity of chilled water coils — over a range of CFM or face velocity values, entering wet-bulb temperatures, entering dry-bulb temperatures, entering water temperatures, temperature rise values and water flow rate values.
- Total (sensible and latent) capacity (TONS) of water chillers — over a range of leaving water temperatures, temperature rise values, evaporator GPM values, condensing temperatures and condensing GPM values.
- Heating capacity of electric heating coils — over a range of CFM or face velocity values, entering dry-bulb temperatures and temperature rise values.
- Heating capacity of heat pump indoor (condensing) DX coils and outdoor (air-source) or remote (water-source) evaporator units — over a range of CFM or face velocity values, entering dry-bulb temperatures, and evaporator temperatures.
- Input and output heating capacity of fossil fuel furnaces — for one or more firing rates with information about temperature rise limits.
- Heating capacity of hot water coils — over a range of CFM or face velocity values, entering dry-bulb temperatures, entering water temperatures, temperature rise values and water flow rate values.
- Input and output heating capacity of hot water boilers — for one or more firing rates with information about temperature rise limits.

Section 6
Selecting Cooling Equipment

Cooling equipment must be selected to satisfy the design sensible load and the design latent load. Gross oversizing is not recommended. Manufacturer's application data can be used to certify equipment performance for the operating conditions that are expected to exist at the site. Equipment capacity must track changes in the cooling load under all operating conditions. Accessories, safety and operating controls must be specified. Fan performance must be specified. Water-side performance must be noted. Dimensions and clearances must be verified.

6-1 Sizing Requirements

When cooling equipment is selected, it is sized so that it has the capacity to neutralize the design load. As explained in Section 4-3 and 4-4 of this manual — the design load for a single-zone system is based on the peak room or zone load and the design load for a multizone system is calculated according to the rules that are summarized by Table 4-2. For convenience Table 4-2 is repeated below.

Combining Zone Loads

System Type	Cooling Capacity	Coil CFM	Fan CFM	Duct Diver-sity
TRH	Sum Zones	Sum Zones	Sum Zones	No
M-VAV	Block	Block	Block	Yes
BP-VAV	Block	Sum Zones	Sum Zones	No
DT-DD	Sum Zones	Sum Zones	Sum Zones	Yes
BT-DD	Block+XS	Block+XS	Sum Zones	Yes
DT-Mult	Sum Zones	Sum Zones	Sum Zones	No
BT-Mult	Block+XS	Block+XS	Sum Zones	No
XS = An allowance to compensate for air that is diverted through the hot deck — approximately 10 percent.				
TRH — Terminal reheat M-VAV — Full modulating variable air volume BP-VAV — Bypass variable air volume DT-DD — Draw-through dual-duct BT-DD — Blow-through dual-duct DT-Mult — Draw-through multizone BT-Mult — Blow-through multizone				

Table 4-2 (Repeated)

Sensible Capacity
The sensible capacity of the cooling equipment must be equal to or greater than the (sensible) design load. As noted above, the sensible design load may be based on a peak block load, a peak zone load or on the sum of a group of peak zone loads. This load should include the sensible component of the ventilation load (if any) and all other sensible "system loads." Refer to Section 4-5 and 4-6 for more discussion about ventilation loads and system loads.

Latent Capacity
The latent capacity of the cooling equipment must be equal to or greater than the (latent) design load. As noted above, the latent design load may be based on a peak block load, a peak zone load or on the sum of a group of peak zone loads. This load should include the latent component of the ventilation load (if any) and all other latent "system loads."

Design Conditions
Equipment capacities should be verified for the operating conditions that will exist at the site. These conditions, which consist of a set of design temperatures and fluid flow rates, are summarized below. Refer to Section 5 of this manual for information about how the values for the design temperatures and flow rates are determined.

Component	Conditions
DX Coils	CFM or face velocity Entering dry-bulb Entering wet-bulb
Water Coils	CFM or face velocity Entering dry-bulb Entering wet-bulb Entering water temp. Water temperature rise
Water Chillers	Leaving water temp. Water temperature drop Evaporator GPM
Air-Cooled Condenser	Outdoor air temperature
Water-Cooled Condenser	Well water temperature City water temperature Cooling tower water temp. Max. earth-loop temp. Max. water-loop temp. Condenser GPM

Sizing Limits — Cooling

Excessive cooling capacity does not have an effect on humidity control if an application has a negligible latent load — sensible cooling in an arid climate for example. Otherwise, grossly oversized cooling equipment tends to "short cycle" with a resulting loss of humidity control and a reduction in comfort, especially during part-load operation. Short operating cycles always increase the wear and tear on the equipment and increase the operating cost. A large amount of surplus capacity increases the first cost of the installation and increases the electrical power demand (KW) on the electrical utility service. However, a small amount of excess capacity is acceptable or even desirable, depending on the application and on the type of equipment. Note that the following sizing limitations apply to either air-cooled or water-cooled equipment.

- The capacity of "cooling-only" equipment should not exceed the total design load (sensible plus latent) by more than 15 percent.
- The cooling capacity of heat pumps that are installed in hot, mild or moderate climates should not exceed the total load (sensible plus latent) by more than 15 percent.
- In very cold climates, heat pump heating performance may be improved by installing a unit that has some excess cooling capacity. A cold climate heat pump installation can be oversized by as much as 25 percent if heating is the primary concern.

Summary of Sizing Limits — Cooling

Cooling-Only or Heat Pump in Mild Climate

Sensible capacity exceeds sensible load

Latent capacity exceeds latent load

Total capacity less than 1.15 x total load

Heat Pump in Cold Climate

Sensible capacity exceeds sensible load

Latent capacity exceeds latent load

Total capacity less than 1.25 x total load

Limits apply to air-source and water-source equipment

6-2 Minimum Capacity Requirement

Most commercial applications require cooling during cool and in many cases, during cold weather. (Refer to Sections 4-8 through 4-11 for more information about part-load analysis.) This requirement is referred to as "low ambient cooling." The minimum cooling load on the refrigeration equipment depends on the method that is used to provide low ambient cooling. As explained in Section 4-16, the minimum load on the cooling equipment will depend on whether an air-side economizer is used to provide low ambient cooling. Figure 4-12 illustrates this point. For convenience, Figure 4-12 and excerpts from Section 4-16 are repeated below.

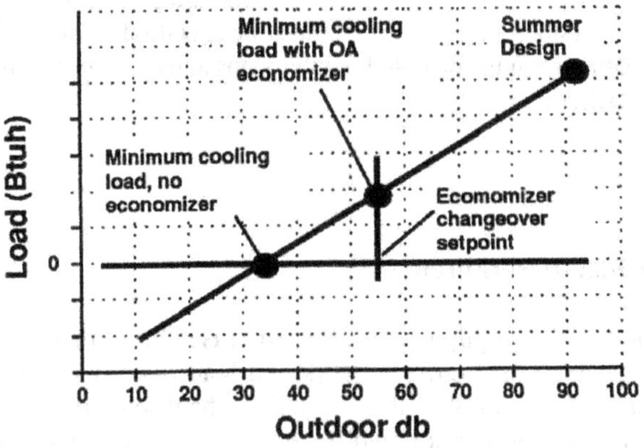

If no air-side economizer is used, the refrigeration equipment must operate when the outdoor temperature is below 55 degrees and the minimum load would be equal to zero at the heating/cooling changeover point. (Low ambient kits will be required for systems that are not equipped with an air-side economizer.)

If an economizer is used, the minimum refrigeration load will occur just before the system switches from mechanical cooling to economizer cooling. In this case, the minimum refrigeration load will most likely be greater than zero.

Figure 4-12 - (Repeated)

In any case, the cooling equipment must be able to adjust its capacity to match any load condition. Small residential equipment, say 3 TONS or less, adjusts capacity by cycling the compressor on and off. Larger equipment requires a more sophisticated form of capacity control. In this case, both the compressor and the cooling coil must have the ability to track changes in the cooling load.

Compressor capacity can be controlled by staging multiple compressors, equipping compressors with unloaders, installing variable speed compressors and equipping compressors with hot gas bypass accessories. (Centrifugal compressors have adjustable inlet vanes, but they are not discussed in this manual because they are not used in small commercial work.) DX cooling coil capacity can be controlled by using solenoid valves to control row-split or face-split coils. Chilled water coil capacity can be controlled with modulating, two-way or three-way water valves.

The minimum cooling load dictates what type of capacity control devices will be required. Compressor staging typically provides steps that reduce the cooling capacity by 25, 33, 50, 66 or 75 percent, depending on the number of compressors installed. Unloaders can be installed if smaller increments of capacity reduction are required. Variable speed compressors provide modulating capacity control. Hot gas bypass accessories can be installed if the minimum cooling load is less than the first step of cooling capacity. Appendix 5 provides more information about controlling cooling capacity.

6-3 Fan Performance

Packaged commercial cooling equipment usually includes a supply fan that can operate over a wide range of speed (RPM) and motor power (HP). A return fan is often offered as an accessory item. This allows the designer to select a fan configuration that will provide the flow (CFM) and pressure (IN.WG.) that is required to move the air through the duct system (supply and return) and the air-side economizer.

Supply Fan Selection

The supply CFM requirement is determined by the space sensible cooling load and the sensible heat equation. This calculation is discussed in Section 5-9 of this manual. The pressure requirement that is associated with the supply side of the system is determined by making a duct system pressure loss calculation. This calculation should include the pressure losses that are associated with the straight runs, fittings and the air-side devices that are external to the air handler. Losses that are associated with any air handler component that are not accounted for in the blower table also should be included — the pressure losses associated with a damper, mixing box or filter for example. (Refer to Appendix 3 in this manual and to ACCA **Manual Q** for more information about air-side design.)

Return Fan Selection

A return fan (or some type of pressure relief damper, installed in the conditioned space) will be required if the air handler is equipped with an air-side economizer. Even with no economizer, a return fan will be required unless the resistance that is associated with the return duct system is negligible. The return CFM requirement is determined by the air balance in the conditioned space and at the air handler. (Refer to ACCA **Manual Q**, Section 17, for a comprehensive discussion of this subject.) The return-side pressure loss includes the losses that are associated with the straight runs, fittings, the air-side devices that are external to the air handler and any air handler component loss that is not accounted for in the blower table.

Variable Volume Systems

Variable volume fans must be selected to provide satisfactory performance over a wide range of operating conditions. In this case, the minimum flow requirement at part load is just as important as the full load flow requirement. In addition, the fan pressure requirement will be affected by the controls and devices that are used to modulate air flow and to adjust the fan performance. It also is important to make sure that the return fan performance has the ability to track the changes in the supply fan performance. Refer to ACCA **Manual Q**, Sections 13, 16 and 17, for more information about VAV fans.

Fan Heat

Supply and return fans create a "system load" on the cooling coil. In some cases, this load is already accounted for in the manufacturer's cooling performance data. In other cases, the effect of the fan heat is not accounted for and the cooling performance must be adjusted for this load. Always read the manufacturer's engineering literature carefully to determine whether or not the fan heat adjustment is necessary. (This point is also discussed in Section 4-6 of this manual.)

6-4 Water-Side Performance

The water flow through (GPM) and the pressure drop across (Feet of Water) hydronic cooling equipment is usually documented in the manufacturer's performance data. (Some examples of this type of equipment include chilled water coils, fan-coil units, unit ventilators and water chillers. This information should be located, inspected and recorded for future reference because it will be required for the pipe sizing and pump selection calculations. (Refer to Section 5-10 of this manual for more information about how water-side flow rates are established. Refer to Appendix 4 for more information about pipe design and pump selection.)

6-5 Air-Side Accessories

Air-side accessories are usually available as part of a package offered by the equipment manufacturer. Examples of accessories that might be required include filters, dampers, mixing boxes, economizers, fan inlet vanes, belt guards and so forth. Some accessories could be available as standard equipment, but others have to be included as options when the equipment requirements are specified.

6-6 Controls

Safety and operating controls are usually available as part of the package offered by the equipment manufacturer. Examples of the controls that might be required include refrigerant-side pressure switches, motor protection, anti-short-cycle relays, motor starters, oil switches, crankcase heaters, low temperature limit switches, fan interlocks, damper interlocks, economizer controls, capacity control devices, central control panels and so forth. Some controls and safety devices could be available as standard equipment, but others have to be requested when the equipment requirements are specified.

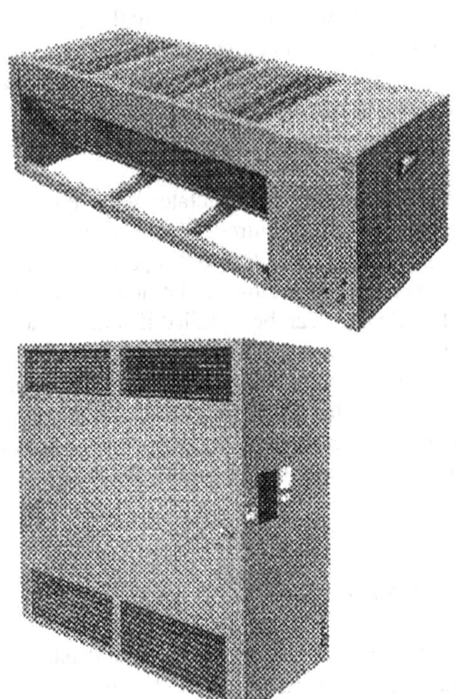

Split-System (Cooling or Heat Pump)

6-7 Physical Dimensions and Arrangement

The designer must verify that the dimensions and the arrangement of the equipment is compatible with the available space. There must be ample room to service the equipment — including room to remove critical parts such as compressors, fans and coils. There also must be ample room to install aerodynamically efficient fan discharge and inlet connections.

> System effect losses associated with fan inlet and discharge connections can be critical. Refer to ACCA **Manual Q**, Section 6, 7 and 8, for more information on this subject.

6-8 Packaged, Air-Cooled, DX Cooling Equipment

The following paragraphs provide a detailed example that illustrates the equipment sizing and selection procedures. This example is designed to show how to select a packaged, rooftop, air-cooled, DX unit. ***A similar procedure would be used to select any type of packaged, air-cooled, direct expansion, cooling unit or heat pump.*** Single-package rooftop, cooling units and heat pumps; self-contained cooling units and heat pumps; ducted split-system cooling units and heat pumps; PTAC cooling units and heat pumps, DX fan-coil units and ductless split-system cooling units and heat pumps are typical of the type of equipment that falls into this category.

Single-Package Heat Pump

Single-Zone or VAV Rooftop Cooling Unit

PTAC (Cooling or Heat Pump)

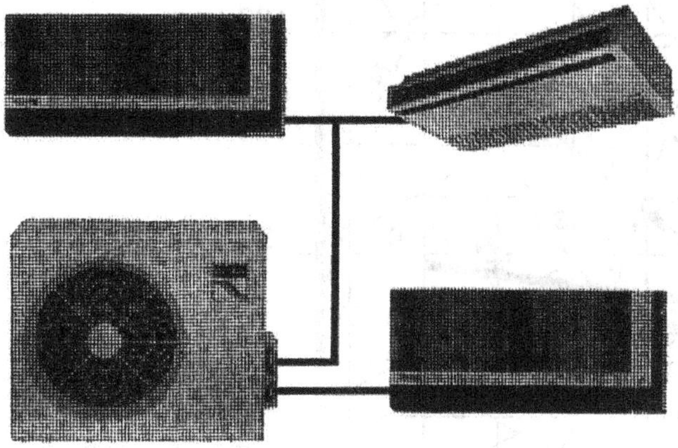

Ductless Split (Cooling Unit or Heat Pump)

Interior, Split or Self-Contained Multizone

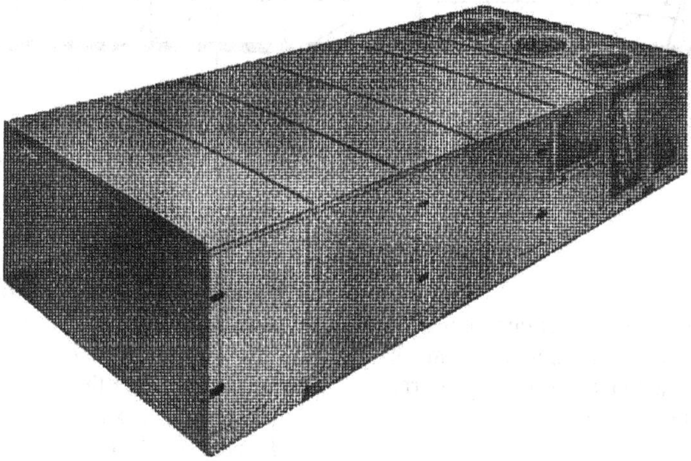

Rooftop Multizone

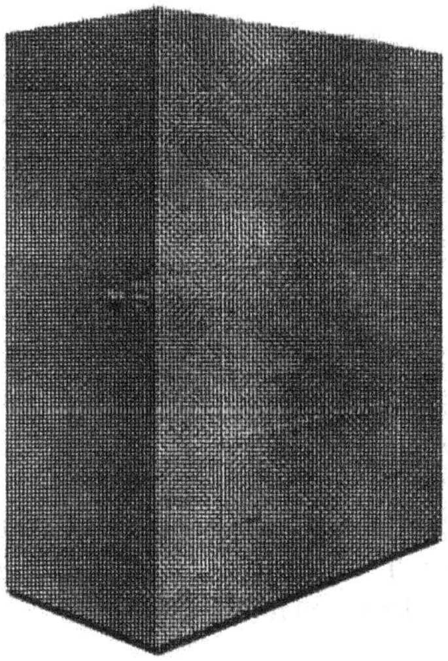

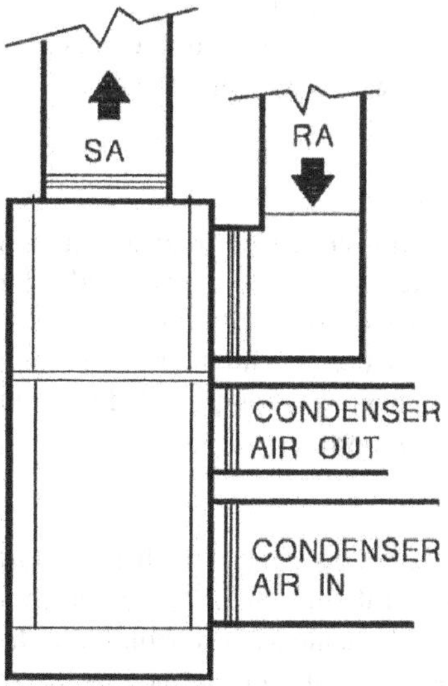

Interior Cooling Unit or Heat Pump with Integral Condenser

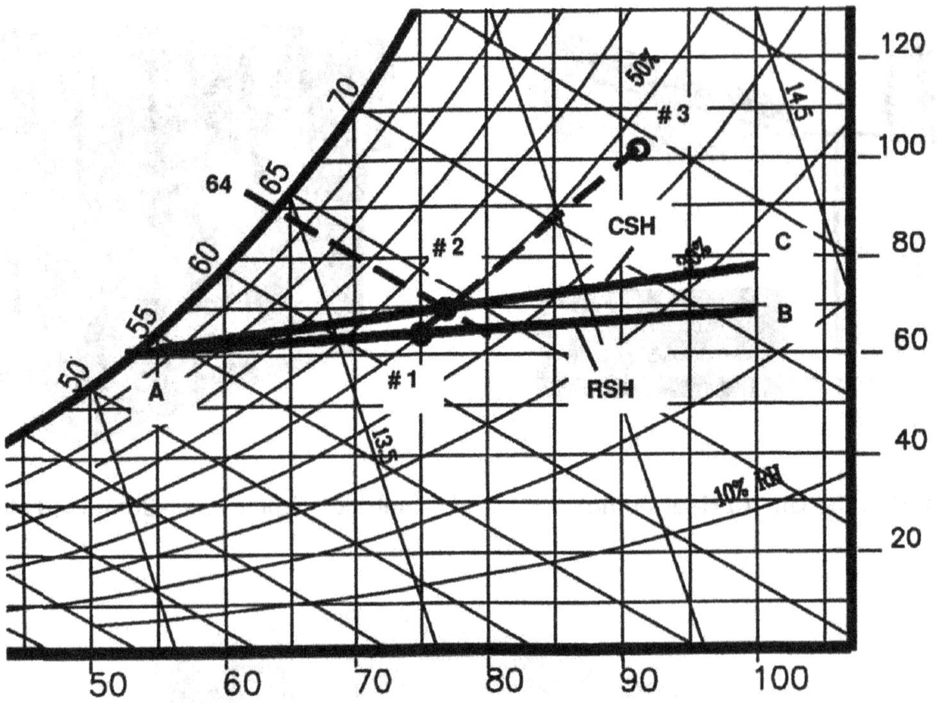

Figure 6-1

Manual N Load Information

A **Manual N** load calculation is always required before the cooling equipment can be selected. For example, assume that this task has been completed and that the following information applies to a single-zone DX cooling unit.

- Indoor design condition is 75°F and 50% RH
- Outdoor design condition is 92°F DB and 74°F WB
- Form N-1, line 12, sensible load is 207,851 BTUH
- Form N-1, line 12, latent load is 42,853 BTUH
- Form N-1, line 8, supply CFM is 7150 (20°F TD)
- 900 CFM of outdoor air is required for ventilation
- No losses associated with return air system

Psychrometric Analysis

The psychrometric chart should be used to determine the condition of the air entering the cooling coil and the acceptable range of conditions of the air leaving the cooling coil. Figure 6-1 summarizes the psychrometric solution to the example problem. The information provided by Figure 6-1 can be interpreted as follows:

Mixed Air Condition

Since there are no losses associated with the return duct system, the condition of the return air at the unit (point #1) will be approximately the same as the condition of the room air (75 DB & 50% RH). Point #3 is defined by the summer outdoor design condition (92 DB & 74 WB).

The mixed air temperature (point #2) falls on a line that connects points #1 and #3. The location of point #2 depends on the percentage of outdoor air. This percentage can be estimated by using the 900 CFM ventilation value and the 7145 CFM (approximate) supply air flow value that is found on line 8, Form N-1.

Percent outdoor air = 900/7145 = 12% to 13%
MAT = 75 + (92-75) x .125 = 77 °F

Entering Condition

Point #2 also represents the condition of the air that enters the cooling coil. The psychrometric chart shows that this condition corresponds to a 77°F dry-bulb temperature and a 64°F wet-bulb temperature.

Room Sensible Heat Ratio

Line A-B represents the room sensible heat ratio. A supply air condition that falls anywhere between point A and point #1, will have the exact combination of sensible and latent capacity that is required to offset the loads that are associated with the room. The room sensible heat ratio is determined from the loads found on line 8, Form N-1.

RSH = 157201 / (157201 + 22027) = 0.90

Coil Sensible Heat Ratio

Line A-C represents the coil sensible heat ratio. A supply air condition that falls anywhere between point A and point #2, will have the exact combination of sensible and latent capacity that is required to offset the loads that are associated with the coil. The coil sensible heat ratio is determined from the loads found on line 12, Form N-1.

CSH = 207851 / (207851 + 42835) = 0.83

Ideal Leaving Condition

The point where the RSH line crosses the CSH line is the *ideal leaving condition.* Air at this condition will have the exact combination of sensible and latent capacity that is required to offset the loads that are associated with the room and the coil. However, some compromise may be necessary because packaged equipment may not be able to produce the ideal leaving condition.

Acceptable Leaving Condition

The dry-bulb temperature of the air off the coil can be at any temperature that is less than the room temperature (75°F). Usually, the supply air dry-bulb temperature is about 20°F less than the room temperature, but it can range from 15 °F to 25°F below the room temperature. In any case, if humidity is to be controlled, the condition of the air off the cooling coil — as defined by the dry-bulb and wet-bulb temperatures — has to fall on or below the RSH line and the CSH line, whichever is lower. (Any leaving condition that falls below the lowest sensible heat ratio line will produce a design that will control the latent load at the coil and the latent load in the occupied space.)

Equipment Selection Data Summary

The designer should summarize the equipment performance requirements before referring to the manufacturer's performance data. This summary should include the indoor and outdoor design conditions, the calculated loads, the entering condition at the cooling coil and an estimate of the required cooling CFM. The information that is required for this example is summarized below.

Indoor Design = 75°F and 50% RH
Outdoor Design = 92°F DB and 74°F WB
Sensible Load = 207,851 BTUH (N-1, line 12)
Latent Load = 42,835 BTUH (N-1, line 12)
Entering dry-bulb = 77°F (from Figure 6-1)
Entering wet-bulb = 64°F (from Figure 6-1)
Approximate CFM = 7,145 (N-1, line 8)

C-25 Gross Cooling Capacities (25 Ton Unit)

CFM STD. AIR	Ent Db (F)	AMBIENT TEMPERATURE																							
		85						95						105						115					
		ENTERING WET BULB																							
		61		67		73		61		67		73		61		67		73		61		67		73	
		MBh	SHR	MBh	SHR	MBh	SHR	MBh	SHR	MBh	SHR	MBh	SHR	MBh	SHR	MBh	SHR	MBh	SHR	MBh	SHR	MBh	SHR	MBh	SHR
5000	75	246	77	276	58	309	42	233	78	261	58	293	42	218	79	246	59	276	42	201	81	228	59	257	42
	80	246	87	276	67	309	50	233	88	262	68	293	47	218	88	246	69	276	52	202	93	228	70	257	51
	85	248	96	276	76	303	58	235	98	262	77	293	59	221	100	246	79	276	60	206	100	228	81	257	61
	90	253	100	276	85	308	66	243	100	261	87	293	67	231	100	248	88	276	68	218	100	229	91	257	70
7000	75	266	81	298	60	331	42	251	83	282	60	314	41	234	85	264	61	295	41	216	88	244	62	274	41
	80	267	93	298	67	331	51	253	95	281	72	314	52	237	98	263	73	295	52	220	100	244	75	274	53
	85	273	100	297	82	331	61	260	100	281	84	314	62	247	100	263	86	295	63	232	100	244	89	274	65
	90	287	100	298	92	331	71	274	100	283	94	313	72	261	100	266	97	294	74	246	100	248	99	274	77
8750	75	278	86	310	61	344	41	262	87	292	63	325	41	244	90	273	64	305	41	225	93	253	65	283	41
	80	280	100	309	74	344	53	266	100	292	78	325	53	250	100	273	78	305	54	234	100	253	80	283	55
	85	292	100	310	87	343	64	279	100	293	89	325	65	264	100	274	81	305	67	248	100	254	95	283	69
	90	308	100	313	98	343	75	295	100	297	100	325	77	280	100	280	100	305	79	263	100	263	100	283	82
10000	75	284	89	316	63	331	41	267	91	298	64	311	41	250	93	279	65	288	41	230	97	257	67	288	41
	80	289	100	316	77	350	54	274	100	298	77	331	55	259	100	278	81	310	55	242	100	257	84	288	57
	85	304	100	316	90	350	66	290	100	299	93	331	68	275	100	280	95	310	70	258	100	261	98	288	72
	90	320	100	322	100	350	78	306	100	306	100	331	81	291	100	290	100	310	83	273	100	273	100	288	87
11000	75	288	91	320	64	355	41	271	94	302	65	335	41	253	96	282	67	314	41	234	99	260	68	291	42
	80	295	100	320	79	355	55	281	100	302	80	335	55	265	100	282	83	314	57	248	100	260	87	291	58
	85	312	100	321	93	354	68	297	100	304	95	335	69	281	100	285	98	314	71	264	100	265	100	291	74
	90	329	100	329	100	354	81	314	100	314	100	335	83	298	100	298	100	314	86	280	100	280	100	292	89

NOTES:

1. All capacities are gross and have not considered indoor fan heat. To obtain net cooling capacities, subtract indoor fan heat
2. MBh - Total Gross Cooling Capacities, SHR - Sensible Heat Ratio, Sensible Heat Capacity (SHC) MBh x SHR

Figure 6-2

Cooling Performance Data
After the equipment performance requirements are tabulated, the manufacturer's performance data can be used to select the equipment. Carefully inspect the performance data for the unit that provides the closest match to the equipment performance requirements. An example of comprehensive cooling performance data is provided by Figure 6-2, on the previous page.

The data in Figure 6-2 provides information on total cooling capacity, sensible capacity and latent capacity. [Sensible capacity equals the total capacity multiplied by the sensible heat ratio (SHR). Latent capacity equals the difference between total capacity and sensible capacity.] Also note that performance is tabulated for a full range of entering conditions, outdoor temperatures and supply CFM values.

> Each manufacturer publishes equipment performance data in a different format. Take the time to become familiar with the data that is published by a particular manufacturer.

> Equipment performance tables typically correspond to operation at sea level. Correction factors are required to compensate for the effects of altitude.

It is always necessary to read the fine print and footnotes that accompany the performance data. The footnotes below the cooling capacity table that was used for this example (Figure 6-2), inform the designer that the effect of the fan heat has not been subtracted from the sensible capacity values. This means that the designer must adjust the sensible cooling capacity values to compensate for this load.

Check Cooling Capacity at 7000 CFM
The equipment selection data summary indicates that the cooling unit should have about 251,000 BTUH of total capacity when it operates at 7,145 CFM. A casual inspection of the data presented by Figure 6-2 indicates that the C25 unit may satisfy the cooling requirements. The unit has 251,000 BTUH of total capacity when it operates at 7,000 CFM (if the outdoor dry-bulb is 95°F, the entering dry-bulb is 75°F and the entering wet-bulb is 61°F). A double interpolation is required to determine what the sensible and latent capacity will be if the entering dry-bulb is equal to 77°F and the entering wet-bulb is equal to 64°F. These calculations are summarized below.

Performance at 7000 CFM, 95/75/64

At 75/61	At 75/67	At 75/64
TC SHR	TC SHR	TC SHR
251 83	282 60	267 71.5

Performance at 7000 CFM, 95/80/64

At 80/61	At 80/67	At 80/64
TC SHR	TC SHR	TC SHR
253 95	281 72	267 83.5

Performance at 7000 CFM, 95/77/64

At 75/64	At 80/64	At 77/64
TC SHR	TC SHR	TC SHR
267 71.5	267 83.3	267 76.3

Capacity at 7000 CFM, 95/77/64
Sensible capacity = 267000 x 0.763 = 203721 Btuh
Latent capacity = 267000 - 203721 = 63279 Btuh

Sensible and Latent Loads
Sensible = 207851 Btuh
Latent = 42853 Btuh

At 7,000 CFM, the unit is slightly short of sensible capacity and it has about 20,400 BTUH of extra latent capacity. These capacities are close to the required capacities, but the fan motor heat has not been accounted for.

Fan Heat
Figure 6-3, at the top of the next page, summarizes supply fan performance. At 7,000 CFM, the fan that is installed in the C25 model has a motor power draw that varies between 2 HP and 5 HP, depending on the external resistance that the fan has to work against. For now, assume that a single-zone duct system with accessory devices will not produce more than 1.5 IN.WG. of external resistance. (This assumption should be verified later, when the duct system is designed.) Figure 6-3 indicates that the fan motor will draw approximately 4 horsepower when the fan moves 7,000 CFM against 1.5 IN.WG. of external pressure. About 5 horsepower will be required as the external resistance approaches 2.0 IN.WG. (or as the supply air flow is increased to more than 8000 CFM).

Figure 6-4, on the next page, indicates that the heat load that is associated with a 5 horsepower fan motor is approximately equal to 14,000 BTUH. This means that the C25 model will be about 14,000 BTUH short of sensible capacity when it operates at 95/77/64. However, the unit has about 24,000 Btuh of excess latent capacity and the actual outdoor design temperature is 92°F, not 95°F. These factors might be enough to offset the sensible load that is associated with the fan heat.

An inspection of Figure 6-2 shows that the unit will have about 4,000 BTUH more total capacity when the outdoor is equal to 92°F. About 3,000 BTUH of this amount will be available as sensible capacity.

Also note that the excess latent capacity (20,400 BTUH) will reduce the humidity in the occupied space. Therefore, the entering wet-bulb temperature will be less than 64°F, say 63°F. This, will cause some of the 20,000 BTUH of excess latent capacity to be converted to sensible capacity. A tedious trial and error analysis is required to determine exactly how much of the 20,400 BTUH of excess latent capacity will be converted to sensible capacity, but 50 percent (10,200 BTUH) would be a safe guess.

Supply Air Fan Performance

Model	CFM	TOTAL STATIC PRESSURE															
		0.250		0.500		0.750		1.000		1.250		1.500		1.750		2.000	
		RPM	BHP	RPM	BHP	RPM	BHP	RPM	BHP	RPM	BHP	RPM	BHP	RPM	BHP	RPM	BHP
C20	4000	447	0.48	567	0.80	669	1.18	757	1.61	833	2.05	902	2.52	964	3.00	1018	3.47
	5000	494	0.73	605	1.10	699	1.50	782	1.96	860	2.47	930	3.00	993	3.54	1052	4.11
	6000	547	1.08	648	1.49	736	1.95	815	2.43	886	2.95	955	3.53	1020	4.14	1080	4.77
	7000	605	1.55	695	2.01	778	2.50	853	3.05	922	3.60	985	4.18	1046	4.81	1104	5.49
	8000	662	2.14	745	2.66	824	3.21	895	3.78	961	4.40	1022	5.03	1080	5.67	1135	6.35
	9000	722	2.87	803	3.50	872	4.07	941	4.70	1003	5.43	1062	6.03	1118	6.74	1171	7.45
C25	5000	469	0.67	581	1.03	678	1.43	764	1.88	843	2.37	914	2.89	978	3.44	1039	4.02
	6000	516	0.98	618	1.39	708	1.83	789	2.31	863	2.82	933	3.38	999	3.97	1060	4.59
	7000	569	1.40	660	1.85	744	2.34	819	2.86	890	3.41	957	3.99	1019	4.60	1079	5.25
	8000	625	1.95	706	2.43	783	2.97	855	3.54	922	4.13	985	4.75	1045	5.39	1103	6.07
	9000	683	2.63	757	3.17	827	3.74	895	4.35	958	5.00	1019	5.66	1975	6.34	1130	7.05
	10000	742	3.47	810	4.06	875	4.65	938	5.32	998	6.01	1055	6.73	1110	7.46	1162	8.21
	11000	803	4.48	866	5.13	926	5.79	984	6.47	1040	7.20	1095	7.96	1148	8.75	1198	9.55
C30	6000	493	0.89	597	1.27	691	1.70	776	2.16	852	2.65	922	3.18	988	3.75	1049	4.36
	7000	542	1.27	634	1.68	720	2.14	800	2.65	874	3.18	943	3.74	1006	4.32	1066	4.94
	8000	593	1.75	676	2.20	754	2.70	829	3.24	900	3.82	965	4.41	1028	5.04	1087	5.69
	9000	645	2.35	722	2.86	793	3.39	863	3.97	929	4.58	992	5.22	1052	5.88	1109	6.56
	10000	699	3.08	771	3.66	837	4.23	900	4.83	962	5.48	1022	6.16	1079	6.86	1135	7.59
	11000	753	3.96	822	4.61	884	5.23	942	5.87	999	6.54	1056	7.25	1110	8.00	1163	8.76
	12000	809	5.00	874	5.72	933	6.41	987	7.08	1040	7.79	1093	8.53	1144	9.31	1195	10.11
	13000	866	6.22	926	7.00	983	7.76	1035	8.49	1084	9.23	1143	10.00	1182	10.81	1230	11.65
	13500	895	6.93	953	7.74	1009	8.53	1060	9.30	1108	10.06	1156	10.85	1203	11.67	1249	12.52

Figure 6-3

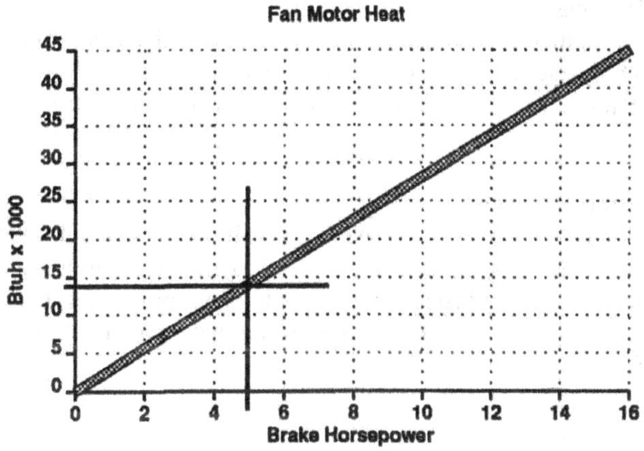

Figure 6-4

Therefore, it appears that at 92/77/63 the unit has about 13,000 BTUH of additional sensible capacity, which should be enough to offset the 14,000 BTUH fan heat load. (Some manufacturers may have computer programs that provide an exact value.) In any case, this unit can be expected to have adequate sensible and latent capacity when it operates at 7000 CFM.

Check Cooling Capacity at 8750 CFM

Another way to deal with a shortage of sensible capacity is to check the performance at a larger supply CFM value. Figure 6-2 indicates that the C25 unit may satisfy the original design requirements if the unit operates at 8,750 CFM. Another double interpolation is required to determine what the sensible and latent capacity will be at the higher CFM value.

Performance at 8750 CFM, 95/75/64

At 75/61		At 75/67		At 75/64	
TC	SHR	TC	SHR	TC	SHR
262	87	292	63	277	75

Performance at 8750 CFM, 95/80/64

At 80/61		At 80/67		At 80/64	
TC	SHR	TC	SHR	TC	SHR
266	100	292	78	279	89

Performance at 8750 CFM, 95/77/64

At 75/64		At 80/64		At 77/64	
TC	SHR	TC	SHR	TC	SHR
277	75	279	89	278	80.6

Capacity at 8750 CFM, 95/77/64

Sensible = 278000 x 0.806 = 224068 Btuh

Latent = 278000 - 224068 = 53932 Btuh

Sensible and Latent Loads

Sensible = 207851 + 14,000 = 221851 Btuh

Latent = 42853 Btuh

At 8,750 CFM the unit has adequate sensible capacity and it has about 11,000 BTUH of extra latent capacity. These capacities are a close match for the 95/77/64 design condition.

Reserve Capacity
If the C25 unit is operated at 7,000 CFM it has no reserve capacity. If the unit is operated at 8,750 CFM it has a few thousand BTUH, perhaps as much as 6,000 BTUH of reserve capacity. The C30 unit would have to be used if more reserve capacity is desired.

Capacity Control — Cooling
A check of the manufacturer's literature indicates that the C-25 unit is equipped with three compressors and three independent refrigeration circuits. The literature also indicates that a hot gas bypass option and an air-side economizer option are available. This means that the cooling capacity can be modulated down to 33 percent of the design capacity before hot gas bypass is required. Point #1, on Figure 6-5, shows that the minimum sensible load on the compressors will be approximately equal to 46,000 BTUH. This load will occur at 55°F, just before the compressors are shut down and the economizer cycle is activated. This load represents 21 percent of the unit's 224,068 BTUH of sensible capacity. Hot gas bypass could be installed, but this would not be necessary if the manufacturer certifies that compressor cycling is an acceptable method of capacity control. (Refer to Appendix 5 for information about controlling the capacity of DX equipment.)

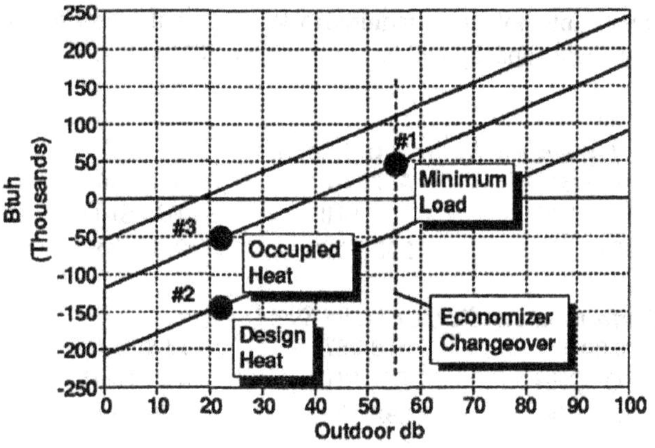

Figure 6-5

Controls and Accessories
The equipment must be specified with the proper controls and accessories. A search of the manufacturer's product literature indicates that the following controls and accessories are either standard equipment or available as options. All the following items will be required with the exception of the hot gas bypass option.

Refrigeration Equipment
Three compressors — standard
Three independent refrigeration circuits — standard
Hot gas bypass — option
Suction accumulator — standard
Motor overload protection — standard
High- and low-pressure cutouts — standard
Oil pressure switch — standard
Anti-short-cycle relay — standard
Crankcase heater — standard

Outdoor Air and Filters
Low leakage dampers — option
100% economizer — option
Lockout on enthalpy — option
Dampers controlled by space temperature — standard
Minimum position setting — standard
Exhaust fan — option
High efficiency 2 inch filter — standard

Electronic Control Panel
Cool-heat-auto-off room thermostat
Heating-cooling deadband 3°F to 30°F
Night Set-back
7-day clock, 18 set-back periods
Cooling lockout
Signal lights (operating mode, malfunction alert)
Low limit on supply temperature
Pneumatic interface

Heating
This single-zone rooftop unit is available with a gas furnace, a hot water coil or a steam coil. Refer to Sections 7-7, -8 and -9 for more information on about selecting heating equipment.

Control Cycles
Control cycles must be specified for all operating conditions. The following control sequences provide an example of a scheme that could be used for a single-zone application. These sequences assume that the unit is equipped with an indirect-fired gas furnace.

Single-Zone Mode — Day Cycle
Mechanical cooling shall be activated during the day cycle, providing the outdoor temperature (or enthalpy) is above the economizer setpoint. The compressors, refrigeration circuits and cooling capacity shall be controlled in response to a signal from the room thermostat. The outdoor air damper shall be open to the minimum position. The supply fan and the exhaust fan shall run continuously. Compressor and fan operation shall be subject to supervision by safeties and interlocks.

Mechanical cooling shall be deactivated when the outdoor temperature (or enthalpy) is below the economizer setpoint. The outdoor air, return air and exhaust air dampers shall be controlled in response to a signal from the room thermostat. A minimum position device shall prevent the outdoor air damper from closing past the minimum position setting. The supply fan and the exhaust fan shall run continuously. Fan and damper operation shall be subject to supervision by safeties and interlocks.

Heat shall be activated when the economizer outdoor air damper modulates to the minimum position. The burner and gas valve shall be controlled by a signal from the room thermostat. The outdoor air damper shall remain in the minimum position as long as heat is required. The supply fan and the exhaust fan shall run continuously. Furnace and fan operation shall be subject to supervision by safeties and interlocks.

Single-Zone — Night Cycle
In the cooling mode, the cooling equipment shall be deactivated during the night cycle. The outdoor air and exhaust dampers shall be closed. The supply fan and the exhaust fans shall be off.

In the heating mode, the burner and gas valve shall be controlled in response to a signal from a night set-back thermostat. The outdoor air and exhaust air dampers shall remain closed. The supply fan shall run intermittently, cycling with the burner. Furnace and fan operation shall be subject to supervision by safeties and interlocks.

Interior, Packaged, Water-Cooled VAV Unit

6-9 Packaged, Water-Cooled, DX Equipment

The procedure that is used for selecting and sizing packaged, water-cooled, DX cooling equipment is almost identical to the procedure that is used for air-cooled equipment. (Refer to Section 6-8 for a comprehensive discussion of this selection procedure.) This basic procedure can be used to select any type of packaged, water-cooled, direct expansion, cooling unit or heat pump. Single-package, interior, cooling units; and water-source heat pumps are typical of the type of equipment that falls into this category.

Water-Source Heat Pump

Interior, Packaged, Water-Cooled Unit

The primary difference between water-cooled and air-cooled equipment is that the water-cooled performance data is based on the entering water temperature and condenser GPM instead of the outdoor air temperature. Figure 6-6 provides an example of water-source heat pump performance data. (The performance data for a cooling-only unit would be similar to this heat pump data.) Note that the cooling capacity depends on the entering water temperature and the condenser GPM and that there is no reference to the outdoor air temperature.

Cooling @ 80 F (EDB) and 2300 CFM													
EWB	EWT	GPM 18.0				GPM 14.0				GPM 10.0			
		TC	SC	KW	HR	TC	SC	KW	HR	TC	SC	KW	HR
71	55	78.0	41.3	4.65	89.1	76.3	40.7	4.81	87.7	74.5	40.1	4.96	86.4
	65	75.3	40.4	5.04	84.9	73.6	39.8	5.21	83.6	71.9	39.3	5.37	82.2
	75	72.6	39.5	5.43	83.2	71.0	38.9	5.61	81.9	69.4	38.4	5.79	80.7
	85	69.9	38.6	5.81	81.4	68.4	38.1	6.01	80.3	66.8	37.6	6.20	79.1
	95	63.0	36.3	6.20	75.5	61.6	35.9	6.41	74.5	60.2	35.4	6.61	73.5
67	55	73.3	50.8	4.53	84.1	71.7	50.2	4.68	82.8	70.1	49.6	4.84	81.6
	65	70.7	49.8	4.91	80.2	69.2	49.2	5.07	78.9	67.6	48.6	5.24	77.6
	75	68.2	48.9	5.29	78.5	66.7	48.3	5.46	77.4	65.2	47.7	5.64	76.2
	85	65.7	47.9	5.67	76.9	64.3	47.4	5.86	75.9	62.8	46.8	6.04	74.8
	95	59.2	45.4	6.04	71.4	57.9	44.9	6.25	70.4	56.6	44.4	6.45	69.5
63	55	68.6	60.2	4.42	79.1	67.1	59.5	4.56	78.0	65.6	58.9	4.71	76.8
	65	66.2	59.1	4.78	75.4	64.7	58.5	4.94	74.2	63.3	57.8	5.10	73.1
	75	63.9	58.1	5.15	73.9	62.4	57.5	5.32	72.8	61.0	56.8	5.49	71.7
	85	61.5	57.1	5.52	72.4	60.2	56.6	5.70	71.4	58.8	55.9	5.89	70.4
	95	55.4	54.4	5.89	67.2	54.2	53.8	6.08	66.4	53.0	53.0	6.28	65.6

Figure 6-6

Psychrometric Analysis

Figure 6-6 indicates that the design conditions that are associated with the indoor coil — namely the entering dry- and wet-bulb temperatures and the CFM across the DX coil — are exactly the same as the conditions that apply to air-cooled equipment. Therefore, the same type of psychrometric analysis that was used for selecting air-cooled equipment also would be used for selecting water-cooled equipment (Refer to Figure 6-1).

Equipment Selection Data Summary

The designer should summarize the equipment performance requirements before referring to the manufacturer's performance data. This summary is similar to the summary that is used for air-cooled equipment, except that the entering water temperature is used in place of the outdoor air temperature. An example of the information that is required is summarized below.

Indoor Design = 75°F and 50% RH
Entering Water = 55°F DB
Sensible Load = 56,450 BTUH
Latent Load = 5,700 BTUH
Entering dry-bulb = 75°F
Entering wet-bulb = 63°F
Approximate CFM = 2,300

Cooling Performance Data

After the equipment performance requirements are tabulated, the manufacturer's performance data can be used to select the equipment. Carefully inspect the performance data for the unit that provides the closest match to the equipment performance requirements. Examples of water-cooled equipment performance data is provided on this page by Figure 6-6 and at the top of the next page by Figures 6-7 and 6-8.

- Some manufacturers publish the performance data in a format that is similar to Figure 6-6. In this case, the cooling capacities are directly referenced to the entering water temperature and the GPM flowing through the condenser — no reference is made to the condensing temperature.

- Some manufacturers publish the performance data in a format that is similar to Figures 6-7 and 6-8. In this case, the cooling capacities are referenced to a condensing temperature in the primary table and a second table is used to establish the relationship between the entering water temperature, the GPM and the condensing temperature.

In either case, the cooling performance data should provide information on total capacity, sensible capacity and latent capacity. In addition, the performance should be tabulated for a full range of entering air conditions, water-side conditions and supply CFM values. Footnotes may indicate that adjustments are required to compensate for operating conditions or system loads that are not included in the basic table — a fan heat adjustment, a different entering wet-bulb or a different entering dry-bulb temperature, for example.

> Each manufacturer publishes equipment performance data in a different format. Take time to become familiar with the data. Also note that the data should be adjusted for the effect of altitude.

Condenser Water Temperature

The temperature of the condenser water is an important design parameter. Condenser water could be supplied by a cooling tower, a well, a closed water loop buried in the earth, a closed reclaim water loop or by a city water connection. Cooling towers are typically sized so that, on a design day, the leaving water temperature is about 85°F. The temperature of well water or city water can vary from below 50°F to more than 60°F, depending on local conditions. The temperature of the water in a closed earth-coupled loop can exceed 90°F during late summer or early fall. The temperature of the water in a closed reclaim loop may exceed 90°F during the most severe operating condition. Refer to Sections 5-6, 5-7 and 5-8 for more information about water temperatures.

Air-Side Performance Data

Air-side performance data consists of a blower table and (possibly) a table or equation that provides information about

Gross Cooling Capacities @ 8000 CFM

Model	ENT Db	ENT Wb	95 F CONDENSING TEMP.					105 F CONDENSINGTEMP.					115 F CONDENSING TEMP.				
			ENT Wb					ENT Wb					ENT Wb				
			61	64	67	70	73	61	64	67	70	73	61	64	67	70	73
200	75	MBh	238	251	265	279	293	231	244	258	271	285	224	236	249	263	277
		SHR	0.82	0.69	0.57	0.46	0.36	0.83	0.70	0.58	0.46	0.36	0.84	0.71	0.58	0.47	0.36
	80	MBh	241	251	265	278	293	235	244	256	271	285	228	237	249	263	276
		SHR	0.95	0.83	0.70	0.59	0.48	0.96	0.84	0.71	0.59	0.48	0.97	0.85	0.72	0.60	0.49
	85	MBh	251	254	265	278	293	246	247	258	271	284	240	241	249	262	276
		SHR	1.00	0.95	0.83	0.71	0.60	1.00	0.96	0.84	0.72	0.61	1.00	0.97	0.85	0.73	0.62
	90	MBh	265	265	268	278	293	259	259	261	271	284	252	252	254	265	276
		SHR	1.00	1.00	0.94	0.83	0.72	1.00	1.00	0.96	0.84	0.73	1.00	1.00	0.96	0.85	0.74

Figure 6-7

Condenser Water Requirements (GPM)

Model	Net Load Tons	95 F Condensing Temp.			105 F Condensing Temp.				115 F Condensing Temp.			
		Ent. Water Temp. F			Ent. Water Temp. F				Ent. Water Temp. F			
		65	70	75	70	75	80	85	80	85	90	95
200	18	34	45	66	28	35	46	69	29	36	48	73
	20	38	52	79	31	39	53	-	32	41	55	-
	22	44	59	-	35	45	61	-	36	46	64	-

Figure 6-8

a fan heat adjustment. (Refer to Figures 6-3 and 6-4.) Other air-side tables may provide information on the air-side pressure drop across components that were not accounted for in the basic blower table — coil, filter or mixing box pressure drops for example. (Refer to ACCA **Manual Q**, Section 9, for more information about how air-side devices and accessories affect the duct sizing and fan selection calculations.

Water-Side Pressure Drop

Manufacturer's data should include tables or charts that provide information about the water-side pressure drop across the condenser and any associated components. This information is required for the pipe sizing and pump selection calculations. (Refer to Appendix 4 for more information about piping design.) Figure 6-9 provides an example of this type of information.

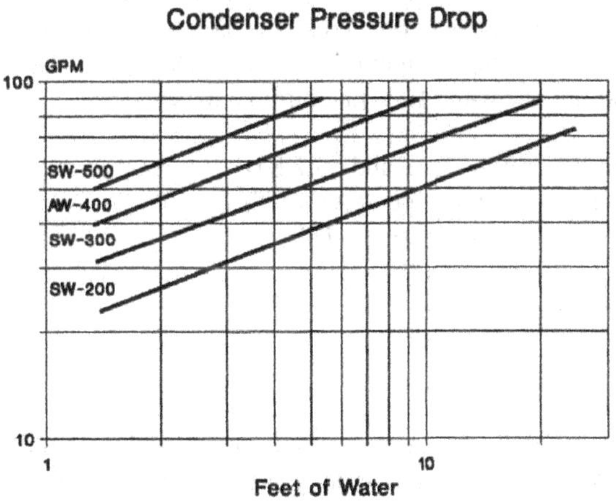

Figure 6-9

6-10 Interpreting Water-Cooled Performance Data

Note that Figure 6-6, on the previous page, only provides information about the cooling capacities that are associated with operation at 2,300 CFM and 80°F dry-bulb entering air. Capacities at other CFM values or entering dry-bulb temperatures are not listed, although they are usually required. For example, Figure 6-1, on page 6-6 — which is typical of comfort cooling applications — indicates a 77°F entering dry-bulb temperature. Therefore, the information presented by Figure 6-6 is not totally adequate for selecting equipment. In this case, the designer must consult with the manufacturer to determine the performance that is associated with entering conditions or CFM values that are not listed in the table.

Also note, there is only a marginal increase in the cooling capacity as the water flow rate increases from 10 GPM to 18 GPM (see Figure 6-6). This behavior is typical of water-cooled equipment. Therefore, the performance advantage associated with operation at the higher water flow rate may not offset the disadvantages associated with pumping more water. (Earth-coupled water-loop systems may require a higher flow rate to insure turbulent flow in the water-loop pipe runs.)

Finally, note that the cooling capacity at 95 EWT is substantially smaller than the cooling capacity at 55 EWT (see Figure 6-6). Therefore, a larger unit will be required for a closed loop, high water temperature system and a smaller unit can be used for a well water or city water system.

Figures 6-7 and 6-8, on the previous page, provide performance data that is more comprehensive than the Figure 6-6 data. In these tables, the data covers a wide range of entering dry-bulb conditions. However, a correction will be required if the air flow rate does not match the nominal CFM value. (Air flow adjustment factors are often included along with the basic performance table.)

To use the more comprehensive performance data (Figure 6-7), the designer must choose a value for the "condensing temperature." Then, this value, plus the entering dry-bulb and the entering wet-bulb temperatures can be used to determine the cooling capacity of the equipment. Once the equipment capacity is verified, the designer can use the companion table (Figure 6-8) to determine the condenser GPM requirement.

Example
Check to see if Model SWUA-200 has the required capacity at the design conditions listed below. Design the system for a 95°F condensing temperature. If the capacity is adequate, determine the water flow rate through the condenser.

Indoor Design = 75°F and 50% RH
Entering Water Temperature = 85°F
Sensible Load = 193,000 BTUH
Latent Load = 52,000 BTUH
Entering dry-bulb = 77°F
Entering wet-bulb = 64°F

Figure 6-7 indicates (with some interpolation) that the unit has 251,000 BTUH of total capacity and that about 75 percent of this capacity is available as sensible capacity. At 77 DB and 64 WB this translates into 188,000 BTUH of sensible capacity and 63,000 BTUH of latent capacity. The unit is 5,000 BTUH short of sensible capacity, but it has 11,000 BTUH of excess latent capacity. This extra latent capacity will cause the entering wet-bulb temperature to drop below 64°F and some (about one half) of this extra latent capacity will be converted to sensible capacity. Therefore, the unit appears to have adequate capacity, but no excess capacity.

251,000 BTUH is about equal to 20 TONS of cooling capacity. Figure 6-8 indicates (at 95°F condensing temperature and 85°F entering water temperature) that 52 GPM of water must flow through the condenser.

6-11 Chilled Water Equipment

The major components of a chilled water system consist of a water chiller, one or more chilled water coils and some type of heat sink. The heat sink can be provided by an air-cooled condenser or by some source of cooling water. Cooling water is usually provided by a cooling tower, but any abundant, inexpensive source of cooling water can be used.

Many of the steps in the procedure that is used to select and size chilled water equipment are similar to those that were used to select and size DX equipment. Equipment size is based on the results of a load calculation and a psychrometric analysis; a summary of the equipment selection data is required and manufacturer's performance data must be used to verify performance at the design conditions. (Refer to Section 6-8 for more information about load calculations and psychrometrics.)

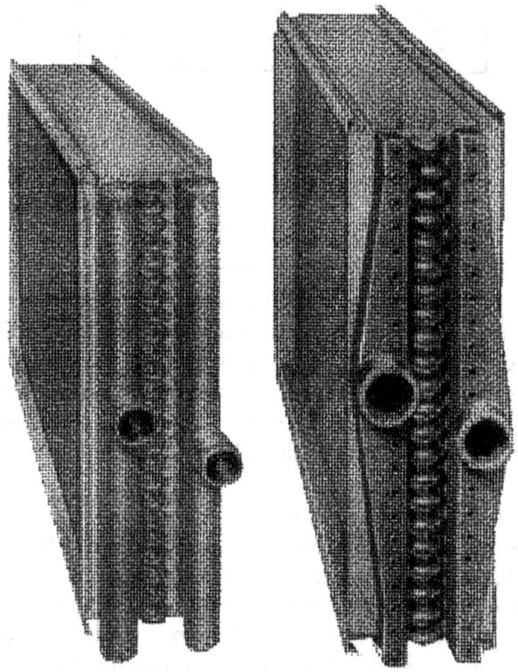

Chilled Water Coils

6-12 Chilled Water Coil Selection

Chilled water coil performance depends on many variables and parameters. An example of one manufacturer's performance data is provided by Figure 6-9, at the top of the next page. Note that the coil performance — which includes MBTUH per square foot of coil face area, leaving air dry-bulb temperature (LDB) and leaving air wet-bulb temperature (LWB) — depends on the following parameters . . .

CHILLED WATER COOLING COIL CAPACITIES
42 EWT
EDB/EWB = 80/67

WTR = 8

FPS	Row	FPF	400 FPM MBH	400 FPM LDB	400 FPM LWB	500 FPM MBH	500 FPM LDB	500 FPM LWB	600 FPM MBH	600 FPM LDB	600 FPM LWB	700 FPM MBH	700 FPM LDB	700 FPM LWB
2	4	80	12.3	59.7	57.3	13.6	61.3	58.5	14.7	62.5	59.5	15.6	63.6	60.2
		100	13.7	57.7	56.1	15.2	59.3	57.4	16.5	60.6	58.5	17.5	61.7	59.3
		120	14.8	56.1	55.1	16.5	57.8	56.5	17.9	59.1	57.6	19.1	60.2	58.5
		140	15.7	54.9	54.3	17.5	56.6	55.8	19.1	57.9	56.9	20.4	59.1	57.9
		160	16.4	54.0	53.6	18.4	55.7	55.2	20.0	57.1	56.4	21.3	58.3	57.4
	6	80	16.1	54.9	53.9	18.2	56.6	55.4	19.9	58.0	56.5	21.3	59.1	57.4
		100	17.6	53.0	52.5	20.0	54.7	54.0	22.0	56.1	55.2	23.7	57.3	56.2
		120	18.8	51.6	51.4	21.4	53.3	52.9	23.6	54.7	54.2	25.5	55.9	55.3
		140	19.7	50.6	50.5	22.5	52.3	52.1	24.9	53.7	53.4	26.9	54.9	54.6
		160	20.4	49.8	49.7	23.4	51.5	51.4	25.9	53.0	52.8	28.0	54.2	54.0
	8	80	19.0	51.6	51.2	21.7	53.3	52.7	24.0	54.7	54.0	26.0	55.9	55.1
		100	20.4	49.9	49.7	23.6	51.6	51.3	26.2	52.9	52.6	28.5	54.2	53.7
		120	21.5	48.8	48.7	25.0	50.3	50.2	27.9	51.7	51.5	30.5	52.9	52.7
		140	22.3	47.9	47.8	26.0	49.4	49.3	29.2	50.6	50.7	31.9	52.0	51.9
		160	22.9	47.3	47.2	26.8	48.8	48.7	30.1	50.2	50.1	33.0	51.4	51.3
4	4	80	13.7	58.4	56.1	15.3	60.0	57.4	16.7	61.4	58.4	17.8	62.5	59.1
		100	15.3	56.1	54.6	17.3	57.8	56.0	19.0	59.1	57.0	20.3	60.3	57.9
		120	16.7	54.3	53.4	18.9	56.0	54.8	20.8	57.3	55.9	22.4	58.5	56.9
		140	17.8	52.9	52.3	20.3	54.6	53.8	22.3	56.0	55.0	24.1	57.2	56.0
		160	18.7	51.8	51.4	21.4	53.5	53.0	23.6	54.9	54.3	25.4	56.2	55.3
	6	80	17.7	53.4	52.4	20.1	55.1	53.9	22.2	56.6	55.1	24.0	57.5	56.1
		100	19.3	51.3	50.8	22.2	53.0	52.3	24.8	54.4	53.5	26.9	55.6	54.6
		120	20.6	49.8	49.5	24.0	51.3	51.0	26.8	52.7	52.2	29.3	53.9	53.3
		140	21.6	48.6	48.5	25.3	50.1	50.0	28.4	51.5	51.2	31.0	52.7	52.4
		160	22.5	47.6	47.7	26.3	49.2	49.1	29.6	50.6	50.4	32.4	51.8	51.6
	8	80	20.4	50.1	49.7	23.7	51.8	51.2	26.4	53.2	52.5	28.8	54.4	53.6
		100	22.0	48.3	48.2	25.7	49.8	49.6	29.0	51.1	50.8	31.9	52.3	51.9
		120	23.1	47.1	47.0	27.3	48.4	48.3	31.0	49.7	49.5	34.2	50.8	50.6
		140	23.9	46.2	46.1	28.4	47.4	47.3	32.4	48.6	48.5	35.9	49.7	49.6
		160	24.6	45.5	45.4	29.3	46.7	46.6	33.5	47.9	47.8	37.2	49.0	48.9

WTR = 10

FPS	Row	FPF	400 FPM MBH	400 FPM LDB	400 FPM LWB	500 FPM MBH	500 FPM LDB	500 FPM LWB	600 FPM MBH	600 FPM LDB	600 FPM LWB	700 FPM MBH	700 FPM LDB	700 FPM LWB
2	4	80	11.8	60.1	57.7	13.1	61.7	58.9	14.1	62.9	59.8	15.0	63.9	60.4
		100	13.2	58.2	56.6	14.7	59.7	57.8	15.9	60.9	58.8	16.9	62.0	59.5
		120	14.3	56.6	55.6	16.0	58.2	56.9	17.3	59.4	58.0	18.5	60.5	58.8
		140	15.2	55.4	54.8	17.0	57.0	56.2	18.5	58.3	57.3	19.7	59.4	58.2
		160	15.9	54.5	54.1	17.8	56.2	55.6	19.4	57.5	56.8	20.7	58.6	57.7
	6	80	15.6	55.4	54.4	17.5	57.1	55.8	19.2	58.4	56.9	20.6	59.5	57.8
		100	17.1	53.6	53.0	19.3	55.2	54.5	21.2	56.6	55.7	22.9	57.7	56.6
		120	18.2	52.2	51.9	20.8	53.8	53.4	22.9	55.2	54.7	24.7	56.3	55.7
		140	19.1	51.1	51.0	21.9	52.6	52.6	24.2	54.2	53.9	26.1	55.4	55.0
		160	19.9	50.4	50.3	22.7	52.0	51.9	25.1	53.5	53.3	27.2	54.7	54.4
	8	80	18.4	52.1	51.7	21.0	53.8	53.2	23.2	55.2	54.6	26.1	56.4	55.5
		100	19.9	50.5	50.3	22.9	52.1	51.8	25.4	53.5	53.1	27.6	54.6	54.2
		120	21.0	49.3	49.2	24.3	50.8	50.7	27.1	52.2	52.0	29.5	53.4	53.2
		140	21.8	48.4	48.3	25.3	50.0	49.9	28.4	51.3	51.2	31.0	52.5	52.4
		160	22.4	47.8	47.7	26.2	49.3	49.2	29.3	50.7	50.6	32.1	51.9	51.8
4	4	80	13.1	58.9	56.6	14.7	60.5	57.8	16.0	61.8	58.7	17.1	62.8	59.5
		100	14.7	56.7	55.2	16.7	58.3	56.4	18.2	59.6	57.5	19.6	60.7	58.3
		120	16.1	54.9	53.9	18.3	56.6	55.3	20.1	57.8	56.4	21.6	58.9	57.3
		140	17.2	53.5	52.9	19.6	55.1	54.3	21.6	56.4	55.5	23.3	57.6	56.4
		160	18.1	52.4	52.0	20.7	54.0	53.5	22.8	55.4	54.7	24.6	56.6	55.8
	6	80	17.0	54.0	53.0	19.3	55.7	54.5	21.3	57.1	55.6	23.1	58.2	56.5
		100	18.7	52.0	51.5	21.5	53.5	52.9	23.9	54.9	54.1	26.0	56.1	55.1
		120	20.0	50.4	50.1	23.2	51.9	51.6	26.0	53.2	52.8	28.3	54.4	53.8
		140	21.1	49.2	49.1	24.6	50.7	50.5	27.5	52.0	51.8	30.1	53.2	52.9
		160	21.9	48.3	48.2	25.6	49.8	49.7	28.7	51.1	51.0	31.5	52.3	52.1
	8	80	19.8	50.6	50.3	22.9	52.4	51.8	25.5	53.7	53.0	27.9	54.9	54.1
		100	21.4	49.0	48.8	25.0	50.4	50.2	28.2	51.7	51.4	30.9	52.9	52.4
		120	22.6	47.6	47.5	26.6	49.0	48.9	30.1	50.2	50.1	33.2	51.4	51.2
		140	23.4	46.7	46.6	27.8	48.0	47.9	31.6	49.2	49.1	35.0	50.3	50.2
		160	24.1	46.0	45.9	28.7	47.2	47.1	32.7	48.4	48.3	36.3	49.5	49.4

Figure 6-9

- Entering water temperature (EWT)
- Entering air dry-bulb temperature (EDB)
- Entering air wet-bulb temperature (EWB)
- Water temperature rise (WTR)
- Coil face velocity (FPM)
- Velocity of the water in the tubes (FPS)
- Number of rows (ROW)
- Number of fins per foot (FPF) or fins per inch (FPI)
- Equipped or not equipped with turbulators

Even when accompanied by detailed instructions, it is not a simple task to use chilled water coil performance data. The designer must refer to many tables and charts; and because there are many parameters to consider, there are numerous coil configurations that will provide the required performance. The best way to proceed is to list the design conditions and let the manufacturer or the manufacturer's sales representative use a computer to select the coil. At a minimum, these design conditions should be listed as follows . . .

- Cooling CFM
- Sensible capacity required (MBTUH)
- Latent capacity required (MBTUH)
- Entering dry-bulb temperature (EDB)
- Entering wet-bulb temperature (EWB)
- Leaving dry-bulb temperature (LDB)
- Leaving wet-bulb temperature (LWB)
- Approximate coil GPM
- Entering water temperature

The information listed above is obtained from the results of the load calculation and the psychrometric analysis. The design loads can be found on Form N-1, line 12. The design CFM can be found on Form N-1, Line 8. The entering and leaving conditions can be determined from the psychrometric analysis (see Figure 6-1). Refer to Section 5-10 for a formula that can be used to estimate the coil GPM.

Note that the leaving dry-bulb temperature must be compatible with the *Supply TD* that was used to establish the design CFM on line 8 on Form N-1. For example, if the room is maintained at 75°F and the line 8 cooling CFM is based on a 19 °F TD, the LDB must be equal to 56°F.

Once the leaving dry-bulb (LDB) temperature is determined, the condition of the leaving air — as indicated by the leaving dry-bulb and leaving wet-bulb (LWB) temperatures — must fall on or below the RSH or the CSH line, whichever is lower (refer to Figure 6-1).

Some secondary design conditions can also be specified. These conditions, which are listed below, are optional. They serve to place limits or restrictions on some aspects of coil performance.

Maximum Air-Side Pressure Drop
A limit on the air-side pressure drop is not usually specified, but one could be requested. The air-side pressure drop is affected by the face velocity, the number of rows and the fin spacing.

Maximum Water-Side Pressure Drop
A limit on the water-side pressure drop is not usually specified, but one could be requested. The water-side pressure drop is affected by the water velocity, the coil circuit arrangement, the number of rows and the turbulator option.

Maximum Face Velocity (FPM)
A face velocity of 500 FPM is typical. It can be higher, but if it is too high, say above 600 FPM, condensate may be blown off the coil or the air-side pressure drop may be too large.

Minimum Entering Water Temperature (EWT)
Entering water temperatures of 42°F and 44°F are typical. Higher water temperatures (up to 60°F) can be desirable because they increase the operating efficiency of the chiller, but they reduce the latent capacity of the coil and increase the face area of the coil. Lower water temperatures (down to 35°F) should only be used if there is no other way to match latent capacity of the coil with the latent load.

Water Temperature Rise
A 10°F water temperature rise across the coil is typical. A larger rise increases the capacity of the coil and reduces the size of the coil. A smaller rise produces the opposite result.

Maximum Number of Rows
A large number of rows increases the coil capacity for a given face area, but there is also an increase in the air-side pressure drop. A small number of rows has the opposite effect.

Minimum Fin Spacing
A close fin spacing increases the coil capacity for a given face area, but there is a corresponding increase in the air-side pressure drop and the coil is more likely to be obstructed by fibers and dust. A wider fin spacing has the opposite effect.

Air-Side Pressure Drop
Part of the work that is associated with the coil selection process involves making an estimate of the air-side pressure drop. If the HVAC system designer selects the coil, the air-side pressure drop can be estimated by using tables and

Air-Side Pressure Drop

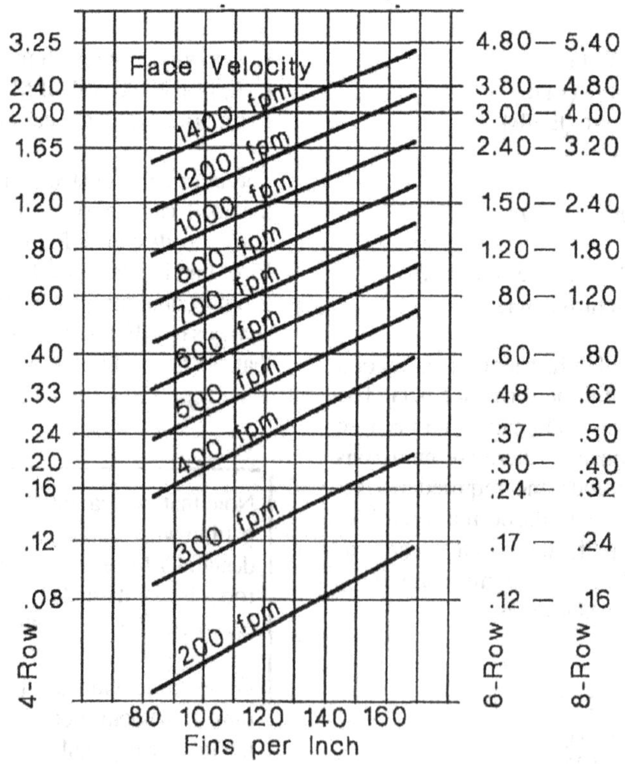

Figure 6-10

charts that are provided by the manufacturer. Or, if the manufacturer selects the coil, the HVAC system designer can get the air-side pressure drop information from the manufacturer. An example of air-side pressure drop data is provided by Figure 6-10, at the bottom of the previous page.

Water-Side Pressure Drop

Another task that is associated with the coil selection process involves making an estimate of the water-side pressure drop. If the HVAC system designer selects the coil, the water-side pressure drop can be estimated by using tables and charts that are provided by the manufacturer. Or, if the manufacturer selects the coil, the HVAC system designer can get the water-side pressure drop information from the manufacturer. An example of water-side pressure drop data is provided in Figure 6-11 below.

Capacity Control

The capacity of a chilled water coil is typically controlled by using two-way (throttling) or three-way (mixing or diverting) valves to control the flow of water (GPM) through the coil. Capacity also can be controlled by using face and bypass dampers to control the flow of air (CFM) across the coil.

6-13 Unit Ventilators and Fan-Coils

Hydronic terminal equipment packages are used for cooling-only, cooling-ventilating and heating-cooling-ventilating applications. Cooling-only equipment is furnished with a single chilled water coil. Heating-cooling units could be equipped with a single coil that is used for both functions or with two coils, one for each function. Packages are available in a wide range of sizes. Each size has a limited number of coil options. In some cases, depending on the make and model, there may only be one type of coil available for a given size unit. This type of equipment is available with floor, wall and ceiling mounted cabinets and depending on the number of coils, it can be installed in two-, three- and four-pipe configurations.

Air Distribution

Hydronic terminal equipment packages are not usually fitted with an air distribution system. When this is the case, the air motion in the room depends on the effectiveness of the discharge grille. This grille can be expected to create a ring shaped air pattern that rolls around the cross-section of the room. Therefore, a number of smaller units, as opposed to one larger unit, should be used to distribute heat over the length of a room.

Header Water Pressure Drop (Feet H20) for Connectors, Headers, and Tube Entrance and Exit

COIL TYPE	FINNED WIDTH	WATER VELOCITY (FPS)								
		0.5	1	2	3	4	5	6	7	8
WC	12", 24"	0.04	0.16	0.66	1.48	2.63	4.11	5.92	8.06	10.53
	18"	0.05	0.22	0.87	1.97	3.50	5.47	7.88	10.73	14.01
	30", 33"	0.03	0.13	0.53	1.20	2.14	3.34	4.81	6.54	8.55
W-1-Row	All	0.05	0.13	0.52	1.20	2.10	3.33	4.79	6.47	8.52
WA-1-Row	6", 9", 12"	0.04	0.15	0.62	1.31	2.43	3.81	5.51	7.41	9.74
	18" thru 33"	0.03	0.13	0.49	1.07	1.92	3.00	4.37	5.88	7.72
W-2-Row	12", 18", 24"	0.02	0.10	0.36	0.80	1.28	2.19	3.15	4.23	5.52
	30", 33"	0.03	0.13	0.53	1.17	2.07	3.22	4.65	6.41	8.25
	36", 42", 48"	-	0.20	0.81	1.82	3.22	5.04	7.25	9.88	12.89
WA-2-Row	12", 18"	0.01	0.04	0.16	0.35	0.63	0.98	1.41	1.92	2.50
	24"	0.01	0.05	0.22	0.49	0.87	1.36	1.96	2.66	3.48
	30", 33"	0.02	0.09	0.34	0.78	1.38	2.16	3.11	4.23	5.52
TT All Row	All	0.04	0.14	0.57	1.29	2.30	3.60	5.18	7.05	9.21

Tube Water Pressure Drop (Feet)

WC 2-Row W	VELOCITY WITH TURBULATORS (FPS)					VELOCITY WITHOUT TURBULATORS (FPS)								
Finned Length	0.5	1.0	2.0	3.0	4.0	0.5	1.0	2.0	3.0	4.0	5.0	6.0	7.0	8.0
12	.016	.056	0.192	0.40	0.68	0.006	0.016	0.06	0.12	0.20	0.30	0.42	0.54	0.68
24	0.03	0.11	0.39	0.80	1.36	0.012	0.032	0.12	0.24	0.40	0.60	0.84	1.08	1.36
36	0.05	0.17	0.58	1.20	2.04	0.018	0.048	0.18	0.36	0.60	0.90	1.26	1.62	2.04
48	0.07	0.23	0.77	1.60	2.72	0.024	0.064	0.24	0.48	0.80	1.20	1.68	2.16	2.72
60	0.08	0.28	0.95	2.00	3.40	0.030	0.080	0.30	0.60	1.00	1.50	2.10	2.70	3.40
72	0.10	0.34	1.15	2.40	4.08	0.036	0.096	0.36	0.72	1.20	1.80	2.52	3.24	4.08
84	0.11	0.39	1.35	2.80	4.76	0.42	0.112	0.42	0.84	1.40	2.10	2.94	3.78	4.76
96	0.13	0.45	1.54	3.20	5.44	0.048	0.128	0.48	0.96	1.60	2.40	3.36	4.32	5.44
108	0.15	0.51	1.74	3.60	6.12	0.054	0.144	0.54	1.08	1.80	2.70	3.78	4.86	6.12
120	0.16	0.56	1.92	4.00	6.80	0.060	0.160	0.60	1.20	2.00	3.00	4.20	5.40	6.80
132	0.18	0.62	2.11	4.40	7.48	0.066	0.176	0.66	1.32	2.20	3.30	4.62	5.94	7.48
144	0.20	0.68	2.30	4.80	8.16	0.72	0.192	0.72	1.44	2.40	3.60	5.04	6.48	8.16

NOTE: For 1-row W, WA coils (all fin spacings), multiply above values by 0.5.

Figure 6-11

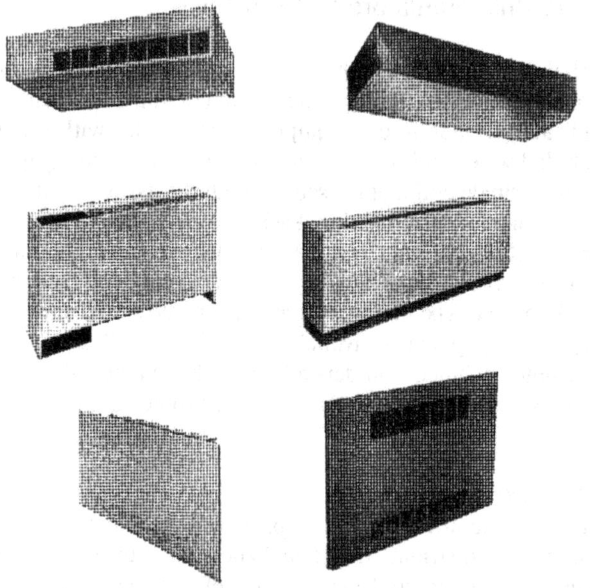

Chilled Water Terminal Equipment

Cooling Coil Performance

The physical characteristics (size, rows, fins per inch etc.) of the coil are fixed by the manufacturer's design. Therefore, the coil performance depends on the entering dry-bulb temperature, the entering wet-bulb temperature, the entering water temperature, the CFM flowing across the coil and the GPM flowing through the coil. An example of this type of coil performance data is provided by Figure 6-12.

SIZE 12 — COIL CAPACITIES

EWB	EDB	EWT	GALLONS PER MINUTE								
			4			6			10		
			SC	TC	TR	SC	TC	TR	SC	TC	TR
64	70	42	16.4	28.4	14.4	18.7	33.9	11.4	21.1	39.4	7.9
		45	15.0	24.9	12.6	16.9	29.6	10.0	19.0	34.5	7.0
		50	18.9	19.0	9.7	14.0	22.3	7.5	15.4	26.1	5.3
	75	42	22.5	29.8	15.1	23.9	34.2	11.5	26.0	39.1	7.9
		45	21.3	26.6	13.5	22.4	30.3	10.2	23.9	34.3	6.9
		50	19.6	21.3	10.9	20.3	24.0	8.1	21.1	26.8	5.4
	80	42	29.3	32.2	16.3	30.5	36.1	12.1	31.3	39.6	8.0
		45	28.4	29.4	14.9	29.3	32.6	11.0	30.1	35.8	7.2
		50	25.3	25.3	12.8	27.4	27.4	9.2	28.3	29.2	5.9
67	70	42	—	—	—	—	—	—	—	—	—
		45	—	—	—	—	—	—	—	—	—
		50	—	—	—	—	—	—	—	—	—
	75	42	19.5	33.0	16.7	21.9	39.3	13.2	24.7	45.8	9.2
		45	18.2	29.5	14.9	20.2	35.0	11.8	22.6	40.9	8.2
		50	16.1	23.6	12.0	17.4	27.7	9.3	19.2	32.4	6.5
	80	42	25.5	34.5	17.4	27.2	39.6	13.3	29.5	45.5	9.1
		45	24.4	31.2	15.8	25.7	35.7	12.0	27.5	40.7	8.2
		50	22.7	25.9	13.1	23.6	29.3	9.9	24.3	32.6	6.6

Figure 6-12

Each manufacturer publishes equipment performance data in a different format. Study manufacturer's literature carefully to familiarize yourself with the data tables and the corresponding design procedures.

Some hydronic terminal equipment packages are designed to accommodate a modest air distribution system. In this case, the fan performance and the CFM that flows through the coil will be affected by external resistance that is associated with the duct system. Cooling capacity will be reduced if the air flow is throttled by this resistance. In some cases, the loss of capacity is only a few percent. In other cases, the loss of capacity may be significant. In any case, the manufacturer should provide a table or a procedure that allows the designer to estimate the change in capacity.

Entering Condition

The condition of the air entering the coil depends on the condition of the return air, the condition of the outdoor air and the percentage of outdoor air. Refer to Sections 5-1 and 6-8 for information on how the psychrometric chart is used to determine the entering condition.

Entering Water Temperature

Any entering water temperature that is compatible with the manufacturer's performance data and which provides the required cooling capacity is acceptable. As noted in Section 5-3, this temperature could vary from 35°F to 65°F, but a range of 42°F to 48°F is much more common. Lower temperatures — say 44°F, 42°F or lower — are required for applications that have high latent loads. Warmer temperatures are suitable when latent loads are relatively small.

Flow Rate and Temperature Rise

Any flow rate (GPM) that is compatible with the manufacturer's data and which provides the required performance is acceptable. The cooling capacity (BTUH), flow rate (GPM) and water-side temperature rise (WTR) are interrelated. If two values are known, the third can be calculated by using the water-side equation.

$$GPM = \frac{\text{Cooling Load BTUH}}{500 \times WTR}$$

Water-Side Pressure Drop

A value for the water-side pressure drop will be required for the pump selection calculations. Manufacturers usually publish this information in their data tables.

Air-Side Pressure Drop

Information about the air-side pressure drop across the coil is not required because the coil is an integral part of the unit. Manufacturers usually do not publish this information.

Fan Performance

Manufacturers usually publish values for the "free delivery" CFM at each fan speed. No other information is required unless the equipment is designed to accommodate a duct system. In this case, a more comprehensive blower table is required to document the relationship between external resistance and the supply CFM.

Unit Sizing

Use the manufacturer's performance data and equipment selection procedures to find a unit size — and in some cases, a coil size — that provides the required cooling performance. If the unit is equipped with a multiple speed fan, use the performance data that corresponds to the medium speed.

If the same unit is used for cooling and for heating, the unit size is based on the cooling requirements. If the unit has a single coil, the coil performance should be matched to the cooling requirements first, and then the heating requirements are addressed. (Refer to Section 7-13 for information on the procedure that is used to size heating equipment.)

Capacity Control

The capacity of a heating unit can be controlled by using two-way (throttling) or three-way (diverting or mixing) valves to modulate the flow of water (GPM) through the coil. If the unit has a multi-speed fan, the capacity can be controlled by changing the fan speed. Capacity also can be controlled by face-and-bypass dampers.

Accessories

Valve packages, damper operators and remote operating controls may be available as standard equipment or as an option. Various types of electrical and motor control equipment also may be available. Other accessory equipment could include filter options, discharge grilles, duct collars, wall sleeves and other installation accessories.

Example — Reading Performance Data

Use Figure 6-12 to find the cooling performance of a Size-12 unit for the following operating conditions.

Entering Water Temperature = 42°F
Entering air dry-bulb = 80°F
Entering air wet-bulb = 67°F
Desired water temperature drop = 8°F to 12°F

Performance data read from Figure 6-12 . . .
Total capacity = 45.5 BTUH
Sensible capacity 29.5 BTUH
Temperature rise 9.1 °F

6-14 Water Chiller Sizing

Chiller packages are available with a water-cooled condenser or an air-cooled condenser. Water-cooled equipment requires a cooling tower or some source of water that can be used as a heat sink. Air-cooled equipment is more suited for small commercial applications because there is no work associated with providing a heat sink.

Water-Cooled Chiller

Packaged Air-Cooled Chiller

The size of a water-cooled or an air-cooled chiller is expressed as TONS of cooling capacity. For a given size unit, this tonnage depends on the temperature of the water or the temperature of the air entering the condenser. An example of one manufacturer's (water-cooled chiller) performance data is provided on the next page by Figure 6-13. A second example, which shows air-cooled chiller performance data, is provided by Figure 6-14 at the bottom of the next page. In either table, the compressor input KW also is tabulated. These types of tables can be used to select a chiller that will satisfy

Leaving Chilled Water Temperature = 42 F

UNIT SIZE	ENTERING CONDENSER WATER TEMPERATURE (F)														
	75			80			85			90			95		
	TONS	KW	EER	TONS	KW	EER	TONS	KW	EER	TONS	KW	EER	TONS	KW	EER
C10M	10.1	9.2	12.8	9.8	9.5	12.1	9.5	9.8	11.4	9.3	10.2	10.7	9.0	10.5	10.0
C15M	15.6	14.6	12.6	15.2	15.1	11.9	14.7	15.6	11.2	14.3	16.1	10.5	13.9	16.6	9.9
C20M	20.9	19.1	13.0	20.4	19.7	12.3	19.8	20.3	11.6	19.3	20.8	11.0	18.7	21.4	10.4
C25M	25.9	21.7	14.2	25.2	22.6	13.2	24.5	23.4	12.4	23.7	24.3	11.6	23.0	25.1	10.9
C30M	31.1	27.4	13.5	30.2	28.4	12.6	29.3	29.4	11.9	28.4	30.3	11.1	27.5	31.3	10.5
C40R	41.0	33.3	14.6	39.9	34.6	13.7	38.8	35.8	12.9	37.6	37.0	12.1	36.4	38.1	11.4
C40M	42.1	38.4	13.1	41.0	39.4	12.4	39.8	40.6	11.7	38.7	41.8	11.0	37.5	43.0	10.4
C50R	51.8	39.5	15.6	50.3	41.2	14.6	48.8	42.8	13.6	47.3	44.3	12.7	45.7	45.8	11.9
C50M	52.0	43.4	14.3	50.5	45.0	13.4	49.1	46.8	12.5	47.6	48.6	11.7	46.1	50.2	10.9
C60R	63.2	47.1	16.0	61.4	49.3	14.9	59.6	51.4	13.8	57.8	53.4	12.9	55.9	55.4	12.0
C60M	62.3	54.8	13.6	60.6	56.8	12.7	58.8	58.8	11.9	57.0	60.8	11.2	55.2	62.6	10.5
C75E	75.0	59.2	15.1	72.0	61.4	14.2	70.9	63.3	13.3	68.8	65.7	12.5	66.6	67.8	11.7
C80R	82.3	67.0	14.6	80.1	69.6	13.7	77.8	72.0	12.9	75.4	74.4	12.1	73.0	76.6	11.4
D10E	100.3	78.4	15.3	97.8	81.3	14.4	95.2	84.2	13.5	92.5	87.0	12.7	89.7	89.7	12.0
D10R	103.5	79.8	15.5	100.5	83.0	14.4	97.4	86.2	13.5	94.3	89.3	12.6	91.1	92.2	11.8
D12R	126.2	94.2	16.0	122.6	98.4	14.9	119.0	102.8	13.8	115.4	106.8	12.9	111.6	100.8	12.0
D15E	147.8	117.6	15.0	143.9	122.0	14.1	139.9	126.2	13.2	135.8	130.4	12.4	131.7	134.6	11.7
D18E	175.4	138.0	15.3	170.9	143.0	14.3	166.3	148.0	13.5	161.6	152.9	12.7	156.7	157.7	11.9
D20E	196.9	154.8	15.2	192.2	160.6	14.3	187.2	166.2	13.5	182.1	171.6	12.7	176.8	177.0	11.9

NOTES:

1. Ratings based on evaporator temperature drops of 10 F. Ratings are applicable for evaporator temperature drops of 8-12 F.
2. Ratings based on a 0.0005 fouling factor for evaporator.
3. Interpolation between points is permissible, but extrapolation is not permitted.
4. KW input is for compressor only.
5. EER-energy efficiency ratio (BTU/watt-hour). Power inputs include compressor and control KW.

Figure 6-13

a set of design conditions. These design conditions include the following information . . .

- Total capacity required (TONS)
- Entering condenser water or air temperature
- Temperature of water leaving the chiller
- Water temperature drop across the chiller
- Chilled water GPM

The total capacity (required tonnage) can be estimated by adding the design sensible load to the design latent load and dividing by 12,000. (One ton is equal to 12,000 BTUH of total capacity.) If water-cooled equipment is used, cooling towers can be selected to supply condenser water over a range of temperatures — say from 75°F to 95°F — with 85°F being the most commonly used value. If air-cooled equipment is used, the design value for the temperature of the air entering

Leaving Chilled Water Temperature = 42 F

UNIT SIZE	AIR-COOLED CONDENSER MODEL NO.	TEMPERATURE OF AIR ENTERING CONDENSER (F)														
		85			90			95			100			105		
		TONS	KW	EER	TONS	KW	EER	TONS	KW	EER	TONS	KW	EER	TONS	KW	EER
C10M	CAUC B10	9.0	11.3	9.4	8.8	11.5	8.9	8.5	11.8	8.5	8.2	12.1	8.0	7.9	12.4	7.5
	CAUC B15	9.6	10.0	11.2	9.3	10.3	10.6	9.0	10.6	10.0	8.7	11.0	9.4	8.5	11.3	8.8
C15M	CAUC B15	14.6	16.0	10.8	14.2	16.5	10.2	13.7	17.1	9.5	13.3	17.6	8.9	12.8	18.2	8.4
	CAUB C20	14.9	15.6	11.3	14.5	16.1	10.6	14.1	16.6	10.0	13.6	17.2	9.4	13.1	17.8	8.8
C20M	CAUB C20	19.6	21.1	11.0	19.0	21.7	10.4	18.4	22.3	9.8	17.8	22.9	9.2	17.2	23.5	8.7
	CAUB C25	20.0	20.6	11.5	19.4	21.2	10.9	18.8	21.8	10.3	18.2	22.4	9.7	17.6	23.0	9.1
C25M	CAUB C25	24.0	24.8	11.5	23.3	25.6	10.8	22.5	26.4	10.2	21.8	27.1	9.6	21.0	27.8	9.0
	CAUB C30	24.6	23.9	12.2	23.9	24.7	11.5	23.1	25.5	10.8	22.4	26.3	10.1	21.6	27.0	9.5
C30M	CAUB C30	28.9	30.8	11.2	28.0	31.7	10.5	27.1	32.6	9.9	26.2	33.5	9.1	25.3	34.4	8.8
	CAUB C40	29.7	29.5	12.0	28.8	30.4	11.3	27.9	31.4	10.6	27.0	32.3	10.0	26.1	33.2	9.4
C40R	CAUB C40	38.1	37.3	12.1	36.9	38.4	11.4	35.7	39.4	10.8	34.4	40.4	10.1	33.2	41.3	9.6
	CAUB C50	38.6	36.6	12.6	37.4	37.7	11.8	36.2	38.8	1.1	35.0	39.8	10.5	33.7	40.8	9.9
C40M	CAUB C40	49.6	42.4	11.1	38.4	43.6	10.5	37.2	44.8	9.9	36.0	46.0	9.3	34.8	47.2	8.8
	CAUB C50	40.2	41.6	11.5	39.0	42.6	10.9	37.8	43.8	10.3	36.6	45.2	9.7	35.4	46.4	9.1

NOTES:

1. Ratings based on evaporator temperature drops of 10 F. Ratings are applicable for evaporator temperature drops of 8-12 F.
2. Ratings based on a 0.0005 fouling factor for evaporator.
3. Interpolation between points is permissible, but extrapolation is not permitted.
4. KW input is for compressor only.
5. EER-energy efficiency ratio (BTU/watt-hour). Power inputs include compressor and control KW.

Figure 6-14

the air-cooled condenser is the same as the summer design temperature (or higher, if located on a roof or near hot masonry materials) that was used to make the load calculation.

Sizing Example
A 207,851 BTUH sensible load and a 42,835 BTUH latent load was calculated at a 95°F outdoor design temperature. Assume that 85°F condensing water will be available for the water-cooled chiller. Use Figure 6-13 to select a water-cooled chiller and Figure 6-14 to select an air-cooled chiller. Size the chillers for a 42°F leaving water temperature and for a 10°F temperature drop across the chiller.

$$TONS = (207851 + 42835)/12000 = 20.9\ TONS$$

Figure 6-13, on the previous page, shows that at an 85°F entering water temperature, the C20M unit is just a little short on capacity (19.8 TONS delivered vs. 20.9 TONS required). If this is not acceptable the C25M unit could be used (24.5 TONS delivered vs. 20.9 TONS required). Figure 6-14, on the previous page, shows that at a 95°F entering air temperature, the C25M/CAUB-C25 unit is the best choice (22.5 TONS delivered vs. 20.9 TONS required).

Evaporator GPM
As shown by the equation below, the flow of chilled water (GPM) through the evaporator is related to the tonnage and the temperature drop (TD) across the evaporator. A 24.5 TON unit (C25M) and a 10°F TD is used to demonstrate this calculation. Chiller performance data should specify the maximum and minimum *allowable* flow rates through the evaporator. An example of this data is provided by Figure 6-15.

$$Evaporator\ GPM = \frac{TONS \times 12000}{500 \times Evaporator\ TD}$$

$$Evaporator\ GPM = \frac{24.5 \times 12000}{500 \times 10} = 58.8$$

Condenser GPM
For water-cooled chillers, the flow of water (GPM) through the condenser is related to the capacity (TONS), the compressor power requirement (KW) and the temperature rise across the condenser (TD). The following equation can be used to estimate the condenser GPM. A 24.5 TON, 23.4 KW unit (C25M) and a 10°F TD rise will be used to demonstrate this calculation. Performance data should specify the maximum and minimum *allowable* flow rates through the condenser. An example of this data is provided by Figure 6-15.

$$GPM = \frac{TONS \times 12000 + KW \times 3413}{500 \times Condenser\ TD}$$

$$GPM = \frac{24.5 \times 12000 + 23.4 \times 3413}{500 \times 10} = 74.8$$

Evaporator Pressure Drop
The water-side pressure drop across the evaporator will be required for the pump sizing calculations. Figure 6-15 indicates that the C25M unit has a C-type evaporator. Figure 6-16, on the next page, provides an example of manufacturer's evaporator pressure drop data. Figure 6-16 indicates that at 58.8 GPM, the evaporator pressure drop (C-type) is about 17 feet of water column or about 7.4 PSI. (One PSI is equal to 2.3 feet of water column.)

Condenser Pressure Drop
The water-side pressure drop across the condenser also will be required for the pump sizing calculations. Figure 6-15 indicates that the C25M unit has a C-type condenser. Figure 6-17, on the next page, provides an example of manufacturer's condenser pressure drop data. Figure 6-17 indicates that at 74.8 GPM, the condenser pressure drop (C-type) also is equal to 17 feet of water column (approximately).

Single Compressor CGWB Models

Compressor	C10M	C15M	C20M	C25M	C30M	C40R	C50R	C60R	C75E	D10E
Model Number	CRHM-150	CRHM-150	CRHM-200	CRHM-250	CRHM-300	CRHR-400	CRHR-500	CRHR-600	2E5G68	2E5G88
Nominal Tons	10	15	20	25	30	40	50	60	75	100
Unloading Option 1	(66.5)	(66.6)	(100-50)	(66.6)	(66.6)	(50)	(66.6)	(66.6)	(50)	(50)
Unloading Option 2	(66-33.3)	(66.6-33.3)	(100-50-25)	(66.6-33.3)	(66.6-33.3)	(50-25)	(66.6-33.3)	(66.6-33.3)	(66.6-50-33.3)	(75-50-37.5)
Evaporator										
Model Designator	A	A	B	C	D	E	F	G	J	K
Storage Capacity (gal)	7.9	7.9	8.5	15.4	13.5	27.2	26.9	23.3	43.0	34.9
Min. Flow Rate (gpm)	12	18	24	30	36	48	60	72	96	120
Max. Flow Rate (gpm)	35	52.5	70	87.5	105	140	175	210	280	350
Condenser										
Model designator	A	A	B	C	D	E	F	G	H	K
Storage Capacity (gal)	1.9	1.9	2.5	3.0	3.6	4.8	5.6	6.7	8.8	10.8
Min. Flow Rate (gpm)	12	18	24	30	36	48	60	72	96	120
Max. Flow Rate (gpm)	37	55	74	92	111	148	185	222	296	370

Figure 6-15

Evaporator Pressure Drop

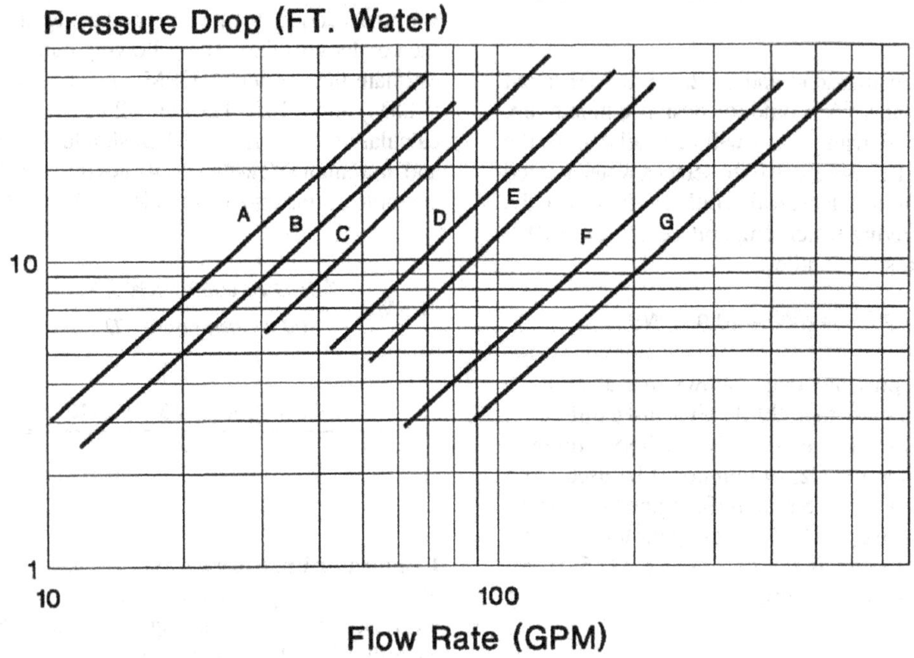

Figure 6-16

Condenser Pressure Drop

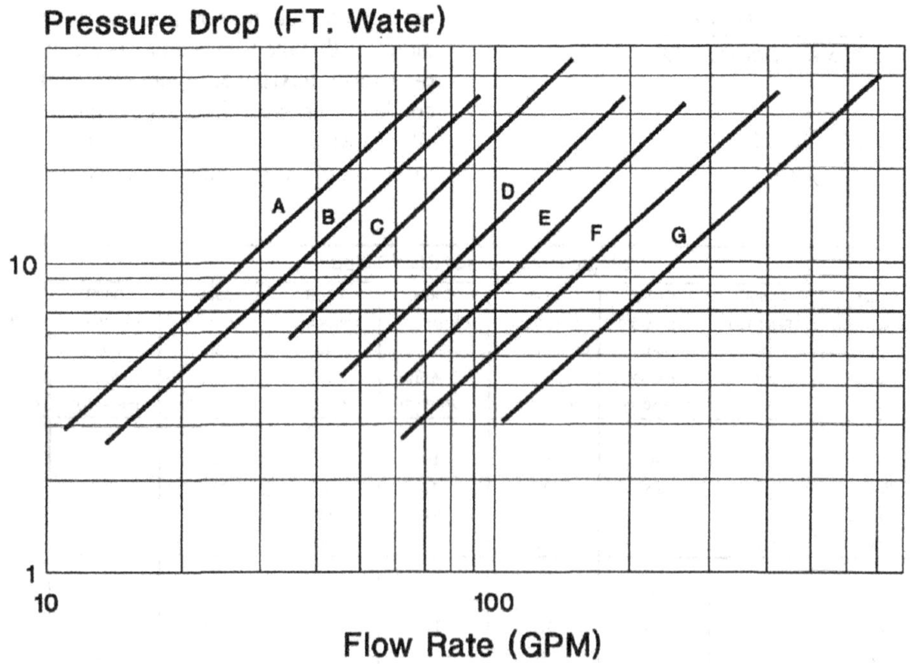

Figure 6-17

Ethylene Glycol Performance

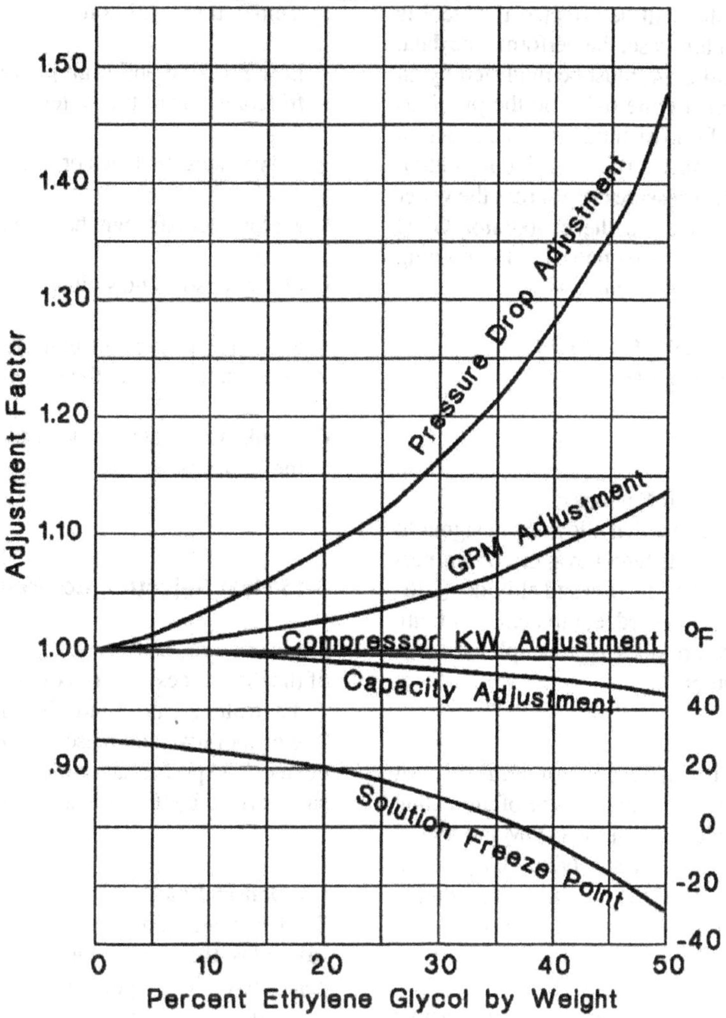

Figure 6-18

Performance Data Adjustment Factors

FOULING FACTOR	CHILLED WATER TD	ALTITUDE							
		SEA LEVEL		2,000 FT.		4,000 FT.		6,000 FT.	
		TONS	KW	TONS	KW	TONS	KW	TONS	KW
0.0005	8	0.993	0.997	0.973	1.007	0.956	1.025	0.935	0.973
	10	1.000	1.000	0.980	1.010	0.960	1.030	0.940	1.040
	12	1.007	1.003	0.987	1.013	0.966	1.034	0.945	0.987
0.001	8	0.973	0.987	0.964	0.997	0.944	1.005	0.925	1.016
	10	0.980	0.990	.0970	1.000	0.950	1.010	0.930	1.020
	12	0.987	0.993	0.975	1.003	0.955	1.015	0.934	1.026
0.0002	8	0.933	0.965	0.925	0.977	0.906	0.987	0.885	0.997
	10	0.940	0.970	0.930	0.980	0.910	0.990	0.890	1.000
	12	0.947	0.974	0.934	0.983	0.915	0.993	0.897	1.003

Figure 6-19

Adjustment for Antifreeze Solution

The performance of the chiller will be affected if glycol is used for freeze protection. In this case, the performance data, as shown by Figures 6-13 and 6-14, must be modified by an appropriate adjustment factor. Figure 6-18, on the previous page, provides an example of one manufacturer's correction factor chart. This chart shows that a 20 percent glycol solution will add about 9 percent to the pressure drop across the water side of the evaporator and increase the evaporator GPM requirement by about 3 percent. Also note that the cooling capacity (TONS) will be about 1 percent lower.

Evaporator pressure drop = 1.09 x 17 = 18.5
Evaporator GPM = 1.03 x 58.8 = 60.6
TONS = 24.5 x 0.99 = 24.3

Other Performance Adjustment Factors

Manufacturers may publish factors that allow the designer to adjust the performance data for different evaporator temperature rises, condenser tube fouling factors and altitude. (Altitude adjustments apply to air-cooled equipment.) Figure 6-19, on the previous page, provides an example of these performance adjustment factors.

Capacity Control

On page 21, Figure 6-15 provides an example of manufacturer's data that shows how many steps of unloading are available. This figure indicates that the C25M model has two unloading options 66.6 percent and 66.6/33.3 percent. Hot gas bypass will be required if the minimum load on the chiller falls below 33.3 percent of the rated capacity.

Accessories and controls

Certain types of devices and controls are required to operate and protect the chiller. In some cases, these controls are standard equipment. In other cases, they must be specified as optional equipment.

- Air-cooled chillers may require controls that allow low ambient or high ambient operation
- Leaving water temperature is controlled by a chilled water temperature controller
- An anti-cycle timer prevents short-cycle damage
- Timed periodic pumpout prevents liquid slugging in the lag compressor (multiple compressor installation)
- Hot gas bypass prevents compressor cycling at light loads
- A load limiting thermostat will prevent nuisance tripouts (Return water temperature may be too warm when a system restart follows an overnight or weekend shutdown.)
- Water flow switches verify flow through the evaporator and the condenser
- Low water temperature controls protect against a low water temperature condition
- Low-pressure and high-pressure switches protect the refrigerant side of the system
- Oil pressure switches protect compressor parts
- Motor overload switches protect the motor
- Phase sensors and switches protect the motor
- Reset relays prevent continued, unattended operation if safety controls shut the system down
- Crankcase heaters prevent the migration of refrigerant to the compressor

6-15 Heat Rejection Equipment

There are three basic types of heat rejection equipment. Two of the three types — air-cooled and water-cooled condensers — have already been introduced in Sections 6-7, -8 and -11. The evaporative condenser is the third type of heat rejection device. The performance characteristics of these devices are summarized by the table at the top of the next page.

Condenser Load

The heat rejected by the condenser is equal to the refrigeration load plus the energy delivered to the compressor. If the compressor KW input is known, the following equation can be used to calculate the condenser load.

Condenser Load (BTUH) = TONS x 12000 + KW x 3413

The preceding equation is a very precise way to estimate the condenser load, but some technical reference manuals use an approximate equation (presented below). This equation is based on the assumption that the compressor heat input is equal to 25 percent of the refrigeration load. (In practice, this value may range from 20 percent for an open compressor to 35 percent for a sealed hermetic compressor.) The use of the 1.25 factor means that the condenser will reject 15,000 BTUH of heat for each TON of cooling load.

Condenser load = 1.25 x TONS of cooling

Air-Cooled Condensers

The capacity of an air-cooled condenser operating at a given condensing temperature varies in direct proportion to the temperature of the ambient air. Therefore, condenser sizing calculations should be based on the cooling season, outdoor design temperature.

Comparison of Heat Rejection Devices					
Type	**Heat Transfer**	**Physical Size and Location**	**Operating Characteristics**	**Water Use**	**Other Comments**
Air-Cooled	Sensible . . . to outdoor air. Depends on outdoor DB temperature.	Relatively large for a given capacity. Locate near compressor and evaporator.	Higher condensing temperatures (110 °F to 140 °F), reduce efficiency and increase compressor KW.	None	Minimum maintenance, no cost associated with using outdoor air as a heat sink.
Water-Cooled	Sensible . . . to cooling water circuit. Depends on condenser water temperature.	Relatively small condenser, but a water tower can be relatively large. Tower location can be remote from the refrigeration equipment.	Once-through systems can have low to moderate condensing temperatures (60 °F to 85 °F), tower systems have mild to warm condensing temperatures (85 °F to 105 °F). Lower condensing temperatures improve efficiency and reduce compressor KW.	Some make up water required for a water tower installation. All water wasted with a "once-through" design. High condenser flow rate (GPM). Towers have significant pumping requirements.	A water tower or an abundant source of inexpensive water is required. Tower and condenser water circuits require maintenance. Water treatment and blow down required. Winter freeze protection required.
Evaporative	Sensible and latent . . . to outdoor air. Depends on outdoor wet-bulb temperature.	Physically smaller than an air-cooled condenser. Locate near compressor and evaporator.	Moderate condensing temperatures, (95 °F to 115 °F) provide better efficiency and lower compressor KW than air-cooled units.	Some make up water required, low condenser flow rate (GPM).	Wet coil design requires more maintenance than air-cooled equipment. Water treatment and blow-down required. Winter freeze protection required.

When the condenser is included as part of a factory assembled package, the combined performance of the compressor, condenser and the DX coil are summarized by a single table. This type of equipment is sized according to the performance the unit would have when the outdoor temperature is equal to the design temperature. Refer to Section 6-8 for an example that shows how to size packaged, air-cooled, cooling equipment.

If the condenser is purchased separately, the refrigerant-side condensing temperature depends on the size of the condenser relative to the size of the compressor (see Appendix 5). The condensing temperature and the compressor power requirements decrease as the size of the condenser increases. But, the first cost and the condenser fan power requirement increase as the size of the condenser increases. Manufacturers usually provide assistance that will help the designer optimize the size and the price performance index of the condensing equipment.

Air-cooled condenser capacity must track changes in the condensing load and changes in the ambient conditions. Capacity can be controlled by varying the air flow across the coil or by flooding part of the coil with refrigerant. Refer to the manufacturer's engineering data for information on capacity control, low ambient operation and refrigerant migration.

> Air-cooled condenser ratings are based on the assumption that the air is at standard conditions. Condenser capacity will be reduced by about 9 percent and system capacity will be reduced by about 3 percent at higher altitudes (5000 feet). Capacity corrections for ambient air temperatures below 125°F are not necessary.

Water-Cooled Condensers

The capacity of a water-cooled condenser varies with the temperature rise of the cooling water. Therefore, the capacity can be increased by increasing the flow rate or decreased by decreasing the flow rate. The condenser water flow rate (GPM) associated with a given temperature rise can be calculated by using the following equation.

$$GPM = \frac{\text{Condenser Load}}{500 \times \text{Rise}}$$

A temperature rise of 15°F to 20 °F is possible when relatively cold city water or well water is used as a source of condensing water. A temperature rise of 10°F is normal when relatively warm, cooling tower water or closed-loop water is used as a source of condensing water.

Condenser sizing calculations should be based on the design cooling load and the desired refrigerant-side condensing temperature. The refrigerant-side condensing temperature depends on the temperature of the water entering the condenser and the temperature rise of the condenser water. Usually, the condensing temperature is 20°F to 30°F above the temperature of the water entering the condenser. The design value for the entering water temperature is equal to the warmest temperature experienced during peak load conditions.

Water-cooled condenser performance is sensitive to scaling and deposits on the water side of the heat transfer surfaces. It is common practice to adjust condenser performance by applying a fouling factor. This fouling factor depends on whether the system is closed or open, once-through or recirculating. Table 6-1 lists fouling factors that can be used for a few specific situations. Manufacturers of water treatment chemicals may be able to make a more precise estimate of the fouling factor that will be associated with a specific application. In any case, fouling can be controlled by water treatment, blow down procedures and by periodically cleaning the water-side heat transfer surfaces.

Equipment Fouling Factors	
Closed Recirculating Systems	0.0005
Open Systems	
Conditioned water	0.0005
Unconditioned water	0.0005 to 0.01
Once-Through Systems	
Sea water	0.0005 to 0.002
Brackish water	0.0005 to 0.006
Great lakes water	0.001 to 0.002
River water	0.001 to 0.003
Sanitary canal	0.006
Muddy or silty water	0.002 to 0.003
Well water	0.001 to 0.003
Surface water	0.001 to 0.003
Consult water treatment chemical manufacturers for advice on the water treatment that is required to obtain the desired fouling factor.	

Table 6-1

When the condenser is included as part of a factory assembled package, the combined performance of the compressor, condenser and the evaporator are summarized by a single table. This type of equipment is sized according to the performance the package would have when it operates at the desired refrigerant-side condensing temperature. Refer to Section 6-9 for an example that shows how to size packaged, water-cooled, cooling equipment. Refer to Section 6-14 for information on how to size packaged water chillers.

If the condenser is purchased separately, the refrigerant-side condensing temperature depends on the size of the condenser relative to the size of the compressor. The compressor power requirements decrease as the refrigerant-side condensing temperature is reduced, but first cost, pumping power and water use tend to increase. Manufacturers usually provide assistance that will help the designer optimize the size and the price performance index of the condensing equipment.

Water-cooled condenser capacity should be adjusted to track changes in the condensing load or changes in the temperature of the condenser water. Capacity can be controlled by varying the water flow through the condenser. Two-way or three-way water valves can be used to maintain the refrigerant-side condensing temperature within the desired range.

Evaporative Condensers

Evaporative (sprayed coil) condensers are smaller than air-cooled units (of the same capacity) because of the efficiency that is associated with the evaporative heat transfer process. With this type of equipment, the capacity of a condenser (operating at a given refrigerant-side condensing temperature) varies with the outdoor wet-bulb temperature. Therefore, evaporative condenser sizing calculations should be based on the cooling season, outdoor design wet-bulb temperature.

When the condenser is included as part of a factory assembled package, the combined performance of the compressor, condenser and the evaporator are summarized by a single table. This type of equipment is sized according to the performance the package would have when the outdoor wet-bulb temperature is equal to the summer design wet-bulb temperature.

If the condenser is purchased separately, the refrigerant-side condensing temperature depends on the size of the condenser relative to the size of the compressor. The condensing temperature and the compressor power requirements decrease as the size of the condenser increases. But the first cost, condenser fan power, water use and pumping power increase as the condenser size increases. Manufacturers usually provide assistance that will help the designer optimize the size and the price performance index of the condensing equipment.

> Since evaporative condenser ratings are based on sea level conditions, they are conservative at higher elevations. Capacity corrections for elevation are not necessary.

Make up water is required to replace water lost by evaporation, carry-over and bleed-off. However, the amount of water required for an evaporative condenser is only about 5 percent of the water that would be required for a once-through water-cooled condenser. Water treatment, bleed-off and periodic cleaning will be required because the water circulating system is open to the atmosphere. Cycling the recirculating pump is not recommended because it promotes the accumulation of scale and deposits.

Evaporative condenser capacity must track changes in the condensing load and changes in the ambient conditions. Capacity can be controlled by varying the air flow through the condenser or by adjusting the entering wet-bulb temperature. Refer to the manufacturer's engineering data for information on capacity control. Also refer to manufacturer's data for information regarding the installation and control requirements associated with winter or year-round operation.

6-16 Cooling Towers

The two basic types of cooling tower designs are atmospheric towers and mechanical draft towers. Mechanical draft towers are usually used with HVAC systems, especially with smaller systems. Mechanical draft towers are subdivided into induced draft designs and forced draft designs. Induced draft designs have the fan located at the leaving air side and forced draft designs have the fan located on the entering air side.

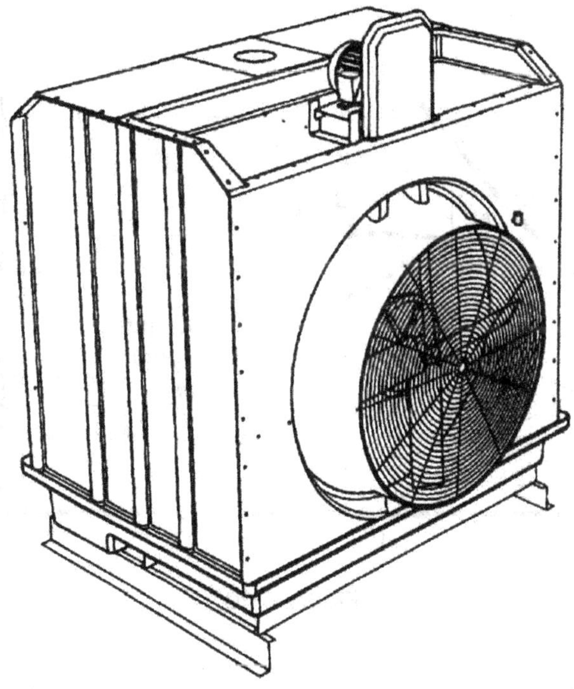

Packaged Cooling Tower

Capacity and Ratings
The capacity of a mechanical draft tower depends on the outdoor wet-bulb temperature, the entering (water on) and leaving (water off) water temperatures, the GPM of water flowing through the tower and the CFM of outdoor air that is forced through the tower. Tower rating tables usually specify tower capacity in terms of GPM for various combinations of wet-bulb temperatures and entering/leaving water temperatures. For example, a tower may be rated at 66 GPM when the water on and water off temperatures are equal to 95°F and 85°F and the wet-bulb temperature is equal to 76°F.

Selection and Sizing
The tower should be sized to meet the condenser GPM requirement during operation at the summer design condition. As explained in Section 6-15, the condenser GPM requirement depends on the condenser load and the temperature rise across the condenser. The design value for the outdoor wet-bulb temperature corresponds to the wet-bulb temperature that is coincident with the summer design dry-bulb temperature. (Refer to **Manual-N**, Table 1 for this value.)

For example, determine the tower sizing parameters if a packaged chiller draws 23.4 KW when it is satisfying a 24.5 TON cooling load. Assume that the chiller performance is based on a 10°F degree rise across the condenser, that the water enters the condenser at 85°F and leaves the condenser at 95°F. Also assume that the coincident outdoor wet-bulb temperature, at the summer design condition, is equal to 76°F.

$$\text{Condenser GPM} = \frac{23.4 \times 12000 + 24.5 \times 3413}{500 \times 10} = 74.8$$

Water on temperature = 95°F
Water off temperature = 85°F
Outdoor wet-bulb temperature = 76°F

Cooling tower ratings, which are based on air at standard (sea level) conditions, are conservative at higher elevations. Capacity corrections for elevation are not necessary.

Capacity Control
Cooling tower capacity must be adjusted to track changes in the condenser load and in the outdoor wet-bulb temperature. This can be done by varying the water flow through the tower or by varying the air flow through the tower. In either case, the objective is to maintain the temperature of the (entering) condenser water within the desired range. (Consult with the refrigeration equipment manufacturer to determine the minimum allowable condensing temperature for a given type of machine.) Refer to manufacturer's engineering data for information on capacity control options.

Winter Operation
Controls are required to make sure that the water off temperature stays well above freezing during winter conditions.

Sumps must be located in a heated space, or exposed sumps must be heated. Fans must not be allowed to operate with ice on the blades.

Direct Free Cooling
During cold weather, the refrigeration equipment can be shut down and the cooling tower water can be used to provide economizer cooling. This can be done by valving the tower water so that it flows directly to the cooling coils or by using a water-to-water heat exchanger to indirectly supply cold water to the cooling coils.

Make Up Water
Make up water is required to replace water lost by evaporation, carry-over, drift and bleed-off. However, the amount of water that is required for a water tower is only about 5 percent of the water that would be required for a once-through water-cooled condenser. Water treatment, bleed-off and periodic cleaning will be required because the water circulating system is open to the atmosphere.

Water Treatment
Water-side performance is sensitive to scaling, to deposits of mineral or biological materials and to deposits of silt. Consult with tower manufacturers and water treatment specialists.

6-17 Evaporative cooling

Under certain circumstances, evaporative cooling systems can be used in place of mechanical cooling systems — or they can be used to reduce the load on the mechanical cooling equipment. The decision on whether to use an evaporative cooling system depends on technical factors and economic factors. Some of the technical factors are associated with maintaining comfort conditions within the occupied space, air-side fan power requirements, water-side pumping power requirements, water use and water treatment. The basic economic factors include installed cost, operating cost and the pay back period.

In general, evaporative cooling should be considered for applications that require a large amount of outdoor air. Obviously, a dry climate provides the most favorable operating conditions, but some designs can provide acceptable performance in more humid climates.

The two basic types of evaporative cooling systems are referred to as direct and indirect. In some designs, both of these two types of evaporative cooling systems are combined into a hybrid indirect-direct system. Either type of evaporative cooling system (or even the hybrid system) may be combined with conventional mechanical cooling equipment. The basic principles associated with direct and indirect equipment will be discussed in the following sections. Contact the Evaporative Cooling Institute, in Las Cruces, NM, for more information about different types of evaporative cooling systems.

6-18 Direct Evaporative Coolers

Direct evaporative coolers can be a spray type or a pad type. In either case, they operate by evaporating water into a stream of air — usually a stream of outdoor air. As the air passes through the cooler it loses sensible heat and exits at a lower dry-bulb temperature, but it gains an equivalent amount of latent heat, so it exits with a higher absolute humidity. This process is illustrated on the psychrometric chart, see line A-B on Figure 6-20 below.

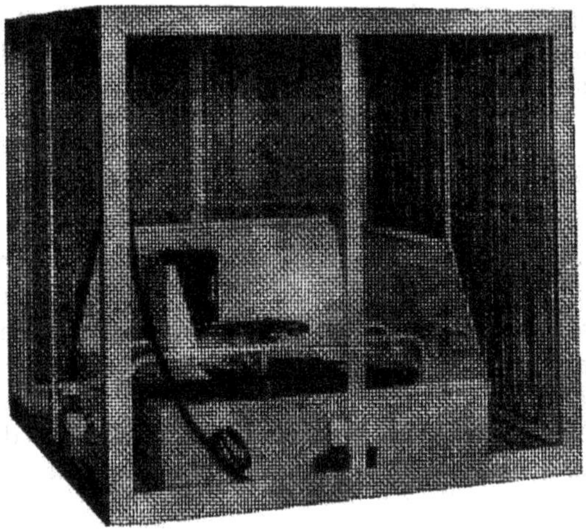

Direct Evaporative Cooler

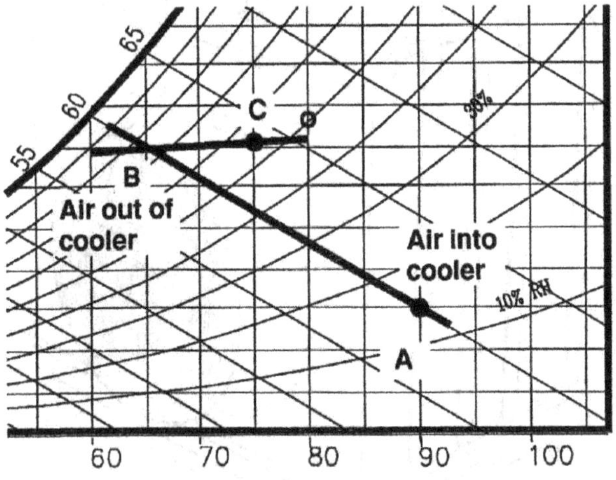

Figure 6-20

When the evaporative cooled air is introduced into the conditioned space, it absorbs the sensible and latent loads. The conditions in the space will depend on the condition of the supply air (point B) and on the size of the sensible and latent loads. This process also is illustrated on the psychrometric chart, see line B-C on Figure 6-20. Comfort conditions in the space will be maintained if point C represents an acceptable combination of temperature and humidity. For comfort applications, an acceptable combination of temperature and humidity is defined by the ASHRAE comfort envelope.

Saturation Effectiveness
The condition of the air leaving the cooler is determined by the *saturation effectiveness* of the device (sometimes referred to as efficiency). If a cooler is 100 percent effective, the air would leave the cooler completely saturated (100% RH). In practice, cooler effectiveness varies from 60 percent to over 95 percent. Pad-type cooler effectiveness depends on the type and geometry of the media and on the velocity of the air passing through the face of the media. Figure 6-21 provides an example of media performance curves. (Spray-type equipment is also rated for effectiveness.)

Media Efficiency & Pressure Drop

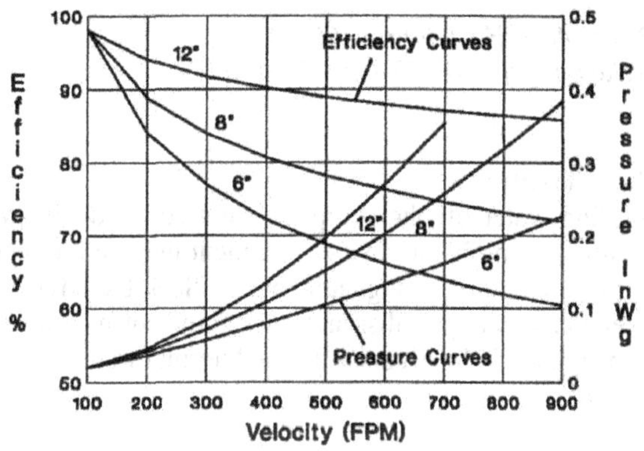

Figure 6-21

Equipment Sizing
Some manufacturers encourage designers to size equipment by using simplified guidelines. These guidelines may be acceptable if humidity control is not required, otherwise a more rigorous sizing procedure is required (see below).

Step 1) Calculate sensible and latent loads as per line 8, Form N-1 and calculate the room sensible heat ratio.

Step 2) Use **Manual N**, Table-1 to find the summer outdoor design condition and locate this point on the psychrometric chart (point A, Figure 6-20).

Step 3) Locate the indoor design condition on the psychrometric chart (point C, Figure 6-20) and draw the room sensible heat ratio line through this point. (Refer to Section 4-17 for more information about sensible heat ratio lines.)

Step 4) Draw the cooler process line by starting at the outdoor design condition (point A, Figure 6-20) and tracing a line along the corresponding wet-bulb line until it reaches the 100 percent relative humidity line.

Step 5) Note where the room sensible heat ratio line crosses the cooler process line (point B, Figure 6-20). The required effectiveness is equal to the relative humidity at this point and the supply air dry-bulb temperature is equal to the dry-bulb temperature at this point.

Step 6) Use the sensible heat equation to calculate the required supply CFM. (Base this calculation on the dry-bulb temperatures at points B and C and on the sensible load.)

Step 7) Refer to manufacturer's literature to find a piece of equipment that has the required air flow and saturation effectiveness.

> Even dry climates have periods of wet weather. Mechanical cooling may still be required during the rainy season.

Example
A single-zone building in Winslow, Arizona (95 DB, 60 WB) has a 105,000 BTUH sensible load and a 11,000 BTUH latent load. The room is to be maintained at 75°F DB and the relative humidity must not exceed 55 %. Figure 6-22 and the following calculations show how the supply CFM and the effectiveness values are determined.

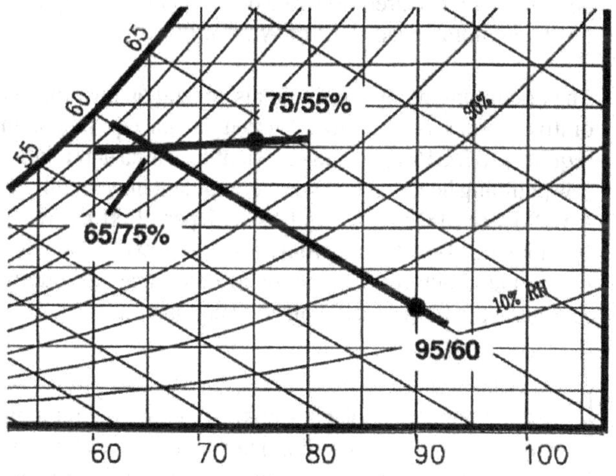

Figure 6-22

Room sensible heat ratio = 105000/116000 = 0.905
Effectiveness = 75 percent (65 DB and 60 WB)
Supply TD = 75 - 65 = 10 °F

$$CFM = \frac{105000}{1.1 \times 10} = 9545$$

Air-Side Design

The evaporative cooler air distribution system consists of a supply duct system and an exhaust system. (A return duct is not required because the evaporative cooling process uses 100 percent outdoor air.) Depending on the application, the supply duct system may consist of a stub duct that feeds one or two large supply outlets or it may be a complete system with multiple duct runs and supply outlets. The exhaust system must be designed to control the pressure in the building. The exhaust system may include duct runs (and even an exhaust fan), or it may simply consist of pressure relief dampers that penetrate the roof or walls.

The design value for the supply CFM can be calculated by the procedure outlined on the previous page. If a blower is included in the evaporative cooler package, it must be able to deliver the required CFM while working against the external resistance that is associated with the air distribution system. If the evaporative cooler does not have an internal blower, the pressure drop across the cooler must be added to the other duct system pressure losses. Refer to manufacturer's literature for blower performance tables or cooler pressure drop data. Refer to ACCA **Manual Q** for information about duct pressure loss calculations, fan selection procedures and building pressure control (make up, relief and exhaust air requirements).

Circulating Water Pump

Water spray equipment and some types of wetted media equipment may or may not include the pump with the cooler package. If an external pump is required the designer must determine the required flow rate (GPM) and the design value for the pumping head (Feet of Water).

The circulating pump requirements depend on how the evaporative cooler is designed. Wetted media equipment may require 1 to 2 GPM per square foot of face area and relatively low pumping heads. Water spray equipment may require from 4 GPM per 1000 CFM to 10 GPM per 1000 CFM, with pumping heads that range from 55 to 100 Feet of Water. Refer to the manufacturer's literature to determine the pumping requirements for a specific piece of equipment.

> Some evaporative cooler designs do not use a recirculating pump. The pump can be eliminated by using "water slingers" to keep the media wet or by mounting the media on a wheel that rotates through a sump that is filled with water.

Make Up Water

Both the pad type coolers and the spray type coolers use recirculated water to condition the air. Some of this recirculated water evaporates into the air and must be replaced. The approximate rate of evaporation in gallons per hour (GPH) can be calculated by using the following formula. In this formula, CFM is the air flow through the device, EDB is the entering dry-bulb temperature and LDB is the leaving dry-bulb temperature.

$$GPH = \frac{1.35 \times CFM \times (EDB - LDB)}{10000}$$

Some water also is required to replace water that is drained off the sump in order to control scale formation (bleed-off). The required amount of *bleed-off* water will usually fall between 50 and 100 percent of the water that is evaporated.

For example, calculate the make up water required for the equipment described on the previous page. Assume that the bleed-off water is equal to 50 percent of the evaporated water.

$$GPH = \frac{1.35 \times 9545 \times (95-65)}{10000} = 38.66$$

Bleed-off GPH = 0.50 x 38.66 = 19.33
Make up GPH = 38.66 + 19.33 = 57.99

Water Treatment

Bleeding water off the sump may not solve all the water conditioning problems. Water treatment may be required to control mineral or biological deposits. Seek the advice of a water treatment specialist and the equipment manufacturer (some types of fill will not withstand normal water treatment methods).

Controls

The sensible capacity of a direct evaporative cooler can be controlled by using bypass dampers to route some outdoor air around the cooler or by installing a preheat or reheat coil. The latent capacity of the cooler depends on the outdoor wet-bulb temperature. (Indoor humidity will not be controlled during humid or wet weather.) During cold weather, the equipment can be decommissioned and an air-side economizer can be used to control the temperature of the supply air.

6-19 Indirect Evaporative Cooling

Indirect evaporative coolers use the evaporative cooling process to provide sensible cooling. No moisture is added to the primary air stream that is supplied to the conditioned space. This is accomplished by using the direct evaporative cooling process to cool the air that is flowing through one side of an air-to-air heat exchanger.

The advantage associated with indirect equipment is that no moisture is added to the air that is supplied to the conditioned space. The disadvantages associated with indirect equipment are that it is more complicated and more expensive than direct evaporative cooling equipment; plus it operates at a lower net effectiveness and higher supply air temperature than the direct equipment. These disadvantages are related to the secondary heat transfer process that occurs at the air-to-air heat exchanger.

Indirect Evapoartive Cooler

Indirect evaporative coolers can be driven by a spray type or a pad type evaporative cooler. Two fans are required to move the air against the resistance that is associated with the primary and secondary air streams. Filters are required on both sides of the system. Pumps and sumps are required for the water-side of the system. Indirect coolers are available as self-contained packages or they can be built up from individual components. Fans and pumps must be capable of providing the desired flow rates when working against the resistance of the duct systems or the piping system.

The psychrometric process associated with the secondary (direct) side of the equipment is the same as the one that is illustrated by Figure 6-20 on page 6-28. The psychrometric process associated with the primary (indirect) side of the equipment is illustrated by Figure 6-23. In this process, the outdoor air enters the primary side of the air-to-air heat exchanger (point A) and it is sensibly cooled (point B).

Effectiveness

The cooling performance of indirect equipment is expressed as a *Performance Factor* (PF) or as the *Indirect Evaporative Cooling Effectiveness.* These performance indexes are defined as — the temperature drop in the primary air stream (DB_{in} - DB_{out}), divided by the difference between the dry-bulb temperature of the air entering the primary side (DB_{in}) and the wet-bulb temperature of the air entering the secondary side (WB_{in}).

$$PF = \frac{DB_{in} - Db_{out}}{DB_{in} - WB_{in}}$$

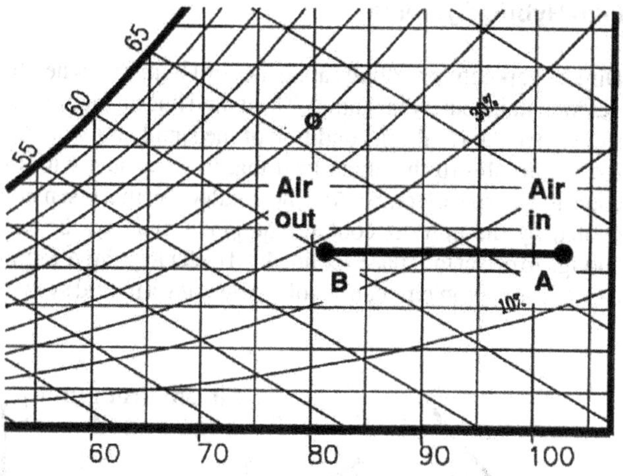

Figure 6-23

Performance

The dry-bulb temperature of the air leaving the primary side of the indirect cooler can be calculated by using the following equation. Before using this equation the designer must refer to manufacturer's data to obtain a value for the performance factor. An example of this data is provided by Figure 6-24. Note that the performance is sensitive to the velocity of the air flowing through the primary side of the equipment.

$$DB_{out} = DB_{in} - (D\,B_{in} - WB_{in}) \times PF$$

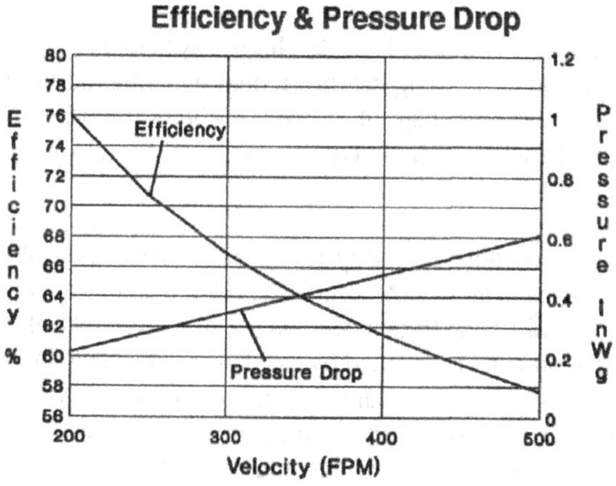

Figure 6-24

Example

The outdoor ambient temperature is 104°F DB and 69°F WB. Use Figure 6-24 to calculate the dry-bulb temperature of 9000 CFM leaving the primary side of the cooler that has a primary-side face area of 26 SQ.FT.

FPM = 9000/26 = 346

PF = 64% (From 6-24)

$DB_{out} = 104 - (104 - 69) \times 0.64 = 81.6°F$

6-20 Hybrid Systems

Direct evaporative cooling cannot control humidity when the outdoor air is somewhat humid. The dotted line on Figure 6-25 shows that the resulting condition of the supply air would be too wet to absorb the latent load that is associated with the conditioned space (room). And, hot outdoor air that is sensibly cooled by the indirect cooling equipment may not be cold enough to offset the space cooling load (refer to point A Figure 6-25). These problems can be solved by using a hybrid system.

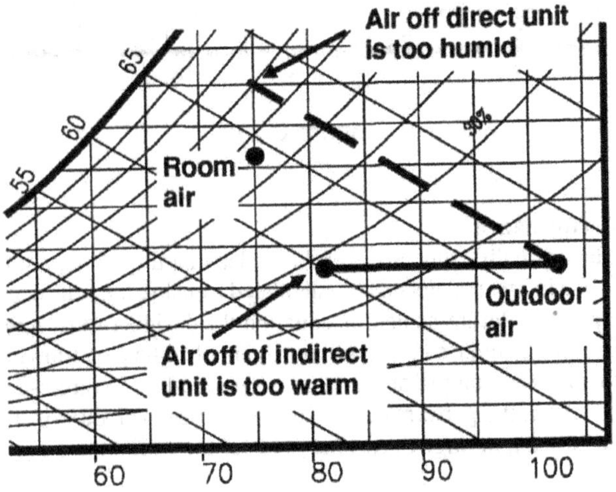

Figure 6-25

If the outdoor humidity is not too high, a system that provides adequate space cooling can be created by using the indirect cooler as the first stage of an indirect-direct hybrid system. The psychrometrics of this type of system are shown by Figure 6-26.

At higher outdoor humidity levels, even the indirect-direct hybrid system may not be able to control humidity in the conditioned space. In this case, an acceptable supply of conditioned air can be obtained by combining first stage indirect cooling with a mechanical refrigeration system. Figure 6-27 shows how direct-indirect cooling plus the mechanical refrigeration system produces an acceptable supply air condition.

There are many other possibilities. Some designs use a three stage system consisting of an indirect stage, followed by a mechanical cooling stage, followed by a direct stage. Other designs use a direct unit to cool exhaust air and a heat wheel to transfer the cooling capacity of the exhaust air to the intake air. It also is possible to use an indirect cooler to enhance the performance of an air-cooled condenser (by using it to cool the outdoor air that flows through the condenser). Many designs add an air-side economizer to the basic system. The air-side economizer is very efficient and it solves most of the problems that are associated with cold weather operation. However, no design is acceptable unless it can be justified by a favorable pay back analysis.

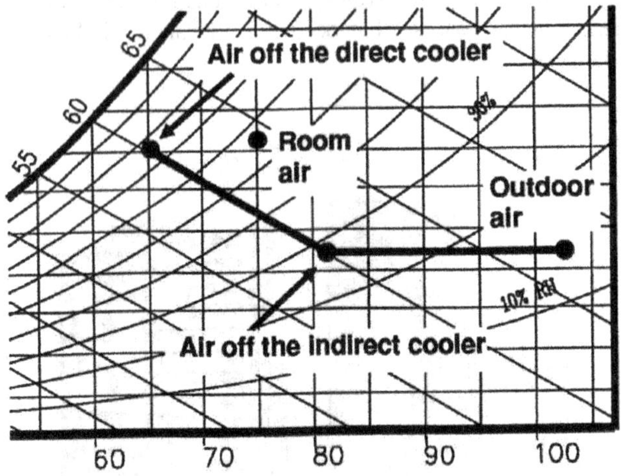

Figure 6-26

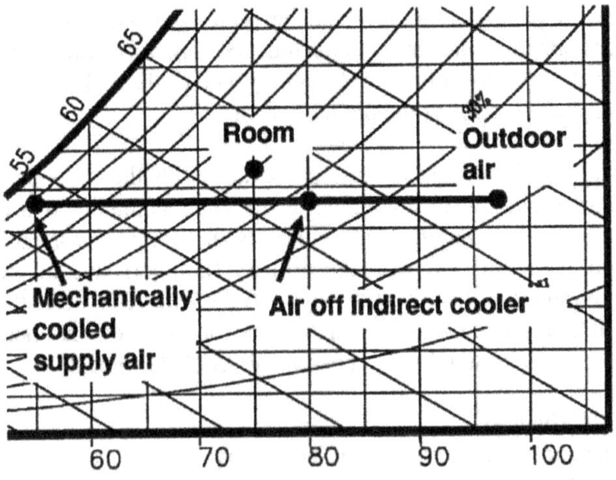

Figure 6-27

Section 7
Selecting Heating Equipment

Heating equipment must be selected to satisfy the design heating load. Gross oversizing is not recommended. Manufacturer's application data can be used to certify equipment performance for the operating conditions that are expected to exist at the site. Equipment capacity must track changes in the heating load under all operating conditions. Accessories, safety and operating controls must be specified. Fan performance must be specified. Water-side performance must be noted. Dimensions and clearances must be verified.

7-1 Sizing Requirements

The design load for a single-zone system is based on the peak room or zone load. The design load for a multizone system is calculated for the largest block load that could occur. In either case, the largest possible load that could (theoretically) occur would be based on the following conditions:

- Outdoor temperature at winter design
- Indoor temperature not set back
- Infiltration estimate assumes exhaust system operation
- Design value for ventilation (outdoor air) CFM
- No credit for internal loads
- No credit for solar gains

These conditions produce the design heating load value found on ACCA Form N1, line 21. This value is traditionally used to size the heating equipment. (Some codes may implicitly or explicitly require that these assumptions be used as the basis for the design heating load calculation.)

It is possible that the load conditions that are outlined above may never occur. The actual day/night (occupied/unoccupied) operating loads would be more likely to conform to the following set of operating conditions:

Day Cycle
- Outdoor temperature at winter design
- Indoor temperature not set back
- Infiltration estimate assumes exhaust system operation
- Design value for ventilation (outdoor air) CFM
- Credit for internal loads
- No credit for solar gains

Night Cycle
- Outdoor temperature at winter design
- Indoor temperature is set back
- Infiltration estimate assumes exhaust systems are off
- Outdoor air CFM based on *closed damper leakage*
- No credit for internal loads
- No credit for solar gains

Usually, the loads that correspond to the day/night conditions will be smaller then the design heating load that is calculated for the traditional (Form N-1) design conditions. If the designer is sure about the day/night operating conditions — and if codes permit — the equipment size could be based on the largest of the day/night loads.

Heating Capacity
The capacity of the heating equipment must be equal to or greater than the design heating load. This load should include the ventilation load (if any) and all other system loads or humidification loads that apply. (Refer to Section 4-5 and 4-6 for more discussion about ventilation loads and system loads.)

Design Conditions
Equipment capacities should be verified for the operating conditions that will exist at the site. These conditions, which consist of a set of design temperatures and fluid flow rates, are summarized below. Refer to Section 5 of this manual for information about how the values for the design temperatures and flow rates are determined.

Component	Conditions
Hot Air Furnaces	CFM Entering dry-bulb Air temperature rise
Hot Water Coils	CFM and face velocity Entering dry-bulb Air temperature rise GPM Entering water temp. Water temperature rise
Electric Coils	CFM and face velocity Entering dry-bulb Air temperature rise

Sizing Limits — Heating
Excessive hot air furnace or hot water boiler capacity does not have a serious effect on operating cost, providing the equipment has a respectable steady state efficiency rating — say 80 percent or better. (This has been documented by tests sponsored by the Department of Energy. Refer to Brookhaven National Laboratory publications BNL-51321 and BNL-51242.) However, grossly oversized heating equipment "short cycles," especially during part-load operation. (Short operating cycles tend to reduce comfort and they increase the wear and tear on the equipment.) Also note that a large amount

of surplus heating capacity increases the first cost of the installation and increases the demand on the utility service. However, a moderate amount of excess capacity is acceptable or even desirable, depending on the application and on the type of equipment. The following sizing recommendations apply to hot air furnaces, hot water boilers and electric coils.

- Hot air furnace short cycling can be minimized and part-load comfort can be enhanced if the output heating capacity of the equipment does not exceed the design heating load by more than 15 percent. Larger capacity margins are permissible, but heating capacities that exceed the design heating load by more than 40 percent are not recommended.

- Hot water boiler short cycling may not have a noticeable effect on comfort. In order to accommodate the "pickup loads" that are associated with the distribution piping, the output heating capacity of the equipment can exceed the design heating load by as much as 15 percent. (Larger capacity margins are permissible, but heating capacities that exceed the design heating load by more than 40 percent are not recommended.)

- Electric resistance heat short cycling can be minimized and part-load comfort can be enhanced if the output heating capacity of the equipment does not exceed the design heating load by more than 15 percent. Larger capacity margins are undesirable because they tend to reduce comfort and increase the connected (KW) load.

- Recovery from set back is best accomplished by using a time clock or a programmable thermostat.

Summary of Sizing Limits — Heating

Hot Air Furnaces

Heating capacity must exceed the design load

Capacity less than 1.15 x design load preferred

Capacity should not exceed 1.40 x design load

Hot Water Boilers

Heating capacity must exceed the design load

Capacity less than 1.15 x design load preferred

Capacity should not exceed 1.40 x design load

Electric Coils

Heating capacity must exceed the design load

Capacity less than 1.15 x design load preferred

Use a time clock or programmable stat for set-back recovery

7-2 Minimum Heating Load and Capacity Control

The heating equipment must adjust its capacity to match any load condition. Usually, heating equipment is required to severely reduce its output capacity so that it can accommodate loads that approach a "no-load" condition.

Small equipment is usually controlled by simple on-off cycling, but larger equipment requires a more sophisticated form of capacity control. Furnace or boiler capacity can be controlled by staging multiple units, by staging the burner or by modulating the burner. Hot water coil capacity can be controlled with modulating, two-way or three-way water valves. Electric heating coils can be controlled by staging coil banks or by using electronic pulse (SCR) controls.

7-3 Fan Performance

Commercial heating equipment is usually available with a wide range of fan speed (RPM) and fan motor power (HP) options. Heating-cooling packages often include a return fan option. This allows the designer to select a fan configuration that will provide the flow (CFM) and pressure (IN.WG.) that is required to move the air through the duct system and maintain the desired pressure in the conditioned space.

Supply Fan Selection

The supply CFM required for heating is determined by the space heating load and the sensible heat equation. This calculation is discussed in Section 5-9 of this manual and applied in Section 7-7 below. However, if cooling is required, the supply CFM is usually determined by the space sensible cooling load and the sensible heat equation. This calculation is discussed in Section 5-9 and applied in Section 6-8 of this manual.

The pressure requirement that is associated with the supply-side of the system is determined by making a duct system pressure loss calculation. This calculation should include the pressure losses that are associated with the straight runs, fittings and the air-side devices that are external to the air handler. Losses that are associated with any air handler component that is not accounted for in the blower table also should be included — the pressure losses associated with a damper, mixing box or filter for example. (Refer to Appendix 3 in this manual and to ACCA **Manual Q** for more information about air-side design.)

Return Fan Selection

A return fan (or some type of pressure relief damper installed in the conditioned space) will be required if the air handler is equipped with an air-side economizer. Even with no economizer, a return fan will be required unless the resistance that is associated with the return duct system is very small. The return CFM requirement is determined by the air balance in the conditioned space and the air balance at the air handler. (Refer to ACCA **Manual Q**, Section 17, for a comprehensive discussion of this subject.) The return-side pressure loss includes the losses that are associated with the straight runs, the fittings, the air-side devices that are external to the air handler and any air handler component loss that is not accounted for in the blower table.

Variable Volume Systems
Variable volume fans must be selected to provide satisfactory performance over a wide range of operating conditions. In this case, the part-load and the minimum flow requirements are just as important as the full flow requirement. In addition, the fan pressure requirement will be affected by the controls and devices that are used to modulate air flow and to adjust the fan performance. (If cooling is required, the operating characteristics of the modulating terminal devices should be reversed when the system changes from the cooling mode to the heating mode.) It also is important to make sure that the return fan performance tracks the changes in the supply fan performance. Refer to ACCA **Manual Q**, Sections 13, 16 and 17, for more information about VAV fans.

Fan Heat
Supply fans and return fans add heat to the air stream. Usually the fan heat that is associated with a piece of heating equipment is not included in the heating capacity data. (Since it is conservative to ignore the fan heat benefit, a fan heat adjustment is not required.)

7-3 Water-Side Performance

The water flow through (GPM) and the pressure drop across (Feet of Water) hydronic heating equipment is usually documented in the manufacturer's performance data. (Some examples of this type of equipment include fin-tube radiation, hot water coils, unit heaters, unit ventilators, heat exchangers and boilers. This information should be located, inspected and recorded for future reference because it will be required for the pipe sizing and pump selection calculations. (Refer to Section 5-10 of this manual for more information about how water-side flow rates are established. Refer to Appendix 4 for more information about pipe design and pump selection.)

7-4 Air-Side Accessories

Air-side accessories are usually available as part of the manufacturer's package. Examples of accessories that might be required include filters, dampers, mixing boxes, fan inlet vanes, belt guards and so forth. Some accessories could be available as standard equipment, but others have to be included as options when the equipment is specified.

7-5 Controls and Accessories

Safety and operating controls are usually available as part of the package offered by the equipment manufacturer. Examples of the controls that might be required include fueltrain operating controls, fueltrain safety controls, high- and low-temperature limit switches, oil pressure switches, pressure relief valves, water level controls, fan interlocks, damper interlocks, motor protection, motor starters, central control panels and so forth. Some controls and safety devices could be available as standard equipment, but others have to be included as options when the equipment is specified.

Many types of accessories are offered by equipment manufacturers. Air-side accessories include stainless steel heat exchangers, discharge nozzles, fresh and return air dampers, economizer packages and filter options. Combustion chamber accessories include stainless steel burners and power venting equipment. Installation accessories include mounting bases, roof curbs, weather hoods, belt guards, and bird screens. Other types of accessories that are associated with the electrical power system and the fueltrain system are usually available.

7-6 Physical Dimensions and Arrangement

The designer must verify that the dimensions and the arrangement of the equipment is compatible with the available space. There must be ample room to service the equipment — including room to remove critical parts such as compressors, fans and coils. There also must be ample room to install aerodynamically efficient fan discharge and inlet connections.

7-7 Indirect-Fired Fossil Fuel Furnaces

The following paragraphs discuss the factors that should be considered when a designer selects an indirect-fired furnace. *These comments apply to any type of indirect-fired heating equipment.* Gas, oil and propane furnaces; combination heating-cooling units; duct furnaces and unit heaters are typical of the type of equipment that falls into this category.

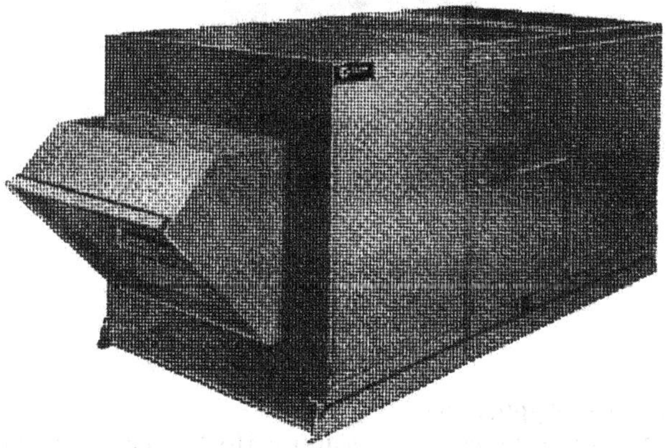

Indirect-Fired Rooftop Furnace

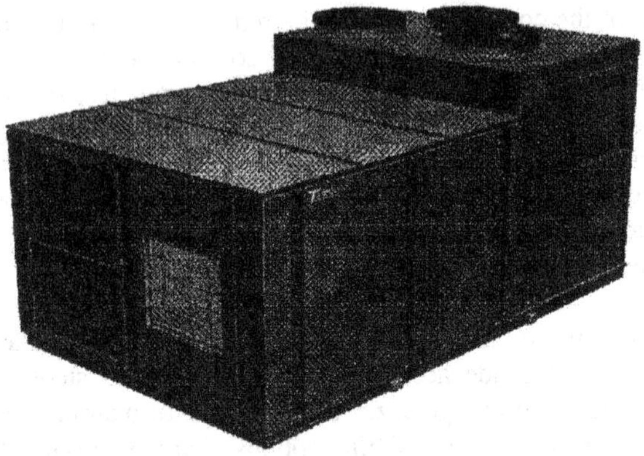

Rooftop Cooling and Heating Unit

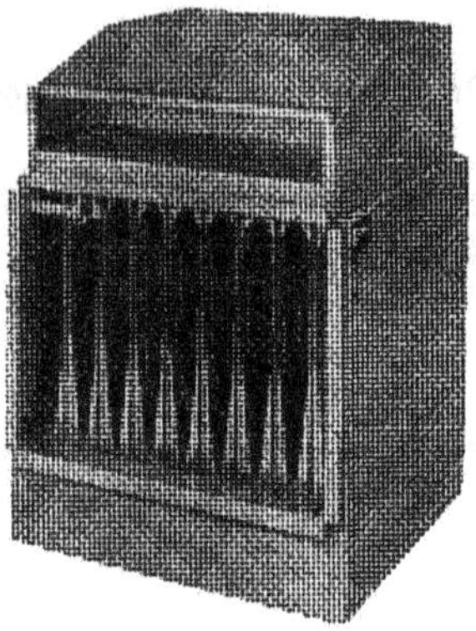

Duct Furnace

Unit Heater

Range of Application
Indirect-fired furnaces are usually used for heating-only or for heating and ventilating applications. This type of furnace isolates the heated space from the (toxic) products of combustion, therefore they should be used when room air is recirculated through the burner. Indirect-fired furnaces that are equipped with stainless steel heat exchangers and modulating burners also can be used for 100 percent make up air applications, but direct-fired designs are more efficient. (Indirect-fired furnaces that have single stage or two stage burners are not recommended for 100 percent make up air applications. In this case, the on-off burner cycling can produce unacceptably large fluctuations in the supply air temperature.)

Manual N Load Information
A **Manual** N load calculation is always a prerequisite to the equipment selection procedure. The following information (which can be extracted from Form N-1) will be required to select the heating equipment.

- Indoor design temperature
- Winter outdoor design temperature
- Form N-1, line 16, heating load in space
- Form N-1, line 21, heating load at unit
- Outdoor air CFM requirement
- Cooling CFM required
- Floor area (for air-loading factor calculation)

Temperature Rise Across the Heat Exchanger
The temperature rise across the heat exchanger must fall within a range that is recommended by the manufacturer. The temperature rise (TR) across the heat exchanger depends on the output capacity (Output BTUH) and the CFM flowing through the heat exchanger. The temperature rise can be computed by using the sensible heat equation.

$$TR = \frac{\text{Output BTUH}}{1.1 \times \text{CFM}}$$

Heating CFM
Mathematically, the design value for the heating CFM is directly related to the space heat load (BTUH) and the temperature difference (HTD) between the heated room (RAT) and the supply air (SAT). The sensible heat equation establishes the relationship between these variables.

$$\text{Heat Load (BTUH)} = 1.1 \times \text{Heating CFM} \times (\text{RAT} - \text{SAT})$$

For indirect-fired, heating-only applications, the CFM value and the heating temperature difference (HTD) value are somewhat arbitrary. A smaller CFM value results in a larger heat exchanger temperature rise and a larger HTD value. A larger CFM value produces the opposite effect. Therefore, from the standpoint of heating ability, it doesn't make much difference what value is used for the heating CFM. However, there are other factors to consider. These factors are listed below. Any design CFM value that satisfies these requirements is acceptable for heating-only applications.

- The temperature rise across the heat exchanger must fall within the range recommended by the manufacturer.
- The supply air temperature should not be too cool or too hot — say between 100°F and 140°F.
- The value for the CFM per square foot of heated space (air-loading factor or ALF) must be compatible with the performance characteristics of the supply outlet. Sidewall and floor diffusers work best when the ALF is between 0.6 and 1.2. Slot diffusers are compatible with ALFs that range from 0.8 to 2.0. Ceiling diffusers are compatible with ALFs that range from 0.9 to 5.0.

> When central equipment provides heating and cooling, the heating CFM value may not be relevant. In this situation, the design value for the supply CFM is based on the larger of the two (heating/cooling) airflow requirements — which is usually the cooling CFM. Different fan speeds should be used for heating and cooling if operation at the cooling CFM value results in an unacceptable temperature rise across the heat exchanger during the heating cycle.

Supply Air Temperature
Ignoring duct losses, the temperature of the air leaving the supply outlets (SAT) is equal to the temperature of the air entering the furnace (EAT) plus the temperature rise (TR) across the heat exchanger. Ideally, the supply temperature should not be too cool or too warm — say between 100°F and 140°F.

SAT = EAT + TR

Ignoring duct losses, the entering air temperature (EAT) depends on the temperature of the outdoor air (OAT), the temperature of the room air (RAT) and the percentage of outdoor air (%OA). The entering air temperature can be estimated by using the following equation.

EAT = RAT - %OA x (RAT - OAT)

When ventilation is required, the entering air temperature (EAT) — and therefore, the supply temperature (SAT) — will vary. This occurs because the air that enters the furnace is a mixture of return air and outdoor air. The entering air temperature will get colder or warmer as the outdoor air temperature changes or as the outdoor air CFM increases and decreases. (If the entering air temperature gets too cold, condensation can form on the heat exchanger. In this case, a stainless steel heat exchanger will be required.) Burner capacity controls are recommended if the operating conditions cause the supply air temperature to exceed 140°F.

Equipment Selection Data Summary
The designer should summarize the equipment performance requirements before referring to the manufacturer's performance data. This summary should include the design heating load, the design value for the outdoor air temperature, the design value for the room temperature, the outdoor air CFM that is required for ventilation, the CFM that is required for cooling and the floor area of the heated space. This information is summarized below.

- Design Heating Load at Unit (N-1, line 21)
- Winter Design Outdoor Air Temperature
- Room Temperature
- Outdoor Air CFM Requirement
- Cooling CFM Required
- Floor Area (for air-loading factor calculation)

Furnace Performance Data — Gas, Oil or Propane
After the equipment performance requirements are listed, the manufacturer's performance data can be used to select the equipment. Carefully inspect the heating performance data and *select the furnace that has an output capacity (BTUH) that is equal to or somewhat greater than the heating load at the unit.* Select an operating CFM and check the temperature rise, the range of possible supply air temperatures and the (supply outlet) air-loading factor.

Inspect the fan performance and air-side accessory data. Verify that the fan will be capable of delivering the design CFM against the (anticipated) resistance created by the duct system and the air-side accessories. (After the duct system is designed, the fan table can be used to specify the fan RPM and motor horsepower.)

Read the information pertaining to controls and accessories. Select burner operating controls (two stage or modulating burner controls are preferred for applications that have wide heating load fluctuations). Select burner safety controls and accessories. Select accessory operating controls, air-side accessories, installation accessories, electrical system devices and fueltrain devices.

> Furnace and fan performance are affected by altitude. Refer to manufacturer's performance data for correction factors that can be used to adjust for the effects of altitude.

Example #1 — Indirect-Fired Rooftop Furnace
An example of manufacturer's performance data is provided by Figures 7-1, -2, -3, -4 and -5. Use figure 7-1 to select a rooftop, single-zone, indirect-fired heating unit. Use Figure 7-2 to check the performance of the fan that comes with the unit. Use Figures 7-3, -4 and -5 to determine the pressure

Performance Data — Indirect-Fired Furnace

	Model Size							
	150	200	250	300	350	351	400	401
BTU/Hr Input	150,000	200,000	250,000	300,000	350,000	350,000	400,000	400,000
BTU/Hr Output	112,500	150,000	187,500	225,000	262,500	262,500	300,000	300,000
Temperature Rise	25-100 °F	25-100 °F	35-100 °F	40-100 °F	45-100 °F	35-45 °F	50-100 °F	35-56 °F
CFM Range	1042-4167	1389-5556	1736-4960	2083-5208	2431-5401	4501-6944	2778-5556	5050-7937
Static Pressure (In.Wg.)	0-1.4	0-1.4	0-1.4	0-1.4	0-1.4	0-1.4	0-1.4	0-1.4

Single stage burner, two stage and modulating burner options available

Figure 7-1

drops that are associated with the accessory equipment, the filter(s), a dampered outdoor air intake and a vertical, downward discharge opening. Assume the following information was extracted from a **Manual** N load calculation.

Indoor Design = 70°F
Outdoor Design = 5°F
Space Heating Load = 212,000 BTUH (N-1, line 16)
Design Heating Load = 240,600 BTUH (N-1, line 21)
Outdoor Air CFM = 400 (N-1, line 18)
Cooling CFM = 0 (No cooling required)
Floor Area = 5,000 SQ.FT.

Capacity
Use Figure 7-1 to find that Model 350 has an output capacity of 262,500 BTUH — (240,600 BTUH required).

Supply CFM
Note that the supply CFM can range between 2431 and 5401 and that the corresponding temperature rise across the heat exchanger ranges from 45°F to 100°F. Proceed by selecting a CFM value that falls somewhere in the cataloged range — say 4000 CFM — and calculate the temperature rise.

$$TR = \frac{262{,}500}{1.1 \times 4000} = 59.6\,°F\ (OK)$$

Supply Air Temperature
Check the value for the coldest supply air temperature (at the winter design condition) and the value for the warmest possible supply air temperature (at a 100 percent return air condition).

%OA = 400/4000 = 0.10

Coldest EAT = 70 - 0.10 x (70-5) = 63.5 °F
Coldest SAT = 63.5 + 59.6 = 123.1 °F (OK)

Warmest EAT = 70° F
Warmest SAT = 70 + 59.6 = 129.6 °F (OK)

Air-loading Factor
Check to see if the air-loading factor is compatible with the supply air outlets.

ALF = 4000/5000 = 0.80 (OK)

Fan Performance
Figure 7-2, at the top of the next page, provides the fan performance data. This graph indicates that at 800 RPM, the blower can move 4,000 CFM of air against a resistance of 1.0 inches of static pressure (water gauge). The graph also indicates that a 2 HP motor will be required.

However, the notes that accompany the fan data indicate that this data only applies to a configuration that consists of a horizontal discharge burner section and a bottom inlet blower section. The pressure losses associated with the outdoor air damper section, filters and a downward discharge plenum — which have to be specified as accessories — must be added to the other external pressure losses.

Accessory Pressure Losses
Figures 7-3, -4 and -5, on the next two pages, provide the accessory pressure loss data. These figures indicate that there is a 0.125 IN.WG. drop across the filters, a 0.13 IN.WG. drop across the downturn plenum and a 0.175 IN.WG. loss across the damper section. This adds up to a total loss of 0.43 IN.WG. This means that 0.57 IN.WG. is available to move the air through the rest of the air distribution system. (The blower data indicates that 1.0 IN.WG. is about all the pressure the blower can produce. This equipment cannot be used if the air distribution system pressure loss exceeds 0.57 IN.WG.)

Controls and Accessories
A search of the manufacturer's product literature indicates that many types of controls and accessories are either standard equipment or available as options. The following items will be included in this design:

Two stage burner controls — option
Two stage room stat and panel — option
Low temperature switch — option
Fire-stat and air flow switch — option

Fan Performance

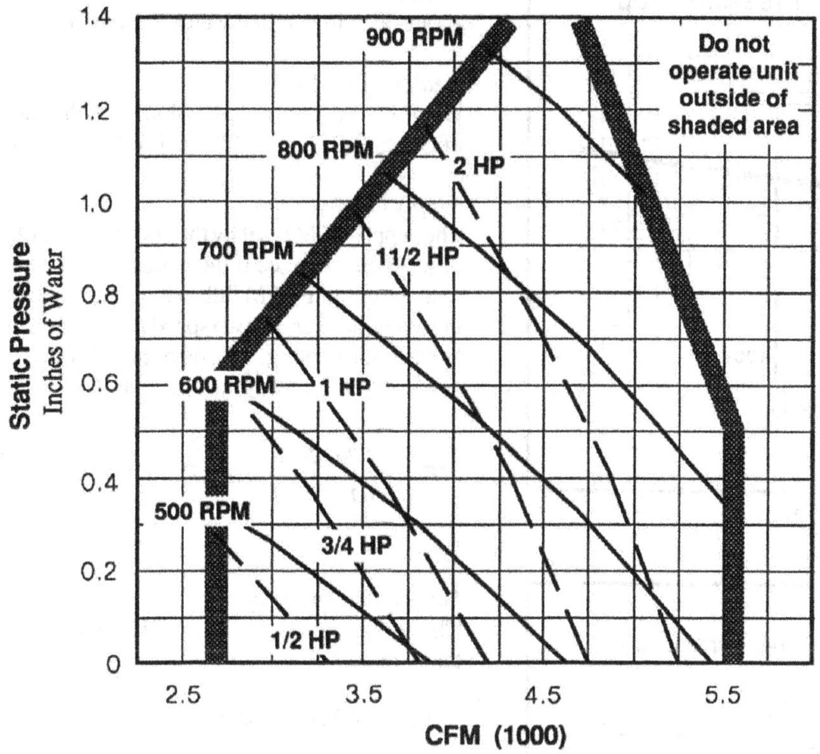

Filter, discharge plenum and damper pressure drops must be added to the duct system pressure losses.

Figure 7-2

Filter Resistance

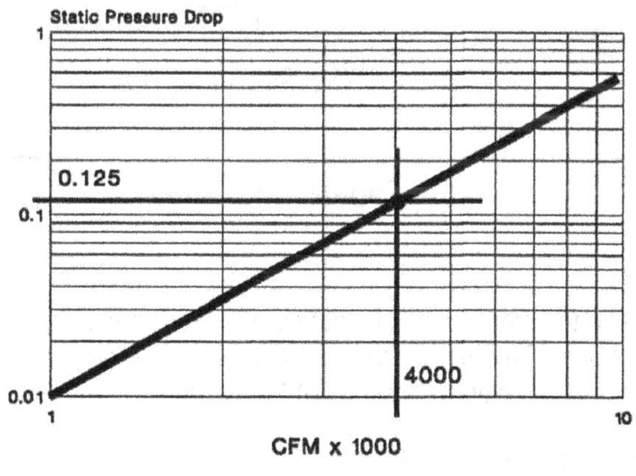

Figure 7-3

Fresh and return air dampers — option
Remote damper position control — option
Filters —) option
Mounting base and roof curb — option
Weather hood and bird screen — option

Control Cycles
Control cycles must be specified for all operating conditions. The following control sequences provide an example of a scheme that could be used for a single-zone application.

Day cycle . . .
The outdoor air damper shall be opened to the minimum position. The supply fan shall run continuously. The burner shall fire at 50 percent capacity in response to a call for heat from the first stage of the thermostat and at 100 percent in response to a second stage call. Fan and burner operation shall be subject to supervision by safeties and interlocks.

Night cycle . . .
The outdoor air and exhaust dampers shall be closed. The supply fan and the burner shall be controlled in response to a signal from a night set-back thermostat. The supply fan shall run intermittently, cycling with the burner. Burner and fan operation shall be subject to supervision by safeties and interlocks.

Example #2 — Rooftop Heating and Cooling Unit
Indirect-fired furnaces are commonly included as part of a heating-cooling equipment package. Except for the design CFM value, the procedure for sizing this type of furnace is the

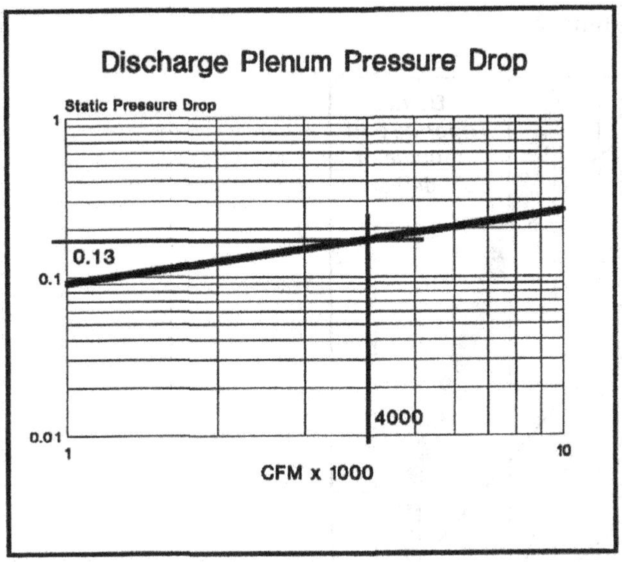

Figure 7-4

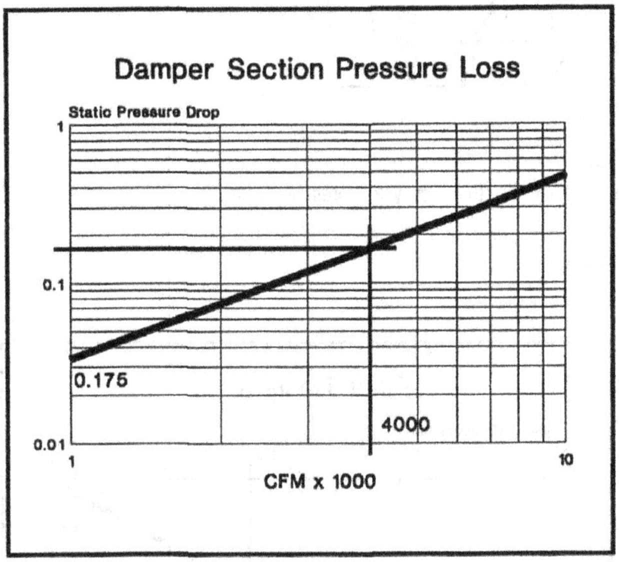

Figure 7-5

same as the procedure that would be used to size heating-only equipment. To illustrate this point, example 1 is repeated for a design that requires 5000 CFM for cooling.

Indoor Design = 70 °F
Outdoor Design = 5 °F
Space Heating Load = 212,000 BTUH (N-1, line 16)
Design Heating Load = 240,600 BTUH (N-1, line 21)
Outdoor Air CFM = 400 (N-1, line 18)
Cooling CFM = 5000 (N-1, line 8)
Floor Area = 5,000 SQ.FT.

Cooling Equipment
Packaged heating and cooling equipment is available with DX cooling coils or chilled water cooling coils. Refer to sections 6-8 and 6-9 for information about selecting DX cooling equipment and refer to sections 6-10, 6-11 and 6-12 for information about selecting chilled water equipment.

Heating Capacity
Use Figure 7-1 to find that Model 350 has an output capacity of 262,500 BTUH — (240,600 BTUH required).

Supply CFM
The supply CFM is set by the 5000 CFM cooling requirement. Always check to see if the temperature rise that is associated with the cooling CFM falls within the range recommended by the manufacturer. (A two speed fan, one speed for heating and another for cooling, will be required if this temperature rise falls outside the recommended range.)

$$TR = \frac{262{,}500}{1.1 \times 5000} = 47.7\,°F \text{ (OK)}$$

Supply Air Temperature
Check the values for the coldest and warmest supply air temperatures

%OA = 400/5000 = 0.08

Coldest EAT = 70 - 0.08 x (70-5) = 64.8 °F
Coldest SAT = 64.8 + 47.7 = 112.5 °F (OK)

Warmest EAT = 70 °F
Warmest SAT = 70 + 47.7 = 117.7 °F (OK)

Air-loading Factor
Check to see if the air-loading factor is compatible with the supply air outlets.

ALF = 5000/5000 = 1.00 CFM/SQ.FT.

Fan Performance
Figure 7-2, on the previous page, indicates that at 900 RPM, the blower can move 5,000 CFM against a resistance of 1.0 IN.WG. of static pressure. (All this pressure may not be available for the duct system. Always refer to the notes that accompany the fan table to determine if there are any components that create external pressure losses.) The graph also indicates that a 3 HP motor will be required.

Accessory Pressure Losses
Figures 7-3, -4 and -5 provide the accessory pressure loss data. At 5,000 CFM, these figures indicate a 0.18 IN.WG. drop across the filters, a 0.20 IN.WG. drop across the downturn plenum and a 0.20 IN.WG. loss across the damper section. This adds up to a total loss of 0.58 IN.WG. This means 0.42 IN.WG. is available to move the air through the rest of the air distribution system. (The blower data indicates that 1.0

IN.WG. is about all the pressure the blower can produce. This equipment cannot be used if the air distribution system pressure loss exceeds 0.42 IN.WG.)

Controls and Accessories
The manufacturer's product literature usually indicates what types of controls and accessories are standard equipment or available as options. Refer above to example 1 for a list of heating cycle devices and refer to Section 6-8 (page 6-10) for a list of cooling cycle devices.

Control Cycles
Control cycles must be specified for all operating conditions. Refer to page 7-7 for an example of a heating mode control cycle. Refer to page 6-10 for an example of a cooling mode and a heating-cooling changeover control cycle.

Example #3 — High Percentage Outdoor Air (Heating)
A high percentage of outdoor air produces a lower mixed air temperature and a colder supply air temperature. Use figure 7-1 to select a rooftop, single-zone, indirect-fired heating unit for an application that requires a large amount of outdoor air. Assume the following information was extracted from a **Manual N** load calculation.

Indoor Design = 70 °F
Outdoor Design = 5 °F
Space Heating Load = 112,000 BTUH (N-1, line 16)
Design Heating Load = 255,000 BTUH (N-1, line 21)
Outdoor Air CFM = 2000
Floor Area = 5,000 SQ.FT.

Capacity
Use Figure 7-1 to find that Model 350 has an output capacity of 262,500 BTUH — (255,000 BTUH required).

Supply CFM
Note that the supply CFM can range between 2431 and 5401 and that the corresponding temperature rise across the heat exchanger ranges from 45°F to 100°F. Proceed by selecting a CFM value that falls somewhere in the cataloged range — say 4000 CFM — and calculate the temperature rise.

$$TR = \frac{262{,}500}{1.1 \times 4000} = 59.6\,°F\ (OK)$$

Supply Air Temperature
Check the value for the coldest supply air temperature.

%OA = 2000/4000 = 0.50

Coldest EAT = 70 - 0.50 x (70-5) = 37.5 °F
Coldest SAT = 37.5 + 59.6 = 97.1 °F (OK)

If a warmer supply air temperature is desired, it can be obtained by reducing the design CFM. However, this strategy will reduce the air-loading factor. The final value for the design CFM may have to be a compromise between a warmer supply air temperature and an acceptable air-loading factor. For example, the following performance will be obtained if the design CFM is reduced from 4000 to 3000.

$$TR = \frac{262{,}500}{1.1 \times 3000} = 79.5\,°F\ (OK)$$

%OA = 2000/3000 = 0.67

Coldest EAT = 70 - 0.67 x (70-5) = 26.5 °F
Coldest SAT = 26.5 + 79.5 = 106 °F (OK)

ALF = 3000/5000 = 0.60 (Barely acceptable)

Example #4 — Indirect-Fired Make Up Air Application
Indirect-fired units can be used for 100 percent make up air applications providing that they are equipped with a stainless steel heat exchanger and a modulating burner. Use figure 7-1 to select a rooftop, single-zone, indirect-fired heating unit for a make up air application. Assume the following information was extracted from a **Manual N** load calculation.

Indoor Design = 70 °F
Outdoor Design = 5 °F
Outdoor Air CFM = 3000 (100 percent)
Outdoor Air Heating Load = 214,500 BTUH
Envelope heat loss = 80,000 BTUH (N-1, line 16)
Design heat loss = 294,500 BTUH (N-1, line 21)

Capacity
Use Figure 7-1 to find that Model 400 has an output capacity of 300,000 BTUH — (294,500 BTUH required).

Supply CFM
In this case, the design CFM is fixed by the make up air requirement. Proceed by checking the performance at 3000 CFM.

$$TR = \frac{300{,}000}{1.1 \times 3000} = 90.9\,°F\ (OK)$$

Supply Air Temperature
Check the value for the coldest supply air temperature.

%OA = 1.00
Coldest EAT = 5 °F
Coldest SAT = 5 + 90.9 = 95.9 °F

If a warmer supply air temperature is desired it cannot be obtained by reducing the design CFM because the design CFM must satisfy the 3000 CFM make up air requirement.

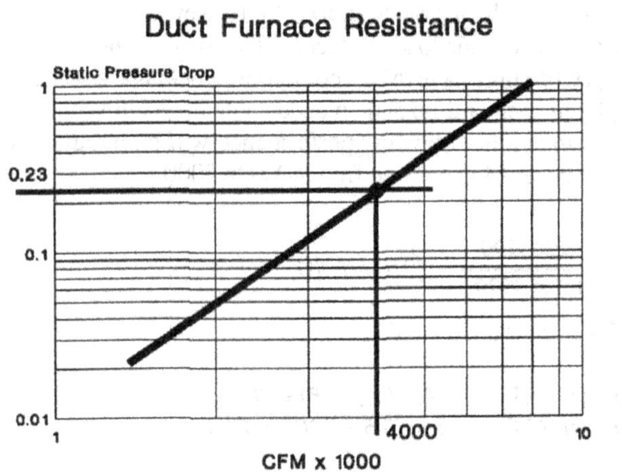

Figure 7-6

Example #5 — Indirect-Fired Duct Furnace
The selection procedure for sizing an indirect-fired duct furnace is similar to the procedure outlined above by example 1. (The main difference between a duct furnace and an indirect-fired heating plant is related to the air-side design. A duct furnace is simply a heat exchanger that is inserted in a duct run. Fans, filters and mixing dampers are not included with the duct furnace package. The duct furnace heat exchanger creates an air-side pressure loss that must be added to the other duct system pressure losses.) Use Figure 7-1 (on page 7-6) to size the duct furnace and use Figure 7-6 to find the pressure drop across the heat exchanger. Select a duct furnace for the following design conditions:

Coldest Entering Air Temperature = 63.5 °F
Heating Load = 240,600 BTUH (N-1, line 21)
CFM = 4000

Capacity
Use Figure 7-1 to find that Model 350 has an output capacity of 262,500 BTUH — (240,600 BTUH required).

Supply CFM
In this case, the design CFM could have been set by some other system requirement, say the cooling CFM requirement. Proceed by checking the heating performance at 4000 CFM.

$$TR = \frac{262{,}500}{1.1 \times 4000} = 59.6\,°F\ (OK)$$

Supply Air Temperature
Use the temperature rise and the entering air temperature to estimate the coldest supply air temperature.

$$Coldest\ SAT = 63.5 + 59.6 = 123.1\,°F$$

Air-Side Pressure Drop
Figure 7-6 indicates that there will be a 0.23 IN.WG. pressure drop across the duct furnace.

7-8 Indirect-Fired Unit Heaters (Gas, Oil or Propane)

Indirect-fired unit heaters are similar to furnaces in that they have an input capacity, an output capacity and a temperature rise across the heat exchanger. They are different from furnaces because they are only used as space heaters, they are not designed to heat outdoor air. Room air is simply forced through a heat exchanger by a fan. Both the fan and the heat exchanger are part of a unit that is located in the heated space. (Some unit heater fans are not designed for use with a duct system and some unit heaters may not be equipped with a filter.) Combustion fumes are expelled through a vent.

> Propeller fan units do not develop much pressure, they are designed for "free delivery" applications. Blower units can develop about 0.40 IN.WG of pressure, which can be equipped with a discharge nozzle or a modest duct system.

Air distribution depends on equipment arrangement, mounting height and throw. Figure 7-7, on the next page, illustrates some arrangements that can be used for various types of buildings. Figures 7-8 and 7-9, on page 7-12, provide examples of propeller fan and blower unit performance. (Blower units can be fitted with nozzles or short duct runs that provide more control over the air distribution pattern.) Use the following guidelines for sizing and locating unit heater equipment.

- Use as few units as possible to blanket the room with warm air. Consider the arrangement, mounting height, throw and spread and try to use an arrangement that minimizes piping, venting and wiring costs.
- Mounting height is critical. Performance will be unsatisfactory unless the installed mounting height matches the mounting height recommended by the manufacturer.
- Propeller fan units are usually used to heat the perimeter of a building. The discharge from this type of heater should wipe the exposed wall, but it should not blow directly toward the wall (see Figure 7-7 on the next page). Arrange the units so that the discharge from one heater feeds the next heater. Space the heaters so that the distance between each unit is about the same as the throw. Try to create a circulating air pattern that blankets the exposed walls.
- Blower-type units that are equipped with discharge nozzles can be used to heat interior areas. Blower units and nozzles also should be used when an application requires a high mounting height. If blower-type units are used exclusively,

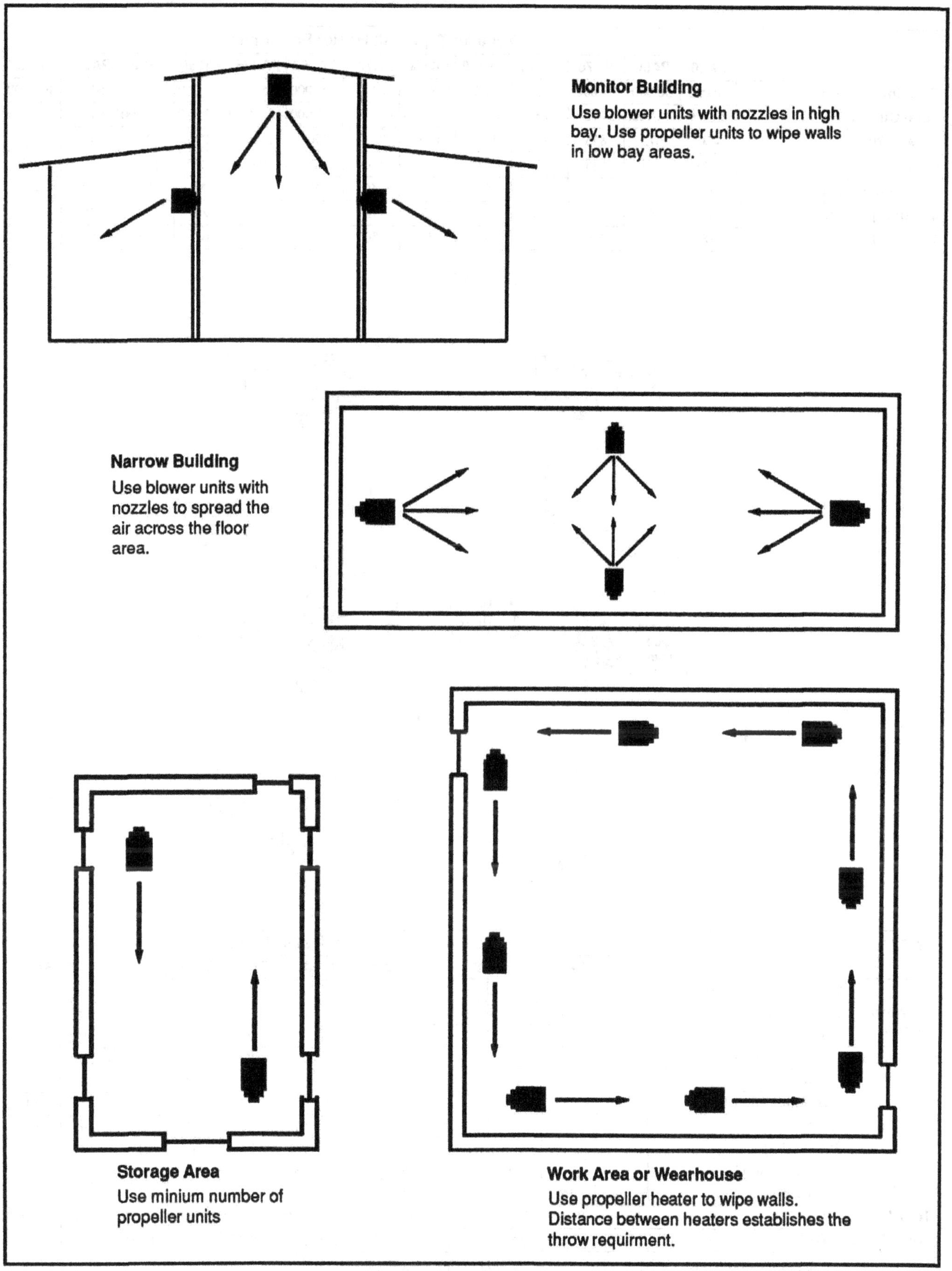

Monitor Building
Use blower units with nozzles in high bay. Use propeller units to wipe walls in low bay areas.
Narrow Building
Use blower units with nozzles to spread the air across the floor area.
Storage Area
Use minium number of propeller units
Work Area or Wearhouse
Use propeller heater to wipe walls. Distance between heaters establishes the throw requirment.

Figure 7-7

	Propeller Type Unit Heater Performance												
	PA30	**PA50**	**PA75**	**PA105**	**PA130**	**PA150**	**PA170**	**PA200**	**PA225**	**PA250**	**PA300**	**PA350**	**PA400**
Total Input BTU/hr.	30,000	50,000	75,000	105,000	130,000	150,000	170,000	200,000	225,000	250,000	300,000	350,000	400,000
Total Output BTU/hr.	22,500	38,500	57,750	80,850	100,100	112,500	127,500	152,000	173,250	192,500	231,000	269,500	304,000
CFM @ 70°F	400	675	855	1400	1725	1900	2540	2900	2900	4275	4275	4400	5300
FPM	470	555	585	825	740	815	955	825	830	1200	1075	975	995
Air Temp. Rise °F	52	53	63	53	54	55	46	49	55	42	50	57	53
Mounting Ht. (Max. Ft.)	8	8	10	10	12	12	12	16	16	20	20	20	20
Throw (Ft.)	18	23	25	40	45	48	50	50	50	65	65	65	65

Figure 7-8

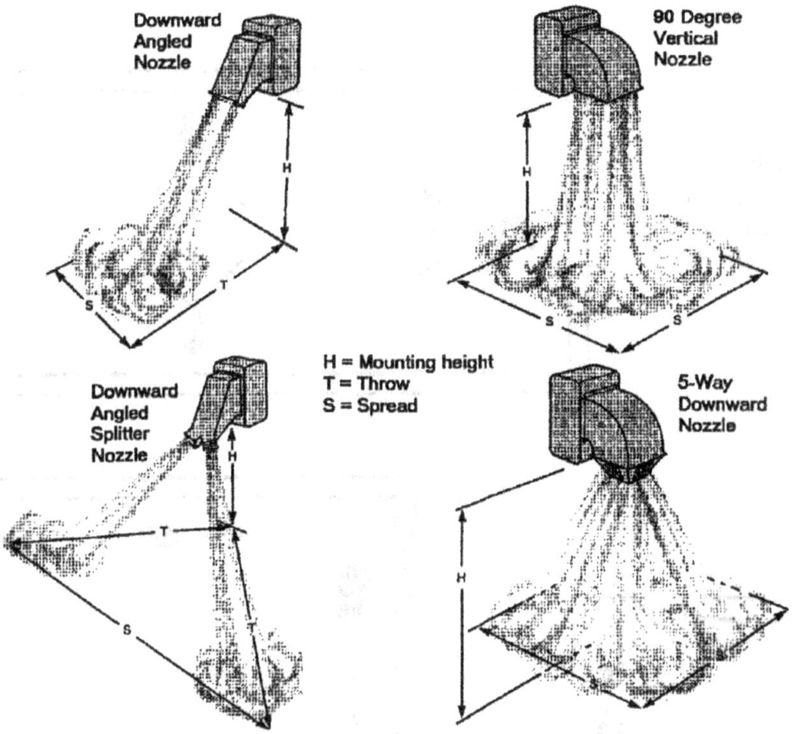

NOZZLE TYPE		Blower Type Unit Heater Performance											
		BA50	**BA75**	**BA105**	**BA130**	**BA150**	**BA170**	**BA200**	**BA225**	**BA250**	**BA300**	**BA350**	**BA400**
40° Downward Nozzle	**Max. Mounting Ht. (Ft.)**	10	13	17	18	21	22	23	24	26	30	33	36
	Throw (Ft.)	30	39	51	54	63	66	69	72	78	90	99	112
	Spread (Ft.)	11	15	19	21	24	25	27	28	31	35	39	44
90° Vertical Nozzle	**Max. Mounting Ht. (Ft.)**	9	14	17	19	21	21	22	24	26	28	30	32
	Spread (Ft.)	9	14	17	19	21	24	24	27	30	30	33	35
40° Splitter Nozzle	**Max. Mounting Ht. (Ft.)**				17	19	21	21	22	24	28	31	34
	Throw (Ft.)				49	52	56	61	65	70	76	84	89
	Spread (Ft.)				85	90	95	100	105	110	115	125	130
5-Way Nozzle	**Max. Mounting Ht. (Ft.)**	8	12	13	14	16	17	18	20	23	25	26	30
	Spread (Ft.)	10	15	17	19	23	26	27	30	34	37	43	47
Total Input BTUH		50,000	75,000	105,000	130,000	150,000	170,000	200,000	225,000	250,000	300,000	350,000	400,000
Total Output BTUH		38,500	57,750	80,850	100,100	112,500	127,500	152,000	173,250	192,500	231,000	269,500	304,000

NOTE: Size door heaters for 90 BTUH per sq. ft. of door area per degree of temperature difference

Figure 7-9

some of them should be arranged so that their air patterns blanket the exterior walls.

- Always place the heaters so the air patterns are not disrupted by columns, machinery, partitions or objects.
- Place the heaters so that they will not blow directly on the occupants.
- Install a larger percentage of the required heating capacity on the windward side of a building that is subjected to a prevailing wind.
- Use a larger percentage of the required heating capacity to offset loads that are associated with large glass areas or large doors.
- Blower units with nozzles can be used as door heaters. In this situation, the units should be mounted so that they produce an air pattern that hits the floor in front of the door. Ideally, this air should hit the floor at about 400 FPM.
- Zoning enhances comfort. Use multiple thermostats to control groups of unit heaters that serve a large room.
- Provide air circulation during warm weather by equipping the unit heaters with summer-winter fan switches.
- Specify stainless steel heat exchangers and burners for applications that are associated with high humidity or salt water atmospheres.
- Never locate atmospheric burners in areas where chlorinated, halogenated or acid vapors are present in the atmosphere.

Example

Use indirect-fired propeller-type unit heaters to offset the transmission and infiltration loads associated with a warehouse. Use blower-type unit heaters near the shipping doors. Base the design on the following **Manual N** information. Note that the 461,000 BTUH envelope heat loss does not include the heat loss associated with the shipping doors.

Outdoor Design = -5 °F
Indoor Design = 55 °F
Envelope heat loss = 461,000 BTUH (N-1, line 16)
Door area (four @ 10x10) = 400 SQ.FT.
Floor Dimensions = 200 Ft. by 100 FT.
Bottom of trusses = 24 FT.

Arrangement

Locate the propeller units around the perimeter of the building and use blower units with discharge nozzles directed toward the shipping doors (refer to Figure 7-10).

Sizing

Figure 7-8, on the previous page, indicates that 10 PA75 propeller units will have a total output capacity of 577,500 BTUH, which is more than enough to offset the envelope loss. The performance table also indicates that a 10 foot mounting height produces a 25 foot throw. However, a throw of 40 to 50 feet, based on the distance between units, is required. The same number of larger units will provide the desired throw — 10 PA105 units for example. In this case, the larger units are preferred because they produce a better wiping action.

The footnote, at the bottom of Figure 7-9 on the previous page, indicates that this particular manufacturer recommends sizing door heaters to offset a load of 90 BTUH per degree temperature difference per square foot of door area. This computes

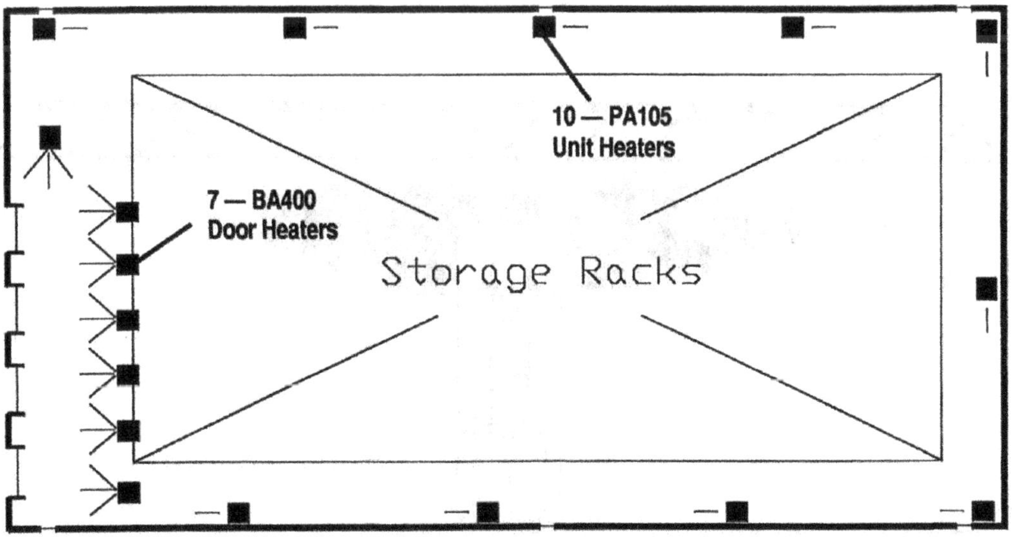

Figure 7-10

to 2,160,000 BTUH of output capacity (90 x 400 SQ.FT. x 60°F). Seven BA400 blower-type unit heaters will have a total output capacity of 2,128,000 BTUH. These units should be equipped with a downward angled nozzle. Ideally, shipping doors should be equipped with air locks (or the warehouse storage area should be isolated from the loading dock).

> The 90 BTUH per degree-temperature-difference per square foot of door area recommended by the manufacturer appears to be a large heat loss, but consider this . . .
>
> A 15 MPH wind blowing through a 1 square foot opening produces 1320 CFM of infiltration. This translates into a load of about 1450 BTUH per degree-temperature-difference per square foot of opening area.

Controls, Accessories and Trim
Some manufacturers offer heaters that can be equipped with a standing pilot or an electronic spark ignition system and one of three types of gas burner operating controls (on-off, two stage and modulating). Two to four types of discharge nozzles are usually available for the blower-type units.

7-9 Large Indirect-Fired Ductless Space Heaters

Large indirect-fired, ductless space heaters are designed to heat industrial buildings and warehouses. These units are suited for applications that would otherwise be handled by installing a larger number of smaller unit heaters. Like unit heaters, they are designed to heat room air. If outdoor air is required for ventilation, it must be tempered by other heating equipment. The advantages that this type of equipment has over unit heaters are summarized below.

- Lower installation costs — no duct system is required and only one or two large units need to be set, piped, wired, vented and controlled.
- Lower maintenance costs — only one or two large units need to be serviced.
- Less stratification — supply air temperatures are only a little warmer than room temperature.
- Less heat loss at the roof — lower temperature under the roof, smaller temperature difference across the roof.
- Uniform temperature throughout the space — high circulation rates thoroughly mix room air with supply air (stratification is not a problem).

The rationale for this design is to use a large, ductless, centrally located unit to create a rotating air pattern that sweeps the roof and the exterior walls. (Refer to Figure 7-11 below.) This requires a large amount of slightly warm air (to minimize stratification) and a high discharge outlet and a low return inlet (to promote circulation). Manufacturers claim that a single centrally located unit can be used to heat a space as large as 500 feet by 300 feet, with less than a 5 degree temperature difference throughout the building. However, this concentration of heating capacity leaves the building without heat if the equipment fails.

Sizing
The unit (or units) must have an output heating capacity (BTUH) that equals or exceeds the **Manual N** transmission and infiltration loads (Form N-1, line 16). The unit also must

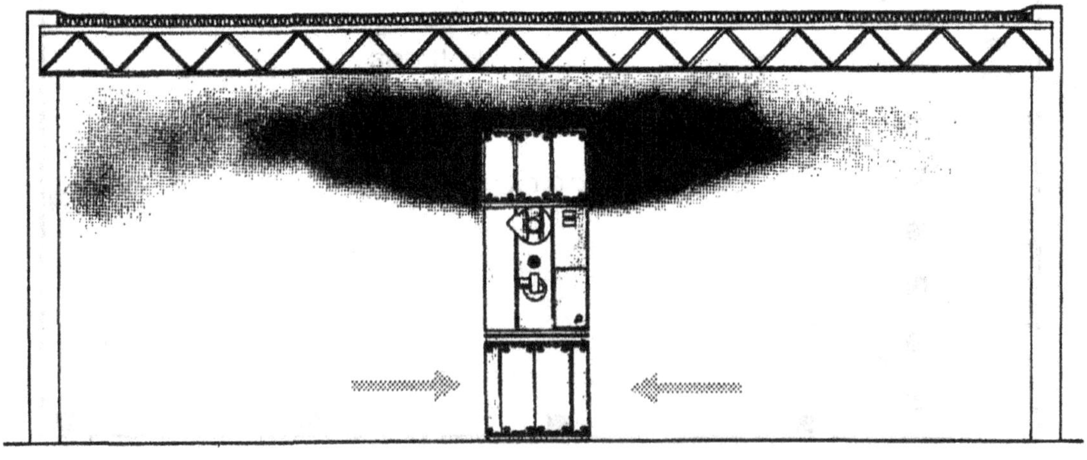

Figure 7-11

provide enough air circulation to insure that a uniform temperature is maintained throughout the heated space. Base the supply CFM requirement on 3 air changes per hour (CFM_1) or on a supply-air to room-air temperature difference of 20°F (CFM_2) — which ever is greater.

$$CFM_1 = (3\ AC/HR \times Volume)/60$$

$$CFM_2 = \frac{Output\ BTUH}{1.1 \times 20}$$

Example
Use one large indirect-fired space heater to offset the transmission and infiltration loads associated with a factory work area. The **Manual N** load information is provided below. Note that the 641,000 BTUH envelope heat loss does not included the heat loss associated with open overhead doors or exhaust fans. Use the manufacturer's performance data provided by Figure 7-12 to select a unit that has the required heating capacity and air circulation capability.

Outdoor Design = -5 °F
Indoor Design = 55 °F
Envelope heat loss = 641,000 BTUH (N-1, line 16)
Floor Dimensions = 200 FT. by 100 FT.
Bottom of trusses = 24 FT.
Volume = 480,000 CU.FT.

Arrangement
Locate the unit near the middle of the room and direct the discharge air in a direction that is parallel to the longest wall.

Sizing Calculations
Start by calculating the CFM that corresponds to 3 air circulations per hour (ACH).

$$CFM_1 = (3 \times 480000)/60 = 24000$$

Use Figure 7-12 to make a tentative selection. This figure indicates that when a 043 unit is equipped with a 42 inch fan, it will deliver 27,000 CFM and provide at least 650,000 BTUH of output capacity.

Use the output burner capacity to calculate the CFM that corresponds to a 20°F temperature rise.

$$CFM_2 = \frac{650000}{1.1 \times 20} = 29545$$

Since 29,545 CFM is larger than 27,000 CFM, use 29,545 CFM as the design CFM. Figure 7-12 indicates that when a 043 unit is equipped with a 48 inch fan, it can deliver 29,545 CFM and 650,000 BTUH of output capacity.

Controls, Accessories and Trim
Some manufacturers offer heaters that can be equipped with different types of burners (gas, propane or oil). Temperature controls, fueltrain safety controls and fueltrain operating controls are usually standard equipment. Various filter and discharge nozzle options also may be available.

UNIT SIZE	BTU OUTPUT	CFM RANGE	FAN QTY	FAN DIA.	FAN HP	UNIT WEIGHT
016	200,000 to 400,000	7,000 to 9,000	2	24	3/4	2,135
		10,000 to 12,000	2	24	1	2,135
		13,000 to 15,000	2	24	1-1/2	2,135
022	650,000 to 750,000	9,000 to 12,000	2	24	1	2,660
		13,000 to 18,000	2	30	1-1/2	2,660
043	650,000 to 1,000,000	19,000 to 27,000	2	42	2	4,425
		27,000 to 40,000	2	48	3	4,425
057	850,000 to 1,750,000	28,000 to 40,000	2	48	3	5,940
		41,000 to 62,000	2	54	5	5,940

Figure 7-12

7-10 Direct-Fired Make Up Air Units

Direct, gas-fired, make up air units are typically used to replace air that is exhausted by HVAC exhaust systems or by process equipment exhaust fans. For example, a direct-fired make up air unit could be used to replace the air that is exhausted from a locker room, kitchen or a foundry. Figure 7-13 on the next page, illustrates this application.

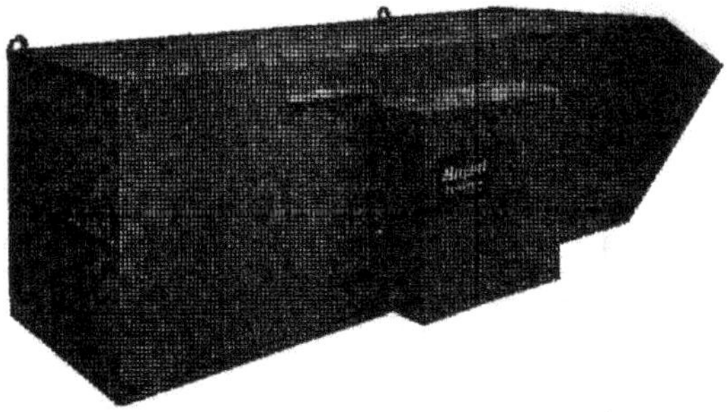

Direct-Fired Make Up Air Unit

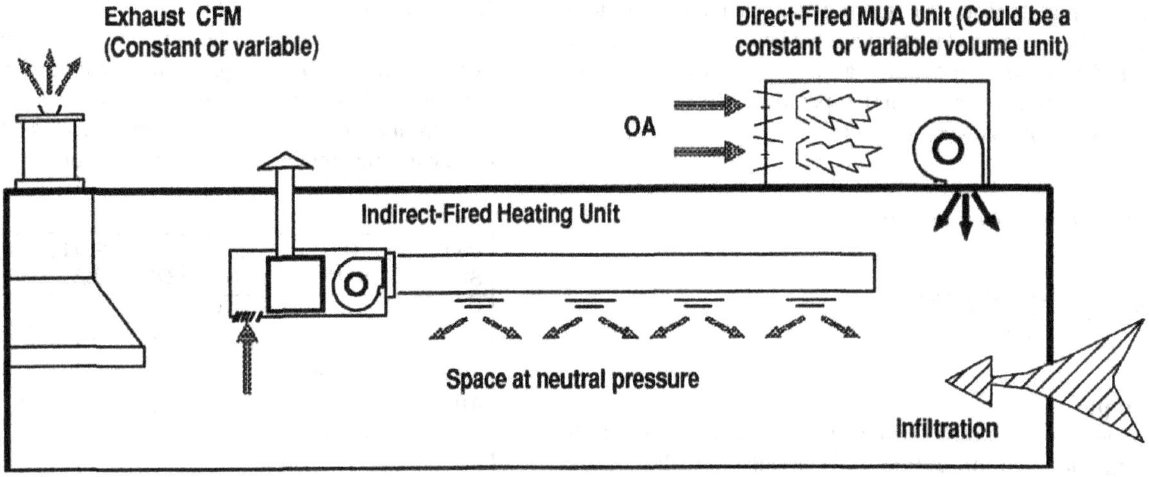

Figure 7-13

Direct-fired equipment also can be used to replace exhaust air and to add heat to a space. There are two important advantages associated with using a direct-fired make up air unit to heat a space. The first is that a direct-fired burner is more efficient than the indirect-fired, heat exchanger design (90 percent *lifetime* efficiency compared to 70 or 80 percent *initial* efficiency). The second is that a duct system is not required — providing that the direct-fired unit is used to pressurize the space. Figure 7-14 below illustrates this application.

Products of Combustion

Direct-fired equipment releases the products of combustion directly into the supply air. However, this is not hazardous because the once-through, 100 percent outdoor air, design insures complete combustion and the resulting concentration of toxic fumes is quite small — well below allowable levels.

If a direct-fired unit is equipped with a return air damper, the damper should be located downstream from, or in parallel with, the burner section (see Figure 7-14). Note that in this arrangement, the full burner capacity is applied to a smaller and smaller amount of outdoor air as the percentage of return air increases. However, the concentration of toxic fumes should not be hazardous, providing that the outdoor air flowing through the burner is never allowed to fall below 20 percent of the design CFM.

> The best way to be sure that the products of combustion are within allowable limits is to use laboratory labeled heaters tested to ANSI Z83.4 or Z83.18 standards. Also make sure that the burner cannot operate if the outdoor air CFM is reduced to less than 20 percent of the design CFM.

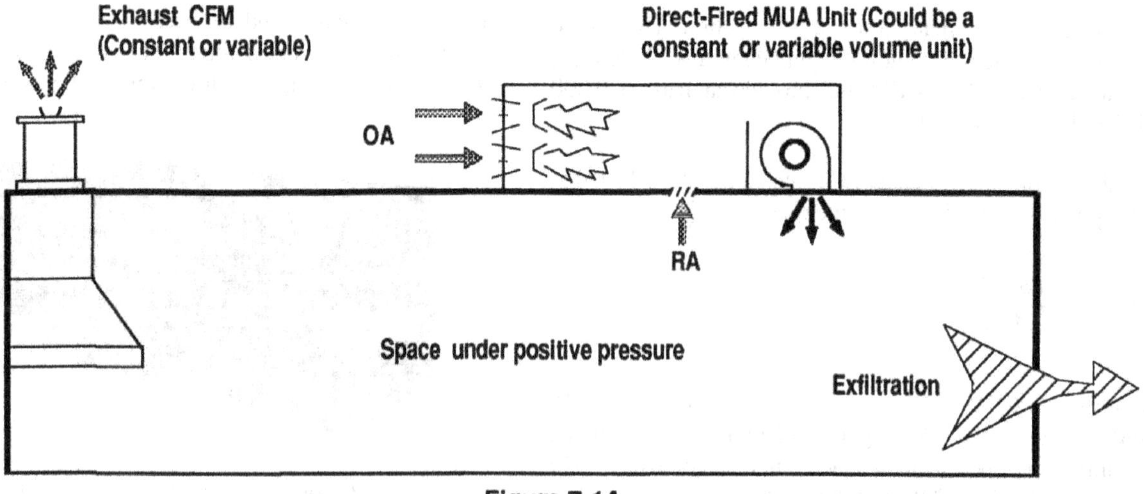

Figure 7-14

Characteristics of Buildings that Require Make Up Air
Make up air is required to replace air that is expelled by the HVAC system or by dedicated process exhaust systems. Buildings that do not have adequate make up air are identified by the following symptoms:

- Exterior doors are hard to open
- Cold drafts around perimeter of building
- Exhaust systems operate below rated capacity
- Relief dampers never open
- Smoke, haze and dust hangs in the air
- Odors do not dissipate
- Vapors condense in exhaust ducts
- Poor combustion at atmospheric burners
- Moisture stains at cracks in exterior masonry walls

Manual N Load Information
A **Manual N** load calculation is always necessary. The following information — which can be extracted from Form N-1 — will be required to select direct-fired heating equipment:

- Indoor design temperature
- Winter outdoor design temperature
- Form N-1, line 16, heating load in space
- Form N-1, line 5, infiltration heating load
- Form N-1, line 18, ventilation heating load
- Exhaust CFM
- Infiltration CFM
- Volume of heated space

Temperature Rise Across the Heat Exchanger
The temperature rise (*TR*) across the direct-fired burner depends on the burner input capacity (Burner BTUH) and the quantity of air (*CFM*) flowing through the burner section. One advantage of the direct-fired burner is that there are no vent losses. The temperature rise (*TR*) can be computed by using the sensible heat equation.

$$TR = \frac{Burner\ BTUH}{1.1 \times CFM}$$

Note that the temperature rise values that are listed in the manufacturer's performance data correspond to the rise that would be obtained if the volume of air flowing through the burner is equal to 100 percent of the rated blower CFM. However, these published temperature rise values — typically 40°F to 120°F — do not represent the maximum rise that can be tolerated by the direct-fired burner. Larger temperature rises can occur when a unit is equipped with a return air damper. In this case, the CFM that flows through the burner can be substantially less than the rated blower CFM and the corresponding rise can exceed the rise listed in the performance data.

Maximum Discharge Temperature
Manufacturers may place a limit on the temperature of the air that leaves the unit. For a simple 100 percent make up air unit, this temperature is the same as the temperature of the air leaving the burner section. For a make up air/heating design, this temperature will be equal to the temperature that results when burner air is mixed with return air. Make sure that the design complies with all of the ANSI requirements.

Blower CFM
Direct-fired make up air units are rated by blower CFM. Units that are used for 100 percent make up air duty should be selected so that the blower CFM is equal to, or slightly greater than, the required make up air CFM value. Ductless units that are used to temper outdoor air, pressurize, and heat the space should be sized so that the blower CFM satisfies the make up air requirement and provides a large amount of air circulation (1/2 to 7 circulations per hour, depending on the building — see Figure 7-16) at a moderate supply air temperature. Ducted units that temper outdoor air and heat the space should be sized so that the blower CFM satisfies the make up air requirement and the supply air outlet air-loading requirement.

Supply Air Temperature
The temperature of the air leaving a 100 percent make up air unit is controlled so that it is equal to the temperature in the heated space. To minimize stratification, the temperature of the air leaving a *ductless* make up air/heating unit should not be more than 15 degrees warmer than the temperature in the heated space. The temperature of the air leaving a make up air/heating unit that is equipped with a complete air distribution system can be substantially warmer than the temperature in the heated space.

Equipment Selection Data Summary
The designer should summarize the equipment performance requirements before referring to the manufacturer's performance data. The information that is required for sizing 100 percent make up air units and for sizing units that will be used for pressurized, ductless, make up air/heating duty is summarized below.

100 percent make up air application
- Indoor design temperature
- Winter outdoor design temperature
- Form N-1, line 18, ventilation heating load
- Exhaust CFM

Make up air plus heating application
- Indoor design temperature
- Winter outdoor design temperature
- Form N-1, line 16, heating load in space
- Form N-1, line 5, infiltration heating load
- Form N-1, line 18, ventilation heating load
- Estimate of building tightness
- Infiltration CFM
- Exhaust CFM
- Volume of heated space

Performance Data
After the equipment performance requirements are tabulated, the manufacturer's performance data can be used to select the equipment. Carefully inspect the performance data and select a unit that has the required CFM rating and burner capacity. Figure 7-15 provides an example of direct-fired make up air unit performance data.

Study the performance information to determine if there are any external pressure losses associated with the filters, discharge nozzles or other air-side accessories. Verify that the fan will be capable of delivering the design CFM against the resistance that is created by these accessories. If a duct system is used, the fan also must produce enough pressure to overcome the resistance that is created by the straight runs, fittings and the air outlets.

Direct-Fired Unit Performance		
CFM	Burner BTUH	Rise ^{o}F
3,000	300,000	92
5,000	500,000	92
7,500	750,000	92
10,000	1,000,000	92
15,000	1,500,000	92
20,000	2,000,000	92
25,000	2,250,000	92
30,000	3,000,000	92
35,000	3,500,000	92
40,000	4,000,000	92
50,000	5,000,000	92
75,000	7,500,000	92
100,000	10,000,000	92
CFM values for operation against an external resistsnce of 0.30 IN.WG. Add the pressure drops associated with filters and air-side accessories to the duct system resistance.		

Figure 7-15

Controls and Accessories
Refer to the manufacturer's literature to determine the features that are offered as standard equipment and what must be ordered as an option. Fan, damper, burner and fueltrain controls (operating and safety) must be compatible with the application. Also specify air-side accessories, installation accessories and electrical system devices. The following list provides examples of features that may be standard equipment or available as an option:

- Ignition devices, flame supervision and safety controls
- Burner capacity controls (turn-down ratio)
- Discharge, space, high- and low-limit temperature controls
- Dual fuel burner (gas/propane) switch-over
- Gas valves (inlet, shut-off, manual) and pressure regulator
- Air flow switch and building pressure control
- Dampers, operators and controls (return air option)
- Control panels, indicating lights, day-night switches
- Time clocks and microprocessor network controls
- Filter options intake hoods and bird screens
- Discharge arrangement and attachments
- Roof curb, service platform, weatherization
- Motor starters, relays, disconnects and motor protection

Design Procedure — Make Up Air Only
In this application, the equipment maintains an air balance while the burner output just neutralizes the outdoor air load (no heat is added to the building). An extensive supply duct system is not required to distribute the make up air. A short duct system or even a simple diffuser can be used to "dump" the make up air into the room. Since no heat is added to the building, a second heating system will be required to offset the transmission and infiltration losses that are associated with the building envelope (see Figure 7-13, on page 7-16).

In a 100 percent make up air application, the equipment size is based on the air flow that is required to replace exhausted air. Select a unit that has a CFM rating that is equal to or greater than the maximum exhaust CFM requirement. Since no net heat is added to the building, the burner capacity is based on the amount of heat that is required to temper the incoming air. This capacity is based on a temperature rise that is equal to the difference between the room temperature (*RAT*) and the winter design, outdoor temperature (*OAT*).

Required CFM = Maximum CFM Exhausted

Required Burner Capacity = 1.1 x CFM x (RAT-OAT)

The make up air unit will have to operate whenever the exhaust fans are energized, but it can be shut down when the exhaust fans are off. A two-position (open/closed) outdoor air damper is all that is required if the exhaust CFM is constant. A variable volume unit is required if the exhaust CFM is not constant. (A number of VAV designs are available. Some just use dampers to control the outdoor CFM and the air flow through the burner, others use dampers plus fan speed or inlet vane control.) If a VAV unit is used, the air-side devices are controlled by a static pressure sensor that is located in the space. (Manufacturers design the equipment so that the outdoor air CFM that is flowing through the burner is never less than 20 percent of the design CFM.) And, burner capacity is controlled to maintain the discharge air temperature setpoint, which is usually the same as the room temperature.

Example — Make Up Air Only
A **Manual** N load calculation indicates that the outdoor winter design temperature is 15°F and the indoor temperature is 60°F. Line 18 on Form N-1 indicates that 4,700 CFM of outdoor air is required to replace air that is exhausted from the

room. Use Figure 7-15 on Page 7-18 to select a direct-fired make up air unit to temper 4700 CFM of exhaust air.

CFM required = 4700
Burner capacity = 1.1 x 4700 x (60-15) = 232650 BTUH
Select the 5000 CFM unit @ 500000 BTUH

Design Procedure — Make Up Air/Heating

Direct-fired make up air units also can be used to replace exhaust air and to satisfy part — or all — of the building heat requirement. This design is illustrated at the bottom of page 7-16 by Figure 7-14. In this application, the direct-fired make up air unit must be equipped with a return air damper that will allow the unit to recirculate the room air without routing it through the burner. A supply duct system is not required, providing that the space is slightly pressurized and providing that the supply air temperature is only slightly warmer than the room temperature. The following guidelines apply to this type of application:

- Usually, the outdoor CFM should exceed the exhaust CFM and the building should be slightly pressurized, say from 0.01 to 0.05 IN.WG. positive.
- A positive pressure in the space will reduce the infiltration. At the winter design condition (15 MPH = 0.11 IN.WG.) the reduction ratio is:

  Reduction ratio = Space pressure / 0.11

- To minimize stratification and obtain a uniform temperature in the space, the supply air should not be more than fifteen degrees warmer than the room air.
- A large air circulation is required to maintain a low supply air temperature and to promote uniform heating throughout the space. The design value for the supply CFM should be based on an air circulation requirement that ranges between 1/2 and 7 air circulations per hour, depending on the size and tightness of the building.
- The outdoor air CFM through the burner should never be allowed to fall below 20 percent of the design CFM. (For safety, this feature should be built into the unit.)
- Barometric dampers should be installed in tight buildings that do not have continuous exhaust systems.
- The amount of heat that is added to the outdoor air that flows through the burner section is equal to the heat that is required to temper the outdoor air CFM plus the heat that is required to offset the envelope loads. This may result in a high temperature rise across the burner, but this is not a problem for a direct-fired burner.
- The unit must be equipped with a modulating burner that has a generous turn-down ratio (25:1).

The make up air unit has to operate whenever heat or make up air is required. The outdoor air damper and the return damper should modulate to maintain an air balance in the space during any operating condition. The dampers can be controlled by a static pressure sensor located in the space. The outdoor air damper should not be allowed to throttle the outdoor air CFM to less than 20 percent of the design CFM. The burner capacity can be controlled by a signal from a discharge air temperature controller that is reset by a temperature sensor located in the heated space.

Pressurized, ductless make up air/heating systems work because pressurization retards infiltration and causes the supply air to blanket the space as it flows from the make up air unit to the openings in the building envelope. Performance also depends on discharging a relatively large amount of slightly warm air into the room. (A ducted air distribution system and carefully selected supply outlets will be required for designs that do not pressurize the space, provide enough circulation or operate at moderate supply air temperature.) The procedure for designing a ductless system is summarized below.

Step 1) Refer to load calculation (Form N-1), and retrieve the values for the transmission load (*TL*) and the infiltration load (*IL*). Calculate the infiltration load reduction ratio (*RR*) and use this ratio to calculate the design heating load (*DHL*).

RR = Space pressure / 0.11
Design Heating Load = TL + RR x IL

Step 2) Calculate the CFM that is required to heat the space if the supply air temperature is 15°F warmer than the room temperature.

$$\text{Heating CFM} = \frac{\text{Transmission load}}{1.1 \times 15}$$

Step 3) Compare the maximum exhaust CFM with the CFM value from Step 2 and select the larger value. Use this value to calculate a *tentative* air circulation rate (*ACR*).

$$ACR = \frac{\text{Larger CFM} \times 60}{\text{Volume of Space}}$$

Step 4) Use Figure 7-16, on the next page, to estimate the air circulation rate that is required for uniform space temperatures. Note that this chart is based on an estimate of the building tightness and the volume of the space.

Step 5) Compare the air circulation rate from steps 3 and 4 and select the largest value. Use this value to determine the design CFM.

$$\text{Design CFM} = \frac{\text{ACIR} \times \text{Volume of Space}}{60}$$

Step 6) Determine if relief air openings are required. Relief will be required if the infiltration CFM is less than 20 percent of the design CFM from step 5. If relief air openings are required, use the following formula to calculate the relief air load (*RL*). In this formula, the TD value is equal to the winter design temperature difference.

RL = 1.1 x 0.20 x Design CFM x TD

Step 7) If relief openings are not required, the burner capacity can be estimated by summing the transmission load (*TL*), infiltration load (*IL*) and ventilation (outdoor air) load (*VL*). If relief openings are required, the burner capacity can be estimated by summing the transmission load (*TL*), relief air load (*RL*) and ventilation air load (*VL*).

Burner Capacity (no relief) = TL + IL + VL
Burner Capacity (with relief) = TL + RL + VL

Example — Make Up Air and Heating
Refer to the **Manual** N load information that is summarized below. Use the performance data, which is provided by Figure 7-15, on page 7-18, to select a direct-fired make up air unit to replace the exhaust air, to pressurize and to heat the space.

- Indoor design = 65°F
- Outdoor design temperature = 5 °F
- Transmission load = 325,000 BTUH
- Infiltration load = 198,000 BTUH
- Building tightness = average
- Infiltration CFM = 3,000
- Exhaust CFM = 15,000
- Ventilation (outdoor air) heating load = 990,000 BTUH
- Volume of heated space = 300,000 CU.FT.
- Pressure in space = 0.05 IN.WG.
- Exhaust CFM is constant

1) Design heating load (DHL)

TL = 325000 BTUH; IL = 198000 BTUH
RR = .05/0.11 = 0.45
DHL = 325000 + 0.45 x 198000 = 413200 BTUH

2) Heating CFM (based on a 15 °F TD)

$$\text{Heating CFM} = \frac{413200}{1.1 \times 15} = 25042$$

3) Compare CFM values, calculate the circulation rate

Heating CFM = 25042; Exhaust CFM = 15000

$$ACR = \frac{25042 \text{ CFM} \times 60}{300000} = 5.01$$

4) Air circulation rate from Figure Fig 7-16

Average @ 300000 CU.FT. = 2.5 to 3.0

5) Select the larger ACR, calculate the design CFM

$$\text{Design CFM} = \frac{5.01 \times 300000}{60} = 25{,}000 \text{ (Rounded)}$$

6) Check relief opening requirement

Infiltration percentage = 3000/25000 = 12% (not required)

7) Compute burner capacity

325000 + 198000 + 990000 = 1513000 BTUH

8) Summary

CFM required = 25000; Burner BTUH required = 1513000
Use 25000 CFM unit — no relief openings required

Air Circulation Rate for Even Temperature Distribution

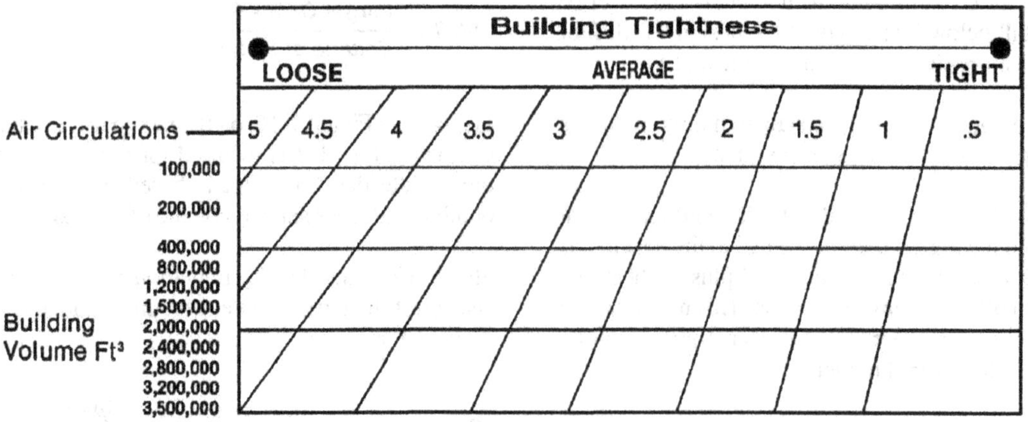

Figure 7-16

7-11 Hot Water Boilers — Gas, Oil and Propane

The following factors should be considered when a designer selects a hot water boiler. *These comments apply to any type of fossil fuel, hot water boiler — gas, oil or propane.*

Hot Water Boiler

Boiler Classification

Hot water, fossil fuel boilers are classified by pressure and temperature. Atmospheric boilers supply water at temperatures that range between 120°F and 212°F. Low pressure boilers operate at a pressure of 15 PSI or less and at temperatures that do not exceed 250°F. Medium and high pressure/temperature boilers operate at much higher pressures and temperatures.

Range of Application

Atmospheric and low pressure boilers are preferred for small commercial applications. They have the right combination of operating temperatures and pressures — and they do not require a licensed operator.

Hot water boilers are used for heating-only or for heating and ventilating applications — depending on the type of terminal equipment. Heating-only terminal equipment consists of finned-tube radiation and various types of unit heaters. The terminal equipment that is used for heating and ventilating applications includes unit ventilators and air handlers that are equipped with a hot water coil and an outdoor air damper. Refer to Sections 7-12, -13 and -14 for information about terminal equipment.

Manual N Load Information

Load calculations are always the basis for boiler sizing. **Manual N** load estimates are required for each piece of terminal equipment (Form N-1, line 21). An estimate of piping losses also is required. If applicable, a pickup load should be estimated. Note that the piping losses are zero if all the piping is located in the heated space. Otherwise, the piping losses must be estimated and added to the other **Manual N** heating loads.

Heating Capacity

The gross output capacity (BTUH) of the boiler must be equal to or greater than the design heating load. This design load should include all of the simultaneous heating loads that are associated with the terminal equipment, plus any piping losses that may apply. The boiler capacity may, under certain circumstances, be increased to allow for a pickup load. Figure 7-17 provides an example of hot water boiler, performance data.

Boiler No.	AGA Rating MBH		CBA	Net Ratings				USA Boiler H.P.	No. & Size of Flue Outlets	Breaching Dia. Inches	Recommended Chimney Sizes	
				Water		Steam						
	Input	Gross Output	Gross Output	Sq.Ft.	1=B=R MBH	1=B=R Sq. Ft.	1=B=R MBH				Round	Rect. In. x In. x Ft.
K-5006-B	401	316.7	320.8	1838	275.7	991	237.8	9.50	1-9"	9	10 x 15	12 x 12 x 15
K-5007-B	468	369.7	374.4	2144	321.7	1157	277.6	11.00	1-10"	10	10 x 15	12 x 12 x 15
K-5008-B	546	431.3	436.8	2504	375.7	1350	324.1	12.88	1-12"	12	12 x 15	12 x 16 x 15
K-5009-B	624	492.9	499.2	2852	427.8	1538	369.1	14.72	1-12"	12	12 x 15	12 x 16 x 15
K-5010-B	702	554.5	561.6	3211	481.7	1732	415.6	16.56	1-12"	12	12 x 15	12 x 16 x 15
K-5011-B	780	616.2	624.0	3571	535.7	1925	462.1	18.41	2-9"	14	15 x 15	16 x 16 x 15
K-5012-B	858	677.8	686.4	3930	589.6	2119	508.6	20.25	1-9",1-10"	14	15 x 15	16 x 16 x 15
K-5013-B	936	739.4	748.8	4290	643.5	2313	555.1	22.09	2-10"	14	15 x 15	16 x 16 x 15
K-5014-B	1014	801.0	811.2	4649	697.4	2507	601.6	23.93	1-10",1-12"	14	15 x 20	16 x 16 x 20
K-5015-B	1092	862.6	873.6	4997	749.6	2694	646.7	25.77	2-12"	14	15 x 20	16 x 16 x 20
K-5016-B	1170	924.3	936.0	5356	803.5	2888	693.2	27.61	2-12"	14	15 x 20	16 x 16 x 20
K-5017-B	1248	985.9	998.4	5716	857.4	3082	739.7	29.45	2-12"	14	15 x 20	16 x 16 x 20
K-5018-B	1326	1047.5	1060.8	6075	911.3	3276	786.2	31.29	2-12"	16	15 x 20	16 x 16 x 20
K-5019-B	1404	1109.1	1123.2	6434	965.2	3470	832.7	33.13	2-12"	16	18 x 20	16 x 20 x 20
K-5020-B	1482	1170.7	1185.6	6782	1017.4	3657	877.7	34.97	2-9",1-12"	16	18 x 20	16 x 20 x 20
K-5021-B	1560	1232.4	1248.0	7142	1071.3	3851	924.2	36.82	1-9",1-10" 1-12"	16	18 x 20	16 x 20 x 20
K-5022-B	1638	1294.0	1310.4	7501	1125.2	4057	973.7	38.66	2-10",1-12"	16	18 x 20	16 x 20 x 20
K-5024-B	1794	1417.2	1435.2	8220	1233.0	4486	1076.7	42.34	1-9",3-10"	18	18 x 20	20 x 20 x 20
K-5026-B	1950	1540.5	1560.0	8927	1339.1	4909	1178.3	46.02	3-10",1-12"	18	18 x 20	20 x 20 x 20

Figure 7-17

A pickup load allowance is not required if night set-back is not used or if the boiler is not shut down for extended periods of time. Even if set back is used, a pickup allowance will not be required unless all of the terminal devices are expected to operate at full capacity during the pickup transient. (For example, air handlers will not operate at full capacity if the outdoor air dampers remain shut during the pickup transient.)

Boiler manufacturers provide a "Net I-B-R" rating. This rating is equal to the gross output capacity minus an allowance for piping and pickup loads. Unfortunately there is no guarantee that the piping and pickup loads that are associated with any particular application will be equal to this Net I-B-R allowance.

Boiler output capacities are rated for operation at elevations that range from sea level to 2000 feet. Above 2000 feet reduce the output capacity by 4% for each 1000 feet above sea level.

Water Temperature Rise

The water temperature rise (TR) across a boiler depends on the output capacity (BTUH) and the GPM flowing through the boiler. This temperature rise can be computed by using the water-side equation.

$$TR = \frac{\text{Output BTUH}}{500 \times \text{GPM}}$$

The temperature rise across a boiler is limited by the potential for causing thermal shock. Cast iron boilers can tolerate a 25°F to 30°F temperature rise. For steel boilers, the temperature rise should be limited to 20°F. Note that the temperature rise across the boiler and the temperature drops across terminal equipment are not independent of each other. Refer to Section 5-4, page 5-5 for more discussion on this subject.

Leaving Water Temperature

Low pressure water boilers can supply water at temperatures that range from 120°F to 250°F. Within this range, the design temperature that is used for a particular application depends on temperatures that are used to select the terminal equipment. Temperatures below 120°F are not recommended because of problems associated with condensation in the boiler flue gas and reabsorption of air into the boiler water. Temperatures above 250°F require boiler operation at pressures that exceed 30 PSI.

Controls and accessories

Capacity control is important. Multiple boilers, two-stage burners or modulating burner controls are preferred for applications that have large heating load fluctuations. Study the manufacturer's literature as it pertains to controls and accessories. Specify the necessary burner operating controls, safety controls and trim (some examples are listed below).

- Burner capacity controls — on-off, two stage, modulating
- Ignition and flame supervision controls
- Master-submaster temperature reset controls
- High limit control
- Pressure temperature gauge
- Pressure relief valve
- Drain cock

Example

Use Figure 7-17, on the previous page, to select a boiler for a building that has three zones. All piping is in the heated space. Unit ventilator and air handler outdoor air dampers will be closed during recovery from night set-back. All zones will be maintained at 70°F during the occupied heating cycle. All zones will be set back to 60°F during the unoccupied heating cycle. All zones will be on the same time clock schedule.

Zone 1 — 124,000 BTUH (baseboard radiation)
Zone 2 — 326,000 BTUH (unit ventilators)
Zone 3 — 840,000 BTUH (four air handlers)
Total — 1,290,000 BTUH

Since the pipes are in the heated space, there are no piping losses. Since all of the zones are on the same operating schedule, there is no load diversity. Since the design load will not be imposed during the recovery cycle (no ventilation load), there is no need for an allowance for a pickup load. Therefore, the required capacity is simply equal to 1,290,000 BTUH. Select model K-5022-B, which has a gross output capacity of 1,310,400 BTUH.

7-12 Hot Water Coils

Hot water coils are used in terminal equipment packages — unit ventilators, cabinet fans and unit heaters, for example. This type of equipment is usually supplied with a single preselected coil, or possibly with one or two coil sizing options. In this type of application, the designer proceeds by selecting a standard size cabinet before checking the details that are associated with coil performance. Refer to Section 7-13 for more information about packaged equipment.

Hot water coils also are used in central packaged equipment — air handlers, rooftop heating and cooling units, packaged interior heating and cooling units, for example. This type of equipment is usually designed so that a hot water coil can be added to the unit as an option. In this case, the designer must select a coil that satisfies a specific set of performance requirements.

Hot Water Coil Selection
Even when accompanied by detailed instructions, it is not a simple task to use hot water coil performance data. The designer must refer to multiple tables and charts and because there are many parameters to consider — there are numerous coil configurations that will provide the required performance. A list of the parameters that affect coil performance is listed below.

- Coil face velocity (FPM)
- Entering water temperature (EWT)
- Water temperature drop (WTD)
- Water-side flow rate (GPM)
- Velocity of the water in the tubes (FPS)
- Number of rows (ROW)
- Number of fins per foot (FPF) or fins per inch (FPI)
- Equipped or not equipped with turbulators

The best way to select a coil is to list the design conditions and let the manufacturer or the manufacturer's sales representative use a computer to select the coil. At a minimum, these design conditions should be listed as follows . . .

- Heating CFM
- Heating capacity required for the space (MBTUH)
- Heating capacity required for reheat (MBTUH)
- Entering air temperature (EAT)
- Leaving air temperature (LAT)

Other primary design conditions are often specified by the designer. These conditions include . . .

- Entering water temperature (EWT)
- Water temperature drop (WTD)
- Water-side flow rate (GPM)

Some secondary design conditions also can be specified. These conditions, which are listed below, are optional. They serve to place limits or restrictions on some aspects of coil performance.

Maximum Face Velocity (FPM)
At the extreme, hot water coil face velocities can range from 200 to 1,500 FPM, but a range of 500 to 1000 FPM is more normal. In practice, face velocities tend to range between 500 and 800 FPM. In some cases, the hot water coil face velocity is selected to match the cooling coil face velocity.

Maximum Air-Side Pressure Drop
A limit on the air-side pressure drop is not usually specified, but one could be requested. The air-side pressure drop is affected by the face velocity, the number of rows and the fin spacing.

Maximum Water-Side Pressure Drop
A limit on the water-side pressure drop is not usually specified, but one could be requested. The water-side pressure drop is affected by the water velocity, the coil circuit arrangement, the number of rows and the turbulator option.

Maximum Number of Rows
A large number of rows increases the coil capacity for a given face area, but there is also an increase in the air-side pressure drop. A small number of rows has the opposite affect.

Minimum Fin Spacing
A close fin spacing increases the coil capacity for a given face area, but there is a corresponding increase in the air-side pressure drop and the coil is more likely to be obstructed by fibers and dust. A wider fin spacing has the opposite affect.

Circuiting
Parallel flow circuiting is recommended if the temperature of the air entering a hot water coil is expected to be less than 32°F.

Coil GPM and Water Temperature Drop
The design value for the GPM flowing through the coil depends on the heating load on the coil (BTUH) and the water temperature drop (WTD) across a coil. Usually, the designer specifies a design value for the temperature drop (typically between 20°F and 40°F) and calculates the corresponding GPM value. The water-side equation provides the relationship between these parameters.

$$GPM = \frac{\text{Heating Load BTUH}}{500 \times WTD}$$

Air-Side Pressure Drop
Part of the work that is associated with the coil selection process involves making an estimate of the air-side pressure drop. If the HVAC system designer selects the coil, the air-side pressure drop can be estimated by using tables and charts that are provided by the manufacturer. Or, if the manufacturer selects the coil, the HVAC system designer can get the air-side pressure drop information from the manufacturer. An example of air-side pressure drop data for cooling coils is provided at the bottom of page 6-16 by Figure 6-10, (similar information is published for hot water coils).

Water-Side Pressure Drop
Another task that is associated with the coil selection process involves making an estimate of the water-side pressure drop. If the HVAC system designer selects the coil, the water-side pressure drop can be estimated by using tables and charts that are provided by the manufacturer. Or, if the manufacturer selects the coil, the HVAC system designer can get the water-

side pressure drop information from the manufacturer. An example of cooling coil water-side pressure drop data is provided at the bottom of page 6-17 by Figure 6-11, (similar information is published for hot water coils).

Capacity Control
The capacity of a hot water coil can be controlled by using two-way (throttling) or three-way (diverting or mixing) valves to modulate the flow of water (GPM) through the coil. Capacity also can be controlled by face-and-bypass dampers or by adjusting the temperature of the water leaving the boiler. Frequently, both strategies — supply water temperature reset control and coil flow control — are combined into the same control system.

7-13 Hydronic Terminal Equipment

Unit heater, unit ventilator and fan-coil packages are used for heating-only, heating-ventilating and heating-cooling-ventilating applications. Heating-only equipment is furnished with a single hot water coil. Heating-cooling units could be equipped with a single coil that is used for both functions or with two coils, one for each function. Packages are available in a wide range of sizes. Each size has a limited number of coil options. In some cases, depending on the make and model, there may only be one type of coil available for a given size unit. This type of equipment is available with floor, wall and ceiling mounted cabinets, and depending on the number of coils — in two-, three- and four-pipe configurations.

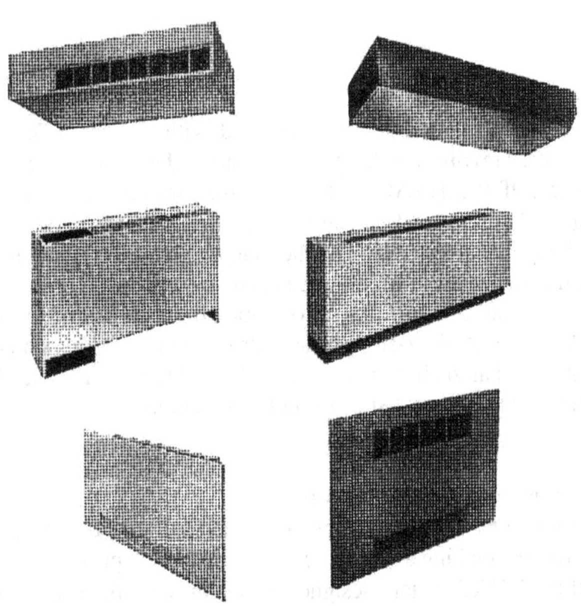

Hydronic Terminal Units

Air Distribution
Hydronic terminal equipment packages are not usually fitted with an air distribution system. When this is the case, the air motion in the room depends on the effectiveness of the discharge grille. This grille can be expected to create a ring shaped air pattern that rolls around the cross-section of the room. Therefore, a number of smaller units (as opposed to one larger unit), should be used to distribute heat over the length of a room.

Heating Coil Performance
The physical characteristics (size, rows, fins-per-inch etc.) of the coil are fixed by the manufacturer's design. Therefore the coil performance depends on the entering air temperature, the entering water temperature and the GPM flowing through the coil. An example of this type of coil performance data is provided by Figure 17-18, at the top of the next page. Note that this data is only valid if the entering air temperature is equal to 60°F and if the entering water temperature is equal to 180°F. However, performance at other conditions can be estimated by using the correction factors provided by the manufacturer — see Figure 17-19, on the next page.

Some hydronic terminal equipment packages are designed to accommodate a modest air distribution system. In this case the fan performance and the CFM that flows through the coil will be affected by external resistance that is associated with the duct system. The heating capacity will be reduced if the air flow is throttled by this resistance. In some cases, the loss of capacity is only a few percent. In other cases, the loss of capacity may be significant. In any case, the manufacturer should provide a procedure that allows the designer to estimate the change in capacity. An example of external resistance correction data is provided on page 7-26 by Figure 7-20.

> Each manufacturer publishes equipment performance data in a different format. Study manufacturer's literature carefully to familiarize yourself with the data tables and the corresponding design procedures.

Entering Air Temperature
The temperature of the air entering the coil depends on the outdoor air CFM. The following equation can be used to estimate the entering air temperature (EAT) associated with the heating cycle. (RAT is the return air temperature, *OAT* is the outdoor temperature and *Percent OA* is the percentage of outdoor air used for ventilation.)

$$MAT = RAT - Percent\ OA \times (RAT - OAT)$$

Entering Water Temperature
If the coil is just used for heating, the coil can be sized for any entering water temperature that is specified by the designer.

Hot Water Coil Capacity, 60°F EAT and 180°F EWT — Free Delivery

UNIT	HIGH SPEED AIR FLOW					LOW SPEED AIR FLOW				
	GPM	WPD	MBH	WTD	FAT	GPM	WPD	MBH	WTD	FAT
02	0.75	0.15	13.6	36.5	112.5	0.75	0.15	11.3	30.2	128.6
	1.00	0.24	15.3	30.9	119.2	1.00	0.24	12.4	24.9	135.3
	1.50	0.50	17.6	23.6	128.0	1.50	0.50	13.7	18.5	143.5
	2.00	0.83	19.0	19.3	133.5	2.00	0.83	14.5	14.7	148.3
	3.00	1.70	20.7	14.0	140.0	3.00	1.70	15.4	10.4	153.7
	5.00	4.20	22.3	9.0	145.0	5.00	4.20	16.2	6.6	158.6
	1.90	0.76	18.8	20.0	132.6	1.35	0.41	13.4	20.0	141.5
03	0.75	0.17	18.4	49.2	109.7	0.75	0.17	15.4	40.9	125.3
	1.00	0.28	20.9	42.0	116.4	1.00	0.26	17.0	34.0	132.1
	1.50	0.58	24.1	32.5	125.2	1.50	0.58	18.9	25.4	140.5
	2.00	0.97	26.2	26.5	130.8	2.00	0.97	20.1	20.2	145.5
	3.00	1.99	28.6	19.3	137.4	3.00	1.99	21.5	14.4	151.2
	5.00	4.90	31.0	12.6	143.7	5.00	4.90	22.7	9.1	156.3
	2.87	1.84	28.4	20.0	136.8	2.03	0.99	20.2	20.0	145.7
04	0.75	0.23	21.4	57.8	104.7	0.75	0.23	18.4	49.5	119.1
	1.00	0.38	24.6	49.9	111.4	1.00	0.38	20.6	41.7	126.1
	1.50	0.77	28.9	39.2	120.5	1.50	0.77	23.4	31.6	135.1
	2.00	1.28	31.7	32.3	126.4	2.00	1.28	25.1	25.5	140.7
	3.00	2.63	35.2	23.9	135.5	3.00	2.63	27.1	18.4	147.1
	5.00	6.49	38.5	15.7	140.5	5.00	6.49	28.9	11.8	153.0
	3.75	3.90	36.7	20.0	136.9	2.71	2.19	26.6	20.0	145.6
	1.46	0.73	28.6	40.0	119.9	1.07	0.42	21.0	40.0	127.6
06	0.75	0.27	27.2	73.1	100.8	0.75	0.27	23.7	63.6	114.6
	1.00	0.45	31.6	64.0	107.5	1.00	0.45	26.8	54.1	121.9
	1.50	0.93	37.6	51.1	116.8	1.50	0.93	30.9	41.7	131.4
	2.00	1.54	41.9	42.6	123.0	2.00	1.54	33.5	34.0	137.3
	3.00	3.16	47.1	31.9	130.7	3.00	3.16	36.6	24.8	144.4
	5.00	7.80	52.2	21.3	138.4	5.00	7.80	39.5	16.0	151.1
	5.38	8.89	52.6	20.0	139.4	3.88	4.97	38.2	20.0	148.1

WPD = Water Pressure Drop WTD = Water Temperature Difference FAT = Final Air Temperature

Figure 7-18

This temperature could vary from 120°F to 250°F, depending on other factors that are associated with the design. If the coil is used for cooling and heating, the coil size and flow rate will be set by the cooling requirement, and a specific value for the entering water temperature will be required to obtain the desired heating performance. These concepts are illustrated by the examples at the end of this section.

Flow Rate

If the coil is just used for heating, the coil can be sized for any flow rate that is compatible with the manufacturer's data. If the coil is used for cooling and heating, the coil size and flow rate will be set by the cooling requirement and the same flow rate value will apply to the heating performance. These concepts are illustrated by the examples at the end of this section.

Flow Rate and Temperature Drop

The heating capacity (BTUH), flow rate (GPM) and water-side temperature drop (WTD) are interrelated. If two values are known, the third can be calculated by using the water-side equation.

$$GPM = \frac{\text{Heating Capacity BTUH}}{500 \times WTD}$$

Water-Side Pressure Drop

A value for the water-side pressure drop will be required for the pump selection calculations. Manufacturers usually publish this information in their data tables, refer to Figure 7-18.

Capacity Correction Factors for Entering Conditions other than 60°F EAT and 180°F EWT

EAT °F	ENTERING WATER TEMPERATURES °F													
	95	100	110	120	130	140	150	160	170	180	190	200	210	220
50	.375	.417	.500	.583	.666	.75	.835	.917	1.00	1.08	1.16	1.25	1.33	1.42
60	.292	.333	.417	.500	.583	.666	.750	.835	.917	1.00	1.08	1.16	1.25	1.33
70	.208	.250	.333	.417	.500	.583	.666	.750	.835	.917	1.00	1.08	1.16	1.25
80	.125	.167	.250	.333	.417	.500	.583	.666	.750	.835	.917	1.00	1.08	1.16

Required capacity (BTUH) = Design Load x Correction Factor

Figure 7-19

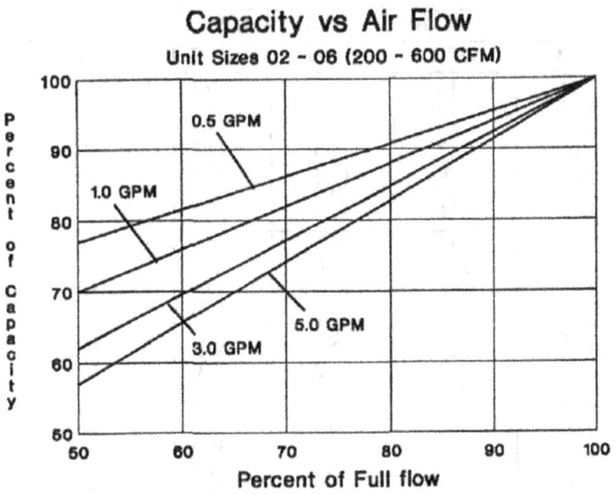

Figure 7-20

Air-Side Pressure Drop
Information about the air-side pressure drop across the coil is not required because the coil is an integral part of the unit. Manufacturers usually do not publish this information.

Fan Performance
Manufacturers usually publish values for the "free delivery" CFM at each fan speed. No other information is required unless the equipment is designed to accommodate a duct system. In these cases, a more comprehensive blower table is required to document the relationship between external resistance and the supply CFM.

Unit Sizing
Use the manufacturer's performance data and equipment selection procedures to find a unit size — and in some cases, a coil size — that provides the required amount of heat. If the unit is equipped with a multiple speed fan, use the performance data that corresponds to the medium speed setting. Note that the details of the sizing procedure will vary, depending on whether the equipment is designed for heating-only or for heating and cooling.

For a heating-only application, the unit and coil size can be based on the heating load. If the design values for the entering water and the entering air temperatures match the performance data temperatures, the unit size can be selected directly from the performance table, otherwise corrections will be required.

If the same unit is used for cooling and for heating, the unit size is based on the cooling requirements first and then the heating requirements are addressed. If the unit has a single coil, the coil performance should be matched to the cooling requirements . (Refer to Section 6-13 for information on the procedure that is used to size cooling coils.) If the unit has a separate heating coil, the coil size can be based on the heating load, as discussed in the previous paragraph.

Capacity Control
The heating capacity can be controlled by using two-way (throttling) or three-way (diverting or mixing) valves to modulate the flow of water (GPM) through the coil. If the unit has a multi-speed fan, the capacity also can be controlled by changing the fan speed. Or, capacity can be controlled by adjusting the temperature of the water leaving the boiler.

Accessories
Valve packages, damper operators and remote operating controls may be available as standard equipment or as an option. Various types of electrical and motor control equipment may also be available. Other accessory equipment could include filter options, discharge grilles, duct collars, wall sleeves and other installation accessories.

Example — Heating-Only
Use Figures 7-18 and 7-19 to select a unit for the following conditions. No cooling is required.

Heat load = 30,000 BTUH
Entering water temperature = 200 °F
Entering air temperature = 70 °F
Maximum water temperature drop = 25 °F

Figure 7-19 correction factor = 1.08
Required capacity from table = 30000/1.08 = 27778
Select size 03 at 3 GPM . . .

Capacity = 28,600 BTUH
WPD = 1.99 FT.
WTD = 19.3 °F

Example — Heating-Cooling Unit
Assume that a size 04 unit (at 3.75 GPM) was selected for cooling (by using the procedures that are presented in Section 6-13). Use Figures 7-18 and 7-19 to check the heating capacity of the unit at 180°F entering water temperature and determine the minimum allowable entering water temperature.

Unit size for cooling = 04
GPM for cooling = 3.75
Heat load = 35,000 BTUH
Entering air temperature = 70 °F

Figure 7-19 correction factor = 0.917
BTUH @ 180 EWT /70 EAT = 36700 x 0.917 = 32715

The entering water temperature must be hotter than 180°F to obtain the required capacity. Figure 7-19 can be used to determine the required temperature.

Water temperature Correction = 35332/32715 = 1.08
EWT required at 70 EAT = 200 °F

7-14 Hot Water Radiation

Hot water radiation equipment is used to add heat to the occupied space. In this capacity, finned-tube (baseboard) radiation and cabinet convectors can either be used as a primary source or as a secondary source of heat. Baseboard units, installed along an outside wall and under windows, produce fairly uniform temperatures from the floor to the ceiling and provide more comfort near the surface of a slab floor. Convection cabinets are commonly used near entrances and in vestibules. If ventilation is required, an independent source of heat is required to temper the flow of outdoor air before it enters the occupied space.

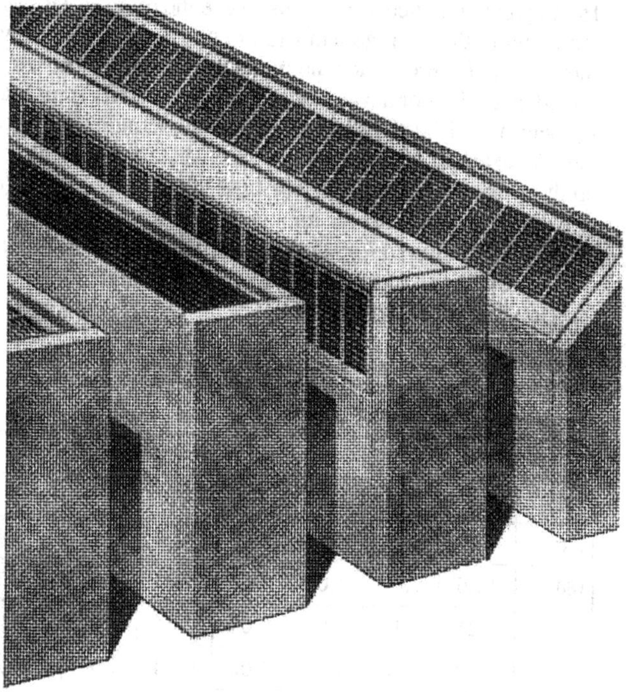

Baseboard Radiation

Air Distribution

Room air motion depends on the effectiveness of convection currents. These currents create a ring shaped air pattern that rolls around the cross-section of the room. Therefore, baseboard units can provide uniform heat throughout the room, providing they are sized so that the heating capacity is distributed along the length of wall — preferably the outside wall.

Baseboard Finned-Tube Performance

Heating performance depends on the physical characteristics (pipe size, rows, fins-per-inch, tube/fin material etc.) of a particular fixture. These parameters are not controlled by the HVAC system designer — they are established by the manufacturer's design. However, manufactures offer this type of equipment in a wide variety of configurations.

Ratings of Wall Fin Elements with Type S Enclosure

Element	Rows	Enclosure	Installed Height	Steam Capacity/Ft. 1 PSI at 65°F Air		Hot Water Capacity-Btu/Hr./Ft. at 65°F Air, Average Water Temp. (IBR Factor-Steam to Hot Water)					
				EDR Sq. Ft.	Btu/Hr.	220°F (1.05)	210°F (0.95)	200°F (0.86)	190°F (0.78)	180°F (0.69)	170°F (0.61)
Steel-1 1/4" Series-40 Fins-2 1/2" x 5 1/4" x .027" Finish-Black	1	12S	15 7/32"	5.45	1310	1380	1240	1130	1020	900	800
		16S	18 7/32"	5.60	1340	1410	1270	1150	1050	920	820
		20S	21 7/32"	5.70	1370	1440	1300	1180	1070	950	840
	2*	16S	18 7/32"	8.75	2100	2210	2000	1810	1640	1450	1280
		20S	21 7/32"	9.00	2160	2270	2050	1860	1680	1490	1320
	2**	20S	21 7/32"	9.40	2260	2370	2150	1940	1760	1560	1380
Steel-1 1/4" Series-52 Fins-2 1/2" x 5 1/4" x 0.27" Finish-Black	1	12S	15 7/32"	6.40	1530	1610	1450	1320	1190	1060	930
		16S	18 7/32"	6.90	1650	1730	1570	1420	1290	1140	1010
		20S	21 7/32"	7.25	1740	1830	1650	1500	1360	1200	1060
	2*	16S	18 7/32"	9.15	2200	2310	2090	1890	1720	1520	1340
		20S	21 7/32"	9.90	2380	2500	2260	2050	1860	1640	1450
	2**	20S	21 7/32"	10.25	2460	2580	2340	2120	1920	1700	1500
Copper-Alumium-1 1/4" Series-60 Fins-2 1/2" x 5 1/4" x .015" Finish-Black	1	12S	15 7/32"	7.65	1840	1930	1750	1580	1440	1270	1120
		16S	18 7/32"	8.55	2050	2150	1950	1760	1600	1410	1250
		20S	21 7/32"	9.35	2240	2350	2130	1930	1750	1550	1370
	2*	16S	18 7/32"	10.15	2440	2560	2320	2100	1900	1680	1490
		20S	21 7/32"	11.25	2700	2840	2570	2320	2110	1860	1650
	2**	20S	21 7/32"	11.45	2750	2890	2610	2370	2150	1900	1680
Copper-Alumium-1" Series-68 Fins-2" x 3 1/4" x .011" Finish-Black	1	10S	14 3/32"	4.70	1130	1190	1070	970	880	780	690
		14S	17 3/32"	5.30	1290	1330	1210	1090	990	880	790
		18S	20 3/4"	5.90	1410	1480	1340	1210	1100	970	860
	2*	14S	17 3/32"	6.15	1470	1540	1400	1260	1150	1010	900
		18S	20 3/4"	6.75	1620	1700	1540	1390	1260	1120	990
	2**	18S	20 3/4"	7.25	1740	1830	1650	1500	1360	1200	1060

*4" CENTERS **8" CENTERS

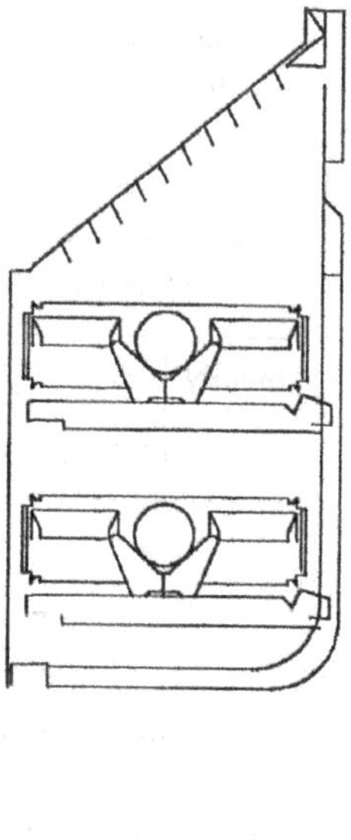

Figure 7-21

Heating performance also is sensitive to the room air temperature and to the average temperature of the water that flows through the fixture. Since the room temperature is predetermined by the indoor design conditions, the capacity is primarily determined by the entering water temperature and flow rate. An example of finned-tube performance data is provided on the preceding page by Figure 17-21. Note that this data is only valid if the room air temperature is equal to 65°F. A correction is required for other room air temperatures. Figure 17-22 provides an example of the correction factor data.

Baseboard Correction Table

	Entering Air Temperature						
EWT	50	55	60	65	70	75	80
170	1.26	1.18	1.10	1.00	0.93	0.87	0.79
180	1.25	1.17	1.09	1.00	0.94	0.88	0.82
190	1.21	1.14	1.08	1.00	0.94	0.88	0.82
200	1.20	1.13	1.07	1.00	0.94	0.90	0.84
210	1.18	1.12	1.06	1.00	0.95	0.89	0.84
220	1.15	1.10	1.04	1.00	0.93	0.89	0.84

Figure 7-22

Hydronic fixtures are usually tested and rated at a 3 foot-per-second (FPS) water velocity. Heating capacity is reduced at lower water velocities. Expect a capacity reduction of five percent or less, if the water velocity in the tubes is less than 3 FPS but greater than 1 FPS. Expect a five to ten percent capacity reduction if this velocity falls below 1 FPS. Table 7-23, below, provides an estimate of these capacity reduction percentages. Manufacturer's engineering data usually provides similar information. Note that there will only be a very slight increase in capacity if the velocity is greater than 3 FPS.

Capacity Reduction Percentage

Velocity	0.50	0.75	1.00	1.25	1.50	1.75	2.00	2.25	2.50
Percent	0.93	0.95	0.96	0.97	0.97	0.98	0.98	0.99	0.99

Figure 7-23

A small capacity adjustment also will be required if the fixture is not mounted at the height that is recommended by the manufacturer. Refer to the manufacturer's engineering data for more information about mounting height recommendations and mounting height correction factors.

Cabinet Convector Performance

Cabinet convector heating performance depends on the physical characteristics of the coil, but these parameters are established by the manufacturer's design. Performance also is sensitive to the room air temperature, the average temperature of the water flowing through the coil and the water temperature drop across the coil. An example of cabinet convector performance data is provided on the next page by Figure 17-24. Note that this data is only valid if the room air temperature is equal to 65°F. A correction is required for other room air temperatures. Figure 17-22 provides an example of the required correction factor data.

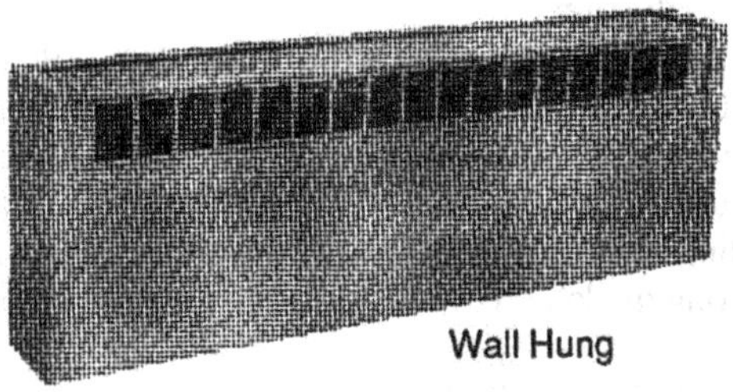

Wall Hung

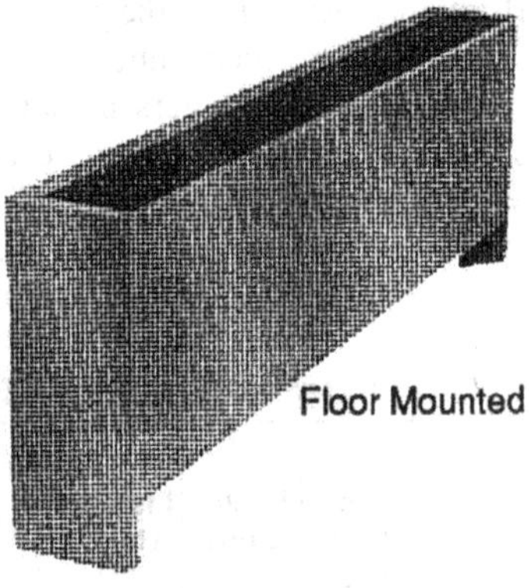

Floor Mounted

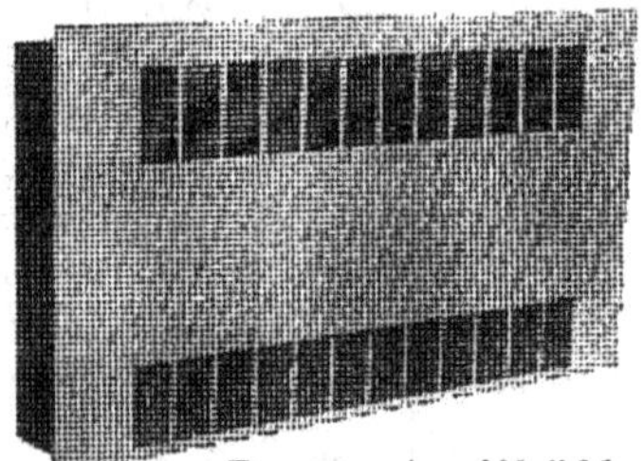

Recessed — Wall Mounted

Cabinet Convectors

Average Water Temperature

Hydronic heating fixtures can be sized for any average water temperature (AWT) that is listed in the manufacturer's performance data. These temperatures usually range between 160°F and 220°F. The average water temperature is equal to

HEIGHT	DEPTH	LENGTH																				
		26			32			38			44			50			56			62		
10	4	2.5	2.3	2.2	3.1	2.7	2.6	3.8	3.4	3.2	4.5	4.1	3.8	5.2	4.8	4.4	5.9	5.5	5.0	6.6	6.2	5.6
12	6	4.3	3.7	3.6	5.5	4.9	4.7	6.7	6.0	5.7	7.9	7.0	6.7	9.1	8.0	7.7	10.3	9.0	8.7	11.5	10.0	9.7
14	4	2.7	2.4	2.3	3.4	3.1	2.9	4.2	3.7	3.5	5.0	4.3	4.1	5.8	4.9	4.7	6.6	5.5	5.3	7.4	6.1	5.9
16	6	4.7	4.2	4.0	6.0	5.3	5.1	7.2	6.4	6.2	8.5	7.6	7.2	9.8	8.8	8.2	11.1	10.0	9.2	12.4	11.2	10.2
	8	6.1	5.4	5.2	7.6	6.8	6.5	9.2	8.2	7.9	10.9	9.7	9.2	12.5	11.1	10.6	14.1	12.6	12.0	15.7	14.0	13.4
18	4	2.9	2.6	2.5	3.8	3.4	3.2	4.5	4.0	3.9	5.2	4.8	4.6	5.9	5.5	5.3	6.6	6.2	6.0	7.3	6.9	6.7
20	6	5.1	4.5	4.3	6.4	5.7	5.5	7.8	6.9	6.6	9.1	8.1	7.8	10.4	9.3	9.0	11.7	10.5	10.2	13.0	11.7	11.4
	8	6.4	5.7	5.5	8.1	7.2	6.9	9.8	8.7	8.3	11.5	10.2	9.8	13.2	11.8	11.2	14.9	13.3	12.7	16.7	14.8	14.2
24	4	3.1	2.7	2.6	3.9	3.5	3.3	4.7	4.2	4.0	5.5	4.9	4.7	6.3	5.6	5.4	7.1	6.3	6.1	7.9	7.0	6.8
26	6	5.3	4.7	4.5	6.8	6.0	5.8	8.2	7.3	7.0	9.7	8.6	8.2	11.0	9.8	9.4	12.5	10.8	10.6	13.5	11.8	11.6
	8	6.8	6.0	5.8	8.6	7.6	7.3	10.4	9.3	8.9	12.2	10.9	10.4	13.9	12.4	11.9	15.7	14.0	13.4	17.6	15.6	14.9
30	4	3.6	3.2	3.0	4.3	3.8	3.6	5.0	4.4	4.2	5.7	5.1	4.9	6.4	5.7	5.5	7.3	6.5	6.2	8.1	7.2	6.9
32	6	5.7	5.1	4.9	7.1	6.4	6.1	8.7	7.7	7.4	10.2	9.1	8.7	11.7	10.4	9.9	13.1	11.7	11.2	14.7	13.0	12.5
	8	7.2	6.4	6.2	9.1	8.1	7.7	10.9	9.7	9.2	12.8	11.4	10.9	14.6	13.0	12.3	16.5	14.7	14.0	18.4	16.3	15.6

BTUH Heating Capacity
190 °F Average Water Temperature

Temperature Drops
Left column - 10 °F
Middle column - 20 °F
Right column - 30 °F

Figure 7-24

the mean value of the entering water temperature (EWT) and the leaving water temperature (LWT). The water temperature drop (WTD) also can be used to compute the average water temperature.

AWT = (EWT+LWT)/2
AWT = EWT - WTD/2 or LWT + WTD/2

Flow Rate and Temperature Drop

The heating capacity (BTUH), flow rate (GPM) and water-side temperature drop (WTD) are interrelated. If two values are known, the third can be calculated by using the water-side equation. In practice, water temperature drops usually range from 10°F to 30°F with 20°F being a popular choice.

$$GPM = \frac{\text{Heating Capacity BTUH}}{500 \times WTD}$$

Water Velocity

The velocity of the water in the tube depends on the inside diameter of the tube and on the GPM flowing through the tube. Manufacturer's usually publish tables or charts that provide velocity values. To insure quite operation, the water velocity should not exceed 5 FPS. Capacity will be degraded by 5 to 10 percent if the water velocity is less than 1 FPS. The following equation shows how the flow rate (GPM) and the nominal pipe diameter (D) can be used to make an estimate of the water velocity (FPS).

$$FPS = \frac{0.37 \times GPM}{D^2}$$

Circuiting

Multi-row convectors should be piped in series. Improper flow, low velocity, loss of capacity or air binding may result if the coils are piped in parallel.

Water-Side Pressure Drop

A value for the water-side pressure drop information will be required for the pump selection calculations. Manufacturers usually publish this information as part of their engineering and design data.

Unit Sizing

Use the manufacturer's performance data and the corresponding equipment selection instructions to find a configuration or size that provides the required heating capacity. If the design values for the average water temperature and entering air temperatures match the performance data temperatures, the unit size can be selected directly from the performance table, otherwise corrections will be required.

Capacity Control

The capacity of a heating device can be controlled by using two-way (throttling) or three-way (diverting or mixing) valves to modulate the flow of water (GPM) through the piping circuit — or capacity can be controlled by adjusting the temperature of the water leaving the boiler. Many types of convector cabinets are equipped with a simple manually operated, damper that throttles the flow of air across the coil.

> Convector performance will be seriously degraded if furniture disrupts the flow through the coil. Do not place file cabinets, bookcases or other barriers on the floor, directly in front of a baseboard convector. One way to work around this problem is to set the file cabinets (etc). on a platform that has about a four inch air space between the platform and the floor.

Accessories and Options

Valve packages, integral manually operated dampers and various cabinet options are usually available. Refer to the manufacturer's engineering data for a complete list of options and accessories.

Example — Baseboard Finned-Tube Heating
Heat is required for a room that has 30 feet of exterior wall. Use Figures 7-21, -22 and -23 to select a hot water baseboard radiation coil for this room if the following operating conditions apply.

Heating load = 60,000 BTUH
Entering water temperature = 200 °F
Water temperature drop = 20 °F
Room (entering) air temperature = 70 °F

1) Calculate the average water temperature

AWT = (200 - 20) / 2 = 190 °F

2) Figure 7-22 entering air temperature correction

Correction = 0.94 (@ 190 AWT and 70 EAT)

3) Calculate the required BTUH for 1 foot of coil at 65 EAT

$$BTUH/FT = \frac{60000}{30 \times 0.94} = 2128$$

4) Select Series 60 —1-1/4", two row coil (Figure 7-21)

Capacity per foot @ 190 EWT & 65 EAT = 2150 BTUH

5) Calculate the GPM

$$GPM = \frac{60000}{500 \times 20} = 6.0$$

6) Calculate the water velocity

$$FPS = \frac{0.37 \times 6}{(1.25)^2} = 1.42$$

7) Adjust the capacity for velocity

Velocity correction (Figure 7-23 interpolation) = 0.97
BTUH/Ft = 0.97 x 2150 = 2086

This capacity is a little short (2086 vs 2128 required), but it is close. (The capacity could be increased by raising the entering water temperature or reducing the temperature drop.)

Example — Wall Convector Sizing
Use Figures 7-22 and 7-24 to select a wall hung convector for a vestibule if the . . .

Heating load = 12,000 BTUH
Entering water temperature = 200°F
Water temperature drop = 20°F
Vestibule (entering) air temperature = 55°F

1) Calculate the average water temperature

AWT = (200 - 20) / 2 = 190 °F

2) Figure 7-22 entering air temperature correction

Correction = 1.14 (@ 190 AWT and 55 EAT)

3) Calculate the required BTUH at 65 EAT

$$BTUH/Ft = \frac{12000}{1.14} = 10526$$

4) Select a 44 x 26 x 8 inch cabinet (Figure 7-24)

Capacity @ 190 EWT & 65 EAT = 10900 BTUH

5) Calculate the GPM

$$GPM = \frac{12000}{500 \times 20} = 1.2$$

Section 8
Selecting Heat Pump Equipment

Heat pump equipment is unique because the same machinery provides cooling and heat. When both of these functions are required, which is usually the case, the equipment size is based on the sensible and latent cooling loads. Occasionally, heat pumps are just used for heat. When this is the case, the equipment size can be based on the heating load. Gross oversizing is not recommended for either type of application.

Manufacturer's engineering data must be used to certify equipment performance for the operating conditions that are expected to exist at the site. Equipment capacity must track changes in the cooling and heating loads during all operating conditions. Accessories, safety and operating controls must be specified. Fan performance must be specified. Water-side performance must be noted. Dimensions and clearances must be verified.

8-1 Sizing Requirements Based on the Cooling Load

Because heat pumps are normally expected to provide cooling, the cooling performance becomes the primary consideration for sizing the equipment. The standard practice is to size the heat pump to satisfy the cooling load and then check the heating balance point and the auxiliary heat requirement.

The procedure that is used to size heat pump equipment for cooling duty is identical to the procedure that is used to select ordinary cooling equipment. This procedure has been fully discussed in Section 6. Refer to Section 6-1 for general information about the factors and sizing limitations that affect equipment selection. Also see Sections 6-8 and 6-9, for detailed information that pertains to air-source (air-cooled) and water-source (water-cooled) equipment.

8-2 Minimum Cooling Capacity

Most commercial buildings require cooling during cool and in many cases, during cold weather. Usually, this means that the cooling equipment must operate at reduced capacity for extended periods of time. The capacity control requirements that are associated with heat pump equipment are the same as the capacity control requirements that are associated with ordinary cooling equipment. Refer to Section 6-2 for more information about this subject.

Also note that heat pumps do not have as many capacity control options as ordinary cooling equipment. Most manufacturers offer dual refrigeration circuits and some manufacturers offer one or more stages of unloading per circuit. Manufacturers do not mention hot gas bypass in their engineering catalogs. However hot gas bypass can be installed, providing that it is locked out during the heating cycle.

8-3 Heating Performance

The *instantaneous heating capacity* of an air-source heat pump depends on the temperature of the outdoor air. Figure 8-1 shows how the instantaneous capacity declines as the outdoor temperature drops. (The instantaneous capacity does not include an allowance for the capacity reduction that is caused by the defrost cycle.)

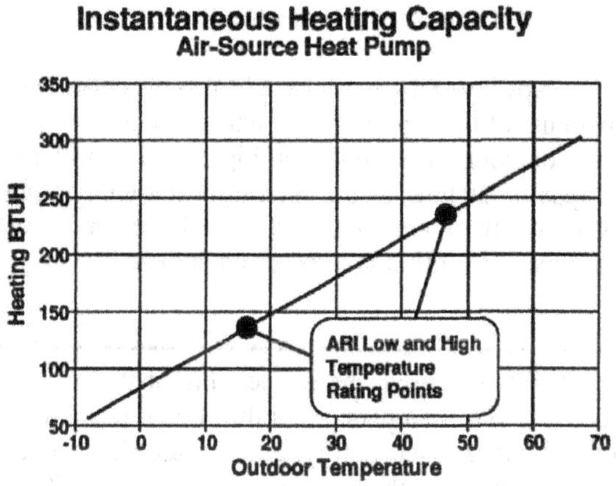

Figure 8-1

The *net heating capacity* of an air-source heat pump includes a correction for the cooling that occurs during the defrost cycle. Figure 8-2, at the top of the next page, shows how the defrost cycle affects the heating capacity as the outdoor temperature drops. The footnotes that accompany the manufacturer's performance tables usually specify whether the defrost penalty has been included in the performance data.

Note that the defrost penalty that is published in the manufacturer's data corresponds to a specific set of test conditions (outdoor temperature and humidity). However, the winter conditions that are associated with a particular city may be quite different than the conditions that are simulated in the test chamber. Therefore, the actual, on-site, defrost penalty will usually be different than the published defrost penalty. For example, the defrost penalty will be larger if a city that has a lot of wet weather when the outdoor temperature is

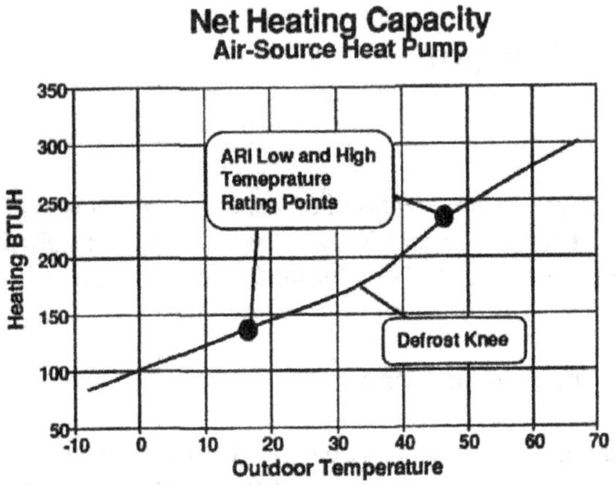

Figure 8-2

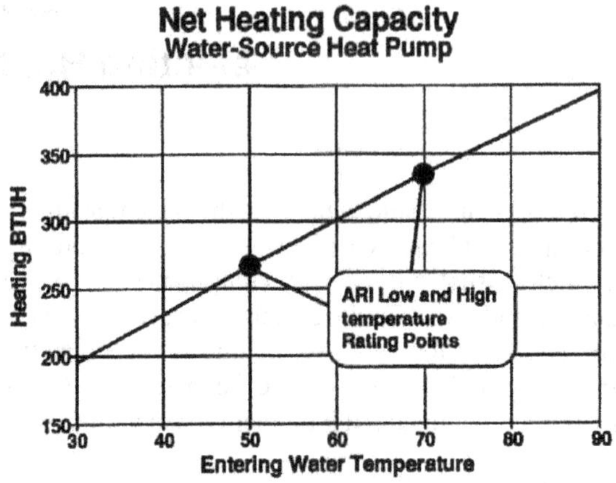

Figure 8-3

between 30°F and 40°F, but smaller in a city that has a very dry climate. (Refer to ACCA **Manual H**, page 39 for more information about defrost cycle penalties.)

The heating capacity of a water-source heat pump depends on the temperature of the supply water. In this case, the temperature of the outdoor air is irrelevant. Figure 8-3 shows how heating capacity declines as the entering water temperature drops. Note that there are no defrost cycle losses or defrost penalties associated with water-source equipment.

> When well water is used, the temperature of the supply water is equal to the temperature of the ground water. If a reclaim heat pump water-loop system is installed, the temperature of the water in the loop can vary from about 60°F to about 90°F. (Refer to Section 5-5 for more information about the water temperatures that are associated with a reclaim system.) When a earth-coupled water-loop system is installed, the temperature of the water in the loop can vary from below 30°F, during late winter, to more than 90°F, during late summer. (Refer to Section 5-7 for more information about the water temperatures that are associated with an earth-loop system.)

8-4 Balance Point Diagram

A balance point diagram can be used to compare the heat requirement of the structure with the heating capacity of the heat pump, at various outdoor temperatures. For residential applications, the heating load is assumed to decrease linearly from the **Manual J** load (at the winter design temperature), to zero load (at approximately 65°F). However, this simplified heating-cooling load-line model does not apply to commercial structures.

During the day, when they are occupied, commercial buildings usually have large internal loads. They also may have substantial solar gains. Therefore, in a commercial building, the cooling-heating changeover usually occurs when the outdoor temperature is well below the 65°F value that is used for residential work. Changeover temperatures between 20°F and 40°F are common and in some cases, they can be lower — even below zero! This means that, during cold weather, cooling will usually be required when the building is occupied.

During the night, when they are not occupied, commercial buildings may not have significant internal loads and there are no solar gains. In this case, the cooling-heating changeover could occur when the outdoor temperature is approximately equal to the indoor temperature. This means that, during cold weather, heat will be required whenever the building is not occupied.

Another important difference between commercial and residential applications is that the commercial heat pump is (usually) subjected to a sizable outdoor air (ventilation) load when the building is occupied. (If the outdoor air damper is closed when the building is not occupied, damper leakage will still cause an outdoor air load.) Therefore, the balance point diagram for a commercial heat pump application must be based on the space loads and the outdoor loads that are associated with the application.

Load lines can be used to evaluate the load characteristics of commercial spaces. Load lines also are used to evaluate the load that is imposed on the equipment. Refer to Sections 4-12, -15 and -16 for more information about the load-line diagrams that are associated with a conditioned space and the equipment.

Two separate balance point diagrams are required, one for an occupied building and one for an unoccupied building. These diagrams can be drawn by plotting the equipment perfor-

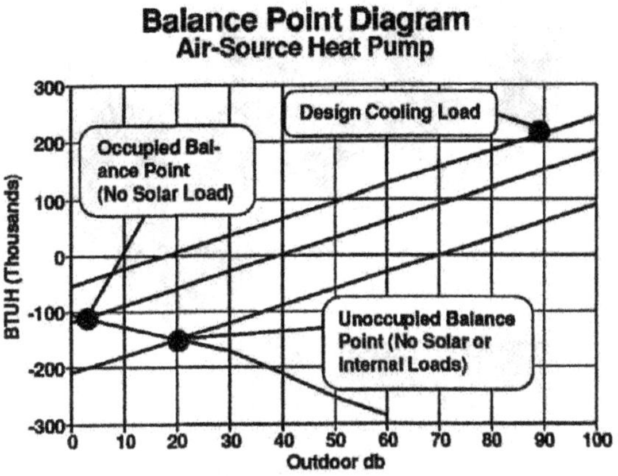

Figure 8-4

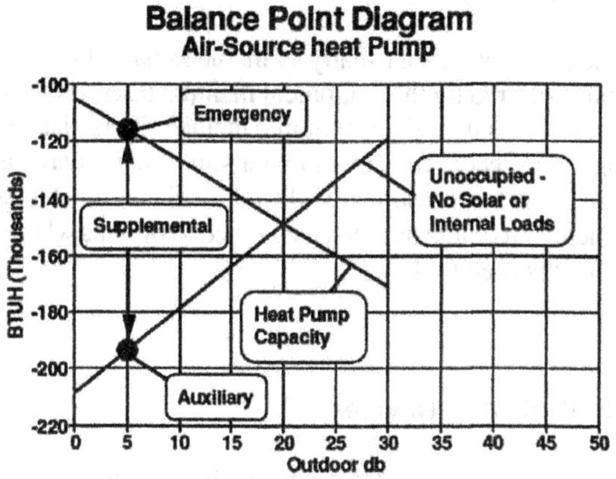

Figure 8-5

mance data on the load-line diagrams. Figure 8-4 provides an example of these balance point diagrams. The occupied balance point is based on the load line that represents all of the loads except the solar load. The unoccupied balance point is based on the load line that includes all the loads except the solar load and the internal load.

8-5 Supplemental, Emergency and Auxiliary Heat

When the outdoor air temperature is below the balance point, a second source of heat is required to make up for the difference between the building load and the capacity of the heat pump. The amount of supplemental heat that must be installed is determined by the differential that is associated with operation at the winter design temperature. Figure 8-5 shows how the balance point diagram is used to evaluate this differential.

In addition to furnishing heat during normal operation, supplemental heat can be used to heat the building if the refrigeration cycle equipment fails. However, the supplemental heat may not have enough capacity to maintain comfort during very cold weather. In this case, another increment of secondary heat will be required. This appendage is designated as emergency heat and it would be locked out during normal operating conditions. Figure 8-5 shows how the balance point diagram is used to evaluate the emergency heat requirement. Note that emergency heat may or may not be installed, depending on codes, utility regulations, owner preference and the designers judgment.

In this manual (and in other ACCA manuals) the term *auxiliary heat* refers to the total amount of electric resistance heat that is associated with the heat pump unit. Therefore, the auxiliary heating capacity is equal to the supplemental heat plus the emergency heat.

8-6 Fan Performance

Commercial heat pump equipment is usually available with a wide range of pulley (fan speed — RPM) and motor horsepower options. This allows the designer to select a fan configuration that will satisfy the flow (CFM) and pressure (IN.WG.) requirements that are associated with moving the air through the duct system. These requirements are similar to the requirements that are associated with ordinary cooling equipment. Also note that a return fan will be required if the unit is equipped with an economizer cycle, or a return duct system. Refer to Section 6-3 for more information about supply fan selection, return fans, variable volume fans and the cooling load that is associated with fan heat.

8-7 Water-Side Performance

The water-side flow rate (GPM) through — and the corresponding pressure drop (Feet of Water) across — water-source heat pump equipment is normally published in the manufacturer's performance tables. Usually, these tables document air-side and water-side performance for three or four different flow rates. Upon inspection, this data shows that the capacity of water-source equipment is not very sensitive to the water flow rate. Therefore, if a once-through well-water system is installed, the design should be based on the smallest recommended flow rate. This strategy will minimize pumping power. However, if an earth-coupled water-loop system is installed, the flow rate will have to be large enough to produce turbulent flow in the water-loop piping system. In any case, the flow rate and pressure drop information should be noted and recorded for future reference because it will be required for the pipe sizing and pump selection calculations. (Refer to Appendix 4 for more information about pipe design and pump selection.)

8-8 Air-Side Accessories

Air-side accessories are usually available as part of the package that is offered by the equipment manufacturer. Examples of accessories that might be required include filters, dampers, mixing boxes, economizers, fan inlet vanes, belt guards and so forth. Some accessories could be available as standard equipment, but others have to be included as options when the equipment is specified.

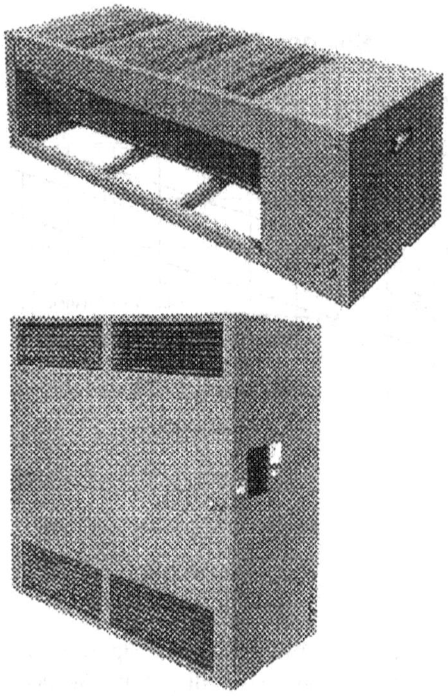

Split-System Heat Pump

8-9 Controls and Devices

Safety and operating controls are usually available as part of the package offered by the equipment manufacturer. Examples of the controls and devices that are peculiar to heat pump equipment include defrost controls, refrigerant reversing valves, accumulators, supplemental heat controls, outdoor thermostats and emergency heat switches. Other controls that are required include indoor thermostats, refrigerant-side pressure switches, motor protection, anti-short-cycle relays, motor starters, oil switches, crankcase heaters, low-temperature limit switches, fan and pump interlocks, damper interlocks, economizer controls, capacity control devices, central control panels and so forth. Some controls and safety devices could be available as standard equipment, but others have to be included as options when the equipment is specified.

Packaged Rooftop Heat Pump

8-10 Physical Dimensions and Arrangement

The designer must verify that the dimensions and the arrangement of the equipment is compatible with the available space. There must be ample room to service the equipment — including room to remove critical parts such as compressors, fans and coils. There also must be ample room to install aerodynamically efficient fan discharge and inlet connections.

8-11 Air-Source Heat Pump Equipment

The following paragraphs provide an example that illustrates air-source heat pump sizing and selection procedures. Initially, the heat pump is sized to satisfy the cooling load. This part of the selection procedure is identical to the procedure that is used to select ordinary cooling equipment. The balance point and the supplemental heat requirement are checked after the equipment has been sized for cooling. Finally, controls and accessories are specified.

Manual N Load Information

A **Manual** N load calculation is always required before the equipment can be selected. For this example, assume that this task has been completed and that the following information is available for a simple single-zone system. Note that the following cooling load data is the same as the data that was used for the packaged DX cooling equipment example (refer to Section 6-8 on page 6-6).

- Summer indoor design condition is 75°F & 50% RH
- Summer outdoor design condition is 92°F DB & 74°F WB
- Form N1, line 2, solar loads are 14,899 BTUH
- Form N1, line 4, internal loads are 90,590 BTUH
- Form N1, line 12, sensible load is 207,851 BTUH
- Form N-1, line 12, latent load is 42,853 BTUH
- Form N-1, line 8, supply CFM is 7145 (20°F TD)
- Winter indoor design temperature is 75°F

- Winter outdoor design temperature is +5°F
- Form N-1, line 16, heating load in space is 130,126 BTUH
- Form N-1, line 21, heating load at unit is 194,000 BTUH
- 900 CFM of outdoor air is required for ventilation
- Outdoor air damper leakage is 5 percent of supply CFM
- No losses associated with return air system

Psychrometric Analysis

The psychrometric chart should be used to determine the condition of the air entering the cooling coil and the acceptable range of conditions of the air leaving the cooling coil. Refer to page 6-6 for a detailed discussion of the psychrometric analysis that applies to this example. This analysis (Figure 6-1) indicates that the dry-bulb temperature of the air entering the coil will be 77°F and that the entering wet-bulb temperature will be 64°F.

Equipment Selection Data Summary

The designer should summarize the equipment performance requirements before referring to the manufacturer's performance data. This summary should include the indoor and outdoor design conditions, the calculated loads, the entering condition at the cooling coil and an estimate of the required cooling CFM. The information that is required for this example is summarized below.

Indoor Design = 75°F and 50% RH
Outdoor Design = 92°F DB & 74°F WB
Sensible Load = 207,851 BTUH (N-1, line 12)
Latent Load = 42,835 BTUH (N-1, line 12)
Entering dry-bulb = 77°F (from Figure 6-1)
Entering wet-bulb = 64°F (from Figure 6-1)
Approximate CFM = 7,145 (N-1, line 8)
Winter indoor temperature is 75°F
Winter outdoor temperature is +5°F
Heating load at unit is 194,000 BTUH

Cooling Performance Data

After the equipment performance requirements are tabulated, the manufacturer's performance data can be used to select the equipment. An example of comprehensive cooling performance data is provided by Figure 6-2 on page 6-7. This data, as explained on pages 6-8 and 6-9, indicates that a 25 TON unit, operating at 8750 CFM will satisfy the cooling load. This selection includes an allowance for 14,000 BTUH of fan heat.

Fan Performance

The fan performance must be compatible with the flow and pressure requirements of the duct system. The cooling equipment sizing procedure indicates that the fan must deliver 8750 CFM of supply air. A value for the resistance (pressure) that the fan must work against will be available after the duct system is designed. At this point, a fan performance table can be used to determine the fan speed (RPM) and the fan motor horsepower. Figure 6-3 on page 6-9 provides an example of a fan performance table.

Capacity Control — Cooling

A check of the manufacturer's literature (not shown) indicates that the 25 TON heat pump is equipped with two compressors and two independent refrigeration circuits. This means that the cooling capacity can be staged from 0 percent, to 50 percent, to 100 percent of the design capacity. The literature also indicates that an air-side economizer is available.

Load-line diagrams can be used to determine the minimum load on the compressors. (Load-line diagrams are discussed in Sections 4-9 through 4-12.) Point #1 on Figure 8-6 shows that the minimum sensible load on the compressors will be approximately equal to 46,000 BTUH. This load will occur at 55°F, just before the compressors are shut down and the economizer cycle is activated. This load represents 21 percent of the unit's 224,068 BTUH of sensible capacity. Hot gas bypass could be installed, but this would not be necessary if the manufacturer certifies that compressor cycling is an acceptable method of capacity control. (Refer to Appendix 5 for more information about controlling the capacity of DX equipment.)

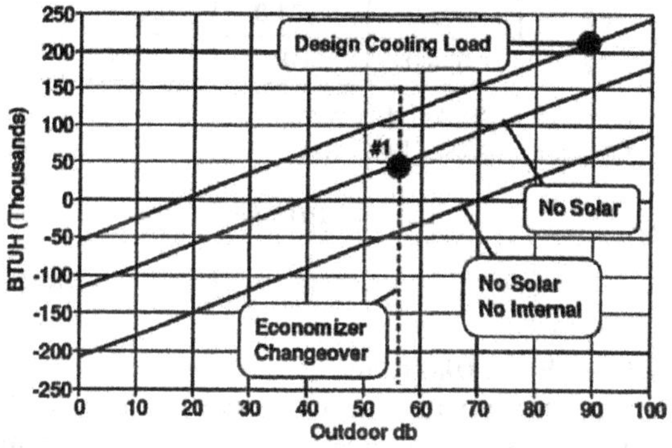

Figure 8-6

Heating Performance

Figure 8-7, at the top of the next page, provides an example of a heat pump manufacturer's heating performance data. The first footnote below the performance table indicates that the capacity reduction that is caused by the defrost cycle has been accounted for. (If the heating performance data does not include a defrost penalty, assume a 10 percent reduction in the heating capacities that are associated with the 17°F to the 47°F outdoor temperature range.) The second footnote indicates that the fan heat has not been added to the heating capacity values. When the equipment was selected for cooling, it was determined that the supply fan will add (about) 14,000 BTUH of heat. Therefore, the heating capacity of the heat pump will be about 14,000 BTUH larger than the table values.

HP-25 Heating Capacities 8750 CFM				
OD Temp. °F	Heating Capacity (MBh) at Indicated Indoor Dry-Bulb Temp.			
	60	70	75	80
-18	62.5	61.4	60.8	60.2
-13	73.7	72.3	71.6	70.9
-8	84.8	83.3	82.5	81.7
-3	96.0	94.2	93.3	92.4
2	107.0	105.0	104.0	103.0
7	118.0	116.0	115.0	114.0
12	129.0	127.0	126.0	125.0
17	141.0	138.0	137.0	135.0
22	152.0	149.0	147.0	146.0
27	163.0	160.0	158.0	157.0
32	174.0	171.0	169.0	167.0
37	190.0	187.0	185.0	183.0
42	216.0	211.0	209.0	207.0
47	241.0	236.0	234.0	231.0
52	257.0	252.0	250.0	247.0
57	274.0	269.0	266.0	263.0
62	291.0	285.0	282.0	279.0
67	307.0	301.0	298.0	295.0
72	324.0	318.0	315.0	311.0

1) Capacities adjusted for defrost cycle.
2) Capacities not adjusted for fan heat.

Figure 8-7

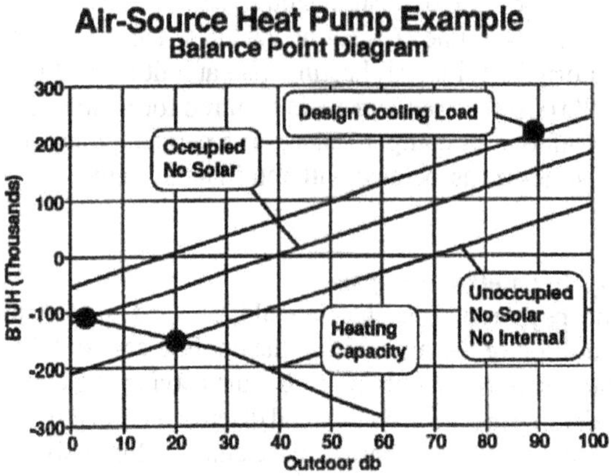

Figure 8-8

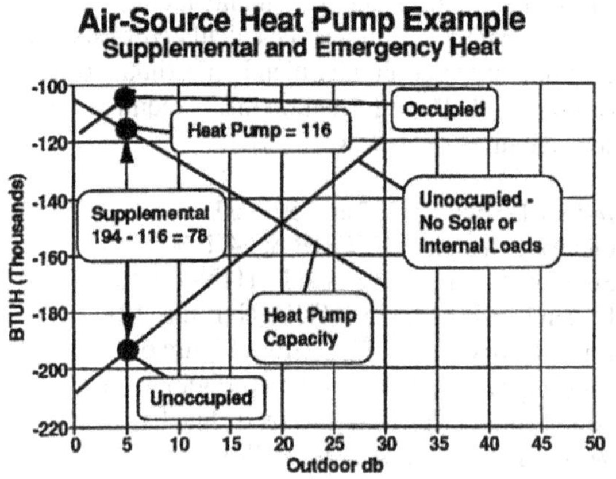

Figure 8-9

Balance Point Diagrams

The balance points that are associated with an occupied building and an unoccupied building can be determined by plotting the equipment performance data on the (equipment) load-line diagram. Figure 8-8 shows this work. In this case — with no credit for solar gains or fan heat — the occupied balance point occurs at 3°F. If the outdoor air damper is not closed (open to the minimum position to satisfy the 900 CFM ventilation requirement), the unoccupied balance point occurs at 20°F. However, the unoccupied balance point could be lower if the outdoor air damper is closed when the building is unoccupied. (In this example, there will still be 5 percent damper leakage.)

Supplemental Heat

At the 5°F winter design temperature, the balance point diagram (Figure 8-9) shows that the heat pump capacity is equal to 116,000 BTUH, the occupied building load is equal to 105,000 BTUH and the unoccupied building load is equal to 194,000 BTUH. Since it is the worst case, the supplemental heating capacity should be based on the unoccupied condition. However, the unoccupied load includes a 900 CFM ventilation load. If the outdoor air dampers are closed, this load should be replaced by a damper leakage load, which would be based on 5 percent (for this example) of the 8,750 supply CFM. The following calculations show that 13.16 KW of supplemental heat is required.

Ventilation load = 1.1 x 900 x (70 - 5) = 64350 BTUH
Leakage load = 1.1 x 0.05 x 8750 x (70 - 5) = 31281 BTUH
Damper load adjustment = 64350 - 31281 = 33069 BTUH
Supplemental = 78000 - 33069 - 116000 = 44931 BTUH
Supplemental KW = 44931/3413 = 13.16

Emergency Heat

At the 5°F winter design temperature, an additional 116,000 BTUH of electric resistance heat is required to satisfy the design heating load if the heat pump is out of service. This represents 34 KW of emergency heat that would be activated by a manual switch.

Controls and Accessories
The heat pump equipment must be specified with the proper controls and accessories. A search of the manufacturer's product literature indicates that the following controls and accessories are either standard equipment or available as options. All the following items will be required with the exception of the hot gas bypass, head pressure control and low ambient cooling options.

Refrigeration Side
Two compressors — standard
Two independent refrigeration circuits — standard
Time-temperature defrost control — standard
Outdoor thermostat(s) — option
Low ambient cooling to 20°F — standard
Low ambient cooling to 0°F — option
Hot gas bypass — option
Suction accumulator — standard
Motor overload protection — standard
High- and low-pressure cutouts — standard
Head pressure control — option
Oil pressure switch — standard
Anti-short-cycle relay — standard
Crankcase heater — standard
Liquid line filter-dryers — standard

Outdoor Air and Filters
Low-leakage dampers — option
100% economizer — option
Dampers controlled by space temperature — standard
Minimum position setting — standard
Exhaust air fan — option
High efficiency 2" filter — standard

Electronic Control Panel
Cool-heat-auto-off room thermostat
Heating-cooling deadband 3°F to 30°F
Night set-back
7-day clock, 18 set-back periods
Cooling lockout
Signal lights (operating mode, malfunction alert)
Low-limit on supply temperature
Pneumatic interface

Control Cycles
Control cycles must be specified for all of the possible operating conditions. The following control sequences provide an example of a scheme that could be used for a single-zone application.

Single-Zone Mode — Day Cycle
Mechanical cooling shall be on-line during the day cycle, provided the outdoor temperature (or enthalpy) is above the economizer setpoint. The compressor(s), refrigeration circuit(s) and the cooling capacity shall be controlled in response to a signal from the room thermostat. The outdoor air damper shall be opened to the minimum position. The supply fan and the exhaust fan shall run continuously. Compressor and fan operation shall be subject to supervision by safeties and interlocks.

Mechanical cooling shall be off-line when the outdoor temperature (or enthalpy) is below the economizer setpoint. The outdoor air, return air and exhaust air dampers shall be controlled in response to a signal from the room thermostat. A minimum position device shall prevent the outdoor air damper from closing past the minimum position setting. The supply fan and the exhaust fan shall run continuously. Fan and damper operation shall be subject to supervision by safeties and interlocks.

Changeover to reversed cycle heat shall be provided in response to a signal from the room thermostat. Supplemental heat shall be activated by the second stage of the room thermostat (subject to lockout by an outdoor thermostat). The economizer outdoor air damper should be at the minimum position when the unit operates in the heating mode. The supply fan and the exhaust fan shall run continuously. Compressor, fan and damper operation shall be subject to supervision by safeties and interlocks.

Single-Zone — Night Cycle
In the cooling mode, the cooling equipment shall be off-line during the night cycle. The outdoor air and exhaust dampers shall be closed. The supply fan and the exhaust fans shall be off.

In the heating mode, the reversed cycle heat shall be controlled in response to a signal from the room thermostat. Supplemental heat shall be activated by the second stage of the room thermostat (but subject to lockout by an outdoor thermostat). The outdoor air and exhaust air dampers shall remain closed. The supply fan shall run intermittently, cycling with the call for heat. Compressor, fan and damper operation shall be subject to supervision by safeties and interlocks.

8-12 Water-Source Heat Pump Equipment

The procedure that is used for selecting and sizing water-source heat pump equipment is almost identical with the procedure that is used for air-source equipment. (Refer to Section 8-11 for a comprehensive discussion of the air-source heat pump selection procedure.) The primary difference between water-source and air-source equipment is that the water-source performance data is based on the entering water temperature and flow rate (GPM), while the air-source performance data is based on the outdoor air temperature. Figure 8-10, at the top of the next page, provides an example of water-source heat pump performance data. Note that the cooling capacity and the heating capacity depend on the entering water temperature and the condenser GPM — there is no reference to the outdoor air temperature.

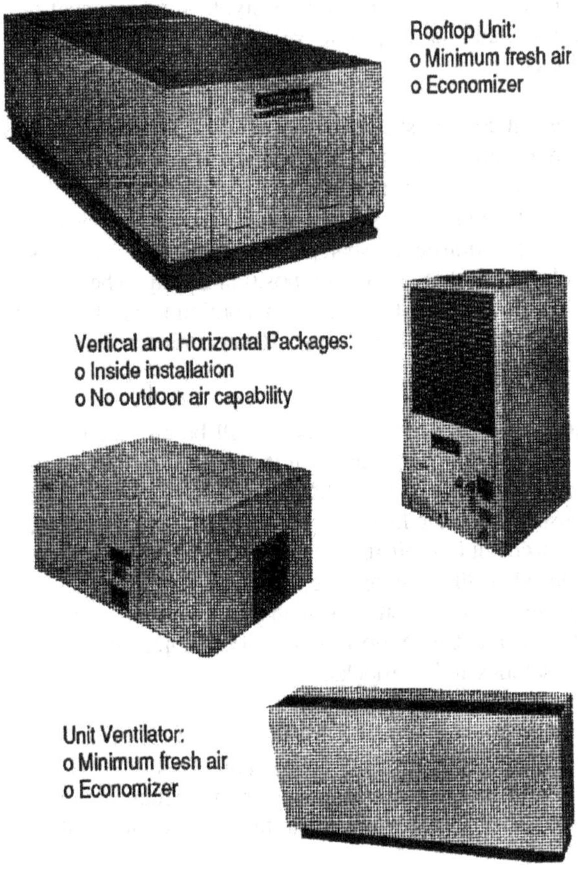

Water-Source Heat Pump Equipment

Water-Source Heat Pump Capacities

Cooling (75 EDB, 63 EWB) and Heating (70 EDB)

EWT	WPD	TC	SC	HC
32 Gallons per Minute; 8000 CFM				
30	7.0	290	231	195
50	5.9	270	222	267
70	5.5	244	207	335
90	5.2	219	194	396
110	4.0	193	180	NA
48 Gallons per Minute; 8000 CFM				
EWT	**WPD**	**TC**	**SC**	**HC**
30	14.2	299	236	206
50	12.0	279	227	279
70	11.1	252	211	351
90	10.5	226	198	411
110	10.0	199	184	NA
64 Gallons per Minute; 8000 CFM				
EWT	**WPD**	**TC**	**SC**	**HC**
30	22.0	308	241	213
50	28.6	287	231	287
70	17.3	260	216	361
90	16.3	233	201	420
110	15.5	205	188	NA

EWT (Entering Water Temperature)
WPD (Water Pressure Drop - Feet of Water)
TC = Total Cooling capacity — 1000's BTUH
SC = Sensible Cooling capacity — 1000's BTUH
HC = Heating Capacity — 1000's BTUH
Fan heat not included in capacity values.

Figure 8-10

Load Information and Psychrometric Analysis

The same set of load calculations can be used for air-source or water-source equipment. The design conditions that are associated with the indoor coil — namely the entering dry- and wet-bulb temperatures and the CFM across the DX coil — are exactly the same as the conditions that apply to air-source equipment. Therefore, the psychrometric analysis that could be used for selecting air-source equipment also would be used for selecting water-source equipment. For this example, assume that the load calculations and the psychrometric analysis have been completed and that the following information is available for a single-zone (rooftop) system.

- Summer indoor design condition is 75°F & 50% RH
- Summer outdoor design condition is 92°F DB & 74°F WB
- Form N1, line 2, solar loads are 14,899 BTUH
- Form N1, line 4, internal loads are 90,590 BTUH
- Form N1, line 12, sensible load is 207,851 BTUH
- Form N-1, line 12, latent load is 42,853 BTUH
- Form N-1, line 8, supply CFM is 7145 (20°F TD)
- Winter indoor design temperature is 70°F
- Winter outdoor design temperature is +5°F
- Form N-1, line 16, heating load in space is 130,126 BTUH
- Form N-1, line 21, heating load at unit is 194,000 BTUH
- 900 CFM of outdoor air is required for ventilation
- Outdoor air damper leakage is 5 percent of supply CFM
- No losses associated with return air system

Entering Water Temperature

The design value for the temperature of the supply water is an important design parameter that affects equipment capacity. Water could be supplied by a well, a closed "reclaim" water loop or a closed water loop that is buried in the ground. For the following example assume that 50°F well water is circulated through the heat pump.

The temperature of well water can vary from below 40°F to more than 60°F, depending on local conditions. The temperature of the well water stays reasonably constant throughout the year. Use the same temperature to determine the cooling capacity and the heating capacity.

The temperature of the water in a closed reclaim-loop will vary from about 60°F to about 90°F. Use the lower temperature to determine the heating capacity and the higher temperature to determine the cooling capacity.

The temperature of the water in a closed earth-coupled loop can vary from below 30°F to more than 90°F, depending on local conditions. Use the lower temperature to determine the heating capacity and the higher temperature to determine the cooling capacity.

Equipment Selection Data Summary
The designer should summarize the equipment performance requirements before referring to the manufacturer's performance data. This summary is similar to the summary that would be used for air-cooled equipment, except the entering water temperature is used in place of the outdoor air temperature. An example of the information that is required to size the (well water) heat pump is summarized below.

Indoor Design = 75°F and 50% RH
Entering Water = 50°F DB (summer)
Sensible Load = 207,851 BTUH
Latent Load = 42,853 BTUH
Entering dry-bulb = 77°F
Entering wet-bulb = 64°F
Approximate supply CFM is 7145
Indoor Design = 70°F (winter)
Entering Water = 50°F db (winter)
Heating load at unit is 194,000 BTUH
900 CFM of outdoor air is required for ventilation
Outdoor air damper leakage is 5 percent of supply CFM
No losses associated with return air system

Cooling Performance Data
The water-source heat pump performance data should provide information on total capacity, sensible capacity and heating capacity. In addition, the performance should be tabulated for a full range of entering air conditions, entering water temperatures and supply CFM values. Footnotes may indicate that adjustments are required to compensate for operating conditions or system loads that are not accounted for in the basic table — a fan heat adjustment, a different entering wet-bulb or a different dry-bulb temperature, for example.

Each manufacturer publishes equipment performance data in a different format. Take time to become familiar with the data. Also note that the data may have to be corrected for the effect of altitude.

Water-Source Equipment is Sized for Cooling
Oversized water-source equipment can lead to the same humidity control and short cycling problems that are associated with oversized air-source equipment. Therefore, water-source equipment size should be based on the cooling load. If additional heating capacity is desired, the unit can be oversized (for cooling) by about 25 percent of the total capacity that is required.

After the equipment performance requirements have been tabulated, the manufacturer's performance data can be used to select the equipment. Carefully search the manufacturer's performance data for the unit that provides the closest match to the cooling performance requirements. Since water-source equipment is not very sensitive to flow rate (GPM), use the lowest flow rate that is compatible with the type of supply water system (as explained in Section 8-7).

An inspection of the example performance data (Figure 8-10, page 8-8) shows that the documented entering conditions (75 EDB and 63 EWB) do not match the desired entering conditions (77 EDB and 64 EWB). However, these differences are small and the sensible and latent capacities of the unit are about right. For this example, assume that the unit will have adequate capacity (at 32 GPM) after the cooling performance is corrected for the entering conditions and fan heat.

Capacity Control and Economizer Cycle
The capacity control requirements that are associated with water-source heat pump equipment are the same as the capacity control requirements that are associated with air-source heat pumps or with ordinary cooling equipment. Load-line diagrams can be used to determine the minimum load on the compressor. (Load-line diagrams are discussed in Sections 4-9 through 4-12. Also, refer to Appendix 5 for more information about controlling the capacity of DX equipment.)

Economizers are used with water-source equipment because they are more efficient than mechanical cooling and because they eliminate the problems that are associated with operating compressors during cold weather. Refer to manufacturer's literature for capacity control and economizer options.

Air-Side Performance Data
Air-side performance data consists of a blower table, and possibly a table or equation that provides information about a fan heat adjustment. (Refer to Figures 6-3 and 6-4 on page 6-9.) Other air-side tables may provide information on the air-side pressure drops across components that were not accounted for in the basic blower table — coil, filter or mixing box pressure drops for example. (Refer to ACCA **Manual Q**, Section 9, for more information about how devices and accessories affect the duct sizing and fan selection calculations.)

Water-Side Performance Data
Information about the water flow rate (GPM) through, and the corresponding pressure drop (Feet of Water or PSI) across, water-source heat pump equipment is usually included in the

manufacturer's performance data. This flow rate and pressure drop information should be recorded for future reference because it will be required for the pipe sizing and pump selection calculations. (Refer to Appendix 4 for more information about pipe design and pump selection.)

Balance Point Diagrams

The balance points that are associated with an occupied building and an unoccupied building can be determined by drawing the heating capacity line on the (equipment) load-line diagram. Figure 8-11 shows this work. In this example, even without credit for fan heat, the capacity of the heat pump is greater than the heating loads that are associated with the occupied or the unoccupied conditions. Therefore (in this particular example), the balance points are not relevant.

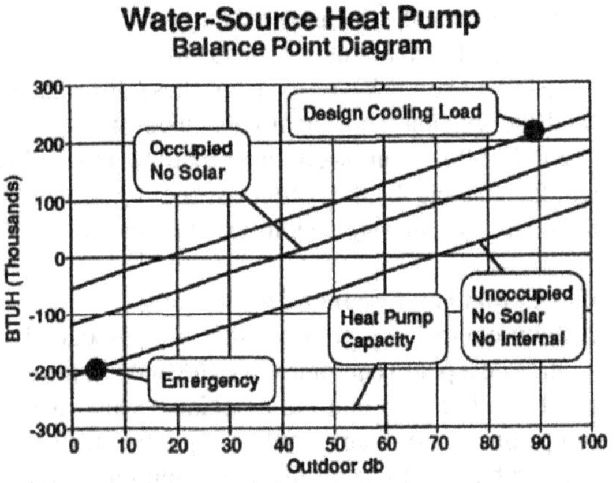

Figure 8-11

Supplemental Heat

At the 5°F winter design temperature, the balance point diagram (Figure 8-11) shows that even without credit for fan heat, the 267,000 BTUH capacity of the heat pump is greater than the heating loads that are associated with the occupied or the unoccupied conditions. Therefore, in this example, no supplemental heat is required. However, this is not always the case. Applications with relatively small cooling loads and large heating loads may require supplemental heat.

Emergency Heat

At the 5°F winter design temperature, the balance point diagram (Figure 8-11) shows that the unoccupied heating load is equal to 194,000 BTUH. Since it is the worst case, the emergency heating capacity should be based on the unoccupied condition. However, the unoccupied load includes a 900 CFM ventilation load. If the outdoor air dampers are closed, this load should be replaced by a damper leakage load, which would be based on 5 percent of the supply CFM. The following calculations show that 46.37 KW of emergency heat that would be required if the heat pump is out of service.

Ventilation load = 1.1 x 900 x (70 - 5) = 64350 BTUH
Leakage load = 1.1 x 0.05 x 8000 x (70 - 5) = 28600 BTUH
Damper load adjustment = 64350 - 28600 = 35750 BTUH
Adjusted heat load = 194000 - 35750 = 158250 BTUH
Emergency Heat = 158250 / 3413 = 46.37 KW

Controls and Accessories

The equipment must be specified with the proper controls and accessories. Manufacturer's product literature will indicate if the following controls and accessories are available as standard equipment or as options.

Capacity control options
Motor overload protection
High- and low-pressure cutouts
Low water-temperature cutout
Anti-short-cycle relay
Liquid line filter-dryers
Crankcase heater
Low-leakage dampers
Minimum outdoor air damper
100% economizer
Return and exhaust air fan options
Filter options
Water-side valves, hoses and fittings
Cool-heat-auto-off control panel
Air-side interlocks and safeties
Water-side interlocks and safeties

Control Cycles

Control cycles must be specified for all of the possible equipment operating conditions. The following control sequences provide an example of a scheme that could be used for a single-zone application.

Single-Zone Mode—Day Cycle

Mechanical cooling shall be on-line during the day cycle, provided the outdoor temperature (or enthalpy) is above the economizer setpoint. The compressors, refrigeration circuits and cooling capacity shall be controlled in response to a signal from the room thermostat. The outdoor air damper shall be opened to the minimum position. The supply fan and the exhaust fan shall run continuously. Compressor, fan and pump operation shall be subject to supervision by safeties and interlocks.

Mechanical cooling shall be off-line when the outdoor temperature (or enthalpy) is below the economizer setpoint. The outdoor air, return air and exhaust air dampers shall be controlled in response to a signal from the room thermostat. A minimum position device shall prevent the outdoor air damper from closing past the minimum position setting. The supply fan and the exhaust fan shall run continuously. Fan and damper operation shall be subject to supervision by safeties and interlocks.

Changeover to reversed cycle heat shall be provided in response to a signal from the room thermostat. Supplemental heat shall be activated by the second stage of the room thermostat (subject to lockout by an outdoor thermostat). The economizer outdoor air damper should be at the minimum position when the unit operates in the heating mode. The supply fan and the exhaust fan shall run continuously. Compressor, fan, pump and damper operation shall be subject to supervision by safeties and interlocks.

Single-Zone—Night Cycle
In the cooling mode, the cooling equipment shall be off-line during the night cycle. The compressor and the water pump shall be shut off. The outdoor air and exhaust dampers shall be closed. The supply fan and the exhaust fan shall be switched off.

In the heating mode, the reversed cycle heat shall be controlled in response to a signal from the room thermostat. Supplemental heat shall be activated by the second stage of the room thermostat (subject to lockout by an outdoor thermostat). The outdoor air and exhaust air dampers shall remain closed. The supply fan shall run intermittently, cycling with the compressor on a call for heat. Compressor, fan, pump and damper operation shall be subject to supervision by safeties and interlocks.

Section 9
Radiant Heat

Infrared heating equipment can be used to partially or to completely heat a space. Area or spot heating is appropriate if occupancy is restricted to a relatively small part of a large room or building. The entire room or building must be heated if the occupants are not confined to a specific area. Area and spot heating reduces operating costs because radiant heat can provide comfort at a reduced indoor air temperature. Radiant heat also provides an efficient way to maintain a uniform design temperature in an entire room or building.

9-1 Air Motion

Radiant heat does not create much air motion in the room. In fact, excessive air motion, say velocities that exceed 50 FPM (1/2 MPH) reduce the sensation of comfort that is provided by radiant heating systems. Additional radiant capacity may be required to offset the effects of excessive air motion.

9-2 Temperature, Air Motion, Clothing and Comfort

Ambient temperature, clothing and air velocity affect the heat loss from a person's body. Activity level determines the amount of heat generated within a person's body. Comfort can be maintained if the radiant heat absorbed by the body is equal to the difference between the heat lost from the body and the heat generated within the body. Figure 9-1, below, shows the relationship between heat loss, temperature, air velocity, clothing and activity level. This diagram can be used to estimate how much radiant heat must be received by each square foot of body surface area to insure comfort.

9-3 Comfort and Operating Cost

If the amount of radiant energy received by the occupant is sufficient, the occupant will feel comfortable, even if the ambient air temperature is relatively cold. Therefore, limited areas of a given facility can be made comfortable even though the ambient temperature in the facility is normally considered uncomfortable. The lower indoor temperatures associated with area and spot heating strategies translate into reduced energy consumption and lower operating costs. There also is evidence that lower indoor design temperatures are appropriate when the whole facility is heated. In this case, the indoor temperature can be reduced by 5 or 10 degrees without sacrificing comfort.

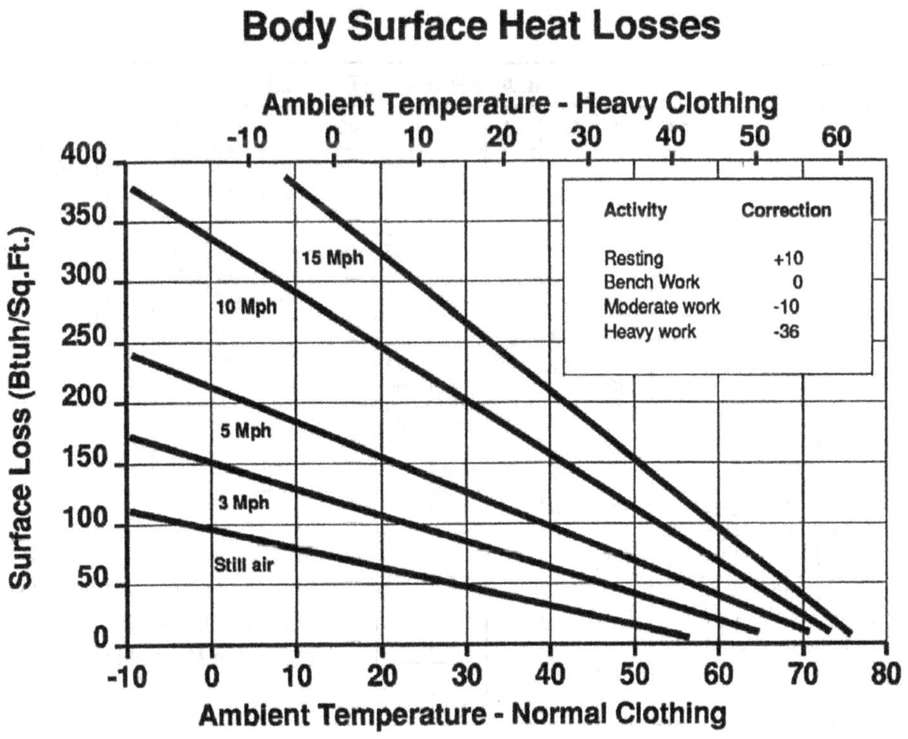

Figure 9-1

At part-load conditions, oversized equipment can have an adverse effect on comfort — especially if the designer is expecting the direct radiation that is received by the occupants to compensate for a lower indoor design temperature.

9-4 Stratification and Operating Cost

Radiant heating devices transfer heat from a hot radiating surface directly to an occupant, the floor and to the low-level surroundings — the air is not directly heated. The floor and the surrounding materials heat up as they absorb radiation and they release this heat to the air. During this process the air is never heated to more than a few degrees above the desired room temperature. When compared with hot air systems, radiant systems are found to cause less stratification — temperatures are more uniform throughout the room and are not as hot near the roof and high wall areas. In high bay structures, reduced stratification results in lower heat losses, lower fuel consumption and reduced operating cost. Figure 9-2 below provides one manufacturer's estimate of the energy savings potential. Note that the potential savings increase as the distance between the heater and the roof is increased. Heaters mounted near the roof do not reduce energy consumption.

9-5 Sizing Adjustments

Figure 9-2, which is based on manufacturer's field data, indicates that radiant systems consume less fuel than warm air heating systems. These claims seem to be justified — refer to the explanations provided in the preceding paragraphs. However, the designer should question any suggestion that fuel consumption factors can be used to reduce the design heating load value.

- Fuel consumption is reduced if stratification is reduced. But, this does not necessarily affect the calculated heating load because this calculation is usually based on an assumption of uniform indoor temperature. In fact, if stratification does occur, an unadjusted load estimate would underestimate the stratified load.
- Fuel consumption is reduced if the radiant heat absorbed by the occupants provides comfort at a lower indoor design temperature. But, a consumption-based load correction is not necessary because this effect can be accounted for by adjusting the indoor design temperature that is used to estimate the heating load.

9-6 Reflectors

Reflector shape has a significant effect on the amount of direct radiation that strikes the floor and the lower levels of the occupied space. Some reflector shapes spread the radiation over a wider area than others. Shapes that provide a wide spread are more suitable for low mounting heights. Shapes that concentrate the radiation into a smaller area are more suitable for spot heating or high mounting heights. (Some reflectors can be rotated, concentrating a maximum amount

BUILDING HEIGHT — FEET	HEATER HEIGHT — FEET																				
	8	10	12	14	16	18	20	22	24	26	28	30	32	34	36	38	40	42	44	46	48
50									-	.77	.81	.84	.87	.89	.91	.93	.95	.96	.98	.99	1.0
48									.75	.79	.83	.86	.88	.91	.93	.95	.96	.98	.99	1.0	-
46									.77	.81	.85	.88	.90	.92	.94	.96	.97	.99	1.0	-	-
44							-	.75	.80	.83	.87	.89	.92	.94	.96	.97	.99	1.0			
42							-	.78	.82	.85	.89	.91	.93	.95	.97	.99	1.0	-			
40							.75	.80	.84	.88	.90	.93	.95	.97	.99	1.0	-	-			
38						-	.78	.83	.87	.90	.92	.95	.97	.99	1.0						
36						.75	.81	.85	.89	.92	.94	.96	.98	1.0	-						
34						.78	.83	.86	.91	.94	.96	.98	1.0	-	-						
32				-	.75	.81	.86	.90	.93	.96	.98	1.0									
30				-	.79	.84	.89	.93	.96	.98	1.0	-									
28				.75	.82	.88	.92	.95	.98	1.0	-	-									
26			-	.79	.86	.91	.95	.98	1.0												
24			.75	.83	.89	.94	.97	1.0	-												
22			.80	.88	.93	.97	1.0	-	-												
20	-	.75	.85	.92	.97	1.0															
18	-	.81	.90	.96	1.0	-															
16	.75	.87	.95	1.0	-	-															
14	.83	.94	1.0																		
12	.91	1.0	-																		

The factors in this table apply to fuel consumption calculations. Do not use these factors to adjust the calculated **Manual N** heat load.

Figure 9-2

of radiant energy on the floor while minimizing the radiant energy impinging on the walls. This feature allows installation near an outside wall, without increasing the heat loss through the wall.)

9-7 Zoning

When radiant heating devices are installed, zones can be established by using a group of heaters to radiate a specific area — partitions are not required. Each zone and the corresponding group of heaters can be controlled by a separate thermostat. Spot heat can be controlled by a thermostat or by a manual switch.

9-8 Controls

Normally, all the safety and operating controls, except the thermostat, are built into the heating device. Manufacturers usually offer a variety of ignition and flame sensing options. Observe manufacturer's recommendations regarding thermostat location.

9-9 Part-Load Control

Radiant devices are either on, supplying full capacity, or off. The space temperatures may hold steady, say within a few degrees when the units are off, but the occupants will not receive any direct radiation. Manufacturers do not say much about this problem. However, it would seem that if direct radiation is used to compensate for a reduced indoor design temperature, the sensation of comfort would be reduced when the units are off. One solution to this problem might be to use an outdoor thermostat to stage the number of units that are on-line during part-load conditions. In any case, oversized equipment will only aggravate this problem.

9-9 Fossil Fuel Equipment

A variety of fossil fuel radiant heating devices are available — some of these devices are shown on this page. Indirect-fired, low intensity unit heaters and indirect-fired, low intensity radiant-tube systems may be vented or non vented. Natural gas, propane or oil can be used to fuel indirect-fired units. Oil fired units must be vented. Direct-fired, high intensity, gas and propane heaters also are available. Direct-fired equipment is not vented and it operates with what amounts to an open flame at the radiating surface.

9-10 Burner Efficiency, Venting and Operating Cost

Nonvented heaters operate at a relatively high burner efficiency — about 90 to 92 percent. This higher efficiency translates into lower fuel costs. However, these units require a supply of outdoor air to dilute the products of combustion. The net efficiency will decrease and the fuel cost will increase if a supplemental supply of outdoor air is required to maintain air quality or to control humidity. Nonvented units are suitable for heating loose, leaky structures or spaces that would otherwise require a generous amount of outdoor air.

Vented unit heaters operate at a lower burner efficiency — about 80 percent. The burner efficiency of radiant-tube systems varies, depending on the length of the pipe, it can approach 93 percent. Vented devices do not require outdoor air to dilute the products of combustion. Vented units should be used to heat tight structures and spaces that do not require much outdoor air.

Indrect-Fired Unit

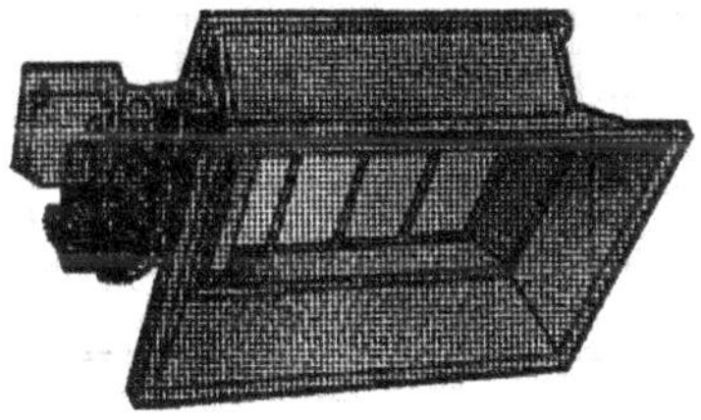

Direct-Fired Unit

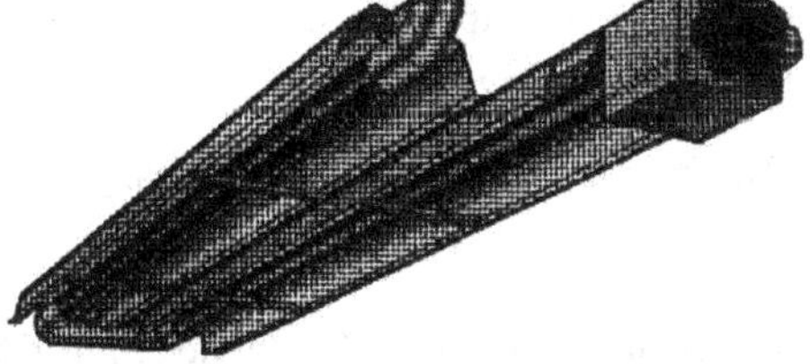

Radiant-Tube Equipment

9-11 Venting and Dilution Air Requirements

Vented units expel the products of combustion to the outdoors and therefore, they do not require dilution ventilation. *The information and procedures discussed in this section only apply to nonvented equipment.*

Dilution CFM

Since unvented units release the products of combustion into the occupied space, outdoor air will be required to dilute toxic fumes and to help control humidity. The factors in Table 9-1 can be used to estimate the amount of outdoor air that will be required to dilute toxic fumes. These factors are based on an ANSI code requirement that specifies 4 CFM of outdoor air for each 1000 BTUH of natural gas input capacity and 5 CFM of outdoor air for each 1000 BTUH propane input capacity.

Fume Dilution Factors Unvented Heaters		
Design TD	**Natural Gas (4 CFM)**	**Propane (5 CFM)**
30	0.276	0.359
35	0.284	0.372
40	0.291	0.385
45	0.299	0.399
50	0.308	0.414
55	0.317	0.430
60	0.326	0.448
65	0.336	0.467
70	0.347	0.488
75	0.358	0.511
80	0.370	0.536
85	0.383	0.563
90	0.397	0.594

Table 9-1

In loose buildings, the dilution air may be provided by infiltration. To estimate the minimum amount of infiltration required for dilution (DCFM) multiply the transmission load by a factor from Table 9-1 and divide by 60.

DCFM = (Dilution factor x Transmission Loss) / 60

Compare the expected infiltration CFM (as determined by the load calculation) with the dilution CFM (DCFM). A powered ventilation system will be required if infiltration does not provide enough outdoor air for fume dilution. Note that some codes may require that 100 percent of the dilution CFM be supplied by a powered ventilation system, even if the expected infiltration is adequate.

Powered Ventilation Systems

Some manufacturers suggest that a simple powered ventilation system can be provided by installing an exhaust fan and fresh air intake openings. An example of how this suggestion might be phrased is provided below.

> *Small exhausters should be distributed around the roof or at the high points of the building walls. Outdoor air intake openings are also required. Install one square inch of free inlet area for each 1000 BTUH of input capacity. A single opening should not be more than 2 square feet in size. Multiple openings should be located on the side walls at the same level as the heaters and distributed around the perimeter of the room.*

However, if this recommendation is followed, the total amount of untreated outdoor air flowing through the space will be equal to the infiltration CFM plus the ventilation CFM — therefore, the total heating load will be increased, larger heaters will be necessary and even more dilution air will be required. In this situation, the designer should consider using a make up air unit to temper the incoming ventilation air, or simply switch to vented equipment.

Example — Dilution CFM Estimate

The calculated infiltration for a building is equal to 5,300 CFM. The winter design temperature is equal to 5°F and the room design temperature is equal to 65°F. Estimate the amount of outdoor air required to dilute the products of combustion expelled from unvented natural gas heaters if the transmission load is equal to 1,500,000 BTUH. Determine if powered ventilation is required.

TD = 65 - 5 = 60 °F
Factor from Table 9-1 = 0.326
DCFM = (0.326 x 1500000) / 60 = 8150

The 8,150 CFM dilution value exceeds the 5,300 infiltration value. Either a power ventilation system (8,150 CFM) or vented equipment is required. If exhaust fans are installed, a make up air unit should be used to temper the incoming ventilation air.

Infiltration Estimates

The ACCA **Manual N** load calculation procedure produces an estimate of the expected infiltration. This procedure considers the tightness of the building envelope, the pressure in the space, door traffic and the influence of exhaust systems. The air change estimates provided by Figure 9-3, on the next page, can be used to supplement the **Manual N** infiltration tables. Note that Figure 9-3 considers envelope leakage and door traffic, but it does not consider the effect of exhaust

Manufacturer Air Change Recommendations				
Type of Building	**Construction**	**Door Usage**	**ACH**	**ACH**
			Under 30,000 SQ.FT.	Over 30,000 SQ.FT
Warehouse	Poor	Minor	1.50 to 3.00	1.00 to 2.50
		Considerable	2.00 to 4.50	1.50 to 4.00
	Average	Minor	1.00 to 2.00	0.75 to 1.50
		Considerable	1.50 to 3.50	1.25 to 3.00
	Good	Minor	0.75 to 1.50	0.50 to 1.25
		Considerable	1.25 to 3.00	1.25 to 2.75
			Under 15,000 SQ.FT.	Over 15,000 SQ.FT.
Light Manufacturing	Poor	Minor	2.00 to 3.50	1.50 to 3.00
		Considerable	2.50 to 4.00	2.00 to 2.50
	Average	Minor	1.00 to 2.50	0.75 to 2.00
		Considerable	1.50 to 3.00	1.25 to 2.50
	Good	Minor	0.50 to 1.00	0.25 to 0.75
		Considerable	1.00 to 1.50	0.75 to 1.25
			Under 15,000 SQ.FT.	Over 15,000 SQ.FT.
Aircraft Hangar	Poor	Minor	3.00 to 4.00	2.00 to 3.50
		Considerable	4.00 to 5.50	3.00 to 5.00
	Average	Minor	2.00 to 3.00	1.50 to 2.50
		Considerable	3.00 to 4.50	2.50 to 4.00
	Good	Minor	1.50 to 2.00	1.00 to 1.50
		Considerable	2.50 to 3.50	2.00 to 3.00
Tennis Courts	Poor	Minor	2.00 to 3.00	30,000 SQ.FT.
	Average	Minor	1.00 to 2.00	
	Good	Minor	0.50 to 1.00	
Garages (Engine exhaust vented)	Poor	Minor	3.00 to 4.00	5000 to 7500 SQ.FT.
		Considerable	4.00 to 6.00	
	Average	Minor	1.50 to 3.00	
		Considerable	2.50 to 5.00	
	Good	Minor	0.50 to 1.00	
		Considerable	1.50 to 3.00	
Vehicle Storage	Poor	Min/Max . . .	1.50/3.50 or 2.50/5.00	5000 SQ.FT.
	Average	Light or heavy	1.00/2.00 or 2.00/4.00	
	Good	door traffic	0.50/1.00 or 1.50/3.00	

Figure 9-3

systems. If exhaust systems are required, the information provided by Figure 9-3 can be used with the **Manual N** procedure as follows:

Step 1) Calculate the infiltration CFM.
Use the Figure 9-3 air change values to calculate the infiltration CFM associated with envelope leakage and door traffic.

Infiltration CFM = (ACH x Volume) / 60

Step 2) Determine if make up air equipment is required.
Compare the exhaust CFM value with the estimated infiltration CFM value. Use make up air equipment to maintain the space at neutral pressure if the exhaust CFM value is greater than 10 percent of the infiltration CFM value.

> Make up air equipment may not be available in a suitable size if the amount of air exhausted from the space is small. Direct-fired make up air units are available in sizes that are suitable for tempering as little as 1200 CFM of outdoor air. Indirect-fired make up air units that have turn-down ratios of 20:1 or better are available in sizes that are suitable for tempering as little as 600 CFM of outdoor air. Below 600 CFM electric heating coils can be staged or pulsed to maintain the temperature of the make up air; or raw, untempered air can be used if the make up air load is small.

Step 3) Determine the design value for the infiltration CFM.

- If make up air equipment is not used, add the exhaust CFM to the infiltration CFM.
- If make up air equipment is used, do not add the exhaust CFM to the infiltration CFM.

Condensation
If unvented heaters are installed, the water vapor created during the combustion process is released into the room. Condensation will not occur if the dew point temperature of the indoor air is lower than the temperature of the coldest surface. Therefore, the maximum allowable dew point temperature is equal to the temperature of the coldest surface.

The temperature (Ts) of the inside surface of a structural component depends on the U-value of that component, the indoor temperature (Ti) and the outdoor temperature (To). The equation below can be used to estimate the inside surface temperature of an exposed surface, which establishes the maximum allowable room dew point temperature (RDPT).

$$T_s = T_i - 0.65 \times U \times (T_i - T_o) = RDPT$$

Also consider cold surfaces that are associated with process equipment or with materials that are moved into the room from outside storage areas. Condensation will form on these surfaces if the temperature of the surface is lower than the dew point temperature of the indoor air.

The room dew point temperature depends on the amount of outdoor air flowing through the space (OCFM), the moisture content of the outdoor air (OAGR), the internal latent loads associated with people and processes (ILH) and the latent heat released to the space by unvented combustion gases (LHC). Values for these factors can be determined as follows:

Outdoor Air - OCFM
The amount of dilution air flowing through the space is equal to the total amount of outdoor air entering the space through cracks and outdoor air intake openings *plus any outdoor air that is introduced to the space through direct or indirect-fired make up air heating units.*

Outdoor Air Moisture Content - OAGR
The worst possible situation occurs when the outdoor air entering the space is saturated (100 percent relative humidity). Figure 9-4, above, can be used to estimate the grains of moisture associated with saturated outdoor air. (For any given outdoor air temperature, just move vertically to the reference line and then horizontally to the grains value. For example, at 5°F, saturated outdoor air holds 7 grains of moisture.)

Grains of Moisture in Saturated Air

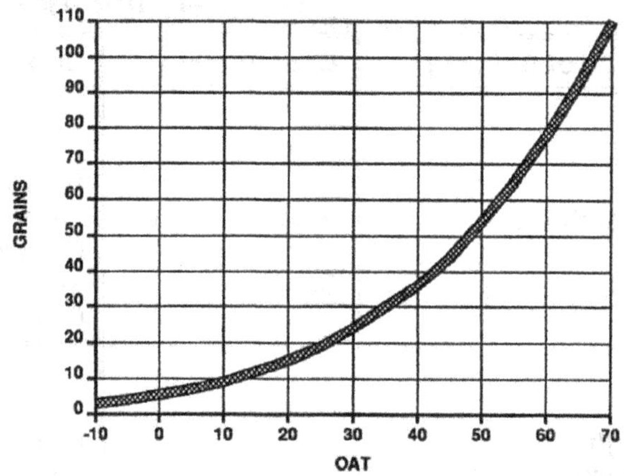

Figure 9-4

Internal Latent Load - ILH
The internal latent load associated with people and some types of equipment can be estimated by using **Manual N**, Tables 10 and 12. The latent loads associated with undocumented industrial process or manufacturing equipment will have to be determined from information gained at the site. In many applications, the internal latent loads are relatively small. Whenever these loads are small, they can be ignored.

Latent Heat From Combustion - LHC
The following equations can be used to estimate the latent heat added by unvented radiant heaters and direct-fired make up air equipment. In these equations, the *outdoor air load* is the ventilation load associated with unvented equipment. This load is based on the total amount of outdoor air entering the space through cracks and free intake openings, plus the outdoor air that is processed by direct-fired make up air heating units. Do not include any outdoor air loads that are associated with indirect-fired make up air units.

For natural gas
LHC = 0.115 x (Transmission load + Outdoor air load)

For propane
LHC = 0.89 x (Transmission load + Outdoor air load)

Figure 9-4 can be used to estimate the room dew point temperature (RDPT), but first it is necessary to calculate the grains of moisture that will be expected in the room (RMGR). The following equation can be used for this purpose.

$$RMGR = \frac{(ILH + LHC)}{0.68 \times (OCFM)} + OAGR$$

Use the calculated room grains value to determine the room dew point temperature. Enter Figure 9-4 at the RMGR value and move horizontally to the reference line and then down to the outdoor temperature value. For example, a value of 30 grains corresponds to a 35°F dew point temperature.

When condensation is anticipated, the design will have to be modified. In this case, the designer should consider using one or more of the following options.

1) Increase the indoor temperature.

2) Add insulation to the cold surfaces. (Note that when insulation is installed on the inside of the building, the inside surface of the insulation must be impermeable to moisture — that is, a vapor barrier is required. Windows and skylights can be particularly troublesome because they must either be eliminated, replaced with better windows, fitted with a second glazing or covered with insulated panels.)

3) Increase the amount of outdoor air. (The following equation can be used to estimate how many CFM of outdoor air will be required to keep the room dew point temperature below the temperature of the coldest surface.)

$$OCFM = \frac{(ILH + LHC)}{0.68 \times (RMGR - OAGR)}$$

4) Use a vented heater. (Vented equipment expels the products of combustion to the outdoors.)

A single comparison of the surface temperature and dew point, made at the winter design condition, is sufficient to insure that condensation will not occur at any part-load operating condition providing that:

- The internal latent loads are negligible.
- Outdoor air CFM flowing through the space is greater than the dilution CFM value (DCFM) that was established by applying the factors provided in Table 9-1.

If internal latent loads cannot be ignored, a dew point comparison must be made at the winter design condition and at each outdoor temperature that falls between the winter design condition and the room temperature (5 degree increments will be sufficient). All values including the inside surface temperature, the transmission load, the outdoor air load, the outdoor air grains, the room grains and the expected dewpoint in the room must be recalculated for each outdoor air temperature. If the calculations indicate that condensation is possible in the lower range of outdoor temperatures, one of the recommendations listed above may solve the problem. Vented equipment should be used if the calculations indicate that condensation will occur at moderate outdoor temperatures. If the internal latent loads are large enough, condensation may occur even if vented equipment is installed.

Example — Condensation Check, No Make Up Air
The largest U-value associated with the opaque surfaces is equal to 0.30, the glass U-value is equal to 0.70. The calculated infiltration for a building is equal to 8300 CFM. The winter design temperature is equal to 10°F and the room design temperature is equal to 65°F. The transmission load is equal to 1,500,000 BTUH. **The internal latent loads are negligible.** Determine if unvented, natural gas heaters are likely to cause condensation on the various surfaces.

1) Calculate the surface temperatures.

Opaque Ts = RDPT = 65 - 0.65 x 0.30 x (65 - 10) = 54 °F
Glass Ts = RDPT = 65 - 0.65 x 0.70 x (65 - 10) = 40 °F

2) Calculate the outdoor air CFM.

Factor from Table 9-1 = 0.317
DCFM = (0.317 x 1500000) / 60 = 7925
Infiltration CFM = 8300
OCFM = 8300 (8300 larger than 7925)

3) Calculate the outdoor air load on unvented equipment.

OCFM = 8300
OA-Load = 1.1 x 8300 x (65 - 10) = 502150 BTUH

4) Calculate the combustion latent heat to the space.

LHC = 0.115 x (1500000 + 502150) = 230247 BTUH

5) Use Figure 9-4 to find the grains value for the outdoor air.

OAGR = 9 (at 10 °F OAT)

6) Determine the grains of moisture in the room.

$$RMGR = \frac{0 + 230247}{0.68 \times 8300} + 9 = 50$$

7) Use Figure 9-4 to find the room dew point temperature.

RDPT = 48 °F

8) Compare the dew point with the surface temperatures.

No condensation on opaque surfaces — the 54°F surface is warmer than the 48°F dew point.

Condensation on glass — the 40°F surface is colder than the 48°F dew point.

Example — Condensation Check, Direct-Fired MUA Unit
Solve the previous problem if a 5000 CFM process equipment exhaust system and a 5000 CFM direct-fired make up air heating unit are added to the building. (Since the space will

be under neutral pressure the infiltration will still be equal to 8300 CFM.)

1) From previous example . . .

Opaque Ts = 54 °F
Glass Ts = 40 °F
Required dilution CFM = 7925
Outdoor grains = 9

2) Calculate the outdoor air CFM.

OCFM = 8300 + 5000 = 13300
(Use 13300 since it is larger than 7925)

3) Calculate the outdoor air load on unvented equipment.

OCFM = 13300
OA-Load = 1.1 x 13300 x (65 - 10) = 804650 BTUH

4) Calculate the combustion latent heat to the space.

LHC = 0.115 x (1500000 + 804650) = 265035 BTUH

5) Determine the grains of moisture in the room.

$$RMGR = \frac{265035}{0.68 \times 13300} + 9 = 38$$

6) Use Figure 9-4 to find the room dew point temperature.

RDPT = 41 °F

8) Compare the dew point with the surface temperatures.

No condensation on opaque surfaces — the 54°F surface is warmer than the 41°F dew point.

Condensation on glass — the 40°F surface is colder than the 41°F dew point.

Example — Condensation Check, Indirect-Fired MUA
Solve the previous problem if a 5000 CFM indirect-fired make up air unit is substituted for the direct-fired unit.

1) From previous example . . .

Opaque Ts = 54 °F
Glass Ts = 40 °F
OCFM = 13300
Outdoor grains = 9

2) Calculate the outdoor air load on unvented equipment.

CFM = 8300 (indirect-fired MUA unit is vented)
OA-Load = 1.1 x 8300 x (65 - 10) = 502150 BTUH

3) Calculate the combustion latent heat to space.

LHC = 0.115 x (1500000 + 502150) = 230247 BTUH

4) Determine the grains of moisture in the room, (ILH = 0).

$$RMGR = \frac{0 + 230247}{0.68 \times 13300} + 9 = 34$$

5) Use Figure 9-4 to find the room dew point temperature.

RDPT = 38 °F

6) Compare the dew point with the surface temperatures.

No condensation on opaque surfaces — the 54°F surface is warmer than the 38°F dew point.

No condensation on glass — the 40°F surface is warmer than the 38°F dew point.

Example — Condensation Check, Internal Latent Loads
The largest U-value associated with the opaque surfaces is equal to 0.30. The calculated infiltration for a building is equal to 8300 CFM. The winter design temperature is equal to 0°F and the room design temperature is equal to 65°F. The transmission load is equal to 1,500,000 BTUH. The internal latent loads are equal to 55,000 BTUH. Determine if unvented, natural gas heaters are likely to cause condensation on the various surfaces.

The solution to this problem requires separate calculations (as shown above) for each operating condition that occurs as the outdoor air temperature varies from 0°F to 65°F. The following graph shows the results of these calculations. Note that condensation will occur when the outdoor temperature is below 12°F and when it is above 48°F.

Room Dew Point with Internal Latent Loads

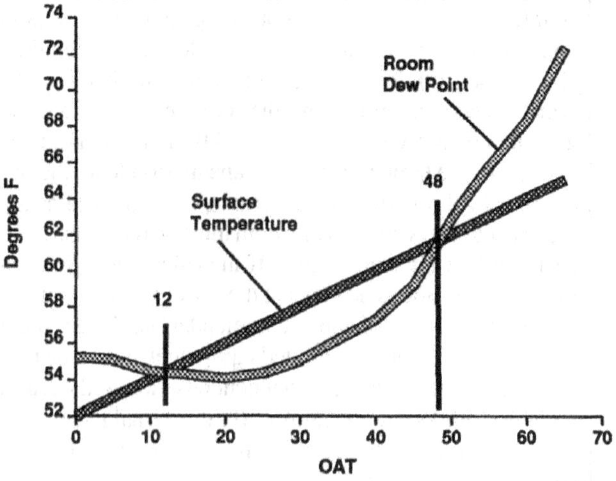

9-12 Combustion Air

All direct-fired, unvented equipment uses indoor air for combustion. Indirect-fired, vented equipment also may use indoor air for combustion. (Some indirect-fired equipment can use outdoor air for combustion.) When combustion air is extracted from the heated space, the designer must verify that there is an adequate supply of combustion air.

- If unvented equipment is installed, the dilution ventilation will more than satisfy the combustion air requirement.
- If vented equipment is installed, infiltration or mechanical ventilation must replace the air that is used by the burner. However, the amount of air required for combustion is not large — 0.50 CFM of air for each 1000 input BTUH will provide a liberal amount of excess air. Refer to local codes, BOCA, ANSI and the NFPA codes for more guidance regarding combustion air requirements.

Some indirect-fired, vented equipment is designed to take combustion air directly from the outdoors. This type of equipment is required if the indoor air is not a suitable source of combustion air.

- Combustion air should not be extracted from the heated space if the air contains corrosive vapors or fumes.
- Combustion air should not be extracted from the heated space if the space contains vapors or fumes that degenerate into toxic or corrosive gases when exposed to an open flame — fluorocarbon compounds for example.
- Combustion air should not be extracted from the heated space if the air contains chlorine or chlorinated compounds.
- Combustion air should not be extracted from the heated space if the air contains an excessive amount of dust, grit, dirt or painting over-spray.

9-13 Precautions

All types of fossil fuel infrared heaters have high surface temperatures and direct-fired devices expose the air in the heated space to an open flame. Observe the following precautions.

- Heaters should not be installed close to combustible materials. Observe codes and manufacturer's recommendations regarding clearances. (Refer to local, state, ANSI and NFPA codes; also ask insurance underwriters and manufacturer's for guidance.)
- Signs that specify maximum stacking heights may have to be posted in storage areas.
- Sufficient clearance must be provided between the heater and any vehicle (including the load) that may be parked under or near the heater.
- Sufficient clearance must be provided between the heater and sprinkler heads.
- Sufficient clearance must be provided between the heater and the structural components of the building.
- Infrared heaters should not be used in a combustible or explosive environment.
- Infrared heaters should not be used in a situation where the hot surface or exposed flame will cause fumes and vapors to decompose into a toxic pollutant.
- Infrared heaters should not be used in swimming pools or other areas that have a chlorinated atmosphere if they are susceptible to corrosion caused by chlorine. Vented heaters that use outdoor air for combustion may be suitable for this type of application — consult with the manufacturer.
- Unvented heaters require the proper amount of dilution ventilation.
- Dilution air will not be effective if it stratifies before it is thoroughly mixed with room air.
- Unvented heaters may cause humidity and condensation problems.
- If the heaters extract combustion air from the heated space, an adequate supply of outdoor air is required to replace the combustion air.
- Combustion air should not be extracted from the heated space if the space contains corrosive vapors or fumes.
- Combustion air should not be extracted from the heated space if the air in the space contains an excessive amount of dust, dirt or mists.
- Improperly installed heaters may cause the occupants to be too cold or too hot.
- Winds or drafts degrade heater effectiveness.

9-14 Application of Fossil Fuel Equipment

Infrared, fossil fuel equipment is commonly used to heat warehouses, hangers, transportation terminals, parking garages, indoor tennis courts, gymnasiums, swimming pools, sports facilities, grandstands, greenhouses, workshops, foundries, factory areas and loading docks. In these applications, radiant heating equipment may either be used to provide total building heat, area heat or spot heat.

9-15 Direct-Fired Heaters — Total Building Heat

Direct-fired heaters operate at a high surface temperature, they expose the indoor air to an open flame, they use indoor air as the source of combustion air and they are not vented. Direct-fired radiant heaters can be used in situations where an unvented, atmospheric burner would not create a problem. For example, they could be used in:

- Loosely constructed buildings or buildings that have a lot of door traffic — infiltration provides the required dilution CFM.
- Well insulated, well ventilated buildings that would not have problems associated with condensation or fumes.
- Buildings with atmospheres that are compatable with atmospheric burners.
- Buildings with adequate ceiling heights.
- In situations where there is no problem providing adequate clearances between the heater and the structure or between the heater and combustible materials.

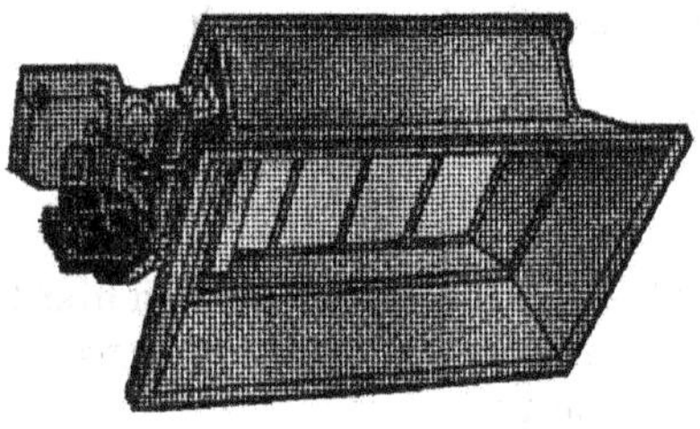

Direct-Fired Unit

The heating performance of direct-fired units will depend on the physical characteristics (size, geometry, radiating surface, reflecting surface, operating temperature, etc.) of a particular fixture. These parameters are not controlled by the HVAC system designer, they are established by the manufacturer's design. However, manufacturers offer direct-fired equipment in a variety of sizes and designs.

Heating performance, as it applies to comfort, also is sensitive to the room air temperature, room air motion, mounting height, distance between fixtures and the fixture pattern. These parameters are under the control of the system designer.

Design Procedure — Total Building Heat

To design a total building heating system, the designer must determine the required input capacity and then decide how many heaters to use. Placement of the heaters also is important. **Because direct-fired heaters are not vented, the designer has to perform the calculations that are associated with fume dilution and condensation.** Controls and accessories also should be considered. This design procedure is explained below.

Step 1) Make a **Manual** N Load Estimate.
Perform a **Manual N** load estimate. Make a load summary that tabulates the values of the transmission loads, the infiltration load, and if required, the ventilation load. It also may be necessary to modify the standard **Manual** N load estimate to account for the following effects:

- If cold materials are brought into the building and heated to warmer temperature, they will produce an additional heating load on the equipment. This "materials load" must be added to the **Manual** N design load.
- The design value for the heating load can be reduced if a substantial amount of heat is added to the space by process equipment, process piping or electrical devices. However, do not subtract these loads from the design heating load unless the operating hours of these devices coincide with the operating hours of the heating system.

Step 2) Evaluate the dilution ventilation requirement.
The dilution ventilation requirement is based on the transmission load. If the dilution requirement is not satisfied by the expected infiltration, a powered ventilation system or vented equipment will be required. Refer to Section 9-11 for a complete discussion of this subject and the associated calculation procedures.

Step 3) Evaluate the condensation potential.
Condensation must not be allowed to occur on interior surfaces. Refer to Section 9-11 for a complete discussion of this subject and the associated calculations.

Step 4) Determine the input heating capacity.
The combined input heating capacity of the heaters should equal the design heating load (DHL). Include all the transmission and outdoor air loads that will actually be imposed on the radiant heaters. Do not include outdoor air loads that are neutralized by make up air equipment. If make up air units or other type heating sources add heat to the building, reduce the design heating load accordingly. Door heaters and air curtains should be used to control the heat loads associated with shipping doors.

Step 5) Determine the number of heaters.
Try to use the same size heater throughout the building, but this is not mandatory. Try to mount the heaters as low as possible. Larger heaters require higher mounting heights. Ignore heater sizes that have recommended mounting heights that are incompatible with the desired mounting height. For a

specific heater model, the required number of heaters (N) can be calculated by dividing the Step 4 design heating load (DHL) by the input capacity of the heater.

N = DHL / Input Capacity

Try to use as few heaters as possible. The final number of heaters will be a compromise between installed cost and performance. Consider the recommended mounting height, the effectiveness of the heating pattern, the required clearances associated with the structure, the materials that are stored in the structure and any obstructions that would restrict heater placement options. It may be necessary to evaluate two or more designs, depending on the spacings and heating patterns that are produced by using larger or smaller heaters.

Step 6) Layout a heating pattern.
The preferred heating pattern is produced by a layout that distributes the heaters along the outside walls. The number of heaters required along a particular wall (N_w), can be determined by multiplying the total number of heaters (N) by the ratio of the loss associated with the wall, divided by the total loss that is associated with all the walls.

N_W = N x (Wall Loss / Total of Wall Losses)

It may not always be possible to use the perimeter layout pattern, but other patterns will produce acceptable performance. Heaters can be uniformly distributed to blanket the floor area or they can be lined up below the ridge line of the roof. Combinations of these three basic patterns also can be used. The following factors should be considered when a pattern is evaluated.

- A larger number of smaller heaters will provide a more uniform heating pattern, but this must be balanced against the installation cost.
- Narrow buildings are ideal candidates for the perimeter pattern because the floor area in the middle of the building can be radiated by the perimeter units.
- A combination of perimeter units and ridge-mounted units can be used to radiate the entire floor area of a wide building.
- Distribute most of the heaters in areas that are normally occupied. Concentrate more heat in areas that have higher loads, use less heat in storage areas.
- Zone control becomes more precise as the number of heaters increase.
- Use spot heat for isolated work stations. Use spot or area heat in areas that have unusually high loads.
- The radiant heating patterns should always be directed toward the open floor areas; do not radiate stored materials.
- Reflectors can either be used to spread a pattern over a wider area or to concentrate the pattern. Reflector geometry also affects the recommended mounting height and the minimum distance between the heater and a wall.
- Units can be installed at an angle. This reduces the recommended mounting height and allows the heater to be installed closer to the wall.
- Layouts must be compatible with safety. Consider clearances required for combustible materials and hazardous fumes or processes.
- Layouts must accommodate obstructions such as cranes, lights, piping, sprinklers and manufacturing equipment.

Step 7) Verify clearances and spacings.
The heating pattern layout must be compatible with the manufacturer's recommendations regarding clearances and spacings. Refer to the manufacturer's performance data and installation manuals for information about clearances and spacing.

- Minimum recommended mounting height — mounting at a lower height could cause discomfort.
- Maximum recommended mounting height — mounting at a higher height reduces the intensity of the radiation striking the floor.
- Clearances between the heater and the ceiling — overall efficiency decreases and fuel consumption increases if the heaters are mounted close to the ceiling.
- Clearances between the heater and the wall — wall heat losses increase and fuel consumption increases if walls are radiated.
- Maximum distance between heater rows — excessive row spacing reduces comfort.
- Maximum distance between heaters in a given row — excessive spacing reduces comfort.
- Reflector geometry has an effect on the recommended mounting heights, clearances and spacings.
- Clearances between the heater and combustible materials.

Step 8) Select controls and accessories.
Refer to the manufacturer's product literature for information about controls and accessories. Select a burner control package (a variety of ignition and flame supervision control op-

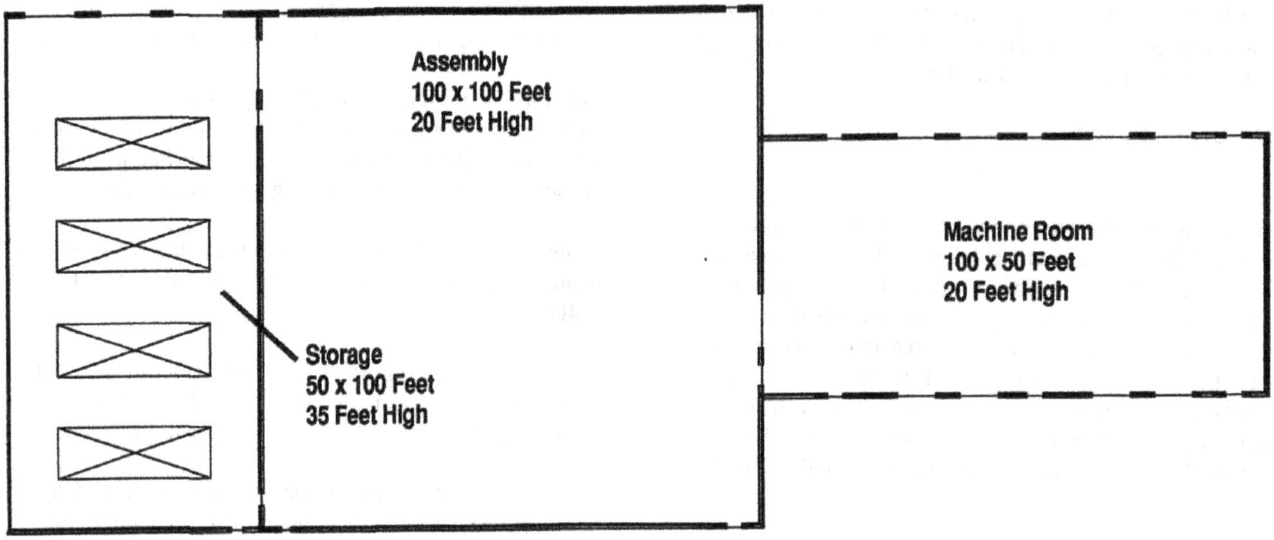

Figure 9-5

tions are usually available). Remote thermostats can be used to control space temperature. The number of heaters controlled by one thermostat will depend on the total number of units, the heating pattern layout and the zoning requirements; refer to the manufacturer's installation instructions for specific recommendations. Locate thermostats in areas of greatest heat loss. Unless the manufacturer advises otherwise, locate thermostats so they are not radiated by the heaters. In some cases, a concentrating reflector or reflector extensions, must be ordered as an accessory item.

Example — Total Building Heat

Figure 9-5, above, shows a manufacturing building that could be heated by **direct-fired, unvented** radiant heating units. The building has a machine tool room, a materials assembly room and a warehouse. The tool and assembly rooms have process exhaust systems that require make up air units. Direct-fired make up air units will be used to temper the outdoor air that is required to replace exhausted air. The internal latent loads are negligible. Use this information and information extracted from the load calculation to design the radiant heating system.

1) Load calculation

The table on this page summarizes pertinent information that was extracted from the load calculation. The heating loads are based on a 0°F outdoor design temperature.

Load Calculation Summary — Industrial Building			
	Machine	Assembly	Storage
Room Temp.	70	60	40
Design TD	70	60	40
Inf. CFM	833	2,500	5,833
Exh. CFM	2,000	2500	0
MUA CFM	2,000	2,500	0
MUA Load	154,000	165,000	0
U-Windows	0.50	NA	NA
U-Skylights	NA	1.12	NA
U-Roof	0.10	0.10	0.67
U-Wall	0.08	0.08	1.124
Trans. Load	129,000	173,000	455,000
Infilt. Load	64,000	165,000	257,000
Trans. + Inf.	193,000	338,000	712,000

2) Dilution ventilation requirement

The fume dilution factors found in Table 9-1 were used to calculate the minimum dilution CFM requirements. The design value for the dilution CFM is based on this minimum requirement or the infiltration CFM, whichever is greater.

	Machine	*Assembly*	*Storage*
Dilution CFM	*746*	*940*	*2210*
Infilt. CFM	*833*	*2500*	*5833*
Design Value	*833*	*2500*	*5833*

3) Evaluate condensation possibilities.

Condensation is not expected to occur in the machine tool room. The machine room calculations should be based on the window U-value because the U-values associated with the other surfaces are smaller than the window U-value. The calculations also should allow for dilution air and the latent heat produced by a 2000 CFM direct-fired make up air unit. The key details of the machine room condensation calculations (as explained on page 9-8) are summarized below.

Outdoor air = 0 °F (5 Grains)
Room temperature = 70 °F
Largest U-value = 0.50
Transmission load = 129000 BTUH
Basis for the outdoor air load = 833 + 2000 = 2833 CFM
Outdoor air load = 1.1 x 2833 x 70 = 218000 BTUH
Basis for LHC load = 129000 + 218000 = 347000
Latent load = LHC = 0.115 x 347000 = 39900 BTUH
Outdoor air = OCFM = 833 + 2000 = 2833
Room grains = RMGR = 25.7
Dew point = DPT = 32 °F (Figure 9-4)
Surface temperature = Ts = 47 °F -OK

Condensation is expected to occur in the assembly room. The assembly room calculations should be based on the skylight U-value because the U-values associated with the other surfaces are smaller than the skylight U-value. The calculations also should allow for dilution air and the latent heat produced by a 5000 CFM direct-fired make up air unit. The key details of the assembly room condensation calculations (as explained on page 9-8) are summarized below.

Outdoor air = 0 °F (5 Grains)
Room temperature = 60 °F
Largest U-value = 1.12
Transmission load = 173000 BTUH
Basis for the outdoor air load = 2500 + 5000 = 7500 CFM
Outdoor air load = 1.1 x 7500 x 60 = 495000 BTUH
Basis for LHC load = 173000 + 495000 = 668000 BTUH
Latent load = LHC = 0.115 x 668000 = 76820 BTUH
Outdoor air = OCFM = 2500 + 5000 = 7500
Room grains = RMGR = 20
Dew point = DPT = 26 °F (Figure 9-4)
Surface temperature = Ts = 16 °F — Not OK

However, condensation in the assembly room can be avoided by making a design change. (Add a plastic glazing to the skylights.) These calculations are summarized below.

Outdoor air = 0 °F (5 Grains)
Room temperature = 60 °F
Largest U-value = 0.45
Transmission load = 163500 BTUH (Glass U = 0.45)
Basis for the outdoor air load = 2500 + 5000 = 7500 CFM
Outdoor air load = 1.1 x 7500 x 60 = 495000 BTUH
Basis for LHC load = 163000 + 495000 = 658000 BTUH
Latent load = LHC = 0.115 x 658000 = 75670 BTUH
Outdoor air = OCFM = 2500 + 5000 = 7500
Room grains = RMGR = 20
Dew point = DPT = 26 °F (Figure 9-4)
Surface temperature = Ts = 42 °F — OK

Condensation is expected to occur in the storage room when the overhead doors are shut. The storage room calculations should be based on the wall U-value because the U-values associated with the other surfaces are smaller than this U-value. The details of the storage room condensation calculations (as explained on page 9-7) are summarized below.

Outdoor air = 0 °F (5 Grains)
Room temperature = 40 °F
Largest U-value = 1.124
Transmission load = 455000 BTUH
Basis for the outdoor air load = 5833 CFM
Outdoor air load = 1.1 x 5833 x 40 = 257000 BTUH
Basis for LHC load = 455000 + 257000 = 712000 BTUH
Latent load = LHC = 0.115 x 712000 = 81880 BTUH
Outdoor air = OCFM = 5833
Room grains = RMGR = 26
Dew point = DPT = 32 °F (Figure 9-4)
Surface temperature = Ts = 11 °F — Not OK

However, condensation in the storage room can be avoided by improving the wall U-value. (The U-value can be reduced to about 0.42 by adding an inside surface.) These calculations are summarized below.

Outdoor air = 0 °F (5 Grains)
Room temperature = 40 °F
Largest U-value = 0.42
Transmission load = 258000 BTUH (Wall U = 0.42)
Basis for the outdoor air load = 5833 CFM
Outdoor air load = 1.1 x 5833 x 60 = 257000 BTUH
Basis for LHC load = 258000 + 257000 = 515000 BTUH
Latent load = LHC = 0.115 x 515000 = 59220 BTUH
Outdoor air = OCFM = 5833
Room grains = RMGR = 20
Dew point = DPT = 26 °F (Figure 9-4)
Surface temperature = Ts = 29 °F — OK

4) Determine the input heating capacity.
The required input capacity is equal to the transmission load plus the infiltration load. (The direct-fired make up air units will temper the outdoor air that replaces the exhausted air.) These loads, which are based on the revised design, are summarized below.

	Machine	*Assembly*	*Storage*
Transmission	*129000*	*163500*	*258000*
Infiltration	*64000*	*165000*	*257000*
Total	*193000*	*328500*	*515000*

5) Determine the number of heaters.
Heater sizes range from 30,000 BTUH to more than 150,000 BTUH per unit. It's desirable to use as few heaters as possible, but it is also necessary to consider the minimum mounting height and to have a heating pattern that provides good

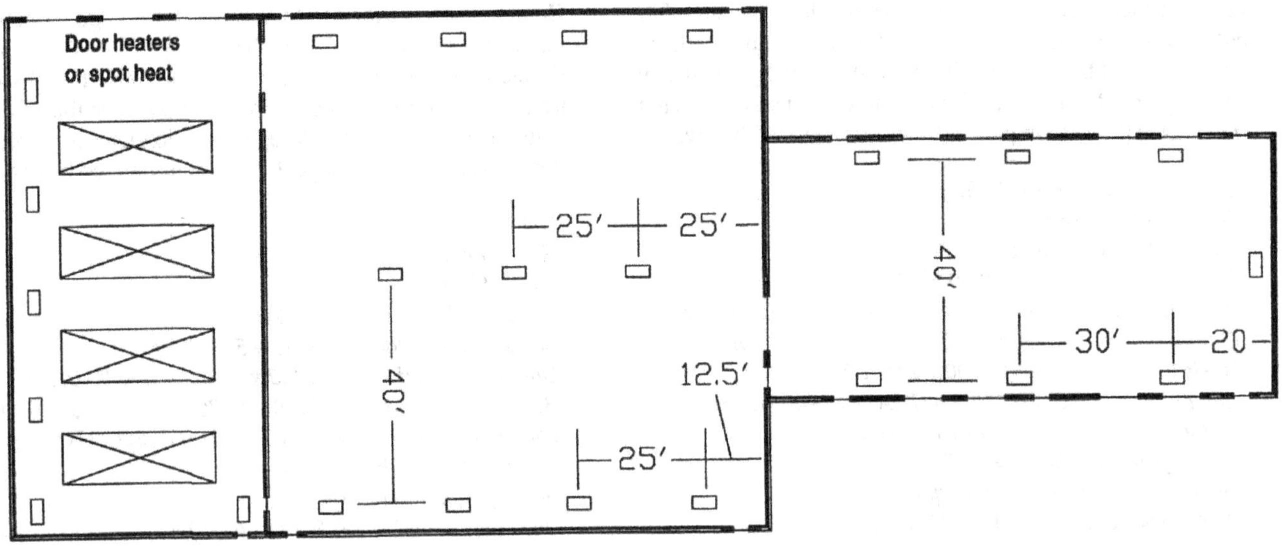

Figure 9-6

coverage. Based on the manufacturer's recommendations regarding mounting height and the distance between heaters, a 30,000 BTUH unit was selected for the machine and assembly rooms and a 75,000 BTUH unit was selected for the storage area. The number of heaters required for each area is calculated by dividing the design load by the input capacity.

	Machine	*Assembly*	*Storage*
Load	*193000*	*328500*	*515000*
Capacity	*30000*	*30000*	*75000*
Number	*7*	*11*	*7*

6) Lay out a heating pattern.
The machine room is long and narrow, so the heaters can be uniformly distributed along the outside walls. The assembly room is wide, so some heaters should be installed in the middle of the room and the rest can be distributed along the outside walls. Materials storage racks cover a large portion of the warehouse floor area. The warehouse heaters can be distributed along the outside walls, but they should be positioned so that as much radiation as possible is directed to the floor. Spot heaters or door heaters will be required near the large overhead doors. A preliminary layout is shown above, by Figure 9-6.

7) Verify clearances and spacings.
An example of manufacturer's data is provided on this page. A comparison of this data with the proposed layout verifies that the unit sizes and the heating patterns are acceptable.

8) Controls and accessories.
Use the manufacturer's catalog to select an ignition and flame supervision control package. Specify the number of thermostats that will be required. Also specify any reflector options that may be appropriate.

Performance Data — Direct-Fired Heaters

Unit Size	Input BTUH	Recom. Mount. Height	Distance to Wall	Distance Between Units	Distance Between Rows
30	30000	10 - 12	6	8 - 30	10 - 70
50	50000	12 - 14	10	14 - 40	14 - 80
75	75000	15 - 17	12	16 - 50	20 - 100
100	100000	17 - 20	12	20 - 60	20 - 120

9-16 Indirect-Fired Heaters — Total Building Heat

Indirect-fired units are available in a vertical panel design and in unitized radiant tube configurations. Indirect-fired heaters operate at a lower surface temperature than direct-fired heaters, they do not expose the indoor air to an open flame, they can use outdoor air as a source of combustion air and they can be vented. Indirect-fired radiant heaters can be used wherever direct-fired heaters would be used and they can be used in situations where a direct-fired unit would not be appropriate. For example, they should be used in:

- Tight buildings or poorly insulated buildings where condensation could be a problem.
- Buildings with contaminated atmospheres that would cause atmospheric burners to malfunction.
- Buildings with contaminated atmospheres that would cause toxic fumes to be generated by an atmospheric burner.

- Situations that require a highly controlled (indoor) atmosphere.
- Situations where direct-fired equipment is unlawful.
- Buildings with low ceiling heights.
- Situations where there are limited clearances between the heater and the structure or between the heater and combustible materials.
- Situations where a direct-fired burner would be exposed to high winds.

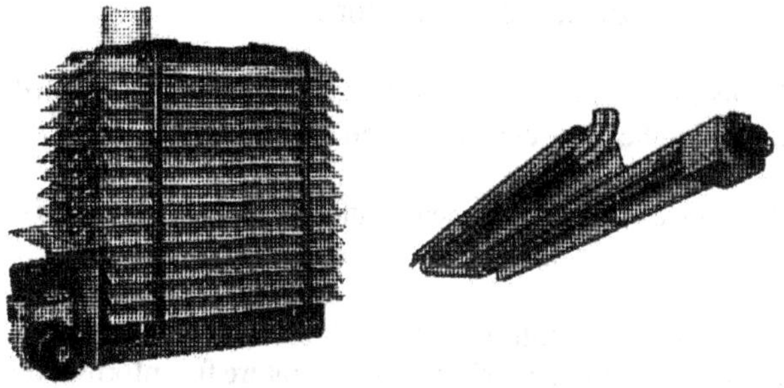

Indirect-Fired Equipment

Providing that the products of combustion are vented, indirect-fired radiant unit heaters do not require dilution ventilation. Therefore, the procedure that is used to size and position vented, indirect-fired unit heaters is less complicated than the procedure that is used for unvented, direct-fired equipment. This procedure is summarized below. **Note that when the heater is vented, the design is based on the output capacity of the indirect-fired unit.** *(If the products of combustion are not vented, the procedure that is used to size and position direct-fired unit heaters — as described in Section 9-15 — applies to the indirect-fired equipment.)*

Design Procedure — Total Building Heat
To design a total building heating system, the designer must determine the required output capacity and then decide how many heaters to use. Placement of the heaters also is important. If vented equipment is used, the calculations that are associated with fume dilution and condensation are not required. Controls and accessories also should be considered.

Step 1) Make a **Manual** N load estimate.
Perform a **Manual N** load estimate and list the values of the transmission loads, the infiltration load, and if required, the ventilation load. It also may be necessary to modify the standard **Manual N** load estimate to account for heating load associated with cold materials that are brought into the building from the outdoors.

Step 2) Determine the output heating capacity.
The combined output heating capacity of the heaters should equal the design heating load (DHL). Include all the transmission and outdoor air loads that will actually be imposed on the radiant heaters. Do not include outdoor air loads that are neutralized by make up air equipment. If make up air units or other types of heat sources add heat to the building, reduce the design heating load accordingly. Door heaters and air curtains should be used to control the heat loads associated with shipping doors.

> Some manufacturers may not list the output capacity of vented, indirect-fired equipment. When this is the case, either ask the manufacturer for this information or assume that the output capacity is equal to 75 percent of the input capacity.

Step 3) Determine the number of heaters.
Try to use the same size heater throughout the building, but this is not mandatory. Try to mount the heaters as low as possible. Larger heaters require higher mounting heights — ignore heater sizes that have recommended mounting heights that are incompatible with the desired mounting height. For a specific heater model, the required number of heaters (N) can be calculated by dividing the Step 2 design heating load (DHL) by the output capacity of the heater.

$$N = DHL / Output\ Capacity$$

Try to use as few heaters as possible. The final number of heaters will be a compromise between installed cost and performance. Consider the recommended mounting height, the effectiveness of the heating pattern, the required clearances associated with the structure, the materials that are stored in the structure and any obstructions that would restrict heater placement options. It may be necessary to evaluate two or more designs, depending on the spacings and heating patterns that are produced by using larger or smaller heaters.

Step 4) Layout a heating pattern.
The preferred heating pattern is produced by a layout that distributes the heaters along the outside walls. The number of heaters required along a particular wall (N_w) can be determined by multiplying the total number of heaters (N) by the ratio of the loss associated with the wall in question divided by the total loss associated with all the walls.

$$N_W = N \times (Wall\ Loss / Total\ of\ Wall\ Losses)$$

It may not always be possible to use the perimeter layout pattern, but other patterns will produce acceptable performance. Heaters can be uniformly distributed to blanket the floor area or they can be lined up under the ridge line of the roof. Combinations of these three basic patterns also can be used. The following factors should be considered when a pattern is evaluated.

- A larger number of smaller heaters will provide a more uniform heating pattern, but this must be balanced against the installation cost.
- Narrow buildings are ideal candidates for the perimeter pattern because the floor area in the middle of the building can be radiated by the perimeter units.
- A combination of perimeter units and ridge-mounted units can be used to radiate the entire floor area of a wide building.
- Distribute most of the heaters in areas that are normally occupied. Concentrate more heat in areas that have higher loads and use less heat in storage areas.
- Zone control becomes more precise as the number of heaters increase.
- Use spot heat for isolated work stations. Use spot or area heat in areas that have unusually high loads.
- The radiant heating patterns should always be directed toward the open floor areas, do not radiate stored materials.
- Reflectors can either be used to spread a pattern over a wider area or to concentrate the pattern. Reflector geometry also affects the recommended mounting height and the minimum distance between the heater and a wall.
- Units can be installed at an angle — this reduces the recommended mounting height and allows the heater to be installed closer to the wall.
- Layouts must be compatible with safety. Consider clearances required for combustible materials and hazardous fumes or processes.
- Layouts must accommodate obstructions such as cranes, lights, piping, sprinklers and manufacturing equipment.

Step 5) Verify clearances and spacings.
The heating pattern layout must be compatible with manufacturer's recommendations regarding clearances and spacings. Refer to the manufacturer's performance data and installation manuals for information about clearances, mounting height and the spacing between heaters.

- Minimum recommended mounting height — mounting at a lower height could cause discomfort.
- Maximum recommended mounting height — mounting at a higher height reduces the intensity of the radiation striking the floor.
- Clearances between the heater and the ceiling — overall efficiency decreases and fuel consumption increases if the heaters are mounted close to the ceiling.
- Clearances between the heater and the wall — wall heat losses increase and fuel consumption increases if walls are radiated.
- Maximum distance between heater rows — excessive row spacing reduces comfort.
- Maximum distance between heaters in a given row — excessive spacing reduces comfort.
- Reflector geometry has an effect on the recommended mounting heights, clearances and spacings.
- Clearances between the heater and combustible materials.

Step 6) Select the controls and accessories.
Refer to the manufacturer's product literature for information about controls and accessories. Select a burner control package. (A variety of ignition and flame supervision control options are usually available.) Remote thermostats can be used to control space temperature. The number of heaters controlled by one thermostat will depend on the total number of units, the heating pattern layout and the zoning requirements; refer to the manufacturer's installation instructions for specific recommendations. Locate thermostats in areas of greatest heat loss. Unless the manufacturer advises otherwise, locate thermostats so they are not radiated by the heaters. In some cases, special reflectors or reflector extensions may be available as an accessory item.

Example — Total Building Heat
Figure 9-5, on page 9-12, shows a manufacturing building that could be heated by indirect-fired, vented, radiant heating units. The building has a machine tool room, a materials assembly room and a warehouse. The tool and assembly rooms have process exhaust systems that require make up air units. Direct-fired make up air units will be used to temper the outdoor air that is required to replace exhausted air. The internal latent loads are negligible. Use this information and information extracted from the load calculation (see Step 1 below) to design a vented, indirect-fired radiant heating system.

1) Load calculation
The table on page 9-12 summarizes pertinent information that was extracted from the load calculation. The heating loads are based on a 0°F outdoor design temperature. These loads are summarized as follows:

	Machine	Assembly	Storage
Transmission	129000	173000	455000
Infiltration	64000	165000	257000
Total	193000	338000	712000

In the previous example, the direct-fired, unvented equipment caused condensation to form on the assembly room skylights and on the storage room walls. This condensation problem was neutralized by adding a second glazing to the skylights and by adding an inside surface to the warehouse walls. In this example, building envelope modifications are not necessary because condensation will not be a problem when the products of combustion are vented directly to the outdoors.

Even though envelope modifications are not necessary to prevent condensation, it still may be a good idea to improve the U-value of the warehouse walls. As in the previous example, the addition of an inside surface will almost cut the transmission loss in half. This simple modification will substantially reduce the first cost (less installed capacity) and the operating costs (less fuel consumption).

2) Determine the output heating capacity.
The required output capacity is equal to the transmission load plus the infiltration load. (The direct-fired make up air units will temper the outdoor air that replaces the exhausted air.) These loads, which are based on an improved wall in the storage area , are summarized below.

	Machine	Assembly	Storage
Transmission	129000	173000	258000
Infiltration	64000	165000	257000
Total	193000	338500	515000

3) Determine the number of heaters.
It's desirable to use as few heaters as possible but it also is necessary to consider the minimum mounting height and to have a heating pattern that provides good coverage. Based on the manufacturer's recommendations regarding mounting height and the distance between heaters, a 40,000 BTUH (30,000 output) unit was selected for the machine and assembly rooms and a 100,000 BTUH (75,000 output) unit was selected for the storage area. The number of heaters required for each area is calculated by dividing the design load by the output capacity.

	Machine	Assembly	Storage
Load	193000	338500	515000
Capacity	30000	30000	75000
Number	7	11	7

4) Layout a heating pattern.
The machine room is long and narrow, so the heaters can be uniformly distributed along the outside walls. The assembly room is wide, so some heaters should be installed in the middle of the room and the rest can be distributed along the outside walls. Materials storage racks cover a large portion of the warehouse floor area. The warehouse heaters can be distributed along the outside walls, but they should be positioned so that as much radiation as possible is directed to the floor. Spot heaters or door heaters will be required near the large overhead doors. A preliminary layout is shown by Figure 9-6, on page 9-14.

5) Verify clearances and spacings
An example of radiant tube heater performance data is provided below. A comparison of this data with the proposed layout verifies that the unit sizes and the heating patterns are acceptable.

Performance Data — Indirect-Fired Heaters						
Unit Size	Output Btuh	Recom. Mount. Height	Distance to Wall 0°	45°	Distance Between Units	Distance Between Rows
40	30000	9 - 15	15	5	9 - 30	10 - 50
60	45000	11 - 18	17	7	12 - 38	14 - 70
75	56000	11 - 20	20	8	15 - 43	16 - 85
100	75000	13 - 25	20	10	13 - 40	16 - 85

6) Controls and accessories
Use the manufacturer's catalog to select an ignition and flame supervision control package. Specify the number of thermostats that will be required. Also specify any reflector options that may be appropriate.

9-17 Indirect-Fired, Vacuum-Tube Systems

Indirect-fired, vacuum-tube, low intensity radiant heating systems may consist of a customized, field fabricated system, or a system that is pieced together from pre-engineered, factory assembled components. In either case, one or more burners inject combustion gases into a "radiant pipe" and then these hot gases are pulled through downstream "tail pipes" by a vacuum pump, which ultimately expels the products of combustion to the outdoors. **Since the combustion gases are vented, the problems associated with dilution ventilation and condensation are avoided.**

These systems can be very efficient. A properly designed system will extract almost all of the available heat from the combustion gases before they are expelled to the outdoors. Efficiencies of 90 percent or higher are possible.

Vacuum-tube systems can be used wherever direct- or indirect-fired unitary heaters would be used and they can be used in situations where direct-fired equipment would not be appropriate. For example, they should be used in:

- Tight buildings or poorly insulated buildings where condensation could be a problem.
- Buildings with contaminated atmospheres that would cause atmospheric burners to malfunction.
- Buildings with contaminated atmospheres that would cause toxic fumes to be generated by an atmospheric burner.
- Situations that require a highly controlled (indoor) atmosphere.
- Situations where direct-fired equipment is unlawful.
- Buildings with low ceiling heights.
- Situations where there are limited clearances between the heater and the structure or between the heater and combustible materials.
- Situations where a direct-fired burner would be exposed to high winds.

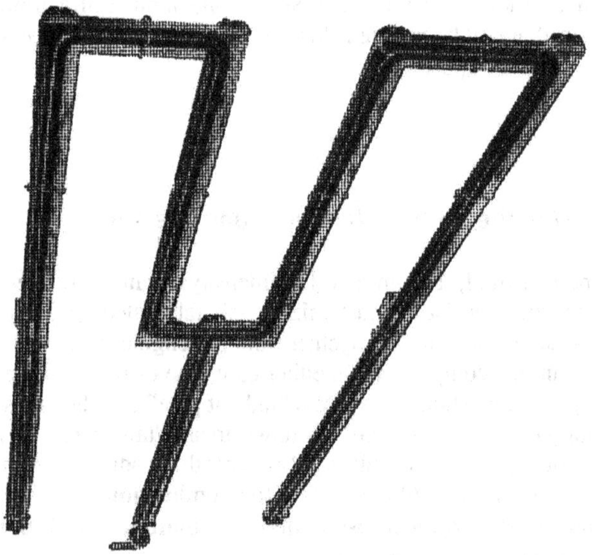

Radiant Vacuum-Tube System

Vacuum-tube radiant heating systems consists of one or more gas burners, end vents, sections of steel tubing, a vacuum pump, a condensate trap and reflectors. The burners can extract combustion air from the heated space or from the outdoors. (Manufacturers offer optional, powered, outdoor air combustion systems.) Reflectors can be adjusted to project the radiant energy straight down or at an angle.

The steel tubing is divided into radiant sections and a tailpipe section. Radiant sections carry the hot gasses that are immediately downstream from the burner. Manufacturers specify minimum and maximum lengths for these sections. A tailpipe connects the radiant sections with the vacuum pump. Manufacturers also specify limitations on the length of the tailpipe section. Figure 9-7, at the top of the next page, shows the primary components of a vacuum-tube system.

A burner and an end vent are always located at the starting end of a radiant pipe. Multiple burners can be located at the starting ends of a system of parallel radiant pipes. Parallel burner arrangements are typical because burners usually cannot be installed in series. However, one manufacturer offers an exclusive, patented design that allows the burners to be installed in series. In any case, the radiant pipes are exhausted by a single vacuum pump located at the end of the tailpipe. A condensate drain also is located at the end of the tailpipe. Figure 9-8, on page 9-20, provides diagrams that show some typical piping arrangements.

Design Procedure — Total Building Heat

To design a total building heating system, the designer must determine the required heating capacity and then decide what size burners to use and how many burners to use. Placement of the burners, the geometry of the radiant tubes and the geometry of the tailpipe section also are important. Since vacuum systems are always vented, the calculations that are associated with fume dilution and condensation are not required. Controls and accessories also should be specified.

Step 1) Make a **Manual N** load estimate.

Perform a **Manual N** load estimate and list the values of the transmission loads, the infiltration load and if required, the ventilation load. It also may be necessary to modify the standard **Manual N** load estimate to account for a heating load that is associated with cold materials that are brought into the building from the outdoors.

Step 2) Determine the required burner input capacity.

The required burner input capacity (RIC) should be based on the design heating load (DHL). Include all the transmission and outdoor air loads that will actually be imposed on the radiant heaters. Do not include outdoor air loads that are neutralized by make up air equipment. If make up air units or other types of heat sources add heat to the building, reduce the design heating load accordingly. Door heaters and air curtains should be used to control the heat loads associated with shipping doors.

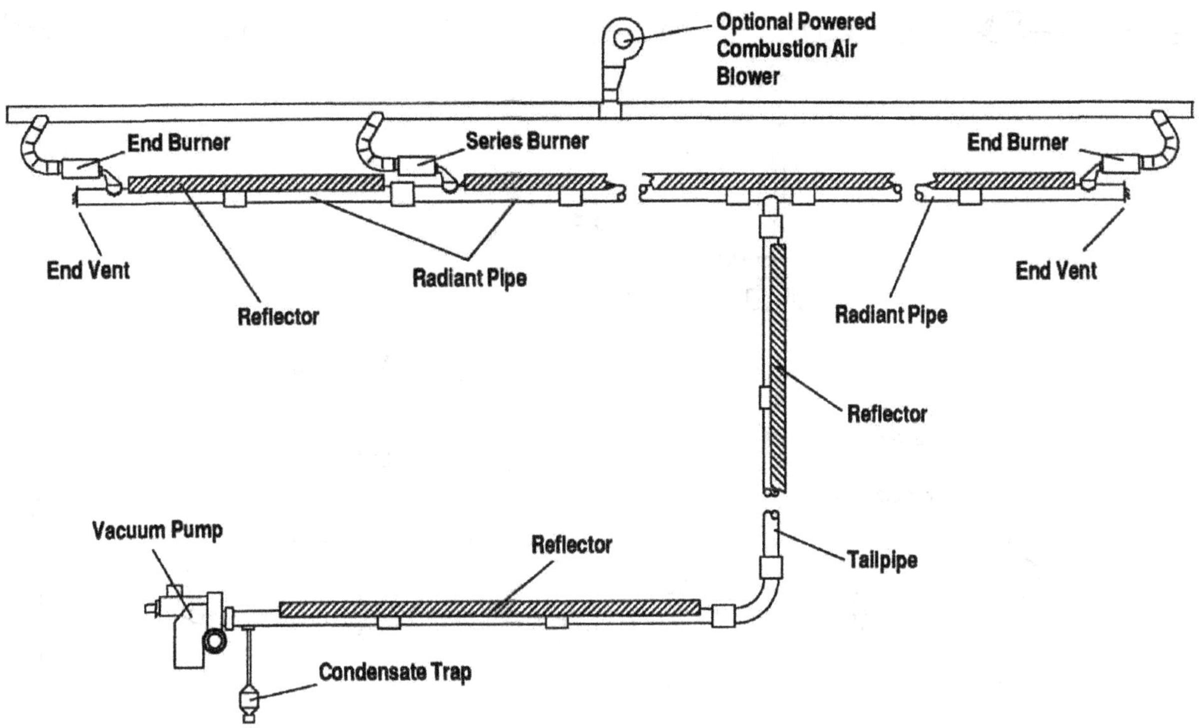

Figure 9-7

Note, the required input capacity may have to be increased if the actual mounting height is higher than the maximum mounting height recommended by the manufacturer. Refer to the manufacturer's engineering data for instructions on how to make a correction for mounting height. For example, one manufacturer requires a 1 percent increase in the input capacity for each foot in excess of the recommended (20 foot) mounting height. (A capacity increase of 10 percent would be required for a 30 foot mounting height.)

RIC = DHL x Height Correction

Step 3) Determine the number of burners.
Try to use the same size burner throughout the system, but this is not mandatory. Larger burners require higher mounting heights — avoid sizes that have recommended mounting heights that are incompatible with the desired mounting height. The number of burners will depend on the required input capacity, the capacity of a single burner and on the geometry of the piping layout. For a specific burner model, the required number of burners (N) can be calculated by dividing the Step 2 input capacity requirement (RIC) by the input capacity of the prototype burner.

N = RIC / Input Capacity

Try to use as few burners as possible. The final number of burners will be a compromise between installed cost and performance. Consider the recommended mounting height, the coverage of the piping pattern, the required clearances associated with the structure and the materials that are stored in the structure, and any obstructions that would interfere with equipment placement. It may be necessary to evaluate two or more designs, depending on the spacings and heating patterns that are associated with different piping layouts.

Step 4) Layout a piping pattern.
The preferred piping pattern is produced by a layout that distributes the radiant tubes along the outside walls. Try to locate the hottest pipes near areas that have the highest heat loss. Tail pipes should be installed in less critical areas. In most cases, it is not essential to cover interior areas with radiant pipes. Center areas, and other areas that have a low heat loss, will be adequately heated if the total system input capacity is equal to or greater than the required input capacity. Interior patterns, or a combination of perimeter and interior patterns, can be used if it is not possible to install the required heating capacity at the perimeter. The following factors also should be considered:

- Install more pipe in the areas that are occupied and less pipe in storage areas.
- Zone control becomes more precise as the number of burners increase.
- Use spot heat for isolated work stations. Use spot or area heat in areas that have unusually high loads.

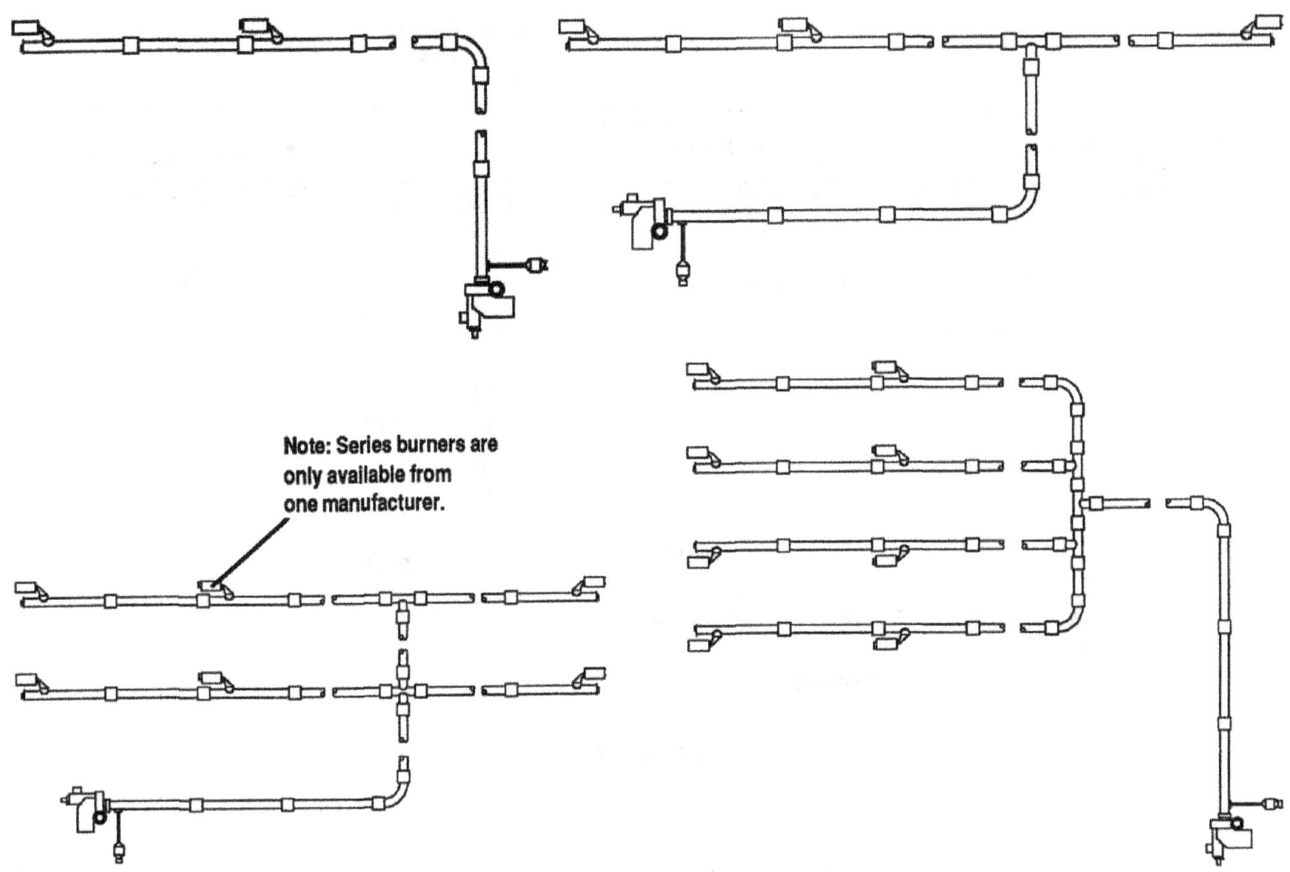

Figure 9-8

- The radiant heating patterns should always be directed toward the open floor areas; do not radiate storage racks or machinery.

- Reflectors can either be used to spread a pattern over a wider area or to concentrate the pattern — reflector geometry also affects the recommended mounting height and the minimum distance between the heater and a wall.

- Reflectors can be installed at an angle — this allows the pipe to be installed closer to the wall.

- Layouts must be compatible with safety — consider clearances required for combustible materials and hazardous fumes or processes.

- Layouts must accommodate obstructions such as cranes, lights, piping, sprinklers, storage racks and manufacturing equipment.

Step 5) Check manufacturer's design guidelines.
Manufacturer's design manuals should provide rules that govern how the various components can be combined to form a complete system. (This step is not required for pre-engineered, factory assembled equipment.) These rules should cover the following items:

- How to match end vents with burner capacity.

- How to determine the tailpipe length. There is a relationship between the number of burners and end vents and the tailpipe length. Or, if the burners and the corresponding end vents are not all the same size, there is a relationship between the maximum input capacity and the tailpipe length.

- How to select a vacuum pump. There are limitations on the maximum number of burners and end vents that can be served by one vacuum pump. If the burners and the corresponding end vents are not all the same size, there is a limitation on the maximum input capacity that can be handled by one vacuum pump. **Note that vacuum pump performance must be adjusted for altitude.**

- For systems that allow burners to be placed in series there is a limit on the number of burners that can be placed in a single branch. If the burners are not all the same size, there is a limitation on the maximum input capacity that can be installed on one branch.

- How to select a combustion air blower. (There are limitations on the maximum number of burners and end vents that can be served by one combustion air blower. If the burners and the corresponding end vents are not all the same size, there is a limitation on the maximum input capacity that can be handled by one blower.)

Step 6) Verify clearances, spacings.
The heating pattern layout must be compatible with manufacturer's recommendations regarding clearances and spacings. Refer to the manufacturer's performance data and installation manuals for information about clearances, mounting height, minimum and maximum pipe lengths and the spacing between burners. For example, consider the . . .

- Minimum mounting height. (Mounting at a lower height could cause discomfort.)
- Maximum recommended mounting height. (Mounting at a higher height may increase the required input capacity.)
- Clearances between the system components and combustible materials.
- Clearances between the system components and the ceiling.
- Clearances between the system components and the wall. (Wall heat losses increase and fuel consumption increases if the walls are radiated.)
- Minimum and maximum distances between burners on the same section of pipe. (This only applies to series systems).
- Minimum length of tubing required directly upstream from a burner.
- Minimum and maximum lengths for radiant pipe sections that are downstream from a burner.
- Minimum and maximum lengths for the tailpipe section.
- Maximum distance between pipe rows. (Excessive row spacing reduces comfort.)
- Reflector geometry. (Because it has an affect on the recommended mounting heights, clearances and spacings.)

Step 7) Select the controls and accessories.
Refer to the manufacturer's product literature for information about controls and accessories. Select a burner control package (a variety of ignition and flame supervision control options are usually available). Remote thermostats can be used to control space temperature. The number of burners controlled by one thermostat will depend on the total number of units, the piping pattern layout and the zoning requirements — refer to the manufacturer's installation instructions for specific recommendations. Locate thermostats in areas of greatest heat loss. Unless the manufacturer advises otherwise, locate thermostats so they are not radiated by the pipes. Also specify the type of reflectors or reflector extensions.

Example — Custom System with Series Burners
Figure 9-5, at the top of page 9-12, shows a manufacturing building that could be heated by a radiant vacuum-tube system that has branch burners arranged in series. The building has a machine tool room, a materials assembly room and a warehouse. The tool and assembly rooms have process exhaust systems that require make up air units. Direct-fired make up air units will be used to temper the outdoor air that is required to replace exhausted air. The internal latent loads are negligible. A combustion air blower is not required because the indoor air can be used for combustion air. The plant is located 600 feet above sea level. Use this information and information extracted from the load calculation (see Step 1 below) to design a radiant vacuum-tube system.

Custom, field assembled systems that use burners in series are only offered by one manufacturer, but they are very common. The manufacturer of this proprietary product introduces the concept of a "flow unit" in their design manual. This concept also is used in the following example. A flow unit is a measure of the amount of air and fuel that flows through a burner, an end plate or the vacuum pump. Flow units also are used to calculate minimum and maximum pipe lengths and to establish a value for the maximum capacity of the vacuum pump.

1) Load calculation.
The table on page 9-12 summarizes pertinent information that was extracted from the load calculation. The heating loads are based on a 0°F outdoor design temperature. These loads are summarized below.

	Machine	*Assembly*	*Storage*
Transmission	*129000*	*173000*	*455000*
Infiltration	*64000*	*165000*	*257000*
Total	*193000*	*338000*	*712000*

Even though envelope modifications are not necessary to prevent condensation, it still may be a good idea to improve the U-value of the warehouse walls. The addition of an inside surface will (almost) cut the transmission loss in half. This simple modification will substantially reduce the equipment cost (less installed capacity) and the operating costs (less fuel consumption).

2) Determine the required input heating capacity.
The required burner input capacity (RIC) is based on the design heating load (DHL), which is equal to the transmission load plus the infiltration load. (The direct-fired make up air units will temper the outdoor air that replaces the exhausted air.) The design heating loads are summarized below. Note that the load in the storage area is based on an improved wall.

	Machine	*Assembly*	*Storage*
Transmission	*129000*	*173000*	*258000*
Infiltration	*64000*	*165000*	*257000*
Total	*193000*	*338500*	*515000*

The manufacturer's product information indicates that a 1 percent per foot height correction is required when the mounting height exceeds 20 feet. No correction is required in the machine room or the assembly room because the walls are not more than 20 feet high. An 8 percent correction is required for the storage room because the mounting height will be equal to 28 feet. The design heating loads for the three rooms are recalculated below.

Machine room = 1.00 x 193000 = 193000
Assembly room = 1.00 x 338000 = 338000
Storage = 1.08 x 515000 = 556200

3) Determine the number of burners.
It's desirable to use as few burners as possible, but it also is necessary to have a piping pattern that provides good coverage. The piping pattern also must be compatible with the manufacturer's recommendations regarding pipe lengths.

Based on the manufacturer's design information, as shown by Figure 9-9, a 40,000 BTUH burner was selected for the machine room, a 60,000 BTUH burner was selected for the assembly rooms and a 100,000 BTUH burner was selected for the storage area. These burner sizes can be used in a perimeter pattern and they should be compatible with the manufacturer's recommendations regarding minimum and maximum pipe lengths. The number of burners required for each area is calculated by dividing the required input capacity by the burner input capacity.

	Machine	*Assembly*	*Storage*
Load	*193000*	*338500*	*556200*
Capacity	*40000*	*60000*	*100000*
Number	*5*	*6*	*6*

4) Layout a piping pattern.
Only five burners are required in the machine room but six burners will be used. This way, three burners can be installed along each side of the room. The same plan also will work for the assembly room, since it requires six burners. The six warehouse burners also can be distributed along the outside walls. Spot heaters or door heaters will be required near the large overhead doors. A preliminary layout is shown at the top of the next page by Figure 9-10.

Design Information					
Input BTUH (1000's)	40	60	80	100	120
Maximum burners per branch	4	4	3	3	2
Flow units per burner	4	6	8	10	12
Flow units per end vent	10	15	20	20	20
Distance between burners					
Mlinimum	12	20	25	30	35
Maximum	25	35	45	60	70
Tailpipe length per flow unit (maximum burner spacing)					
Minimum	1.2	1.2	1.2	1.2	1.2
Maximum	2.5	2.5	2.5	2.5	2.5
Tailpipe length per flow unit (minimum burner spacing)					
Minimum	2.0	2.0	2.0	2.0	2.0
Maximum	3.0	3.0	3.0	3.0	3.0
Burner to elbow spacing					
Minimum to downstream elbow	5	10	10	15	15
Minimum to upstream elbow	2	2	2	2	2
Minimum mounting height	8	8	10	15	15

Notes:
Vacuum pump capacity = 110 flow units (0 to 2000 feet)
Minimum distance from wall = 1 foot (with extension)
Combustion air blower capacity = 20 to 120 flow units depending on the diameter and length of the duct runs.

Figure 9-9

5) Check manufacturer's design guidelines.
The manufacturers design information, as provided by Figure 9-9, indicates that the requirements that are listed below must be satisfied by the piping layouts. Reference to Figure 9-10, at the top of the next page, shows that these requirements are indeed satisfied by the proposed layouts.

Machine room layout:

- A maximum of four burners are allowed per branch. Only three burners per branch are required for the proposed layout. Note that three burners plus one end vent adds up to 22 flow units per branch.
- The vacuum pump can handle 110 flow units. There are only 44 flow units associated with the proposed layout.
- A combustion air system is not required.

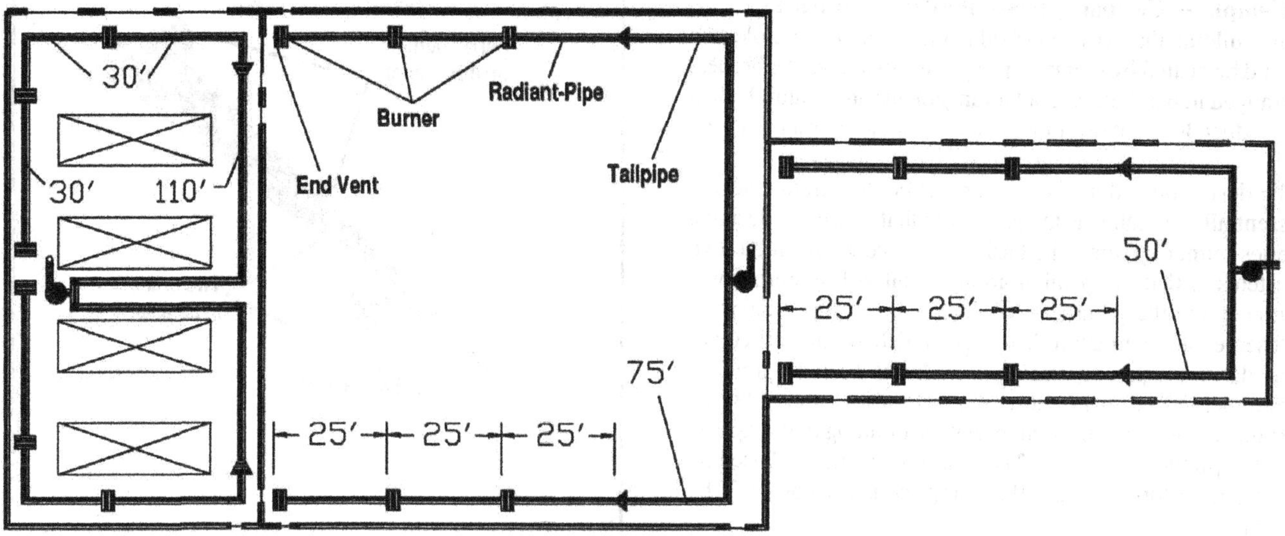

Figure 9-10

Assembly room layout:
- A maximum of four burners are allowed per branch. Only three burners per branch are required for the proposed layout. Note that three burners plus one end vent adds up to 33 flow units per branch.

- The vacuum pump can handle 110 flow units. There are only 66 flow units associated with the proposed layout.

- A combustion air system is not required.

Storage room layout:
- A maximum of three burners are allowed per branch. Only three burners per branch are required for the proposed layout. Note that three burners plus one end vent adds up to 50 flow units per branch.

- The vacuum pump can handle 110 flow units. There are 100 flow units associated with the proposed layout.

- A combustion air system is not required.

6) Verify clearances and spacings.
An example of the manufacturer's data is provided on the previous page by Figure 9-9. A comparison of this data with the proposed layouts, which are shown above, verifies that the layouts satisfy the manufacturer's requirements. Note that reflector extensions should be used so that the burners and pipes can be installed close to the wall.

Machine room layout:
- The minimum distance between burners is 12 feet.

- The maximum distance between burners is 25 feet.

- If the burners are spaced at 25 foot intervals, the minimum length of a branch tailpipe is 26.4 feet (1.2 x 22 flow units) and the maximum length is 55 feet (2.5 x 22 flow units).

- The minimum mounting height is 8 feet.

Assembly room layout:
- The minimum distance between burners is 20 feet.

- The maximum distance between burners is 35 feet.

- If the burners are spaced at 25 foot intervals, the minimum length of a branch tailpipe should be about 66 feet (2.0 x 33 flow units) and the maximum length should be about 99 feet (3.0 x 33 flow units).

- The minimum mounting height is equal to 8 feet.

Storage room layout:
- The minimum distance between burners is 30 feet.

- The maximum distance between burners is equal to 60 feet.

- If the burners are spaced at 30 foot intervals, the minimum length of a branch tailpipe should be about 100 feet (2.0 x 50 flow units) and the maximum length should be about 150 feet (3.0 x 50 flow units).

- The minimum mounting height is 10 feet.

7) Controls and accessories.
Use the manufacturer's catalog to select an ignition and flame supervision control package. Specify the number of thermostats that will be required. Also specify any reflector options that are desired.

Example — Custom System with Parallel Burners
The building that was described in the previous example also could be heated by radiant pipe systems that have the burners arranged in parallel. Parallel arrangements are required when a product does not offer the (patented) series burner feature.

The design procedure for the parallel burner arrangement is essentially the same as the procedure that is used to design a series burner system. The basic difference between the two systems is that the parallel arrangement will require fewer burners, but these burners will have a larger input capacity. Observe all manufacturer's design guidelines and all recommendations regarding spacing — as discussed in the previous example. Figure 9-11 below illustrates how parallel burner arrangements could be used to heat the building that was used for the previous example. These arrangements are based on the same performance data that was provided on page 9-22 by Figure 9-9.

Pre-engineered Vacuum-Tube Systems
Manufacturers also offer pre-engineered vacuum-tube systems. All components are matched at the factory and shipped as a single package. Some of these systems are limited to a parallel burner arrangement, but one manufacturer offers a system that has branch burners installed in series. Either type of arrangement, series or parallel, is available with a variety of burner sizes and pipe geometries. However, there are limits on the number of series burners (usually two) and the number of branches connected to one vacuum pump (usually two). Two examples of pre-engineered systems are provided on this page by Figure 9-12.

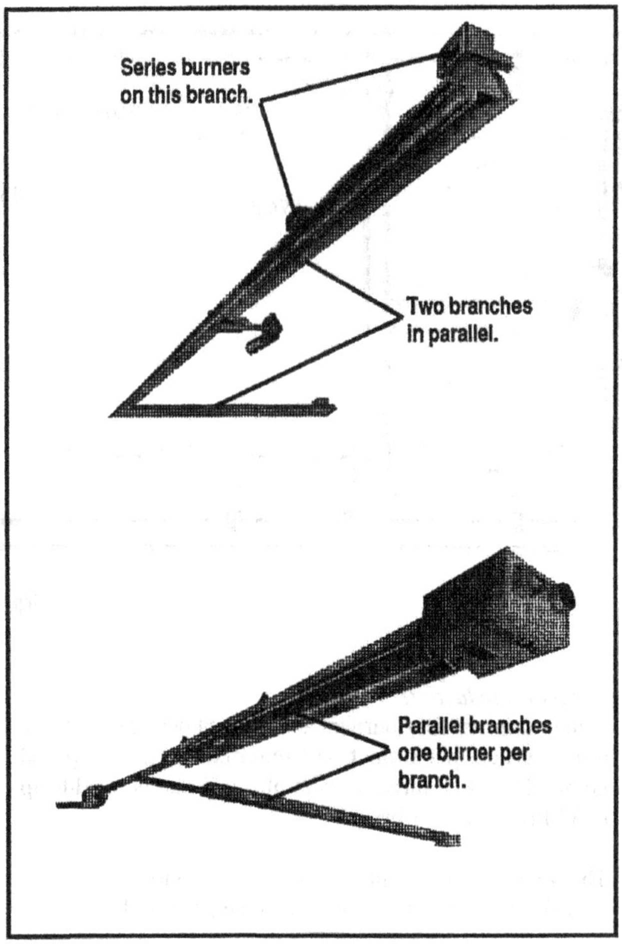

Figure 9-12

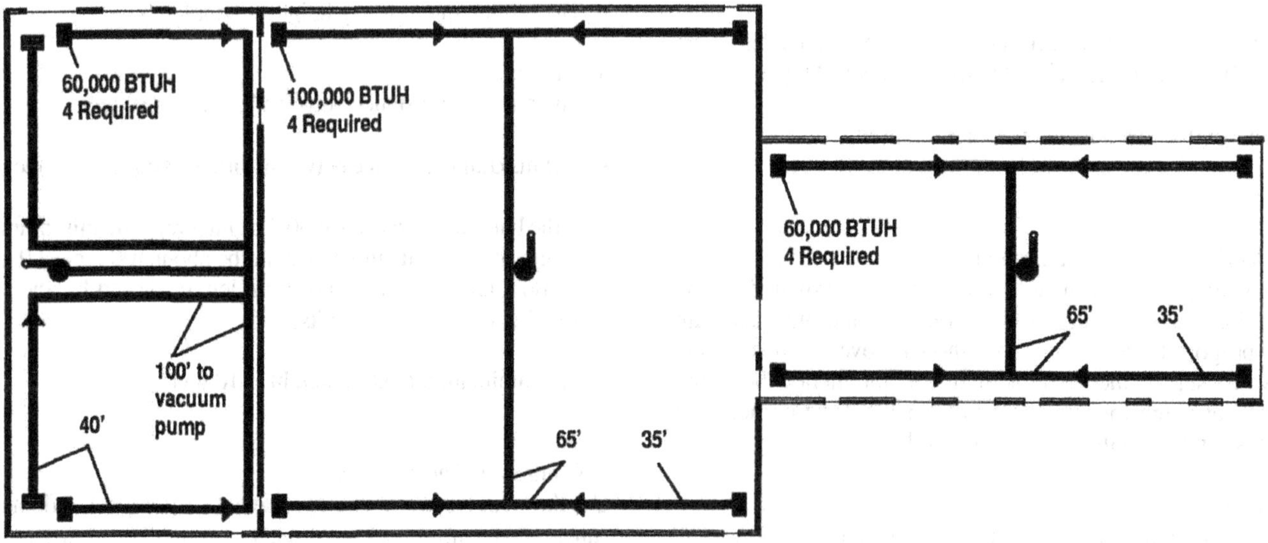

Figure 9-11

9-18 Low-Temperature, Electric, Radiant Panels

Low-temperature, electric, radiant panels operate at a surface temperature that is below 200°F. These panels can be inserted into a T-bar ceiling, surface mounted on a ceiling or wall or suspended by chains. These panels can be used as a primary or secondary source of heat. They also can be used for spot heat. They are frequently used for commercial applications such as offices, motel rooms, schools and condominiums, but there is no reason why they could not be used for residential applications. They are manufactured in sizes that range from 2x2 (feet) to 2x8 (feet). Wattage inputs for a single panel range from 250 Watts (853 BTUH) to 1500 Watts (5120 BTUH). As with any type of electric resistance heat, the installed resistance heating capacity (BTUH) should be equal to or greater than the design heating load.

9-19 Spot Heat and Area Heat

Even though the ambient temperature may be relatively low, spot or area heat can be used to provide a comfortable environment. This is accomplished by heating a persons body with direct radiation. However, this sensation of comfort can only be maintained in the immediate area that is blanketed by a radiant heating pattern.

The equipment sizing procedure consists of predicting the amount of heat that is required at the surface of a person's body and finding a device that will deliver the correct amount of radiant heat. As discussed in Section 9-2, comfort will be maintained if the net heat lost from the body is replaced by direct radiation. Figure 9-1, on Page 9-1, can be used to estimate the amount of radiant heat that is required and manufacturer's performance data can be used to determine the heater size.

Figure 9-13, on this page, provides an example of manufacturer's performance data. Note that the tables indicate the heat flux (BTUH per SQ.FT.) that is provided at a spot that is 3 feet above the floor (Point B). This spot corresponds to the mid-section of a persons body. Also, note that the intensity of the flux at Point B depends on the mounting height (H) and on the offset (D).

Since people loose heat from all sides of their body, it is desirable to deliver the radiant heat from at least two opposing sides. When heat can only be delivered from one side, it is common practice to increase the amount of heat by 20 to 30 percent.

Example — Spot Heat

Use Figure 9-13 to size radiant heaters for a spot heating application. The ambient temperature is 20°F, the velocity of the ambient air is negligible and a heavily clothed person will be performing bench work. The mounting height will be equal to 15 feet and the offset will be equal to 4 feet on either side of the work area.

Heat flux required = 60 BTUH/SQ.FT. (See Figure 9-1)

Use two heaters, one on each side of the workbench.

Required flux per heater = 30 BTUH/SQ.FT.

Heater size = Model K-100 (H=15 and D=4)

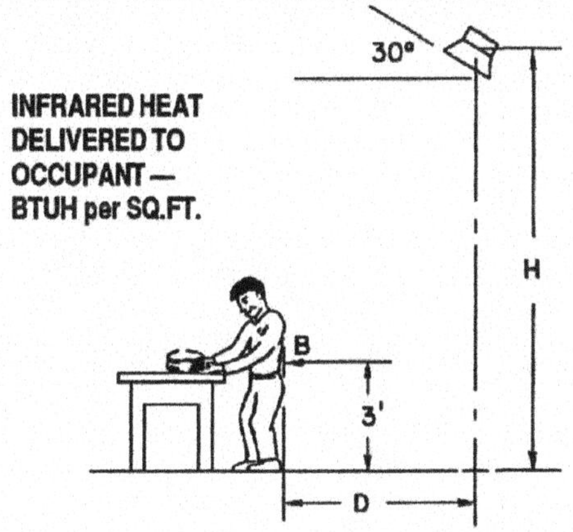

MODEL K-100 TABLE 3

H-FT.	D-FT. 2	4	6	8	10	12	14	16	18	20	22	24
7	354	290	174	110	72	49	34	25	18	14	12	9
9	116	165	129	91	66	49	36	28	22	17	13	11
11	49	88	90	72	56	44	34	27	22	18	15	12
13	25	49	59	56	46	38	31	25	21	18	15	12
15	14	29	39	41	38	32	27	23	19	16	14	12
17	9	18	26	30	30	27	24	20	17	15	13	12
19	6	12	18	22	23	22	20	18	16	14	12	11
21	4	9	13	16	18	18	18	16	14	13	11	10
23	3	6	9	12	14	15	15	14	13	12	10	9
25	2	5	7	9	11	12	12	12	11	10	10	9
27	2	4	5	7	9	10	10	10	10	9	9	8
29	1.4	3	4	6	7	8	9	9	9	8	8	7
31	1.1	2	3	5	6	7	7	8	8	7	7	7
33	0.9	1.8	3	4	5	5	6	6	7	7	6	6

MODEL K-150 TABLE 4

H-FT.	D-FT. 2	4	6	8	10	12	14	16	18	20	22	24
7	531	435	262	164	108	73	51	37	28	22	17	14
9	174	248	193	137	98	73	54	42	32	25	20	16
11	74	133	135	109	84	66	51	41	33	27	22	18
13	38	74	89	84	70	56	46	38	31	26	22	18
15	22	44	59	62	58	48	40	34	29	24	21	18
17	14	28	40	45	45	41	36	30	26	23	20	17
19	9	18	27	33	34	34	31	27	24	21	18	16
21	6	13	19	24	27	28	26	24	22	19	17	15
23	4	10	14	18	21	22	22	21	19	18	16	14
25	3	7	11	14	17	18	18	18	17	16	14	13
27	3	5	8	11	13	15	15	15	15	14	13	12
29	2	4	6	9	11	12	13	13	13	13	12	11
31	2	3	5	7	8	10	11	11	11	11	11	10
33	1.3	3	4	6	7	8	9	10	10	10	10	9

Figure 9-13

Section 10
Selecting Humidification Equipment

Humidity affects comfort and health; it can prevent the buildup of static electricity; and it helps to preserve paper, wood, plaster, fabrics and other materials. Humidity control also is critical for some industrial processes and for some materials storage applications. However, too much humidity is undesirable. Condensation can form on indoor surfaces, particularly on windows and window frames. Concealed condensation is a more insidious problem. Condensation that forms inside the structural components of the building — the walls or ceilings for example — may damage the structure or cause biological growth that can be harmful to the occupants.

Initially, the HVAC system designer must decide how much humidification is desirable for a particular application. Then the designer must balance this requirement with the humidity level that is compatible with the thermal characteristics of the structure. Usually, this work requires some compromise. Humidification equipment can be selected and sized after the design value for the indoor humidity is determined.

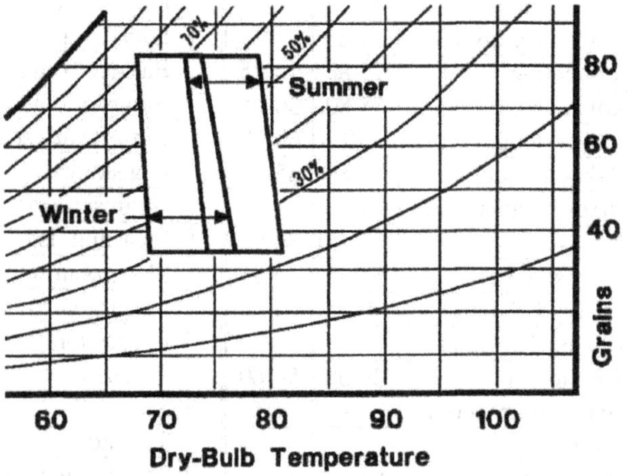

ASHRAE Comfort Chart

10-1 Recommended Humidity Levels

Information about recommended moisture levels, in terms of relative humidity, is provided below. Note that these humidity levels are associated with a specific indoor dry-bulb temperature. However, the maximum indoor humidity level may be limited by the outdoor winter design temperature, the indoor temperature and by the construction details that are associated with the building. Condensation must not be allowed to occur on any exposed or hidden surface.

Humidity Required for Comfort
The ASHRAE comfort chart shows the relationship between comfort, relative humidity and dry-bulb temperature. This chart shows that acceptable humidity levels can range from less than 30 percent to more than 60 percent, depending on the season and the dry-bulb temperature.

Humidity Required for Health
The ability of a persons body, especially the respiratory tract, to resist germs and viruses is optimumized when the relative humidity is between 30 and 60 percent. Health and comfort problems that are associated with the skin and eyes also can be caused by inadequate humidification.

Humidity Required to Control Static Electricity
A humidity level of 45 percent is required to prevent static buildup on people, machinery and on most materials, but higher levels may be required for wool and for some synthetic materials. In explosive environments, the relative humidity should exceed 50 percent.

Humidity Required for Processes and Applications
Manufacturing processes and materials storage applications require specific temperature and humidity levels. Some information about these temperature and humidity levels is provided on the next page by Table 10-1. However, the information that is provided by Table 10-1 is not conclusive or comprehensive. Always ask the people that operate the plant or facility to verify the temperature and humidity levels that are required for a specific process or application. In some cases, it will be necessary to obtain the temperature and humidity requirements from the manufacturer of the process equipment.

10-2 Construction Details Limit Humidity Levels

The upper limit for the indoor humidity might be established by the construction details and the winter design temperatures. Condensation must not be allowed to occur on the building's inside surfaces or within the structural components that are associated with the building envelope. If condensation does occur, the building could be subject to damage from moisture, frost or ice. Condensation also can cause mold, mildew and other types of biological growth that can either damage the building, produce unpleasant odors or compromise the health of the occupants.

Visible Condensation
Condensation on the interior surfaces will not occur if the dew point temperature of the indoor air is lower than the temperature of the coldest inside surface. Therefore, the maximum

Temperature and Relative Humidity for Various Applications

Application	DB	RH
Abrasives Mfg.	78	50
Baking		
Cake mixing	70-75	65
Crackers & biscuits	60-65	50
Fermenting	75-80	70-75
Flour storage	65-80	50-65
Loaf cooling	70	60-70
Make-up room	75-80	65-70
Mixer-bread dough	75-80	40-50
Proof box	90-120	80-90
Yeast storage	32-45	60-75
Bowling Centers	70-75	35-55
Brewing		
Hop storage	29-32	50-60
Yeast culture room	-----	80
Candy		
Chocolate covering	62-65	50-65
Centers for coating	80-85	40-50
Hard candy	70-80	30-50
Storage	30-100	30-90
Ceramics		
Refractory	95-150	50-90
Molding room	80	60-70
Clay storage	60-80	35-65
Decal/decorate room	75-80	48
Cereal Packaging	75-80	45-50
Cheese Curing	45-65	80-90
Computer Room	70-80	40-60
Clean Room	70-74	40-50
Cork Processing	80	45
Data Processing	75	45-55
Distilling		
Grain storage	60	35-40
General manuf.	60-75	45-60
Aging	65-72	50-60
Electrical Products		
Coil/transf. winding	72	15
Tube assembly	68	40
Manufacture & lab.	70	50-55
Calibration room	75	50-55
Close tolerance assm.	72	40-45
Meter assm. & test	76	60-63
Fuse & cutout assm.	73	50
Capacitor winding	73	50
Conductor wrapping	75	65-70
Lightning arrestor	68	20-40
Storage-general	60-80	35-50
Circuit breaker assm.	76	30-60
Thrust runner hopper	70	30-50
Rectifier mfg.	74	30-40

Application	DB	RH
Food		
Apple/apricot/plum grapes/peaches	33-35	85-95
Banana ripening	68	90-95
Banana storage	60	85-90
Butter storage	40	75-85
Citrus fruit storage	40-60	85-90
Egg storage	35-55	75-80
Grain storage	60	30-45
Meat and fish	32-40	80-95
Mushroom storage	32-35	80-85
Mushroom growing	Var.	60-85
Potato storage	40-60	85-90
Produce	Var.	70-95
Sugar	80	30
Tomato storage	34	85
Tomato ripening	70	85
Foundry - Wood pat.	65-70	45-55
Fur Storage	40-50	55-65
Greenhouses	80	70-80
Gum		
Manufacture	77	33
Rolling	68	63
Stripping	72	53
Breaking	74	47
Wrapping	74	58
Hospitals		
Admin. & service	70-80	30-50
Intensive care unit	75	40
Nurseries	75	50-65
Operating rooms	68-76	50-55
Patient rooms	75	40-50
Laboratories		
Animal rooms	70-77	45-50
Clean Rooms	65-75	40-55
Leather		
Drying	70-120	75
Storage-winter	50-60	40-60
Lenses (Optical)		
Fusing	75	45
Grinding	80	80
Library & Museums		
Art storage areas	65-72	50
Stacks, reading room	70-74	40-50
Museum galleries	70-72	50-55
Rare book vaults	70-72	45
Matches		
Drying	70-75	60
Manufacture	72-74	50
Storage	60-62	50

Application	DB	RH
Office Buildings		
Computer areas	70-74	40-50
General (winter)	70-74	30-40
Paint Application		
Air drying lacquers	70-90	60
Air drying oil paint	60-90	60
Pharmaceutical mfg.	75	35
Plastics		
Cellophane wrap	75-80	45-65
Thermoset-molding	80	25-30
Plywood		
Cold pressing	90	15-25
Hot Pressing (resin)	90	60
Printing		
Bind/cut/dry/fold/glue	70-80	45-50
Color sheet lithograph	76-80	50-55
Lithographic press	76-80	43-47
Press rooms	70-80	50
Paper storage	70-80	43-47
Platemaking	75-80	45
Roll storage	73-80	50
Rubber-Dip Goods		
Cementing	80	25-30
Dipping - surgical	75-90	25-30
Storage prior to mfg.	60-75	40-50
Laboratory (ASTM)	73.4	50
School Rooms	72	35-50
Tea Packaging	65	65
Textile		
Cotton, weaving	75-80	80-85
Wool, storage	75-80	50-55
Mixing & blending	75-80	65-70
Carding-worsted	75-80	60-70
Carding-woolen	75-80	60-75
Winding & spooling	75-80	55-60
Weaving, wool/worst	75-80	50-60
Synthetic fiber mfg.	70-75	50-65
All fabrics, knitting	76	45-65
Tobacco		
Cigar/cigarette mfg.	70-75	55-65
Softening	90	85-88
Stemming/stripping	75-85	70-75
Tobacco casing/cond.	75	75
Tobacco store/prep.	78	70
Wrapper storage	75	75
Woodworking		
Finished products	65-70	35-40
Gluing	70-75	40-50
Manufacture	65-75	35-40
Painting & finishing	70-90	60

Table 10-1

allowable indoor dew point temperature is equal to the temperature of the coldest inside surface.

The temperature (Ts) of the inside surface of a structural component depends on the U-value of that component, the indoor temperature (Ti) and the outdoor temperature (To). The following equation can be used to estimate the inside surface temperature of an exposed surface. This temperature establishes the maximum allowable room dew point temperature (RDPT).

$$T_s = T_i - 0.65 \times U \times (T_i - T_o) = RDPT$$

> Note that even if the building envelope is well insulated, localized condensation can occur at points where metal or masonry materials bridge the insulation. Some type or insulation treatment will be required to prevent condensation at these points.

Concealed Condensation

Concealed condensation can occur whenever the temperature of a surface that is within a wall, ceiling or other type of structural component is below the dew point temperature of the air that migrates to that surface. If vapor barriers are not used, the indoor humidity level that can be maintained without producing concealed condensation can be much lower than the humidity level that would produce condensation on the inside surfaces.

The temperature at a concealed surface (Tc) depends on the R-value that is associated with the material between the surface of interest and the outdoors (Rc), the total R-value across the structural component (Rt), the outdoor temperature (To) and the indoor temperature (Ti). The following equation can be used to estimate the temperature at a concealed surface.

$$T_c = \frac{T_o + (T_i - T_o) \times R_c}{R_t}$$

Example

Compare the temperature at the inside surface of a cinder block wall that has insulation installed on the inside of the wall, with a similar wall that has the same amount of insulation installed on the outside of the wall (see the Figure 10-1). An R-value calculation indicates that the thermal resistance of the cinder block is equal to 2, the thermal resistance of the path that connects the outdoor surface and the inside surface of the block is equal to 12 and that the R-value of the entire wall is equal to 13. The heating load design data indicates that the indoor temperature is 70°F and the outdoor temperature is equal to -5°F.

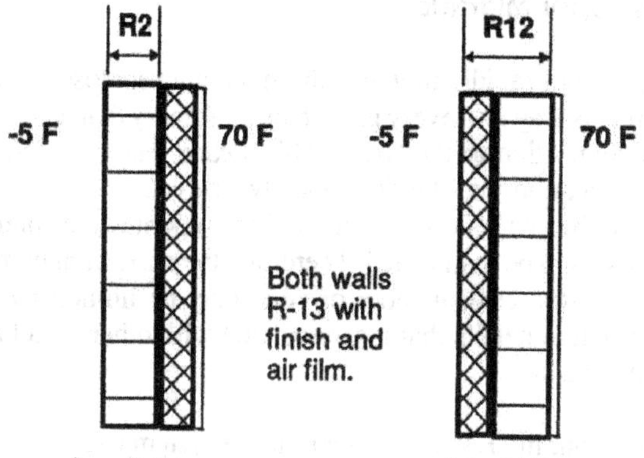

Figure 10-1

Solution

Insulation installed inside:

$T_c = -5 + ((70-5) \times 2)/13 = 5\,°F$

At 5°F saturated air holds 7 Grains of moisture. Therefore condensation could occur if the indoor humidity exceeds 8 percent . . . (70°F and 7 Grains).

Insulation installed outside:

$Tc = -5 + ((70-5) \times 12)/13 = 55\,°F$

At 55°F saturated air holds 64 Grains of moisture. Therefore condensation could occur if the indoor humidity exceeds 60 percent . . . (70°F and 64 Grains).

10-3 Dew Point on the Psychrometric Chart

The psychrometric chart can be used to relate relative humidity, dew point temperature and dry-bulb temperature. Figure 10-2 illustrates these relationships.

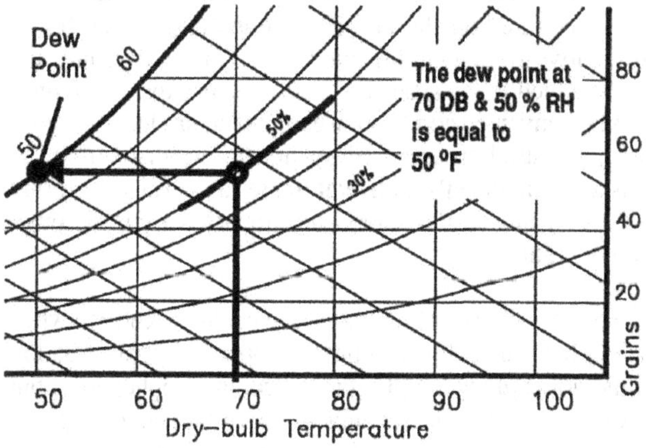

Figure 10-2

10-4 Vapor Migration

Water vapor rapidly migrates throughout an enclosed space, finding its way into every gap, crack and cavity that has an air path connection to the space. This means that even though humidification may be done locally, the entire space — as defined by the exterior walls and/or impermeable interior walls — will be humidified. Therefore, the maximum humidity level in a critical room or zone may be limited by the construction details that are associated with other, unrelated rooms or zones.

In addition, the HVAC air distribution system can mechanically disperse unwanted moisture throughout a building or a large area of a building. If moisture is introduced by a duct-mounted, supply air humidifier, every room or zone that is downstream from the humidifier will receive moisture. Even if the humidifier is located in the room or zone — or in the branch duct that serves a room or zone — the moisture may find its way into other areas by traveling through the return air system. Obviously, vapor barriers and air-side zoning will be required to contain and to control the moisture that is introduced into a space that requires humidification.

10-5 Humidification Load

Infiltration and mechanical ventilation cause the most common types of humidification loads. In this case, the load is created when dry, outdoor air displaces an equal amount of humidified indoor air. The infiltration and ventilation humidification loads can be combined into a single outdoor air humidification load (OAL).

Another type of indoor-outdoor humidification load is produced when no vapor barrier is installed on one side of permeable building materials. In this case, the vapor pressure difference across the structural components causes moisture to migrate through the porous materials. This type of load is called a vapor transmission load (VTL).

Internal *moisture losses* also must be considered. Humidification loads are produced when hygroscopic materials remove moisture from the air. These product regain loads (PRL) are additive to the humidification loads. Conversely, moisture can be added to the space by occupants, occupant activities, cooking, unvented combustion equipment, production equipment, industrial processes or by moisture *released* from hygroscopic materials and wet materials. These internal moisture gains (IMG) should be subtracted from the other humidification loads.

It also is possible that humidification could be required when the mechanical cooling equipment is operating. This situation would not apply to simple comfort cooling applications, but it could occur in some manufacturing or laboratory applications. In this case, more water is removed (from the air) at the cooling coil than is added by the moisture gains to the conditioned space. Therefore, the excessive dehumidification that takes place at the cooling coil could be thought of as a rehumidification load (RHL) in the space.

The design value for the humidification load (DHL) will be equal to the sum of the various types of humidification loads. Depending on the situation, one or more of these loads may be negative (when moisture is added) or equal to zero (because the load is not applicable). The following equation summarizes this calculation.

$$DHL = RHL \pm OAL \pm VTL + PRL - IMG$$

For example, calculate the design humidification load for an application that requires heat and humidification during cold weather. For this work, assume that calculations show that the outdoor air load is equal to 55 pounds per hour, the vapor transmission load is equal to 8.5 pounds per hour, the product regain load is equal to 31.5 pounds per hour and that the internal moisture gains are equal to 27 pounds per hour.

$$DHL = 55 + 8.5 + 31.5 - 27 = 68 \text{ LB/HR}$$

10-6 Outdoor Air Humidification Load (OAL)

The humidification load that is associated with the outdoor air CFM depends on the amount of outdoor air entering the space and the difference in moisture content between the outdoor air and the indoor air. The design value for the minimum amount of outdoor air CFM is equal to the infiltration CFM plus the minimum ventilation CFM. This information can be obtained from the **Manual N**, Form N-1, Lines 15 and 18. The moisture content difference (between the outdoor and indoor air) can be estimated by using Table 10-2, which can be found on the next page. These moisture content values should be evaluated at the winter design condition.

For example, use Table 10-2 to determine how many pounds of water, per hour will be required to humidify 3,500 CFM of ventilation air and 500 CFM of infiltration. For this work assume that the winter design outdoor temperature and relative humidity is 10°F and 80 percent and that the space will be maintained at 70°F and 35 percent.

Outdoor CFM = 3500 + 500 = 4000
LBS/HR/100 CFM = (2.410 - 0.534) = 1.876
LBS/HR = (4000 x 1.876)/100 = 75.04
GAL/HR = 8.99 ... (1 Gallon = 8.35 Pounds)

The humidification load calculations that are associated with outdoor air that is introduced through an air-side economizer are similar to the calculation that is described above. However, when the economizer is used, the largest humidification load may not occur at the winter design condition. To under-

Moisture Content of Air at Various Temperatures and Humidities (Pounds of Water per Hour per 100 CFM)													
DB/RH	**100%**	**85%**	**80%**	**75%**	**70%**	**65%**	**60%**	**55%**	**50%**	**45%**	**40%**	**35%**	**30%**
90	12.84	10.92	10.26	9.60	9.00	8.34	7.68	7.08	6.42	5.78	5.14	4.49	3.85
88	12.06	10.26	9.66	9.06	8.46	7.86	7.26	6.66	6.06	5.41	4.82	4.22	3.62
86	11.40	9.66	9.12	8.52	7.98	7.38	6.84	6.30	5.70	5.13	4.56	3.99	3.42
85	11.04	9.36	8.82	8.28	7.74	7.20	6.60	6.06	5.52	4.97	4.42	3.86	3.31
84	10.68	9.06	8.52	7.98	7.50	6.96	6.42	5.87	5.34	4.81	4.27	3.74	3.20
83	10.38	8.82	8.34	7.80	7.26	6.72	6.24	5.71	5.19	4.67	4.15	3.63	3.11
82	10.08	8.58	8.04	7.56	7.08	6.54	6.06	5.54	5.04	4.54	4.03	3.53	3.02
81	9.78	8.34	7.80	7.32	6.84	6.36	5.87	5.38	4.89	4.40	3.91	3.42	2.93
80	9.48	8.04	7.56	7.08	6.66	6.18	5.69	5.21	4.74	4.27	3.79	3.32	2.84
79	9.18	7.80	7.32	6.90	6.42	5.96	5.51	5.05	4.59	4.13	3.67	3.21	2.75
78	8.88	7.56	7.08	6.66	6.24	5.77	5.33	4.88	4.44	3.99	3.55	3.11	2.66
77	8.64	7.32	6.90	6.48	6.06	5.62	5.18	4.75	4.32	3.89	3.46	3.02	2.59
76	8.40	7.14	6.72	6.30	5.88	5.46	5.04	4.62	4.20	3.78	3.36	2.94	2.52
75	8.10	6.90	6.48	6.06	5.67	5.26	4.86	4.48	4.05	3.64	3.24	2.83	2.43
74	7.86	6.66	6.30	5.89	5.50	5.11	4.72	4.32	3.93	3.53	3.14	2.75	2.36
73	7.62	6.48	6.12	5.71	5.33	4.95	4.57	4.12	3.81	3.43	3.05	2.66	2.29
72	7.38	6.30	5.90	5.53	5.17	4.79	4.43	4.06	3.69	3.32	2.95	2.58	2.21
71	7.14	6.06	5.71	5.35	4.99	4.64	4.29	3.92	3.57	3.21	2.86	2.50	2.14
70	6.90	5.86	5.52	5.17	4.83	4.48	4.14	3.79	3.45	3.10	2.76	2.41	2.07
69	6.72	5.71	5.38	5.04	4.70	4.37	4.03	3.70	3.36	3.02	2.67	2.35	2.02
68	6.48	5.51	5.18	4.86	4.54	4.21	3.89	3.55	3.24	2.92	2.59	2.27	1.94
67	6.24	5.30	4.99	4.68	4.37	4.06	3.74	3.43	3.12	2.81	2.50	2.18	1.87
66	6.06	5.15	4.85	4.54	4.24	3.94	3.64	3.33	3.03	2.72	2.42	2.12	1.82
65	5.87	4.99	4.70	4.40	4.11	3.82	3.52	3.23	2.93	2.65	2.35	2.06	1.76
60	4.97	4.23	3.98	3.73	3.48	3.23	2.98	2.74	2.48	2.24	1.99	1.74	1.49
50	3.52	2.99	2.82	2.64	2.47	2.29	2.11	1.94	1.76	1.58	1.41	1.23	1.06
40	2.45	2.09	1.96	1.84	1.72	1.60	1.47	1.35	1.22	1.10	0.984	0.858	0.74
30	1.67	1.42	1.33	1.25	1.17	1.09	1.00	0.918	0.834	0.750	0.666	0.582	0.50
20	1.07	0.906	0.852	0.798	0.750	0.696	0.642	0.588	0.534	0.480	0.426	0.372	0.32
10	0.667	0.564	0.534	0.498	0.468	0.432	0.396	0.366	0.330	0.30	0.264	0.234	0.20
Zero	0.408	0.348	0.324	0.306	0.288	0.264	0.246	0.222	0.204	0.186	0.162	0.132	0.12
-10	0.246	0.210	0.198	0.186	0.174	0.162	0.150	0.138	0.120	0.108	0.096	0.084	0.074
-20	0.144	0.120	0.114	0.108	0.102	0.096	0.084	0.078	0.072	0.060	0.054	0.048	0.043

Notes:

1) Information about outdoor temperature and coincident humidity levels during the winter can be obtained from the local weather bureau. Otherwise, assume 80 percent relative humidity when the outdoor temperature is below 35 degrees.

2) For temperatures and humidities that are not on this chart, the pounds of water per hour can be estimated by using the following equation. The values for the indoor and outdoor grains can be read off of the psychrometric chart.

Pounds of Water per Hour = Outdoor CFM x 0.00064 x (Indoor Grains - Outdoor Grains)

Table 10-2

stand this, it is necessary to consider what occurs as the outdoor air temperature falls from the economizer changeover temperature to the winter design temperature.

Since the economizer is trying to maintain a constant supply air temperature, the CFM passing through the economizer decreases as the outdoor temperature falls. This action tends to reduce the humidification load. (If there is any infiltration, the infiltration CFM should be added to the economizer CFM.) However, the indoor-outdoor moisture differential increases as the outdoor air temperature drops. This action tends to increase the humidification load. Therefore, the design load could occur at the winter design temperature, the economizer changeover temperature or at any intermediate temperature. This makes it necessary to estimate the humidification load that is associated with each outdoor temperature, starting with the winter design temperature and finishing with the economizer changeover temperature (10 degree increments should provide enough accuracy). Then, by inspection, the design load will be equal to the largest of the various humidification load values.

Example
Use Table 10-2 to determine how many pounds of water, per hour, will be required to humidify the outdoor air that is introduced through an economizer. For this work, assume that the air handler is designed to deliver 12,000 CFM, the economizer changeover temperature is set at 55°F, the winter design temperature is 0°F, the indoor design temperature is 70°F, the desired indoor humidity is equal to 35 percent and that the infiltration is equal to 500 CFM. (The calculations that provide a solution to this problem (70.5 LBS/HR) are summarized by Figure 10-3)

10-7 Vapor Transmission Loads (VTL)

Water vapor flows from a space that has a higher vapor pressure to a space that has a lower vapor pressure. The rate at which this takes place depends on the permeability of the structural materials that separate the spaces. (There will be no vapor flow if the structural material is impermeable or if a vapor barrier is applied to the high-pressure side of a permeable material.)

The humidification load that is associated with vapor transmission (VTL) depends on the square feet of permeable surface area (Area), the grains difference between the indoor air and the outdoor air (Grains Difference) and the permeance of the construction materials that are associated with the transmission path. The following equation can be used to estimate the size of a vapor transmission load.

VTL = Area x Grains Difference x (Permeance/1060)

Values for the grains difference and permeance are required to use this equation. The psychrometric chart can be used to determine the grains difference between the room air and the outdoor air and Table 10-3, which can be found on the next page, can be used to estimate the permeance of some common building materials. (Note, the Table 10-3 permeance values, which are in BTUH, must be divided by 1060 to obtain moisture flow in pounds of water per hour.)

Example
Use Table 10-3 to determine how many pounds of water, per hour will be required to replace the moisture that migrates

OAT	Minimum Outdoor RH	Table 10 Outdoor Air	Table 10 Room Air	Differential LBS/HR/100	Percentage Outdoor Air	Economizer CFM	Outdoor Air CFM	Load LB/HR
0	0.85	0.348	2.410	2.062	0.21	2571	3071	63.3
10	0.80	0.534	2.410	1.876	0.25	3000	3500	65.7
20	0.75	0.798	2.410	1.612	0.30	3600	4100	66.1
30	0.60	1.000	2.410	1.410	0.38	4500	5000	70.5
40	0.55	1.350	2.410	1.060	0.50	6000	6500	68.9
50	0.50	1.760	2.410	0.650	0.75	9000	9500	61.8
55	0.45	1.910	2.410	0.500	1.00	12000	12500	62.5

(1) Minimum relative humidity at various outdoor temperatures can be obtained from local weather data.
(2) Differential LBS/HR/100 = Table 10 value for room air - Table 10 value for outdoor air
(3) Percentage Outdoor Air = (room DB - economizer c/o temperature) / (room DB - outdoor DB)
(4) Economizer CFM = 12000 x Percentage Outdoor Air
(5) Outdoor Air CFM = Economizer CFM + Infiltration CFM
(6) Load LBS/HR = Differential LBS/HR/100 x Outdoor Air CFM / 100

Figure 10-3

Permeance of Structural Components and Building Materials

Structural Component or Material	Permeance BTUH / 100 SQ.FT. / Grains Difference		
	No Vapor Seal	Vapor Sealing Paint - 2 Coats	Foil Backed Vapor Barrier
Walls			
Brick (4") (8") (12") Per inch of thickness	.120 .060 .040 .490	.075 .046 .033	.024 .020 .017
Block — 8"	.240	-	-
Frame — Plaster interior finish Plaster with asphalt coated insulating board	.790 .420	.160 .140	.029 .028
Concrete (6") (12") Per inch of thickness	.067 .034 .400	.050 .029	.021 .016
Clay Tile (4", glazed) (4", common) (4" glazed plus 4" common)	.013 .240 .012	.012 .110 .011	.009 .025 .009
Ceilings and Floors			
Concrete (4") (8")	.100 .051	.067 .040	.023 .019
Plaster on wood or metal lath on joist (no flooring)	2.0	.18	.030
Plaster on wood or metal lath on joist (flooring)	.50	.14	.028
Plaster on wood or metal lath on joist (double flooring)	.40	.13	.028
Partitions			
Insulating Board 1/2" both sides of studding	4.0	.19	.030
Wood or metal lath and plaster both sides of studding	1.0	.17	.029
Roofs			
Concrete 2" to 6" plus 3 layers felt roofing	.02	.018	.012
Shingles, sheathing, rafters-plus plaster on wood/metal lath	1.50	.180	.290
Wood, 1" or 2" plus 3 layers felt roofing	.02	.018	.012
Air Space,			
Still air (3-5/8") (1")	3.6 13.0	-	-
Building Materials			
Gypsum wall board - plain 3/8"	5.2	-	-
Hardboard (standard) (tempered)	.5 to 1.1		
Masonite 1/8"	1.1	.17	.03
Plaster on (wood lath) (metal lath) (gypsum lath)	1.1 1.6 1.9	- - .13	-
Plywood, Douglas fur - (1/4"-3 ply) (1/2"-5 ply)	.63 .27	.10 .06	-
Wood - 1/2" (Pine) (Spruce)	.33 .20	.05 .04	-
Insulating Materials			
Corkboard - 1"	.63	-	-
Interior finish insulating board, 1/2" 1/2" insulating board plus 2 coats water emulsion paint	5.0 to 7.0 3.0 - 4.0	.10 to 1.0	-
Structural insulating board, 1/2" - interior/uncoated	5.2 to 9.4	-	-
Insulating board sheathing, 25/32"	2.6 - 6.1	.05 - 1.0	
Crack			
12 inches long by 1/32 inch wide	5.2 per foot of crack per Grains Difference		

Table 10-3

Permeance of Structural Components and Building Materials — Continued			
Structural Component or Material	**Permeance** BTUH / 100 SQ.FT. / Grains Difference		
	No Vapor Seal	Vapor Sealing Paint - 2 Coats	Foil Backed Vapor Barrier
Packaging Materials			
Cellophane	.01 to .25	-	-
Waxed packaging paper	1 to 3	-	-
Paint Films			
Water based paint - 2 coats	5.0 to 8.0	-	-
Vapor retarder paint, 0.00031" thick	.045	-	
Oil paint - 2 coats	.1 to .6	-	-
Vinyl acetate/acrylic primer, semi-gloss or house paint	.55 to .86	-	-
Asphalt paint or 2 coats of aluminum paint	.05 to .20	-	-
Building Papers			
Duplex or asphalt laminae, 3 LBS/100 SQ.FT.	.15 to .27	-	-
Duplex or asphalt laminae, 4 LBS/100 SQ.FT.	.05 to .09	-	-
Kraft paper (1 sheet) (2 sheets)	8.1 5.1	-	-
Kraft paper (foil on one side) (foil on two sides)	.016 .012	-	-
Plastic and Metal Foils			
Aluminum foil, no penetrations	0	-	-
Polyethylene (.002) (.004) (.006) inches thick	.016 .008 .006	-	-
Polyethylene (.008) (.010) inches thick	.004 .003	-	-
PVC unplasticized, .002 inches thick	.068	-	-
PVC plasticized, .004 inches thick	.080 to .14	-	-
Sheathing and Roofing Materials			
Built-up roofing, hot mopped	0	-	-
Roll roofing, saturated and coated, 65 LB/100 SQ.FT.	.015	-	-
Asphalt impregnated, 7 LBS per 100 SQ.FT.	.02 to .10	-	-
Slaters felt, 6 Lbs per 100 SQ.FT., 50% tar saturation	1.4	-	-
Roofing felt coated with asphalt (25) or (50) LB/SQ.FT.	.015 .011	-	-
Tin sheet, 4 holes @ 1/16" diameter	.17	-	-
Insulating Materials			
Mineral wool - per inch of thickness	11.8	-	-
Expanded polyurethane board, R11 - per inch of thickness	.04 to .16	-	-
Extruded Polystyrene - per inch of thickness	.12	-	-
Polystyrene beadboard - per inch of thickness	.20 to .58	-	-

Notes:
1) Permeance of undocumented construction sandwich will not be greater than the permeance of the least permeable layer.
2) Vapor barrier should always be placed on the side that has the highest vapor pressure.
3) Foil paper or plastic vapor barriers must have joints lapped, taped or sealed with mastic.
4) Multiple by 6.6 to convert *BTUH latent* load values to *Grains of moisture per hour.*
5) Divide *Grains* of moisture values by 7000 to convert to *Pounds of Water* units.
6) Multiply table values by 9.8 to convert to . . . *Grains / (Hour per SQ.FT. per IN.HG. vapor pressure difference)*

Table 10-3

through a bare, 100 foot by 15 foot, 12 inch thick, concrete block wall. For this work, assume that the winter design temperature is 0°F, the indoor design temperature is 70°F and that the desired indoor humidity is equal to 35 percent.

Area = 100 x 15 = 1500 SQ.FT.
Outdoor Grains = 4
Indoor Grains = 38
Grains Difference = 38 - 4 = 34
Permeance = 0.04
VTL = 1500 x 38 x (0.04/1060) = 2.15 LBS/HR

10-8 Product Regain Load (PRL)

An internal moisture loss, or product regain load, occurs when hygroscopic materials absorb moisture from the indoor air. This product regain load only applies if the following conditions are met.

- The moisture content of the product, when it enters the room, is less than it would have been if the product was stored in the room.
- The product stays in the room long enough to absorb a significant amount of moisture.

The size of the product regain load (PRL) depends on the amount of material that is processed in one hour (Pounds/Hour), the regain of the material entering the room (R_{in}) and the regain of the material leaving the room (R_{out}). The following equation can be used to estimate the size of the product regain load.

$$PRL = Pounds/Hour \times (R_{out} - R_{in})$$

As indicated by the equation above, information about the amount of material that is processed in one hour and estimates for the regain of the material — when it enters and leaves the room — will be required. Production rate information may be obtained by talking to plant operating personnel. (Field measurements will be required if the plant operating personnel do not have this information.) Table 10-4, which can be found on the next page, provides values for the moisture holding capacity (regain) of some common hygroscopic materials. These values only apply to material that is in equilibrium with the ambient humidity. Note that these values are expressed as a percentage of the dry weight of the product.

Example
Use Table 10-4 to calculate the product regain load. For this work, assume that 10,000 pounds of absorbent cotton is processed in one hour. Also assume that during the coldest weather, the unprocessed cotton has been stored in a warehouse that has 20 percent relative humidity. And, assume that during the process, the product comes to equilibrium with indoor air that is maintained at 60 percent relative humidity.

Rout = 20.8 percent = 0.208
Rin = 9.0 percent = 0.090
PRL = 10,000 x (0.208 - 0.090) = 1180 LBS/HR

10-9 Internal Moisture Gains (IMG)

Internal moisture gains occur when people, equipment, products or processes release moisture to the indoor air. Many of these loads are identical to the latent loads that are associated with the cooling load calculations. Information about these types of loads can be found in **Manual N**, Tables 10 and 11. Note that this type of latent load information is listed in BTUH units. Divide these values by 1060 to convert the load into Pounds/Water per Hour units.

Pounds/Water per Hour = Latent BTUH / 1060

Regain data also can be used to estimate product related moisture gains. (In the previous example, if the processed cotton is returned to the warehouse, it will release 1180 pounds of water to the ambient air.) Additional information about product and process moisture gains can be obtained by talking to plant operating personal and process equipment manufacturers. (Field measurements may be required if the plant operating personal do not have this information.)

10-10 Rehumidification Load (RHL)

Mechanical cooling is required to maintain the indoor dry-bulb temperature during warm and hot weather. However, moisture also is removed at the cooling coil. Usually this moisture removal is desirable, but this may not be true in every case. Some commercial and manufacturing applications require a relatively high humidity level, with respect to the desired dry-bulb temperature. In this type of application, a cooling coil that is cold enough to satisfy the sensible cooling load will remove too much moisture from the conditioned space, therefore, rehumidification will be required. This rehumidification load is always positive and it only applies if the following conditions are met.

- The humidity level must be maintained at or above some desired value when mechanical cooling (as opposed to economizer cooling) is required.
- The latent capacity of the mechanical cooling system is greater than the net latent load that is associated with the conditioned space.

This rehumidification load (RHL) depends on the moisture content of the room air (GR_{room}), the moisture content of the

Moisture Content of Hygroscopic Materials (Regain)									
Industry and Associated Materials	Percent Relative Humidity of Room or Zone								
	10	20	30	40	50	60	70	80	90
Baking									
Crackers	2.1	2.8	3.3	3.9	5.0	6.5	8.3	10.9	14.9
Flour	2.6	4.1	5.3	6.5	8.0	9.9	12.4	15.4	19.1
White bread	0.5	1.7	3.1	4.5	6.2	8.5	11.1	14.5	19.0
Leather									
Sole Oak, Tanned	5.0	8.5	11.2	13.6	16.0	18.3	20.6	24.0	29.2
Printing									
Common paper, ledger, 75% rag, 1% ash	3.2	4.2	5.0	5.6	6.2	6.9	8.1	10.3	13.9
Newsprint, 24% ash	2.1	3.2	4.0	4.7	5.3	6.1	7.2	8.7	10.6
White bond rag, 1% ash	2.4	3.7	4.7	5.5	6.5	7.5	8.8	10.8	13.2
Writing, 3% ash	3.0	4.2	5.2	6.2	7.2	8.3	9.9	11.9	14.2
Textile									
Cotton - Absorbent	4.8	9.0	12.5	15.7	18.5	20.8	22.8	24.3	25.8
Cotton - American-cloth	2.6	3.7	4.4	5.2	5.9	6.8	8.1	10.0	14.3
Cotton - Sea Isle-roving	2.5	3.7	4.6	5.5	6.6	7.9	9.5	11.5	14.1
Hemp - Manila & Sisal	2.7	4.7	6.0	7.2	8.5	9.9	11.6	13.6	15.7
Jute - Average grade	3.1	5.2	6.9	8.5	10.2	12.2	14.4	17.1	20.2
Linen - Dry Spun-yard	3.6	5.4	6.5	7.3	8.1	8.9	9.8	11.2	13.8
Rayon - Celulose-Acetate-Fibre	0.8	1.1	1.4	1.9	2.4	3.0	3.6	4.3	5.3
Rayon - Cupramonium-Average Skein	4.0	5.7	6.8	7.9	9.2	10.8	12.4	14.2	16.0
Rayon - Viscose Nitrocel	4.0	5.7	6.8	7.9	9.2	10.8	12.4	14.2	16.0
Silk - Raw Chevennes- Skein	3.2	5.5	6.9	8.0	8.9	10.2	11.9	14.3	18.8
Wool - Australian-Morino-Skein	4.7	7.0	8.9	10.8	12.8	14.9	17.2	19.9	23.4
Tobacco									
Cigarette filler	5.4	8.6	11.0	13.3	16.0	19.5	25.0	33.5	50.0
Wood									
Timber - Average	3.0	4.4	5.9	7.6	9.3	11.3	14.0	17.5	22.0
Glue - Hide	3.4	4.8	5.8	6.6	7.6	9.0	10.7	11.8	12.5
Miscellaneous									
Charcoal - Steam	7.1	14.3	22.8	26.2	28.3	29.2	30.0	31.1	32.7
Activated Gelatin	0.7	1.6	2.8	3.8	4.9	6.1	7.6	9.3	11.4
Silica Gel	5.7	9.8	12.7	15.2	17.2	18.8	20.2	21.5	22.6
Soap	1.9	3.8	5.7	7.6	10.0	12.9	16.1	19.8	23.8
Starch	2.2	3.8	5.2	6.4	7.4	8.3	9.2	10.6	12.7

Note:
Product is at equilibrium with the humidity in the room— Moisture content expressed as a percentage of the dry weight.

Table 10-4

air off the cooling coil (GR_{coil}) and the CFM flowing through the cooling coil (CFM_{coil}). The following equation can be used to calculate the RHL load, in Pounds of Water per Hour. As shown by Figure 10-3 on the next page, the GR_{room} and Gr_{coil} values can be obtained from the psychrometric chart.

$$RHL = (0.68 \times CFM_{coil} \times (GR_{room} - GR_{coil})) / 1060$$

Example
Calculate the net humidification load if mechanical cooling and humidification is required on a summer design day. For this work assume that the space is maintained at 70°F dry-bulb and 65 percent RH, there is 500 CFM of infiltration and 1000 CFM of outdoor air for ventilation (95°F, 75 wet-bulb, 40 percent RH), the sensible room load is 82,500 BTUH, 5000

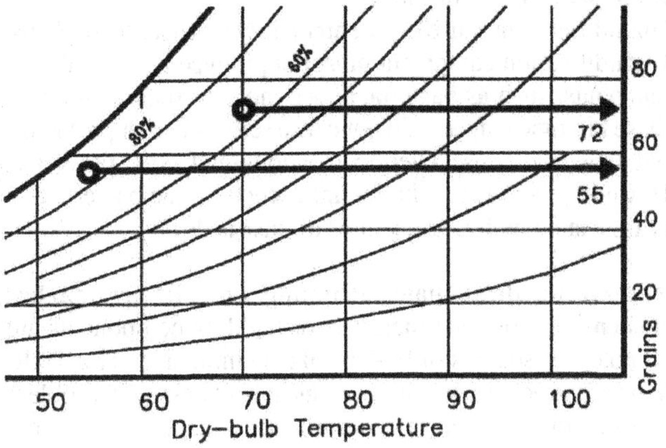

Figure 10-3

CFM leaves the cooling coil at 55 dry-bulb and 85 percent RH and that there is 10.9 LB/Hr hygroscopic product load, but no vapor transmission load. Also, assume that the cooling load calculation indicates that the latent, infiltration and outdoor air load is equal to 15,000 BTUH and that the latent internal load is equal to 10,000 BTUH.

Solution

From Figure 10-3 . . .

$GR_{room} = 72$

$GR_{coil} = 55$

Latent loads . . .

$RHL = (0.68 \times 5000 \times (72 - 55))/1060 = 54.53\ LB/HR$

$OAL = -15000/1060 = -14.15\ LB/HR$

$VTL = 0$

$PRL = 10.9\ LB/HR$

$IMG = -10000/1060 = -9.43\ Lb/Hr$

Design humidification load . . .

$DHL = 54.53 - 14.15 - 0 + 10.9 - 9.43 = 41.85\ LB/HR$

10-11 Moisture Holding Capacity of Air

The psychrometric chart can be used to verify that the air is capable of absorbing the required amount of moisture. For instance, the previous example indicates that about 42 pounds of water per hour must be added to 5000 CFM of air supply air. This corresponds to a latent heat gain of 44,520 BTUH, which is equivalent to the addition of 13 Grains of moisture. (About 1060 BTUs are required to evaporate a pound of water when the temperature is in the 60°F range.)

$$42\ LB/HR \times 1060\ BTU/LB = 44500\ BTUH$$

$$\text{Grains Difference} = \frac{44500}{0.68 \times 5000} = 13$$

As Figure 10-4 shows, this amount of moisture cannot be added to 5000 CFM of 55°F air. In this case, 5000 CFM of air at 85 percent RH will become completely saturated before it can absorb 13 Grains of moisture.

Figure 10-4 also shows, that when 82,500 sensible BTUH is added to the supply air, the temperature of the unhumidified air will rise to 70°F and the moisture content will be about 55 Grains. Air at this condition can easily absorb 13 Grains of moisture without becoming saturated. Therefore some, but not all, of the moisture could be added to the air in the supply duct and the remainder would have to be added to the room air. Or, all the moisture could be added to the room air.

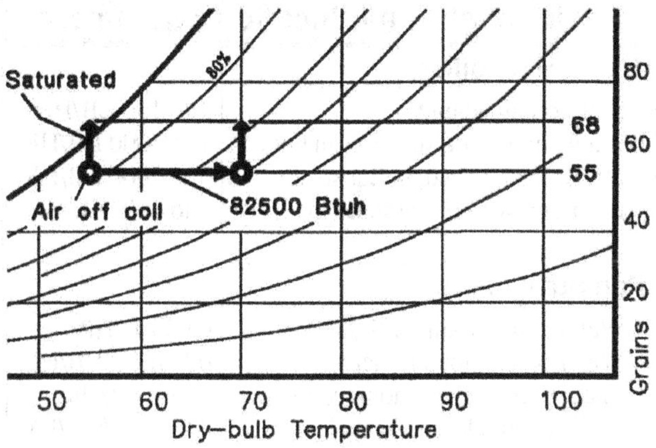

Figure 10-4

10-12 Humidification Equipment

Humidification equipment can either be classified as adiabatic (no heat added by the humidification device) or isothermal (no air temperature rise associated with the humidification process). Adiabatic equipment is also referred to as evaporative equipment. This type of humidifying equipment could be an atomizing device, a wetted element device or an evaporative pan humidifier. Adiabatic devices are not equipped with a source of heat, all the heat that is required to evaporate the water comes from the air. Therefore, humidification with this type of equipment produces a drop in the air temperature. However, there is no temperature change associated with isothermal humidification equipment. In this case, the heat that is required for the evaporation process is provided by an integral heater or by an external heat source. Heated pan humidifiers and steam humidifiers are examples of isothermal humidification equipment.

The performance characteristics that pertain to humidification equipment are discussed below. These characteristics form the basis for selecting a particular type of humidifying device. Figure 10-5 on Page 10-13 compares the performance characteristics of the various types of humidification equipment.

Capacity

Some types of humidifiers are suited for satisfying modest humidification loads and some are designed to handle larger loads. The size of the humidification load is one of the primary factors that should be considered when selecting humidification equipment. Steam humidifiers, depending on the design, can be used to satisfy a wide range of loads and they are the most suitable for very large loads. Commercial atomizing humidifiers are suitable for medium or large loads. Heated pan, wetted media and small atomizing units can be used for small loads.

Capacities of Commercial and Residential Humidification Equipment	
Steam Humidifiers	
Self-contained unit	1 to 150 LB/HR
Single manifold, steam from boiler	1 to 2000 LB/HR
Multiple manifold, boiler steam	3 to 2000 LB/HR
Area unit with or without fan	2 to 400 LB/HR
Pan Humidifiers	
Water pan in room (1 SQ.FT.)	0.03 LB/HR
Water pan in hot air duct	0.5 to 1.5 LB/HR
Plate or disc type - mounted in duct	1 to 3 LB/HR
In duct with electric heat element	3 to 4 LB/HR/KW
In duct with hot water coil (160 F)	4 to 6 LB/HR
In duct with steam coil (2 PSIG)	10 to 20 LB/HR
Self-contained, heated, integral fan	2 to 24 LB/HR
Commercial Wetted Media	
Wet pad (100 CFM - 75 F/35% RH)	1 to 2 LB/Hr
Self contained fan-powered unit	8 to 30 LB/HR
Plenum bypass - no fan	4 to 40 LB/HR
Plenum bypass - with integral fan	24 to 50 LB/HR
Self contained fan-powered, hot air	40 to 310 LB/HR
Residential Wetted Media	
Wet pad (100 CFM - 75 F/35% RH)	1 to 2 LB/HR
Bypass type - no integral fan	1 to 8 LB/HR
Self-contained fan-powered unit	3 to 6 LB/HR
Under the duct - rotating media	2 to 10 LB/HR
Plenum mounted- integral fan	1 to 10 LB/HR
Commercial Atomizing Units	
Per nozzle, with compressed air	1 to 15 LB/HR
Residential Atomizing Units	
Centrifugal plate or cone	1 to 2 LB/HR
Pressure spray nozzle	1 to 10 LB/HR
Splashing ring - plenum mounted	1 to 3 LB/HR

Biological Contamination

Humidifiers can cause indoor air quality and health problems. Humidification equipment must not produce biological contaminants, such as bacteria, algae, mold or fungus. Most of these biological agents do not cause serious health problems, but they can cause allergic reactions and odor problems. However, there is one important exception, the bacteria that is associated with Legionnaires disease is deadly.

Reservoirs, drain pans, dripping or spitting nozzles, condensation on duct materials or building components are all potential sources of biological contamination. The likelihood of the occurrence of a biological contamination problem depends on the type of humidification device. In general, sterile feed water, a high operating temperature, proper sizing and installation, proper water treatment, scheduled inspection and proper maintenance will keep biological growth under control.

- Because of the high temperatures that are associated with steam humidification, there should be no biological contamination — providing that the equipment is properly installed and maintained.

- Atomizing humidifiers do have the potential to cause health and air quality problems. They are the most likely to cause surface wetting, they have the ability to disperse large amounts of bacteria into the air and problems can be caused by dripping nozzles or stagnate water in drip pans.

- Evaporative pan and wetted media humidifiers can cause the same types of problems as atomizing equipment, but to a lesser extent.

- Heated pan humidifiers are less troublesome than evaporative devices because they operate at a higher temperature. The vapor that is produced by these devices is relatively free of biological contaminants, which remain in the pan.

Chemical Contamination

The formation of corrosion, scale and biological growth can be a problem for any type hydronic or steam device. Certain types of chemicals are used to retard, prevent or remove these deposits. However, if these chemicals are present in the humidifier feed water or steam supply, they can be dispersed into the air. Some of these chemicals are known to produce a health hazard. Therefore, a particular type of humidification device should not be used if there is any possibility that water treatment chemicals will contaminate the humidified air.

Maintenance

Isothermal, steam and heated-pan humidifiers discharge water vapor into the air. If these devices are properly installed, there should be no scaling or corrosion on duct or building surfaces. (Steam humidifiers require the least amount of maintenance, but some components of the heated pan equipment will accumulate deposits.) All types of evaporative humidifi-

Characteristics of Commercial Humidification Equipment				
	Steam	**Heated Pan**	**Atomizing**	**Wetted Media**
Capacity Range (LBS-Water/HR)	1 to 2000 per unit	1 to 20 per unit	6 to 15 per nozzle	4 to 300 per unit
Potential for Biological IAQ Problem	Low — hot steam is sterile, no pan, no drain	Moderate — low capacity, bacteria stays in pan	Very high — water/bacteria sprayed into air	High — bacteria can contact and mix with air
Potential for Chemical IAQ Problem	Low to high, depending on feed-water	Low — dissolved solids accumulate on pan and coil	Dissolved solids will precipitate out of spray	Low to moderate, dissolved solids in the sump
Problems with Water Mist or Drips	None if manifold has jacket	None	Severe to none, depends on design	None
Maintenance Requirement	Annually	By week or month, blowdown is important	By week, month or year, depends on the design	By week or month, blowdown is important
Heat Source	Self contained electric heaters or central boiler	Electric, steam or hot water coil; infrared heater	None	None or device to heat water or air heating equipment
Affect on Air Temperature	None	Small rise	Reduction	Reduction
Location	Manifold can be installed in duct or in room	Unit can be installed in duct or in room	Nozzles usually in room but duct mount is possible	Duct, plenum, or self contained room unit
Affect on Air Flow in Duct System	Small pressure drop, minimum height requirement	Not a factor when properly installed	Not usually in duct; consult manufacturer	Pressure drop; bypass units may reduce supply CFM
Control	On-off or throttling with quickest response	On-off with poor response	On-off with good response, throttling is possible	On-off, response depends on design; slow to good
Installation Considerations	Steam generation, piping, strainers, traps, valves, and condensate return	Energy for heat source, water and drain piping	Water and air piping, air compressor, pressure regulators	Water and drain piping, some units may need energy for heat source

Table 10-5

cation equipment, are subject to corrosion. Especially, atomizing equipment, which can be highly susceptible to corrosion and it can cause serious corrosion on duct or building surfaces.

If dissolved minerals are in the feed water, they will precipitate out of the water when atomizer type equipment is used to humidify a space. This causes an accumulation of dust or powder on the surfaces in the room. Even though these deposits do not poise a health hazard, they create a cleaning and maintenance problem.

Heat Source

Evaporative equipment does not have a heat source, it extracts the heat of evaporation from the air that is being humidified. This process causes a drop in the air temperature.

An external heat source supplies the heat of evaporation for isothermal (steam or heated pan) equipment. This heat source can be an integral part of the unit or the heat can be supplied by an external source. For example, a steam humidifier can be self-contained, or steam can be supplied by the boiler that is used to heat the building. (Providing that there are no health hazards created by the chemicals that are added to the boiler feed water).

Location

Some humidification devices are installed in a duct or plenum. Other types of devices are installed within a room or zone. Duct-mounted units humidify all the spaces that are served by the downstream air distribution system. Therefore, the duct system should only accommodate the rooms or zones that actually require humidification. Room mounted units humidify the room where they are located — and intentionally or unintentionally, any adjoining space that is not isolated by a vapor barrier. Both types of devices must be mounted so that clearances are maintained and condensation is prevented.

Control
Since the water is already vaporized, steam humidifiers can provide an immediate response to a call for more humidification, or less humidification. Other types of equipment have a slower response time because of the time lag that is associated with the evaporative process. Capacity control also is important. Steam humidifiers and some pneumatic atomizing units provide the opportunity to use modulating controls. The other types of humidifiers are simple on-off devices.

Cost
Equipment costs vary with the size and type of humidification equipment. Installation costs depend on the work associated with installing equipment, installing piping systems and supplying a heat source. Usually, steam humidifiers provide the most capacity per dollar spent. Since the heat that is required to evaporate a pound of water is the same for every system, the operating cost is determined by the efficiency of the heat source and the cost of the fuel. Steam units tend to have low maintenance costs and atomizing equipment tend to have high maintenance costs.

10-13 Pan Humidifiers

Pan humidifiers are simple devices that are designed to evaporate water from the surface of a heated reservoir. (If the reservoir is not heated, the device is not very effective.) Heat can be supplied by an infrared lamp or by an electric, hot water or steam coil. These devices will not cause a significant change in the temperature of the humidified air because the heat of evaporation is supplied by an external source.

Heated, pan humidifiers are suitable for small loads. They can either be installed in a supply air duct or in the room. Regular blowdown and weekly or monthly maintenance is required to keep the pan and the heating coils free of biological, chemical or mineral deposits. If they are properly installed, there should be no water droplet, wetting, or duct corrosion problems. These humidifiers are on-off devices — and because of dynamics of the evaporation process — they have a sluggish response to a call for humidity. (They are slow to shut down.)

10-14 Wetted Media Humidifiers

Wetted media humidifiers are designed to evaporate water from the surface of a wetted pad. The media can either be wetted by a nozzle or by immersion in a sump. (Nozzle units use considerably more water, but they provide continuous blowdown.) Usually, the heat of evaporation is supplied by the air that is flowing through the media. This psychrometric

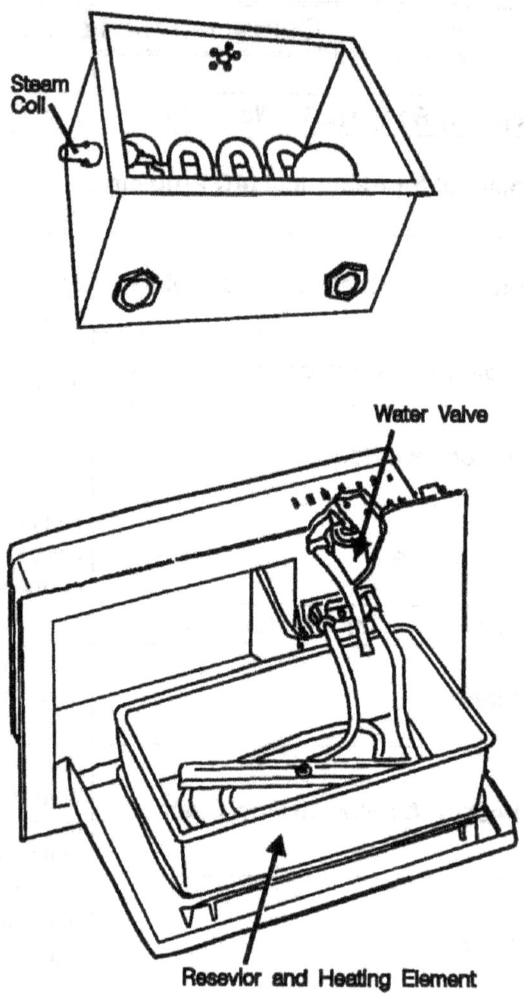

Pan Humidifiers

process causes a drop in the temperature of the humidified air. In some cases, performance may be enhanced by using heated water. Some self-contained room units may use heated air to increase performance.

Most wetted media humidifiers are suitable for small loads but some self-contained units are designed to handle intermediate loads. There are many different types of duct-mounted and self-contained room units to choose from.

The air flow through duct-mounted, bypass units is induced by the pressure difference across the supply air blower. However, bypass units may cause an undesirable reduction in the supply CFM. Other types of duct-mounted devices simply protrude into the air stream. Still other types of duct-mounted devices are equipped with an integral fan that circulates air through the media. Some duct-mounted designs may dissipate a noticeable amount of static pressure, but some designs do not produce much resistance to air flow. Self-contained room units are equipped with a fan and some of these units may have an integral heat source.

Under-Duct, Wet-Media Humidifier

Duct-Wall, Fan-Powered, Wet-Media Humidifier

Residential, Bypass, Wet-Media Humidifier

Heated, Fan-Powered, Evaporative Humidifier

Regular blowdown and weekly or monthly maintenance is required to clear the sump or reservoir of any biological, chemical or mineral deposits. If they are properly installed, there should be no water droplet, wetting or duct corrosion problems. Mineral fallout (dusting) is not a problem with this type of equipment.

Wetted media humidifiers are on-off devices. Depending on the design, they have a slow to good response to a call for humidity and a reasonably quick response to a shut-down signal.

10-15 Atomizing Humidifiers

All types of atomizing humidifiers are designed to spray a fine mist of water droplets into the air. This means that the air must supply the heat that is required to complete the evaporation process. This psychrometric process causes a drop in the temperature of the humidified air.

Residential type, spinning disk and diffusing screen units are designed for very small loads. Pneumatically powered, pressurized water, nozzle spray units can be used to satisfy small, medium and large commercial/industrial humidification loads.

Small residential type units can be installed in a duct, but pneumatic spray nozzles are primarily designed to be installed in the room. There is always the possibility that any type of atomizing equipment could wet nearby surfaces. However, this should not be a problem providing that the equipment is properly sized, controlled and installed. (In some cases, it may be possible to install spray nozzles in a duct without wetting or corroding the duct, but the application should be approved by the manufacturer of the atomizing equipment.)

Regular maintenance may be required to clear clogged nozzles of mineral deposits and to clear biological or mineral deposits from the drip pans. However, some pneumatic nozzle equipment is designed to prevent dripping or spitting; these nozzles are purged with compressed air before and after each operating cycle. This anti-clogging design requires less maintenance than nozzles that are not purged. (Nozzles that are just powered by pressurized water are very likely to cause problems with spitting, dripping and wetting — so they should be avoided.) Mineral fallout (dusting) is a problem that is inherently associated with this type of humidification equipment — anything that is in the water, including biological contaminants, will be dispersed into the air.

Atomizing humidifiers are usually on-off devices, but some manufacturers offer modulating controls. They have a good response to a call for humidity and to a shut-down signal. Since the pneumatic atomizer heads require air and water under pressure, an air compressor, water piping and compressed air piping will be required.

Residential Atomizing Humidifier

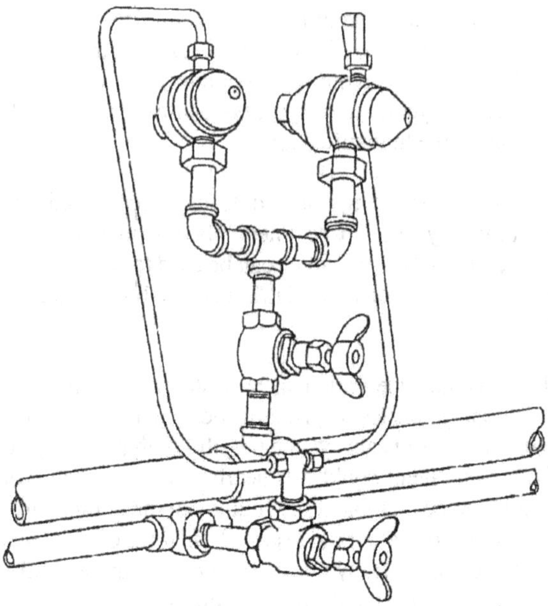

Pneumatic Atomizing Nozzle

10-16 Steam Humidifiers

Because of the high temperatures that are involved, steam humidifiers are potentially the most sterile (no algae, bacteria or odor) type of humidification equipment. And, there should be no problems with chemical contamination, providing that the boiler feedwater is of acceptable quality. However, some of the chemicals that are used to treat central boiler feedwater can make the steam unacceptable as a source for humidification. Self-contained, electrically heated units offer a solution to this problem because they can be supplied with feedwater that is of acceptable quality.

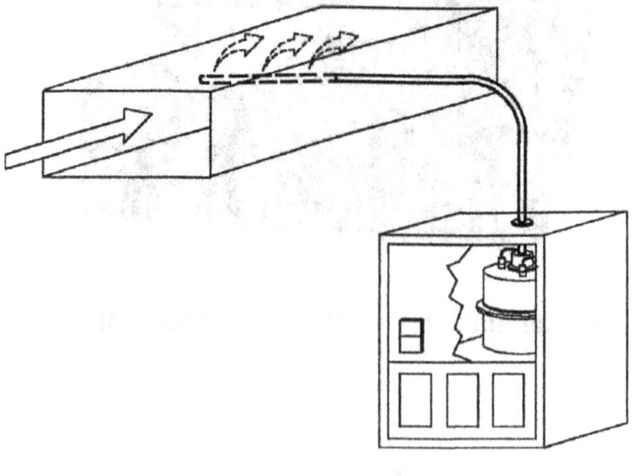

Self-Contained Steam Humidifier

Commercial/industrial steam humidifiers can be used to satisfy small, medium, large and very large humidification loads. This type of humidifier is designed to disperse steam into the air. There is no significant change in the temperature of the humidified air because the heat of evaporation is supplied by an external source.

Conditioned steam humidification equipment can be installed in a duct or a room. Spitting and water droplets will not be a problem providing the unit is equipped with a jacketed manifold and a steam separator. This type of hardware insures that

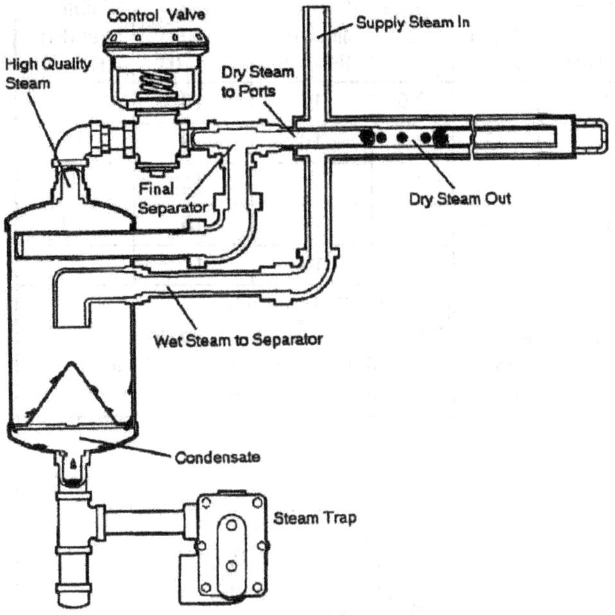

Jacketed, Dry Steam Manifold

the steam will be dispersed as a pure vapor. However, even if the steam is introduced as 100 percent vapor, there is always the possibility that condensation could wet nearby surfaces.

If condensation occurs, it will be a source of corrosion or biological contamination. But, condensation and surface wetting should not be a problem if room units and duct-mounted manifolds are properly sized, controlled and installed. (Units that are installed in the room are often equipped with a self-contained propeller fan that helps disperse the steam before it has a chance to condense.) Note that steam cups and unjacketed steam pipes are not recommended — their performance is characterized by problems associated with droplet fallout, spitting and wetting.

Providing that the feedwater is of good quality, steam humidification equipment will require a minimum amount of maintenance, perhaps as infrequently as once per year. Dusting and fallout is not a problem with steam equipment.

Steam humidifiers offer the best opportunity for using modulating controls, but they also can use two position (on-off) controls. They provide precise control and they have the fast response to a call for humidity and to a shut-down signal. However, the steam valve must be properly sized to obtain optimum performance.

10-17 ARI Ratings

Humidification equipment can be divided into two categories, residential/light-commercial equipment and commercial/industrial equipment. Residential and some small commercial, wetted media and atomizing units are rated according to a ARI standards 610 and 620, which specify the conditions for the rating tests. No specific standards are used to test and rate commercial/industrial steam or pneumatic nozzle equipment.

The ARI capacity rating is for continuous operation. Capacities are listed in *Gallons per Day* for a specific set of operating conditions, which include air velocity, dry-bulb and wet-bulb temperatures, entering water temperature and water flow rate. Some examples of these rating conditions are provided below.

Standard 610 — Central Units
- Capacity listed in gallons per day
- Air velocity in duct = 800 feet per minute
- Supply air temperature = 140°F
- Return air temperature, duct/plenum = 75°F
- Relative humidity, return duct/plenum = 30 Percent
- Entering water temperature = 50 to 60°F

Standard 620 — Self-Contained Units
- Capacity listed in gallons per day
- Entering air temperature = 75 °F
- Entering humidity = 30 Percent
- Entering water temperature = 60°F (no heat)
- Entering water temperature = 140°F (heated)

When the actual operating conditions are different than the ARI rating conditions — a typical situation — the capacity of the humidification equipment will not be the same as the rated capacity. Unfortunately, manufacturers do not usually publish comprehensive application data for small units. For example, Figure 10-5 provides capacity information for different air temperatures, but there are is no correction for the moisture content of the entering air (or the air velocity). Occasionally, it will be necessary to ask the manufacturer for the desired performance data.

Humidifier Capacity

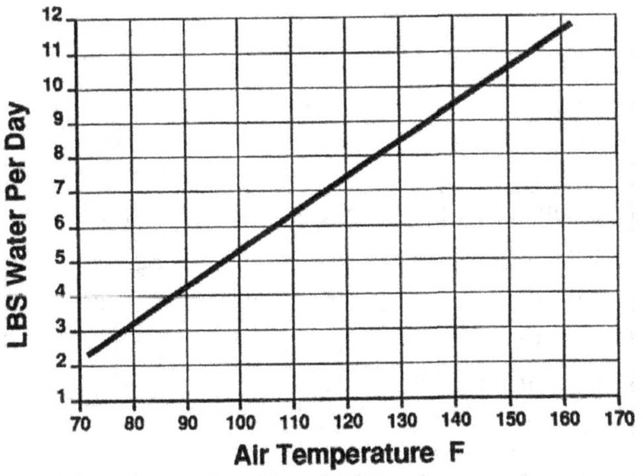

Figure 10-5

10-18 Application Data

When it is available, application (performance) data provides capacity information for a variety of operating conditions. Therefore, application data should always be used to select and size humidification equipment. When this is done correctly, the performance data look-up values will match the operating conditions that are expected to exist at the site. Example of performance data for a steam manifold and a self-contained steam humidifier is provided by Figures 10-6 and 10-7. Figure 10-8 provides an example of pneumatic spray nozzle performance data.

Orifice Size	Steam Pressure - PSIG								
	2	6	10	15	20	30	40	50	60
1/8	5	10	14	17	20	25	30	34	37
3/16	12	23	31	39	46	57	67	74	82
1/4	22	42	55	69	82	102	117	134	147
5/16	33	65	87	108	127	159	187	211	235
3/8	48	93	122	154	181	229	252	300	350

Figure 10-6

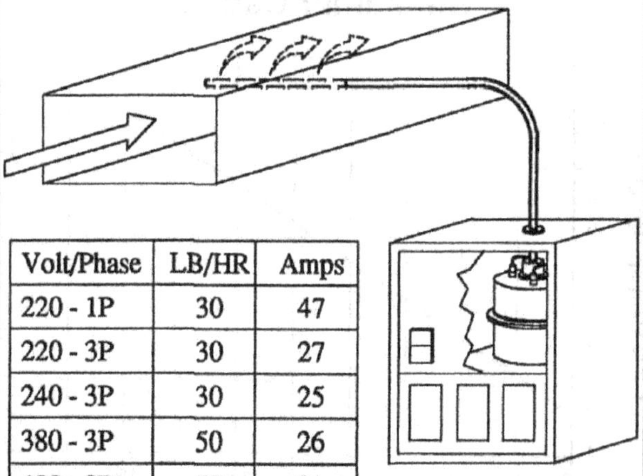

Volt/Phase	LB/HR	Amps
220 - 1P	30	47
220 - 3P	30	27
240 - 3P	30	25
380 - 3P	50	26
480 - 3P	50	21

Figure 10-7

Nozzle Cap. LB/HR	Mount Height (FT)	Spray Diam. (FT)	Visible Mist (Ft) RH < 45%	Visible Mist (Ft) Med RH	Visible Mist (Ft) RH > 60%
06	8	2.5	10	13	16
08	12	2.5	10	13	16
10	14	3.0	11	14	17
12	16	3.5	12	15	18
15	18	4.0	13	17	21

1) For room temperature 68 °F or higher
2) For room 50-67 °F use next smallest nozzle
3) For room below 50 °F drop down two nozzle sizes
4) No air movement near spray pattern

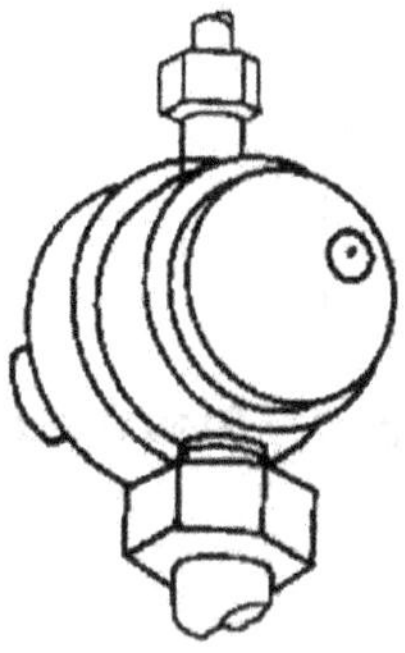

Figure 10-8

10-19 Controls

The output capacity of any type of humidification equipment can be controlled by a humidistat that senses the moisture in a room or in a duct. The humidifier manufacturer may provide the humidistat as standard equipment, as an option or it may be obtained from another source. When a duct-mounted humidifier is used, the humidistat should be interlocked with the blower. This interlock synchronizes operation with the fan and prevents wetting when there is no air flow. In addition, a high-limit humidistat should be installed in a duct or in the room to protect against condensation and wetting. In some designs, a temperature switch is used to lock out the humidification during a morning warmup cycle. If the unit has a sump or reservoir, a high water cut-off valve and a bleed-off valve/drain may be included as standard equipment or as an option. Explosion proof controls are available.

Steam humidification equipment manufacturers offer various types of control valves and valve operators, which may be modulating or on-off devices. All control valves and operators must be sized correctly. An optional, central control panel is normally available for self-contained steam humidification equipment.

Pneumatic nozzle manufacturers usually provide packaged, capacity controls and safety controls. They also sell controls that operate (or interlock) the devices that are used to control the compressed air and pressurized water systems. They also can provide hardware that regulates air pressure and water pressure.

10-20 Installation

Any type of humidifier, whether it is mounted in a duct or in the room, must be installed so that there is adequate clearance between the device and the neighboring surfaces. In addition, duct-mounted units must be installed at a position that is free of upstream and downstream turbulence or backflow. The objective is to make sure that the moisture is completely vaporized and mixed with the ambient air before it contacts any surface. Failure to provide proper clearance can cause wetting and condensation problems. Clearance requirements are usually documented in the manufacturer's literature and installation manuals.

Depending on the type of humidifier, it may be necessary to supply the device with water and compressed air, hot water or steam. Other installation work could involve drain pans, drain lines, condensate return lines, pipe insulation, strainers, steam traps, water treatment equipment, electrical service, electric controls, electronic controls or pneumatic controls. The piping and electric service requirements, piping diagrams, piping details, electrical wiring diagrams and control schematics usually can be found in the manufacturer's literature and installation manuals.

Jacketed Steam Manifolds — Duct Installation
Steam humidifier manifolds should be installed so that the steam is discharged toward the air flow. Always install the manifold so that it spans the widest dimension of the duct and always install the manifold in the middle of the duct. Manifolds can be installed in the horizontal or vertical position. However, if the manifold is vertical, it must point up. Never install a unit with the manifold pointing down. Additional information about manifold installation is provided on the next page by Figure 10-9.

When steam is discharged from the manifold it will immediately begin to condense, forming a visible mist as it releases heat to the air. This mist will extend downstream for some distance before it is completely evaporated. This evaporation process will reabsorb the same amount of heat that was released when the steam condensed into a mist. If temperature is measured along the duct, the temperature in the mist will be higher than the upstream temperature, but the temperature downstream from the mist will be the same as the upstream temperature.

In most duct systems the mist will extend downstream from the manifold for about 10 to 12 feet. However, the actual distance that the mist will extend down the duct will depend on the aspect ratio of the duct, the upstream air temperature, the air velocity and the number of manifolds in the duct. The following factors influence the length of the vapor trail.

- Long manifolds are best for dispersing the steam into the air. Therefore, shorter vapor trails are associated with ducts that are wide and narrow, because they can accommodate a long manifold.
- Since warm air absorbs moisture better than cold air, the vapor trail gets shorter as the air temperature increases.
- Since high velocities carry the water droplets further during a given amount of time, the vapor trail gets shorter as the air velocity decreases.
- Multiple manifolds disperse a given flow of steam better than a single manifold. Therefore, shorter vapor trails are associated with multiple manifolds.

It is very important to make sure that humidity and temperature sensors are not installed in the vapor trail. False readings will be obtained if they are installed in the mist. It also is important to make sure that no high efficiency filters are subjected to the mist because even though they are small, the water droplets will not pass through the filter. Figure 10-9 reinforces these points. Note that multiple manifolds provide an effective way to shorten a vapor trail that would otherwise impinge on a sensing element, air-side device or component. Additional information about the length of the vapor trail that is associated with a specific situation can be obtained from the equipment manufacturer.

Installation — Pneumatic Atomizing Nozzles
The objective is to install the nozzles so that they provide uniform coverage. The nozzle layout and installed height also must be compatible with the spray patterns. The nozzle performance data (see Figure 10-8) will provide information about the spray pattern and the minimum installation height. Other guidelines that affect the nozzle layout are summarized below.

- A spray pattern should not impinge on structural components, machinery, materials, ducts or pipes.
- Install the nozzles as far from the occupants as possible to prevent discomfort caused by droplets or localized evaporative cooling.
- Heads that discharge toward each other must be separated by a distance that exceeds the sum of the spray pattern throws.
- The spray pattern dimensions only define minimum clearances. The heads can be much further apart because the vapor will rapidly diffuse throughout the area.

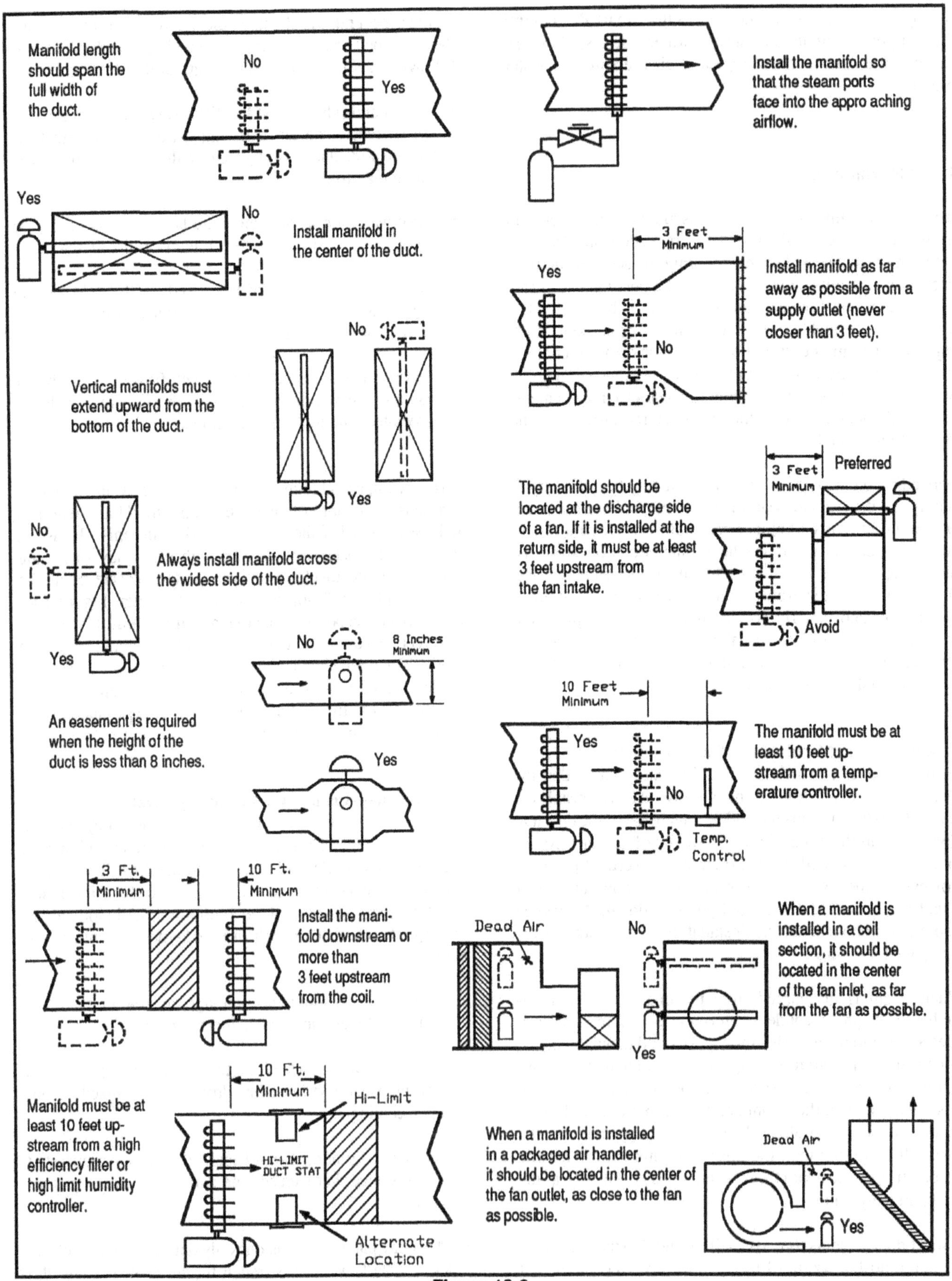
Manifold length should span the full width of the duct.
No
Yes
Install the manifold so that the steam ports face into the appro aching airflow.
Yes
No
Install manifold in the center of the duct.
3 Feet Minimum
Yes
No
Install manifold as far away as possible from a supply outlet (never closer than 3 feet).
No
Vertical manifolds must extend upward from the bottom of the duct.
Yes
3 Feet Minimum
Preferred
The manifold should be located at the discharge side of a fan. If it is installed at the return side, it must be at least 3 feet upstream from the fan intake.
Avoid
No
Always install manifold across the widest side of the duct.
Yes
No
8 Inches Minimum
An easement is required when the height of the duct is less than 8 inches.
Yes
10 Feet Minimum
Yes
No
Temp. Control
The manifold must be at least 10 feet up-stream from a temp-erature controller.
3 Ft. Minimum
10 Ft. Minimum
Install the mani-fold downstream or more than 3 feet upstream from the coil.
Dead Air
No
Yes
When a manifold is installed in a coil section, it should be located in the center of the fan inlet, as far from the fan as possible.
10 Ft. Minimum
Hi-Limit
HI-LIMIT DUCT STAT
Alternate Location
Manifold must be at least 10 feet up-stream from a high efficiency filter or high limit humidity controller.
When a manifold is installed in a packaged air handler, it should be located in the center of the fan outlet, as close to the fan as possible.
Dead Air
Yes

Figure 10-9

- Control stations should be located so that the distance from the station to the most remote head is about the same in all directions.

- Observe the manufacturer's recommendations regarding the relationship between the head location, air piping runs and water piping runs.

10-21 Summary of the Design Procedure

The same design procedure can be used to select and size any type of humidification equipment. The basic steps that are associated with this procedure are summarized below.

1) Establish the humidity and temperature levels that are required for the application — see Section 10-1.

2) Establish the maximum humidity level that can be tolerated by the structure — see Section 10-2

3) Finalize the design value for the humidity level. Base this value on the required humidity level (Step 1), providing that it is compatible with the humidity level that can be tolerated by the structure(Step 2). If there is a conflict between the required humidity level and the humidity level that can be tolerated by the structure, either alter the thermal performance of the structure or design for a lower humidity level.

4) Use the design value for the humidity level to calculate the net load on the humidification equipment — see Sections 10-5 through 10-10.

5) Use the psychrometric chart to make sure that the air will be able to absorb the desired amount of moisture — see Section 10-11.

6) Use manufacturer's performance data to size the humidification equipment — see Section 10-18.

7) Specify controls and accessories.

Section 11
All-Air Systems

All-air systems are very important because they are installed in a majority of commercial buildings. With this type of system, air is the only medium (distribution fluid) that is used to transfer heat between centrally located HVAC equipment and the conditioned space.

11-1 General Discussion

This section provides a summary of the components and performance characteristics that are generic to all types of all-air systems. Specific information about some of the most common types of all-air systems is provided in Sections 11-2 through 11-7.

Function
All-air systems can provide sensible and latent cooling by supplying cool, dry air to the conditioned space or they can provide heat by supplying warm air to the conditioned space. They also can provide outdoor air (for ventilation or make up air), filtration and humidification.

Applications
All-air systems are very popular for small commercial applications and they are commonly used for large commercial, institutional and industrial applications. They can be designed to provide loose control over the space conditions or they can be designed to provide very precise control over the space conditions. They are very competitive from the standpoint of installed cost, but energy efficiency and operating cost vary, depending on the type of system.

Configuration
Packaged equipment is typically used for small commercial applications. Packaged equipment and built-up equipment is used for large commercial, institutional and industrial applications. In any case, all-air systems can either be classified as series-coil systems or as parallel-coil systems. All-air systems also are classified as single-duct or dual-duct systems.

Series-Coil Arrangement
A series-coil system may be a heating-only system, a cooling system, or it may provide heating and cooling. If it is a heating and cooling system the heating coil (or heat exchanger) and the cooling coil are in series. This arrangement allows the system to supply heating, cooling, or cooling with reheat. If reheat is desired, the heating coil must be installed downstream from the cooling coil.

Series-coil systems are commonly used for single-zone and for multizone applications. Zoning can be provided by using reheat coils or by using a variable volume design. Examples of the basic types of series-coil systems are listed below. Other types of systems can be created by combining two of these basic systems, a variable volume, reheat system for example.

- Single-zone, constant volume
- Variable volume
- Reheat
- Variable volume, variable temperature, changeover

Parallel-Coil Arrangement
This arrangement places the heating and cooling coils in parallel, therefore a parallel-coil system is always a heating and cooling system. Parallel-coil systems are used for multiple-zone applications. Zoning is accomplished by supplying each zone with a mixture of hot and cold air. Two examples of parallel-coil systems are listed below:

- Packaged multizone
- Dual-duct

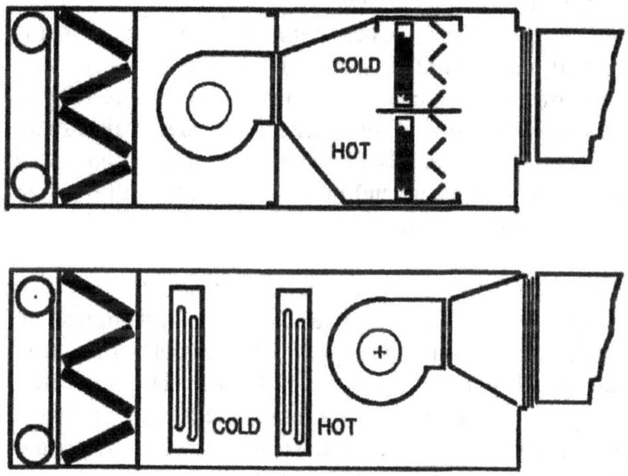

Parallel and Series Coil Arrangements

Single-Duct Systems
Single-duct systems have one supply air duct for each zone. All series-coil systems are single-duct systems. Parallel-coil systems that are designed to mix hot and cold air right at the unit also are single-duct systems. (A packaged multizone system is an example of a parallel coil configuration that uses one supply air duct for each zone.) The diagrams on pages 11-3, -8, -15 and -16 show examples of single-duct systems.

Dual-Duct Systems
Dual-duct systems have two supply air ducts routed to each zone. Parallel coil configurations that use remote mixing boxes are classified as dual-duct systems. The diagram on page 11-29 shows an example of a dual-duct system.

Characteristics
The characteristics that are generic to all types of all-air systems are listed below. The characteristics that are peculiar to specific types of all-air systems are provided in the following sections.

- The equipment is centrally located.
- Maintenance is consolidated at the central unit.
- Equipment noise is confined to the central unit.
- Equipment failure affects a large area or the entire building.
- Outdoor air is available for ventilation.
- Outdoor air is available for makeup air.
- An economizer cycle can be used for free cooling.
- Fans provide positive air distribution.
- Filters remove contaminants
- Adequate clearances are required for duct runs.
- Adequate clearances are required for air-side devices.
- Supply outlets and returns penetrate interior surfaces.
- An air balance is required.
- Fans may consume significant amounts of energy.

Primary Equipment
The primary cooling equipment may consist of a unitary or split DX cooling system, heat pump or an air handler that is equipped with a chilled water coil. Economizers are commonly used for free cooling and return fans are usually required to control the flow of return air and exhaust air. Return fans also provide control over the pressure in the conditioned space. Filters, dampers and in some cases, humidifiers, are installed in the air handler.

The primary heating equipment may consist of a separate heating system, a heat pump or a device that is integrated into the same cabinet as the cooling coil. (An integral heating plant could be a fossil fuel furnace, a hot water coil, an electric heating coil or a steam coil.) If a separate heating system is used, it is typically a perimeter hot water or electric baseboard radiation system, but it could be a hot air or steam system.

Secondary Equipment
Secondary equipment is associated with individual rooms or zones. The type of secondary equipment varies, depending on the type of system. Hot water, electric or steam reheat coils; variable volume devices and mixing boxes provide examples of secondary equipment.

Design Loads
Equipment sizing loads may be based on the peak zone load, the sum of the peak zone loads or the peak block load, depending on the zoning requirements and on the system configuration. Calculations that estimate minimum loads are also important. The calculations that are peculiar to each type of all-air system will be discussed in the following sections. Also refer to Section 4 for a general discussion of the types of load calculations that are used for equipment sizing and for establishing control strategies.

Control
Some all-air systems control the space temperature by cycling the heating or cooling plant on and off, but this type of control is more suited to residential systems. Commercial systems, even small commercial systems, control the space temperature by adjusting the temperature of the supply air, by varying the CFM of supply air or by varying both the temperature and the CFM of the supply air.

The ability of an all-air system to control space humidity and air quality varies, depending on configuration, secondary equipment and controls. The details of how specific types of all-air systems control temperature, humidity and air quality will be discussed in the following sections.

11-2 Single-Zone Constant Volume Systems

This section provides an overview of the performance characteristics, hardware, design procedures and control cycles that are associated with single-zone, constant volume systems. The information provided by Section 11-1 also applies to these systems

Function
Single-zone, constant volume, all-air systems can either provide sensible and latent cooling or they can provide heat. They also can provide outdoor air for ventilation or for make up air and they can provide humidification.

Applications
Single-zone, constant volume all-air systems are used for all types of small to large commercial, institutional and industrial applications. They can be designed to provide loose control over the space conditions or they can be designed to provide very precise control over the space conditions. They are very competitive from the standpoint of installed cost. Systems that do not use reheat are efficient and economical to operate because energy consumption and operating costs are directly related to the space load. Single-zone, constant volume all-air systems also are relatively easy to design because the problems that are associated with zoning and with simultaneous heating and cooling do not have to be addressed.

Configuration
Single-zone, constant volume systems are usually series-coil, single-duct systems. In this configuration, the central unit can provide heating, cooling or cooling with reheat. If reheat is desired, the heating coil must be installed downstream from the cooling coil. A diagram of a single-zone, constant volume system is provided at the top of the next page.

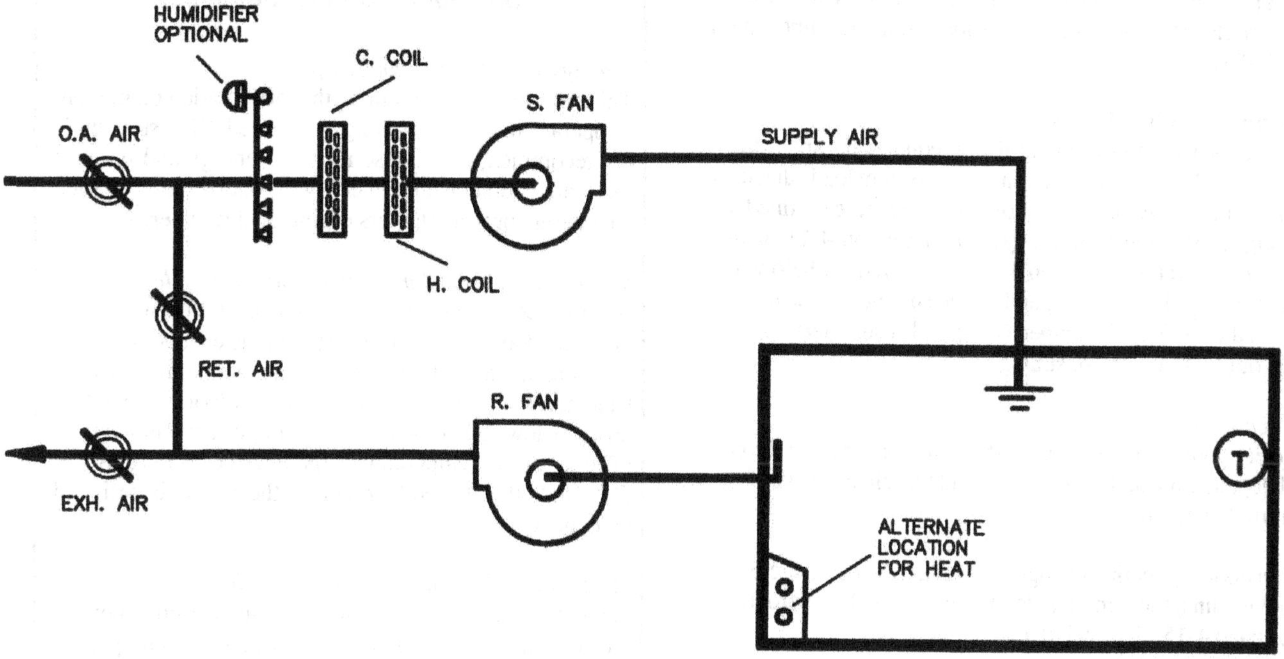

Single-Zone Constant Volume System

Characteristics

The characteristics that are generic to all types of all-air systems are listed in Section 11-1; refer to the paragraph titled *Characteristics*. The characteristics that are peculiar to single-zone. constant volume systems are listed below.

- Each zone requires a separate system.
- The system is easy to design, install and operate.
- The control cycles are relatively simple.
- Control cycles are directly related to the space conditions.
- There is a one-to-one relationship between the percentage of fresh air introduced at the air handler and the percentage of outdoor air required by the zone.
- There is a one-to-one relationship between the condition of the air leaving the cooling coil and the sensible heat ratio that is associated with the zone.
- Seasonal changeover (heating or cooling) is simple.
- Simultaneous heating and cooling is not required.
- Air circulation rates are constant.
- A night cycle can be used when the zone is vacant.
- The equipment can be shut down when the zone is vacant.
- Space temperature and humidity fluctuate if on-off control is used.
- Reheat provides precise control over temperature and places an upper limit on humidity.
- A humidifier is required to maintain exact humidity control during any operating condition.
- The system is efficient and economical to operate, providing there is no reheat.
- It is easy to account for and to meter energy consumption.

Equipment

The type of primary cooling and heating equipment that could be used for a single-zone, constant volume all-air system is listed in Section 11-1; refer to the paragraph titled *Primary Equipment*. Reheat coils are the most common type of secondary equipment that is associated with single-zone, constant volume systems. If required, the system also can be equipped with a humidifier.

Sizing Loads — Cooling

Load estimates should be based on peak load conditions. Estimates of the sensible and latent loads in the conditioned space and at the central equipment are required. The loads that are associated with the conditioned space correspond to the values that are found on line 8 of Form N-1. The loads that are associated with the air handler correspond to the values that are found on line 12 of Form N-1.

Fan heat is a system load that is not included on Form N-1. However, the sensible heat that is added by the supply fan and by the return fan should be accounted for. Sometimes this is accomplished (as per the manufacturer's equipment selection instructions) by subtracting the fan heat from the cooling capacity of the unit. (Refer to Section 6-8 for an example that illustrates this procedure.) Any portion of the fan heat that is not accounted for by a capacity correction should be added to the other system loads that are imposed on the cooling coil.

Pumping heat and piping gains are system loads that are associated with chilled water systems. These sensible loads are not accounted for on the N-1 form, but if these loads exist, they should be added to the other loads that are imposed on the chiller.

Minimum Cooling Loads
The load on the cooling coil is always equal to the space load, plus the outdoor air load, plus any other system loads that may apply. The behavior of the coil load can be examined by making a load-line analysis. (Refer to Section 4-11 of this manual, also refer to Section 4-16.) As explained below, the minimum coil load will depend on the strategy that is used to control the outdoor air damper. (Figure 11-1 summarizes three basic outdoor air control strategies.)

No Economizer
If no economizer cycle is used, the minimum load on the coil will depend on whether or not the load line changes over from cooling to heating.

- If a cooling-heating changeover occurs, (with no credit for solar gains) the minimum load on the coil will be zero (refer to page 4-15, Figure 4-12).
- If a cooling-heating changeover does not occur, (with no credit for solar gains), the minimum load on the coil will be equal to the load that occurs when the outdoor temperature is equal to the winter design temperature.

Simple Economizer
If a simple dry-bulb economizer is installed, the minimum cooling load on the coil (with no credit for solar gains) will occur just before the system switches to the economizer cycle. In this case, the minimum load on the coil will be equal to the load that occurs when the outdoor temperature is equal to changeover temperature (refer to page 4-15, Figure 4-12).

Ultimate Economizer
If an ultimate economizer cycle is installed, it is possible to have the refrigeration equipment operating with no load on the coil. Figure 11-1 illustrates this point.

Night-Cycle (Unoccupied) Cooling
In many applications, the cooling system can be shut down and the outdoor air dampers can be closed when the building is not occupied. This control strategy reduces energy use.

If the cooling system has to operate when the building is unoccupied, a set-up control strategy will reduce energy consumption. Usually, outdoor air is not required during the night cycle, so the outdoor air damper can be closed to obtain additional energy savings. However, if cooling is required, the economizer cycle should still be used whenever the condition of the outdoor air is suitable for free cooling. Internal loads are usually smaller during the night cycle, so the minimum load on the cooling coil could be less than the minimum day-cycle load. An analysis of both operating cycles is required to determine which cycle produces the smallest load on the cooling coil.

Outdoor Air Control Strategies

Strategy 1 — No economizer cycle
This control strategy requires the refrigeration equipment to operate whenever cooling is required. This strategy is not recommended because it wastes energy and because there are problems associated with operating air-cooled condensers or water towers during cold weather.

Strategy 2 — Simple dry-bulb economizer cycle
A simple control cycle, based on the dry-bulb temperature of the outdoor air, can be used to initiate changeover from a mechanical cooling cycle to an economizer cycle. Changeover occurs whenever the outdoor temperature drops below the setpoint of the supply air temperature. (The supply air temperature setpoint usually falls between 50°F and 60°F, depending on how the system is designed and operated.)

Strategy 3 — Ultimate economizer cycle
A more sophisticated control cycle can be used to control the economizer dampers and the refrigeration equipment. This type of control cycle (as shown below) compares the heat content of the outdoor air with the heat content of the return air. If the heat content of the outdoor air is lower than the heat content of the return air, the outdoor air dampers modulate to the 100 percent open position and the return air dampers close. However, the refrigeration equipment will continue to operate, providing that the outdoor dry-bulb temperature is above the setpoint of the supply air temperature controller. Changeover from a mechanical cooling cycle to the economizer cycle still occurs whenever the outdoor temperature drops below the setpoint of the supply air temperature. (The supply air temperature setpoint usually falls between 50°F and 60°F, depending on how the system is designed and operated.)

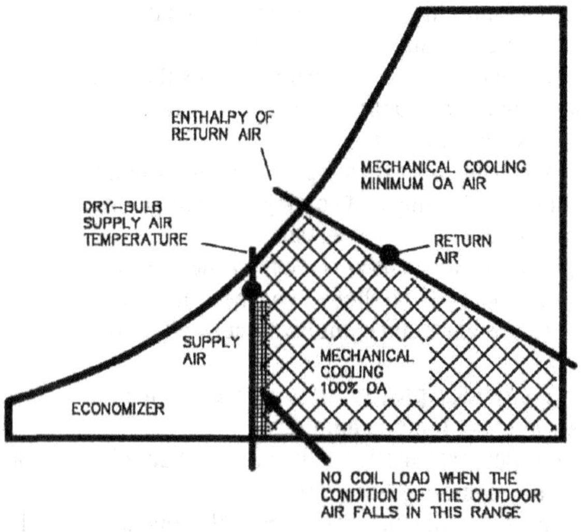

Figure 11-1

Condition of the Air Leaving the Cooling Coil
The performance of the cooling coil must be compatible with the conditioning requirements of the space. This means that the condition of the air leaving the cooling coil must be cold enough to offset the sensible design load and dry enough to neutralize the latent design load. However, the humidity in the space can be expected to vary at part-load conditions. Reheat should be considered if precise, continuous dehumidification is required.

Sizing Load — Heating
The heat load that is associated with the conditioned space corresponds to the sum of the values that are found on lines 16 and 17 of Form N-1. The heat load that is associated with the air handler corresponds to the value that is found on line 21 of Form N-1.

Night-Cycle Heating Loads
The size of the night-cycle space load will depend on whether or not the space temperature is set-back during the night cycle. And, the night-cycle load will depend on whether or not outdoor air is required during the night cycle. If outdoor air is not required, the outdoor air damper should be closed. (Remember to allow for some damper leakage.)

Changeover to Heating
Since there is only one zone, simultaneous heating and cooling is never required. The cooling to heating changeover takes place automatically when the zone thermostat senses that heating is required.

Psychrometrics
An elementary psychrometric analysis of the single-zone cooling cycle should be made because it allows the designer to predict the condition of the air entering the cooling coil. This analysis also defines the range of acceptable conditions for the air that will be supplied to the room during the cooling cycle. This information is directly related to the procedures that are used to select the cooling equipment. Refer to Sections 4-17 and 5-1 in this manual for more information about elementary psychrometrics.

Figure 11-2 shows a comprehensive psychrometric analysis of the single-zone cooling cycle. This diagram also shows how to use the psychrometric chart to estimate the design load on the cooling coil.

Equipment Selection — Primary Equipment
The capacity of the primary equipment should be based on the design values for the cooling and heating loads, and comprehensive engineering data should be used to select the equipment. The equipment performance data should always correspond to the actual operating conditions that are expected to occur at the site. Part-load performance also must be evaluated and an appropriate method of capacity control should be provided. The procedures that are used to select various types of cooling and heating equipment are presented in Sections 6 and 7 of this manual.

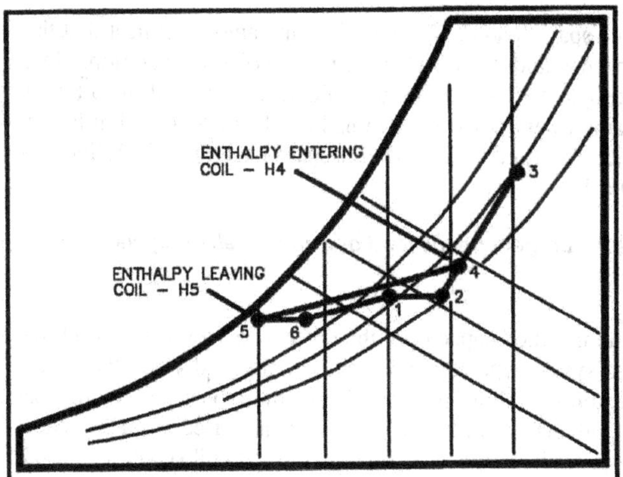

Air leaves the room [1] and picks up return-side heat [2] (the return-side gain could be fan heat, a return air plenum gain or a duct gain). Return air [2] is mixed with outdoor air [3] before the mixed air [4] enters the cooling coil. Cold, dry air off of the cooling coil [5] picks up supply-side heat [6] (the supply-side heat could be fan heat or a duct gain). The cool, dehumidified air [6] enters the room and absorbs the sensible and latent loads and is routed back through the return [1].

The total load on the cooling coil depends on the difference between the enthalpy (*H*) entering the coil [4] and the enthalpy leaving the coil [5]. The design load on the cooling coil can be estimated by using the following equation.

$$Tons = \frac{4.45 \times Coil\ CFM \times (H4 - H5)}{12000}$$

Figure 11-2

Equipment Selection — Reheat Coil
Reheat is not normally used unless close control over humidity and/or temperature is required. If a reheat coil is used, it must have enough capacity to maintain the setpoint temperature during any operating condition. There are two possible types of loads on the reheat coil, a reheat load and a space load.

The load associated with neutralizing relatively cold supply air (raising the temperature of the supply air up to room temperature) is the reheat load. Call this load L1 and use the following equation to calculate this load (*SAT* represents the supply air temperature and *RAT* represents the room air temperature).

$$L1 = 1.1 \times Supply\ CFM \times (RAT - SAT)$$

The space load (*L2*) is the maximum amount of heat that the space would require when the reheat coil is in operation. Most designers do not bother to evaluate the effect of the internal gains when they estimate this load. Instead, they simply use the value for the winter design space load (Form N-1, lines 16 and 17).

L2 = Largest space load during the reheat cycle

The installed capacity of the reheat coil will depend on how the system operates. If reheat is only used when the space requires cooling, the capacity of the reheat coil should be based on the reheat load (*L1*). If reheat is used when the space requires heat, the capacity of the reheat coil should be based on the total of the reheat load and the space load (*L1* + *L2*).

Equipment Selection — Humidifies

If a humidifier is required, its capacity should be equal to or greater than the largest humidification load that is expected to occur during any operating condition. Refer to Section 10 for more information about selecting humidifiers.

Supply CFM

The cooling CFM calculation is based on the peak sensible load that is associated with the conditioned space. This load is determined during the **Manual N** calculation procedure and it appears on Form N-1, line 8.

$$\text{Cooling CFM} = \frac{\text{Peak sensible load for space (BTUH)}}{1.1 \times (\text{Room design DB} - \text{Supply air DB})}$$

The heating CFM calculation is based on the heating load that is associated with the space. This load can be determined during the **Manual N** calculation by adding lines 16 and 17 on Form N-1.

$$\text{Heating CFM} = \frac{\text{Peak heating load for space (BTUH)}}{1.1 \times (\text{Supply air DB} - \text{Room design DB})}$$

If a fossil fuel furnace is used, the heating CFM also must be compatible with the manufacturer's temperature rise (*TR*) or discharge temperature (*TS*) limits. The following equations can be used to check for compliance with these requirements.

$$TR = \frac{\text{Output heating capacity (BTUH)}}{\text{Heating CFM}}$$

$$TS = \text{Entering temperature} + TR$$

If the same air handler provides cooling and heating, the design value for the supply CFM will either be equal to the cooling CFM or the heating CFM, depending on which value is larger. This design CFM value is used to specify blower performance and to design the air distribution system.

Duct Sizing

A single-zone system has no air-side diversity. The CFM flowing through each section of duct and out of each supply outlet is constant during all modes of operation. Therefore, the size of all of the supply and return air ducts (trunks and branches) should be based on the peak load CFM value.

Temperature and Humidity Controls

Equipment capacity is based on peak load conditions, but the peak load condition is not representative of the typical or average operating condition. Therefore, room temperature must be controlled by modulating the temperature of the supply air or by cycling the equipment on and off.

On-Off Control

Simple two-position (on-off) control is commonly used to control residential DX cooling equipment, heat pumps and furnaces. Two-position control also can be used to control some commercial equipment, but there are limitations:

- Space temperature will fluctuate in accordance with the differential that is associated with the thermostat. These fluctuations are usually acceptable for simple comfort conditioning, but they may be unacceptable for applications that require precise temperature control.
- On-off control cycles do a fair job of maintaining space temperature, but the humidity may not be properly controlled. Poor humidity control is experienced when variations in the latent load are not synchronized with the changes in the sensible load.
- On-off cycling can damage compressor motors. Small units may absorb this punishment, but tolerance to cycling decreases as the size of the equipment increases. Never use controls that cycle equipment unless this type of control is approved by the equipment manufacturer.
- On-off control is not appropriate for make up air equipment. Unconditioned or untempered air will be dumped into the occupied zone whenever the thermostat is satisfied.

Temperature Modulation

The method that is used to control the temperature of the supply air will depend on the nature of the equipment. DX coil capacity — providing the coil has face-split or row-split circuits — can be staged by using one or more solenoid valves to stop the flow of refrigerant through one or more of the coil circuits. Or, DX coil capacity can be modulated by using variable-speed fans and variable-speed condenser packages. Fossil fuel burners and electric resistance heating coils can be staged or modulated. Valves or face and bypass dampers can be used to control chilled water coils and hot water coils. (Face and bypass dampers are not recommended for controlling DX coils or electric heating coils because of the danger of not having adequate air flow across the coil.)

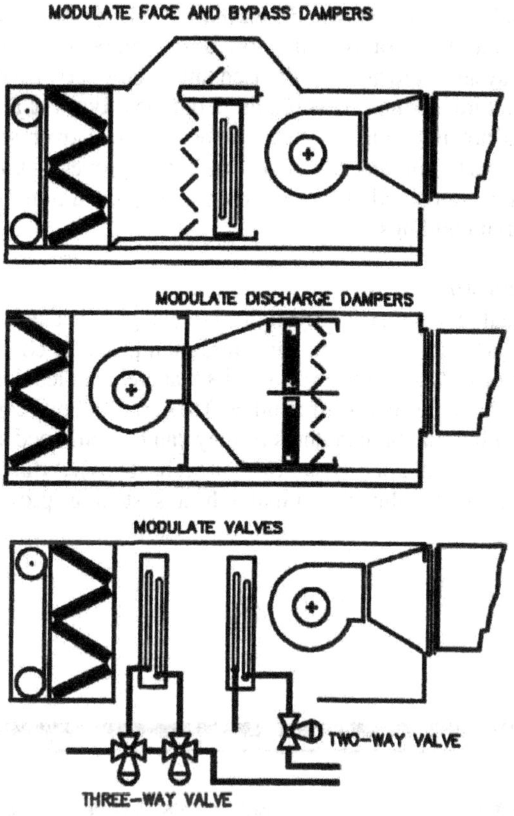

Water-Coil Capacity Control Devices

Some comments about the ability of the various types of heating and cooling equipment to control temperature and humidity are listed below. Note that compressors, chillers and boilers must be capable of adjusting capacity to match the loads that are imposed on the coil. Refer to Appendix 5 for more information about capacity control.

- Single speed DX cooling equipment is not well suited for applications that require close control over temperature and humidity. However, variable-speed DX equipment may provide satisfactory performance.

- Chilled water coils can provide close control over temperature. However, the humidity will not be under close control because the cooling capacity is controlled by sensing temperature. Reheat will be required if close control over temperature and humidity is desired.

- Hot water coils can provide close control over temperature. If required, the heating coil can be supplemented by a humidification system that provides close control over humidity.

- Furnaces and electric heating coils that have modulating controls provide the most accurate type of temperature control. Step controls are less accurate, but they are much better than on-off controls. Step controls can be used when there are large load variations and precise control is not required. Modulating controls are preferred for make up air equipment and for applications that require close control over temperature.

Reheat Control

A reheat coil can be used to provide close control over temperature and humidity. In this case, the temperature of the air entering the reheat coil — which could be the air off of the cooling coil or the air off of the economizer — must be dry enough (based on the grains of moisture per pound of air) to control the humidity in the conditioned space. This supply air temperature control could be a temperature sensor that is located in the supply duct or a humidistat that is located in the room, or in the return air duct. Reheat is required because air that is dry (cold) enough to control humidity may have too much sensible cooling capacity. A room thermostat can control the reheat coil and maintain the desired space temperature.

Controlling a Separate Heating System

In some applications, a separate heating system — such as a baseboard radiation system — is used with a single-zone cooling system. In this case, the two systems must not work against each other. This can be accomplished by using one thermostat to control both systems. This way, the cooling cycle can be de-energized before the heat is activated. Usually, a dead-band between the heating cycle and the cooling cycle prevents short-cycling (between the cooling and heating functions) and reduces energy consumption. However, the fan and damper operating cycles should provide continuous ventilation and air circulation when the indoor temperature is in the dead-band range.

Fan and Damper Control

The fan (or fans if a return air fan is used) should operate continuously when the space is occupied. If outdoor air is required for make up air, the fan (or fans) should operate continuously whenever make up air is required, even if the space is unoccupied.

Outdoor air, return air and exhaust air dampers should be controlled so that the minimum ventilation requirements are met whenever the space is occupied. If outdoor air is required for make up air, the dampers should be controlled so that adequate make up air is always available. (The outdoor air requirement will be dictated by the ventilation requirement or by the make up air requirement, whichever is larger.)

Night-Cycle Control

Control over the entire conditioning system can be transferred to a "night cycle" when the zone is unoccupied — providing that space conditions do not have to be maintained or providing that outdoor air is not required. This usually means that when the night-cycle thermostat is satisfied, the outdoor and exhaust air dampers are closed and the heating (or cooling) plant and the fan (or fans) are de-energized. When the night-cycle thermostat calls for heat (or cooling), the heating (or cooling) plant and the fans (or fans) operate, but the outdoor air and the exhaust air dampers remain closed.

11-3 Terminal Reheat Systems

This section provides an overview of the performance characteristics, hardware, design procedures and control cycles that are associated with constant volume, terminal reheat systems. The information provided by Section 11-1 also applies to these systems

Function
Constant volume, terminal reheat systems can either provide sensible and latent cooling or they can provide heat. They also can provide outdoor air for ventilation or for make up air, and they can provide humidification.

Applications
Before the cost of energy became a primary consideration, terminal reheat systems were very popular — they provide a simple way to zone a building, they are relatively easy to design and they offer simple solutions to the problems that are associated with simultaneous heating and cooling and seasonal changeover. However, escalating energy costs have made other types of systems more desirable. Now, terminal reheat systems are not usually used unless close control over temperature or humidity is required, but these types of applications are not typical. When precise control over space conditions is required, reheat systems can be properly applied to all kinds of small and large commercial, institutional and industrial buildings.

Configuration
Terminal reheat systems are series-coil, single-duct systems. In the basic configuration, the central unit provides cooling; heat is provided by zone reheat coils that are installed downstream from the central air handler. These reheat coils can be inserted into the branch ducts or they can be combined with an air flow regulator (constant volume reheat box). A diagram of a constant volume terminal reheat system is provided below.

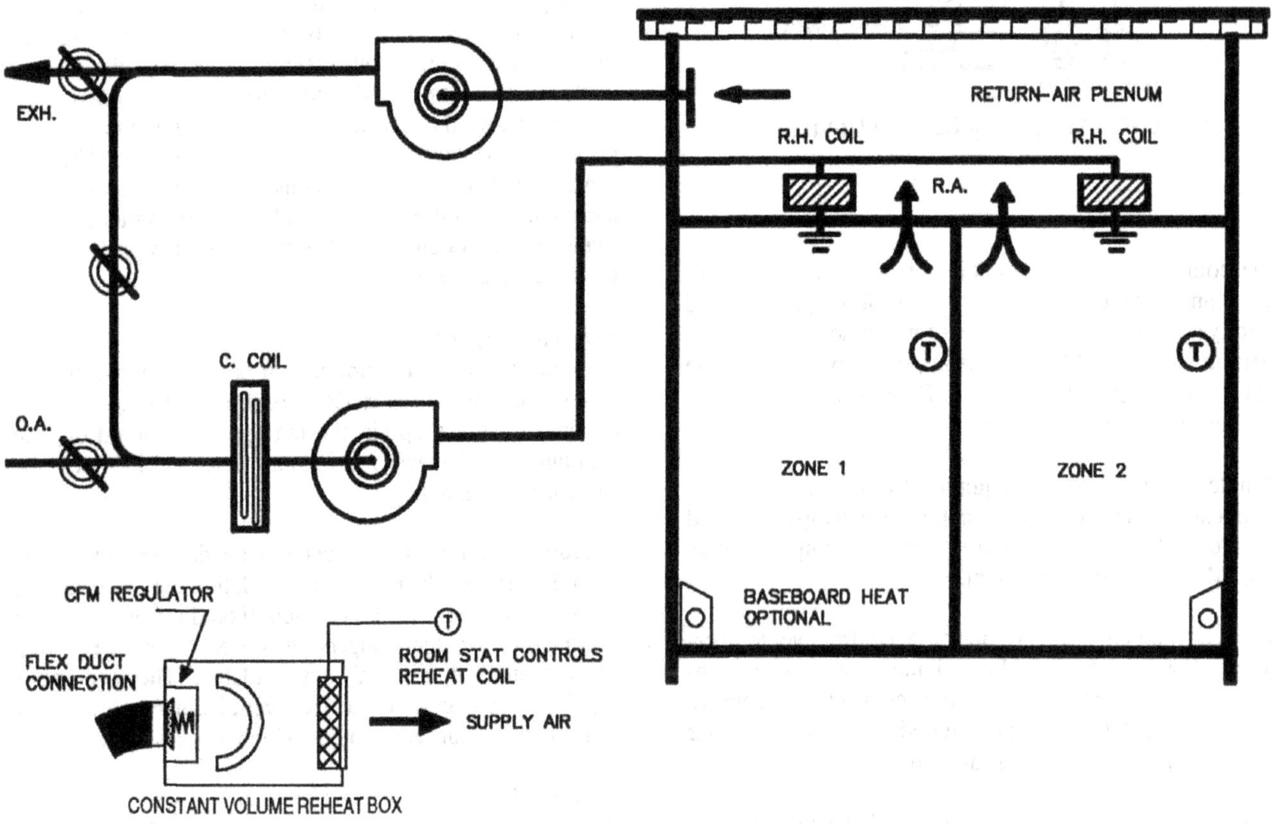

Constant Volume Terminal Reheat System

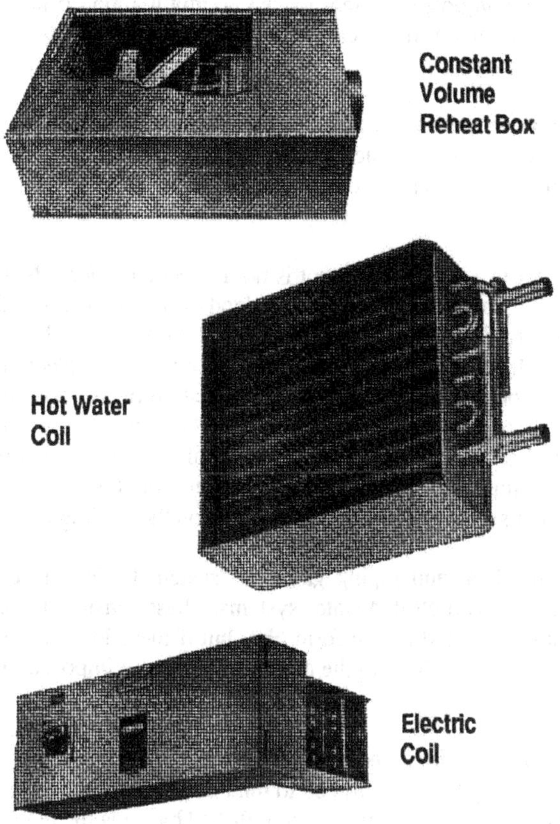

Secondary Equipment

Characteristics

The characteristics that are generic to all types of all-air systems are listed in Section 11-1. The characteristics that are peculiar to terminal reheat systems are listed below.

- The system provides an unlimited number of zones.
- The system provides exact temperature control.
- Air circulation rates are constant.
- Full dehumidification capacity is available at all times.
- The maximum humidity in any zone can be limited to a specified level.
- Humidifiers are required to maintain exact humidity control in every zone.
- Control over the capacity of the cooling coil is dictated by dehumidification requirements.
- Control over the capacity of the cooling coil is not directly related to sensible loads.
- The system is easy to design.
- The system is relatively easy to install and operate.
- The control cycles are relatively simple.
- No seasonal changeover (heating or cooling) is required.
- Simultaneous heating and cooling is always available in every zone.
- Since the fresh air requirements may vary by zone, the percentage of outdoor air introduced at the air handler may not match the outdoor air percentage that is required for each zone.
- Since the room sensible heat ratios may vary by zone, the condition of the air leaving the cooling coil may not be perfectly compatible with all of the room sensible heat ratios.
- The central equipment must be used to condition a single zone even though the rest of the building may be vacant.
- Night-cycle control cannot be applied to the central equipment unless all of the zones are vacant.
- Night-cycle control can be applied to the reheat coils on a zone-by-zone basis.
- The system is characteristically expensive to operate.
- Carefully designed control strategies can produce a significant reduction in operating costs.
- Certain types of reclaim equipment can be used to reduce operating costs.
- It is not easy to account for and to meter energy consumption on a zone-by-zone basis.

Equipment

The various types of primary cooling and heating equipment that could be used for a multizone, terminal reheat system are listed in Section 11-1; refer to the paragraph titled *Primary Equipment*. Reheat coils or constant volume reheat boxes are the most common types of secondary equipment that are associated with reheat systems. If required, the system also can be equipped with a single central humidifier — or individual humidifiers can be provided for each zone.

Sizing Loads — Zone Cooling

The cooling capacity that is required for each zone is equal to the corresponding peak zone load. One **Manual N**, Form N-1 must be filled out for each zone. The sensible and latent loads that are associated with a given zone will correspond to the values that are found on line 8 of Form N-1, (lines 9 through 12 are not required). Note that if there is a supply duct gain, (Form N-1, line 7) it is added to the zone load. This is done so that the duct gain is offset by an increase in the cooling CFM that is supplied to the zone.

Sizing Load — Cooling at the Central Coil

Since the zone reheat coils make up the difference between the peak sensible loads and the off-peak sensible loads, the various zones appear (to the cooling coil) to be simultaneously and continuously subjected to peak load conditions. Therefore, the sensible capacity that is required at the central coil is based on the sum of the peak sensible zone loads.

The latent capacity that is required at the central coil may either be based on the sum of the peak zone latent loads or on a "block" latent load estimate. Use the sum of the peak zone loads if the latent loads in every zone are constant or if humidifiers are used to maintain a constant level of humidity. Use a block latent load estimate if diversity is applicable.

In addition to the sensible and latent zone loads, the cooling coil must have enough capacity to neutralize the outdoor air loads and any other system loads that might apply. Lines 9 through 11 on Form N-1 can be used to estimate the following system loads:

- Sensible and latent outdoor air loads, (line 9).
- Sensible return air plenum load, (line 10).
- Sensible return air duct gain, (line 11).

Fan heat is a system load that is not included on Form N-1. However, the sensible heat that is added by the supply fan and the return fan should be accounted for. Sometimes, this is accomplished (as per the manufacturer's equipment selection instructions) by subtracting the fan heat from the cooling capacity of the unit. (Refer to Section 6-8 for an example that illustrates this procedure.) Any portion of the fan heat that is not accounted for by a capacity reduction should be added to the other system loads that are imposed on the cooling coil.

Pumping heat and piping gains are system loads that are associated with chilled water systems. These sensible loads are not accounted for on form N-1, but if these loads exist, they should be added to the other loads that are imposed on the chiller.

Minimum Cooling Loads — Zones
Zone cooling loads are forced to remain constant during any operating condition. This is accomplished by using the reheat coil to compensate for load reductions that are associated with operation at part-load conditions. Figure 11-3 illustrates this operating characteristic. Note that the actual cooling load plus the reheat load is always equal to the peak cooling load.

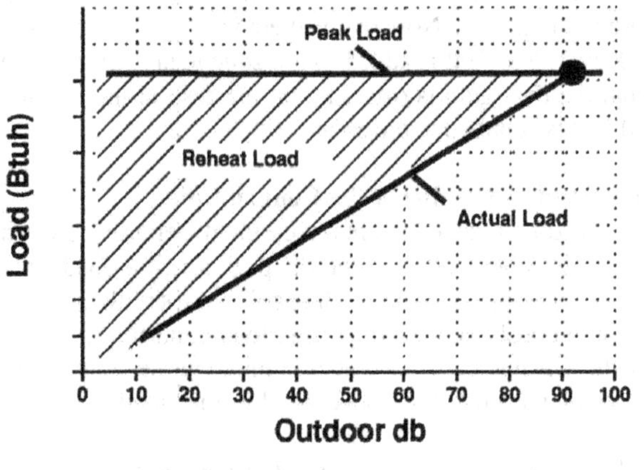

Figure 11-3

Minimum Cooling Load — Central Coil
The minimum load on the central cooling coil is equal to the sum of the peak zone loads, plus the outdoor air load, plus any other system loads that may apply. Since the supply CFM, return air temperature and zone loads are constant (because of reheat), a reasonable estimate of the minimum coil load can be obtained by analyzing the effect of the outdoor air load.

During cool or cold weather, the outdoor air load will be negative because cold outdoor air will be mixed with the return air from the conditioned spaces. As explained below, the net load on the cooling coil will depend on what strategy is used to control the outdoor air dampers. (Refer to Figure 11-1 for a summary of the three basic outdoor air control strategies.)

No Economizer
If no economizer cycle is used, the net load on the coil will depend on the winter outdoor air design temperature and on the percentage of outdoor air that is required to satisfy the minimum outdoor air requirement. In cold climates, it is possible to have a situation where there is no load on the cooling coil. Table 11-1, illustrates this point.

Percentage of the Design Load on the Coil			
Winter Design	**Minimum Outdoor Air Percentage**		
Temperature	**10%**	**20%**	**40%**
0 °F	38	15	0
20 °F	70	38	15
40 °F	77	54	46
Coil load = 4.45 x CFM x (Hin - Hout) Hin and Hout = entering and leaving enthalpy			

Table 11-1

Simple Economizer
If a simple economizer cycle is used, the minimum possible cooling load on the coil will occur just before the system switches to the economizer cycle. An estimate of this load can be made by assuming that the latent load is zero. The psychrometric chart and the following equation can be used to make this calculation.

$$TONS = \frac{1.1 \times CFM \times (T1 - T2)}{12000}$$

In this equation, the tonnage represents the minimum load on the cooling coil. *T1* represents the dry-bulb temperature of the mixed air entering the cooling coil, *T2* represents the economizer changeover temperature and CFM represents the air flow through the coil. Since the CFM and the economizer changeover temperature are constant, the temperature of the air entering the coil is the primary factor that affects the load on the coil.

For example, Figure 11-4, at the top of the next page, shows an analysis of a 10,000 CFM system that switches to the economizer when the outdoor temperature is equal to 55 ^{O}F.

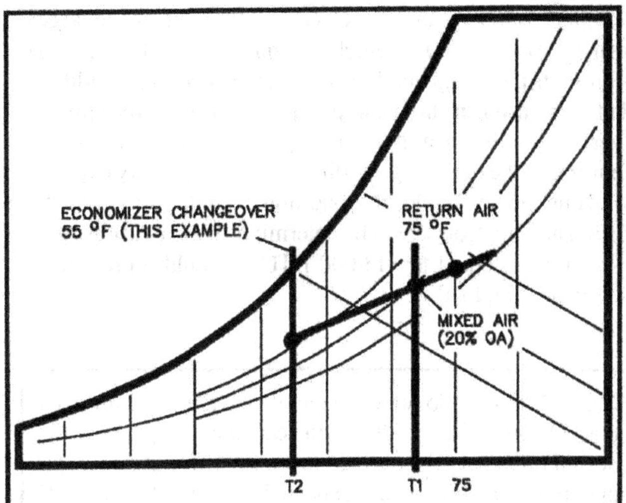

$$T1 = 75 - 0.20 \times (75 - 55) = 71$$

$$Tons = \frac{1.1 \times 10000 \times (71 - 55)}{12000} = 14.67$$

Figure 11-4

Note that the temperature of the entering air (*T1*) depends on the percentage of outdoor air that is required to satisfy the minimum outdoor air requirement. (The minimum percentage of outdoor air varies from job to job. For this example, the outdoor air CFM was arbitrarily set to equal 20 percent of the supply air CFM. Always use the "actual value" for this percentage when performing this calculation.)

Ultimate Economizer

If an ultimate economizer cycle is used, it is possible to have the refrigeration equipment on line, with no load on the coil. Figure 11-1 on page 11-4 provides a diagram that illustrates this point.

Night-Cycle (Unoccupied) Cooling

In many applications, the cooling system can be shut down during the night. This control strategy reduces operating cost. If the system has to operate during the night cycle, the reheat coils will force the zone cooling loads to peak. (Obviously, it would be a mistake to use a "set-up" control strategy during the night cycle. However, a "set-back" control strategy would reduce the amount of energy that is required for reheat.)

A minimum flow of outdoor air may not be required during the night cycle. If outdoor air is not required, the outdoor air damper should be closed when cooling is provided by mechanical refrigeration. However, the economizer cycle should still be used when the condition of the outdoor air is suitable for free cooling.

The internal loads are usually smaller during the night cycle, so when night-cycle cooling is required; the minimum load on the coil is usually less than the minimum day-cycle load. An analysis of both operating cycles is required to determine which cycle produces the smallest load on the cooling coil.

Zone Sensible Heat Ratios

A sensible heat ratio calculation should be made for each zone. These calculations should be based on the sensible and latent loads that are associated with the peak zone loads. There is no reason to expect all of the zones to have the same sensible heat ratio. As explained in the following paragraph, this information is required to establish an acceptable condition for the air leaving the central cooling coil.

Condition of the Air Leaving the Cooling Coil

The performance of the cooling coil must be compatible with the conditioning requirements of each zone. This means that the condition of the air leaving the cooling coil must be cold enough to offset the sensible load in any zone and dry (cold) enough to neutralize the latent load in any zone. Section 4-17 in this manual, which discusses this subject, indicates that the designer will have the following options:

- If humidity control is not critical, the required cooling coil performance can be based on a sensible heat ratio that falls near the mid-range of the various zone sensible heat ratios.
- If humidity control in one or more zones is critical, the cooling coil performance must be based on the lowest sensible heat ratio that is associated with the critical zones.

Sizing Loads — Zone Heating

There are two possible types of zone heating loads, a reheat load and a space load. Figure 11-5 shows these two loads.

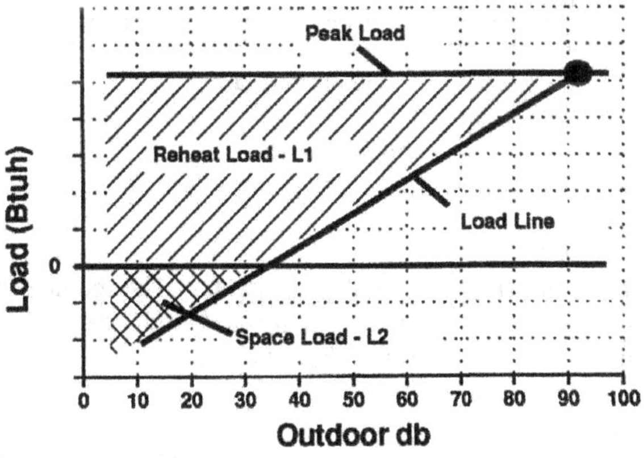

Figure 11-5

The load associated with neutralizing relatively cold supply air (raising the temperature of the supply air up to room temperature) is the reheat load. Call this load *L1* and use the following equation to calculate this load. (*SAT* is the supply air temperature and *RAT* is the room air temperature.)

L1 = 1.1 x Supply CFM x (RAT - SAT)

The space load is the amount of heat that the zone would require at the winter design condition when the envelope losses are balanced against the internal gains. Call this load *L2*. Form N-1 can be used to calculate the envelope losses and the internal gains. (One form is required for each zone. The zone envelope loss is equal to the sum of lines 16 and 17 and the internal gain is equal to the line 4 value.)

L2 = (Envelope losses - Internal gains)

The amount of heat that a zone will require during any operating condition is equal to the reheat load plus the space load (*L1* + *L2*). However, this load will be different for each different operating condition. Figure 11-6 provides a diagram that illustrates this point. This diagram shows how the net heating load is calculated for three typical operating conditions, which are explained below.

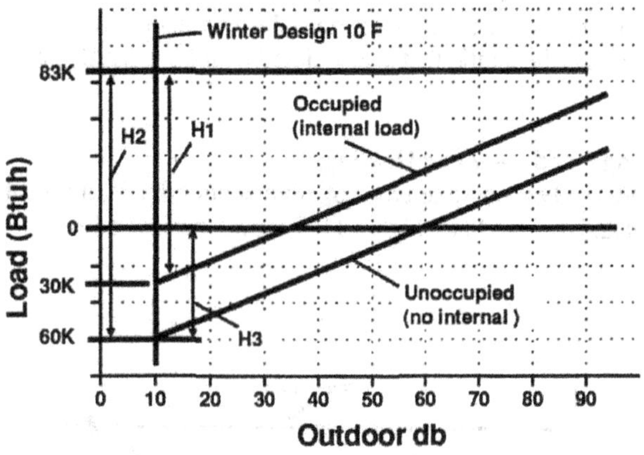

Figure 11-6

Condition 1
H1 represents the amount of heat that would be required if the central system was in the cooling mode and the zone was occupied (internal loads help heat the zone).

Condition 2
H2 represents the amount of heat that would be required if the central system was in the cooling mode and the zone was unoccupied (no internal loads).

Condition 3
H3 represents the amount of heat that would be required if the central system was in the a night-cycle mode (no internal loads, heating-only).

For this particular diagram, Condition 2 produces the largest heating load and *H2*, which is equal to 143,000 BTUH, represents the design load. However, the operating conditions that were used in this example do not necessarily apply to every terminal reheat system. In practice, the analysis must conform to the operating conditions that are actually expected to occur. For example, if Condition 2 could not occur, the design heating load would be determined by Condition 1; and *H1*, which is equal to 113,000 BTUH, would represent the design heating load.

Many designers do not evaluate the heating load for each possible operating condition. Instead, they simply base the design on a "worst case" scenario, even if that particular operating condition is not expected to occur. This type of calculation produces the *H2* value in Figure 11-6. This value is easily determined by adding the standard space heat load (no credit for internal gains) to the reheat load.

Heat Required at the Central Unit
The amount of heat required at the central unit will depend on how the system accommodates the zone loads and the outdoor air load. No central heat will be required, providing that:

- The mixed air temperature (due to the minimum ventilation requirement) does not fall below the cooling cycle, supply air temperature setpoint.
- Adequate heating capacity is installed in each zone.

Night-Cycle Heating Loads
The night-cycle reheat load (*L1*) will depend on how the system is controlled and operated. Consider the following (typical) possibilities:

- If cooling or dehumidification is not required when the space is unoccupied, the cooling system can be shut down during the night cycle. And, if fresh air or make up air is not required when the space is unoccupied, the outdoor air dampers can be closed. Therefore, the reheat load (*L1*) will be almost eliminated because the air leaving the air handler will be close to the temperature of the return air. (The supply air will be a little colder than the return air because of damper leakage).
- If cooling or dehumidification is required when the space is unoccupied, the supply air temperature will be the same as the day-cycle temperature and the night-cycle reheat load (L1) will be the same as the day-cycle reheat load.

The night-cycle space load (L2) will depend on whether or not the zone thermostats are set back during the night cycle and whether or not there are any uninterrupted internal loads. Credit should be taken for night set-back, but credit for internal loads may or may not be applied, depending on the judgment of the designer.

Changeover to Heating
The cooling to heating changeover takes place on a zone-by-zone basis and it automatically occurs as the zone thermostats call for more and more heat. The central equipment does not changeover to a heating cycle, but in some designs the central unit is switched to a night cycle.

Psychrometrics
An elementary psychrometric analysis of the cooling cycle should be made because it allows the designer to predict the condition of the air entering the cooling coil. This analysis also defines the range of acceptable conditions for the air that will be supplied to the room during the cooling cycle. This information is directly related to the procedures that are used to select the cooling equipment. Refer to Sections 4-17 and 5-1 in this manual for more information about elementary psychrometrics.

Figure 11-7 shows a comprehensive psychrometric analysis of the terminal reheat cooling cycle. This diagram also shows how to use the psychrometric chart to estimate the design load on the cooling coil.

Equipment Selection — Primary Equipment
The capacity of the primary cooling equipment should be based on the sum of the peak zone loads (as explained above in the paragraph titled *Sizing Load— Cooling at the Central Coil*) and comprehensive engineering data should be used to select equipment. This engineering data should correspond to the actual operating conditions that are expected to occur at the site. Part-load performance also must be evaluated and an appropriate method of capacity control should be provided. The procedures that are used to select various types of cooling equipment are presented in Section 6 of this manual.

Equipment Selection — Secondary Equipment
Reheat coils must have enough capacity to maintain the setpoint temperature during any operating condition. These capacities should be equal to or greater than the corresponding zone (heating cycle) design load. The procedure that is used for calculating the zone heating loads is provided above, refer to the paragraph titled *Sizing Loads — Zone Heating*. The procedures that are used to select heating coils are presented in Section 7 of this manual.

Equipment Selection — Humidifiers
A single humidifier may be installed in the central air handler or two or more small humidifiers can be installed in two or more zones. If a humidifier is required, its capacity should be equal to or greater than the largest humidification load that is expected to occur during any operating condition. Refer to Section 10 of this manual for more information about selecting humidifiers.

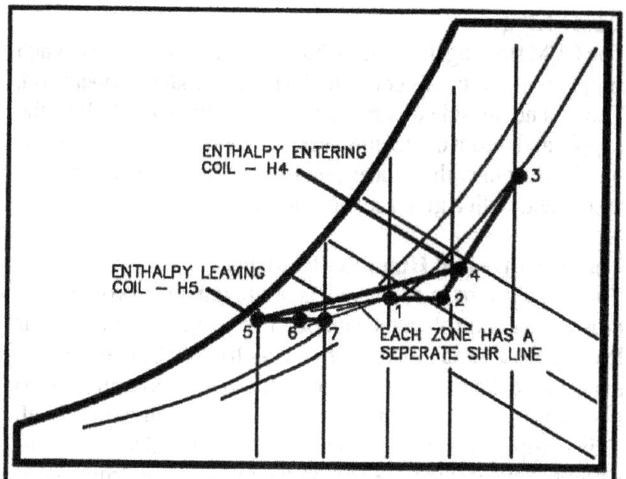

Air leaves the room [1] and picks up return-side heat [2], (the return-side gain could be fan heat, a return air plenum gain or a duct gain). Return air [2] is mixed with outdoor air [3] before the mixed air [4] enters the cooling coil. Cold, dry air off of the cooling coil [5] picks up supply-side heat [6], (the supply-side heat could be fan heat or a duct gain). Air is delivered to the reheat coil [6] and picks up heat [7]. The tempered, dehumidified air [7] enters the room and absorbs the sensible and latent loads and is routed back through the return [1].

The total load on the cooling coil depends on the difference between the enthalpy entering the coil [4] and the enthalpy leaving the coil [5]. The design load on the cooling coil can be estimated by using the following equation.

$$Tons = \frac{4.45 \times Coil\ CFM \times (H4 - H5)}{12000}$$

Figure 11-7

Supply CFM — Zones
The zone CFM values are based on the peak sensible loads that are associated with the various zones. These loads correspond to the value that is found on Form N-1, line 8. One Form N-1 is required for each zone.

$$Cooling\ CFM = \frac{Peak\ sensible\ zone\ load\ (BTUH)}{1.1 \times (Room\ design\ DB - Supply\ air\ DB)}$$

Supply CFM — Central Unit
The supply CFM at the central unit is determined by the peak cooling load requirements. The required CFM is equal to the sum of the various zone CFM values.

Duct Sizing

The CFM flowing through each section of duct and out of each supply outlet remains constant during all modes of operation. There is no air-side diversity. Therefore, the size of all of the supply and return ducts (trunks and branches) should be based on CFM values that correspond to the peak loads that are associated with a given section of duct.

Temperature and Humidity Controls

The temperature in each zone is maintained by a room thermostat that controls a reheat coil. The temperature of the air leaving the central air handler — which could be operating in a mechanical cooling or economizer mode— is controlled so that it has the capacity to absorb the dehumidification load. This (supply air) temperature could be controlled by a temperature sensor that is located in the main supply trunk — or by a humidistat that is either located in a critical zone or in the return air trunk. In any case, this control must be designed to insure that the supply air is dry (cold) enough (based on the grains of moisture per pound of air) to control the humidity in the various zones. (Note that the reheat energy requirements can be reduced by using the warmest possible supply air temperature.) It also is necessary to make sure that the capacity of the refrigeration equipment (chiller or DX unit) tracks the cooling load during any operating condition.

Control of a Separate Heating System

It is possible to use a separate heating system, such as a baseboard radiation system, to supplement or to replace the reheat coil. In this case, it is necessary to make sure that the various devices work in sequence. This can be accomplished by using one thermostat to control all of the devices. Usually, the baseboard heat is kept off if the reheat coil is able to maintain the desired space temperature.

Fan and Damper Control

The fan (or fans if a return air fan is used) should operate continuously when any zone is occupied. If outdoor air is required for make up air, the fan or fans should operate continuously whenever make up air is required, even if all the zones are unoccupied.

Outdoor air, return air and exhaust air dampers should be controlled so that the minimum ventilation requirements are met whenever the building is occupied. If outdoor air is required for make up air, the dampers should be controlled so that adequate make up air is always available. (The outdoor air requirement will be dictated by the ventilation requirement or by the make up air requirement, whichever is larger.)

Night-Cycle Control

Providing that the day-cycle space conditions do not have to be maintained, control over the entire conditioning system can be transferred to a heating-only night cycle when the building is unoccupied. When control is transferred to the night cycle, the outdoor and exhaust air dampers can be closed and the fan (or fans) can be de-energized. When the night-cycle thermostat — which should be located in the zone that has the most rapid rate of temperature drop — calls for heat, the fan (or fans) operate, but the outdoor air and the exhaust air dampers remain closed.

11-4 Variable Volume Systems

This section provides an overview of the performance characteristics, hardware, design procedures and control cycles that are associated with variable volume (VAV) systems. When it is appropriate, the discussion distinguishes between throttling variable volume systems and bypass variable volume systems. The information provided by Section 11-1 also applies to these systems.

Function

Variable volume systems can either provide sensible and latent cooling or they can provide heat. They also can provide outdoor air for ventilation or for make up air, filtration and humidification.

Applications

When precise control over space conditions is not required, basic types of variable volume systems can be properly applied to all kinds of small and large commercial, institutional and industrial buildings. One zone, or any number of zones can be accommodated. Variable volume cooling, with no heat, can be used for interior-zone applications, but some form of heat is usually required for perimeter-zone applications. (Variable volume cooling systems are often used in conjunction with an independent baseboard heating system.) When they are properly designed, these systems are very efficient. Because they are efficient, they are often selected for projects that have budgets that emphasize the importance of operating and installation costs.

Configuration

In their simplest form, variable volume systems are single-coil, single-duct cooling systems. If heat is required, it is provided by an independent system. However, variable volume systems often provide cooling and heat. In this case, heat is usually furnished by zone reheat coils. These reheat coils can be inserted into the branch ducts or they can be combined with a variable volume terminal device, such as a VAV reheat box or a constant volume fan box. It is also possible to use a variable volume system to just provide heat.

VAV system configurations also are identified by the devices that are used to control the flow of the supply air and by a method that is used to control fan performance. Throttling VAV systems use devices that modulate the flow of air to the supply outlets and they have fans that are equipped with inlet vanes or another type of capacity control device. Bypass (or dumping) VAV systems use devices that reduce the flow of air through the supply outlets by dumping excess air into a return air ceiling cavity. Bypass systems do not have devices

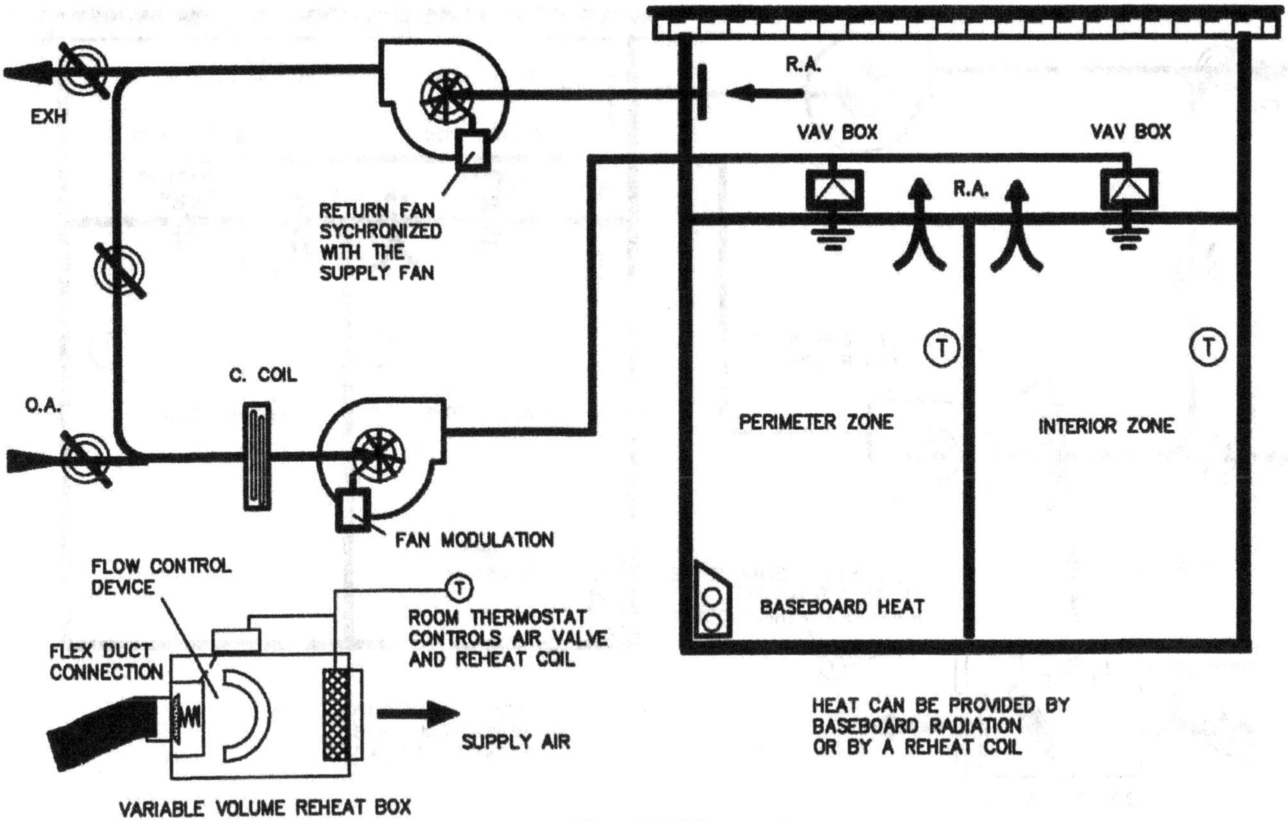

Throttling VAV System

that control the air flow at the fan(s). Therefore, with a bypass system, the air flow through the central equipment remains constant as the dumping devices reduce the air flow to the zones. A diagram of a throttling VAV system is provided above and a diagram of a bypass VAV system is shown at the top of page 11-16.

Characteristics

The characteristics that are generic to all types of all-air systems are listed in Section 11-1. The characteristics that are peculiar to variable volume systems are listed below.

- VAV systems can provide an unlimited number of zones.
- Throttling VAV systems are best suited for large buildings.
- Bypass VAV systems are best suited for small commercial applications.
- VAV systems provide good temperature control.
- VAV systems do not provide exact humidity control, full dehumidification capacity is not always available.
- The maximum humidity in any zone may not be limited to a specific level.
- Air circulation rates will vary, depending on the fluctuation of the space loads.
- VAV systems can take advantage of the diversity that is associated with the zone loads.
- Control over the capacity of the cooling coil is directly related to the sensible space loads.
- Control over the capacity of the cooling coil is not directly related to the latent loads.
- VAV systems require more design calculations.
- VAV systems are relatively easy to install and operate.
- No special provisions for seasonal changeover (heating or cooling) is required.
- Simultaneous heating and cooling can be provided for every zone.
- Since the fresh air requirements may vary by zone, the percentage of outdoor air introduced at the air handler may not match the outdoor air percentage that is required for each zone.
- Without special damper controls, the quantity of outdoor air delivered to a given zone will be throttled in proportion to the reduction in the supply CFM.
- Since the sensible heat ratios may vary by zone, the condition of the air leaving the cooling coil may not be perfectly compatible with all the zone sensible heat ratios.
- Without special controls, the latent capacity of air delivered to a given zone will be throttled in proportion to the reduction in the supply CFM.
- The central equipment must be used to condition a single zone even though the rest of the building may not require conditioning when those zones are vacant.

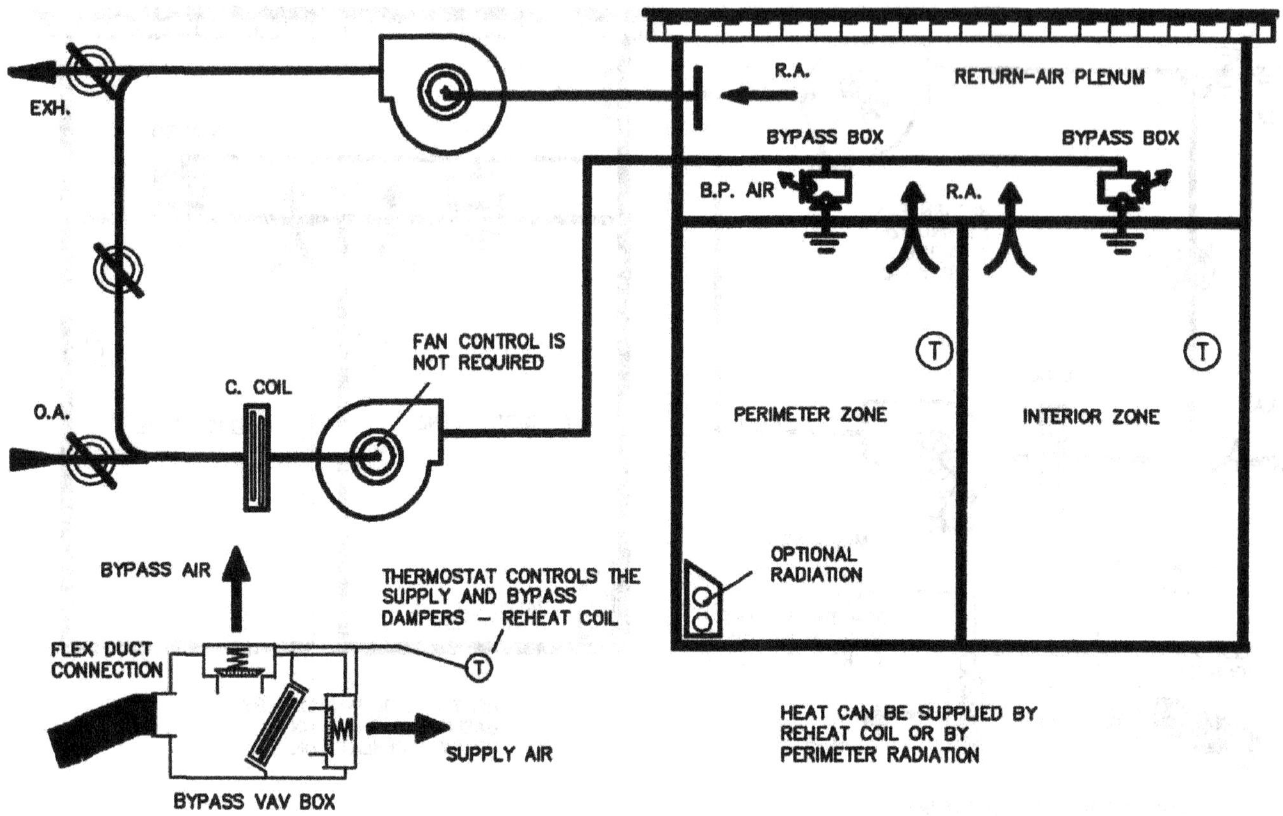

Bypass VAV System

- Night-cycle control cannot be applied to the central equipment unless all the zones are vacant.
- Night-cycle control can be applied to the terminal equipment on a zone-by-zone basis.
- Carefully designed control strategies can improve control over humidity, air circulation and fresh air without causing an unacceptable increase in operating costs.
- VAV system control cycles can be complicated for some of the more sophisticated configurations.
- VAV systems have comparatively low operating costs.
- It is not easy to account for and to meter energy consumption on a zone-by-zone basis.
- VAV systems have the flexibility to accommodate floor plan changes.
- When pressure independent terminal devices are used, the air-side of the system is self balancing.
- VAV systems require special supply air outlets — constant volume outlets are inappropriate.

Primary Equipment

The primary cooling equipment may consist of a unitary or split DX cooling system or an air handler that is equipped with a chilled water coil. Economizers are commonly used for free cooling and return fans are required to control the flow of return air and exhaust air. Return fans also provide control over the pressure in the conditioned space. Filters, dampers and in some cases humidifiers, are installed in the air handler.

If heat is required, it can be supplied by a separate heating system or by heating devices that are installed in the VAV air distribution system. If a separate heating system is used, it is typically a perimeter hot water or electric baseboard radiation system, but it could be a hot air or steam system.

Secondary Equipment

Terminal units, distribution boxes, reheat boxes, fan-powered VAV boxes and fan capacity control devices (next page) are the most common types of secondary equipment that is used with variable volume systems. (Refer to ACCA **Manual Q**, Sections 13 and 14 for more information about terminal devices and refer to Section 16 for information about fans and fan control.) If required, the system also can be fitted with a single central humidifier — or individual humidifiers can be provided for each zone.

Supply Air Diffusers

Special diffusers are required to maintain acceptable room air motion as the supply CFM is throttled from the maximum flow rate to the minimum flow rate — constant volume supply

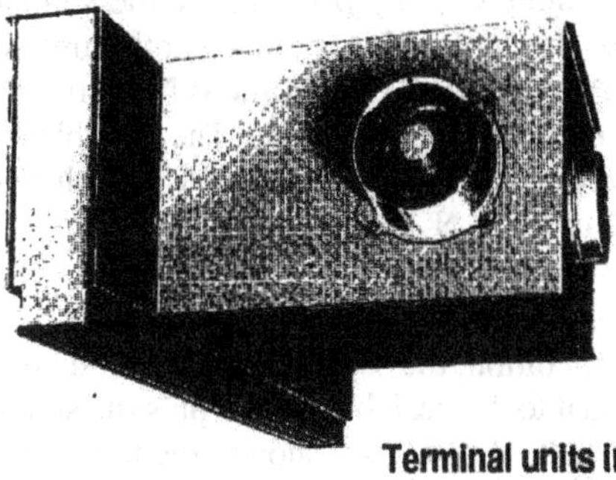

Terminal units integrate a diffuser and a throttling valve into one package.

Distribution boxes throttle air flow to one diffuser or to a group of diffusers.

Inlet vanes provide fan control.

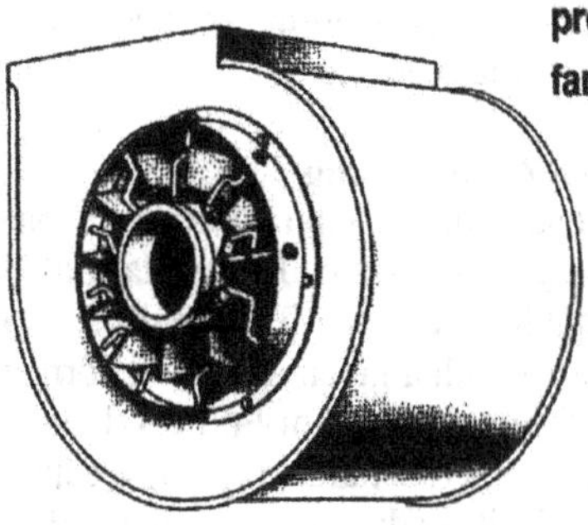

Reheat boxes combine a reheat coil and a throttling valve into one package.

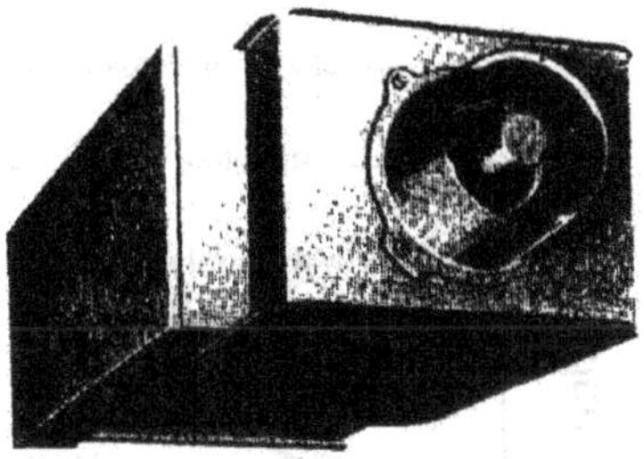

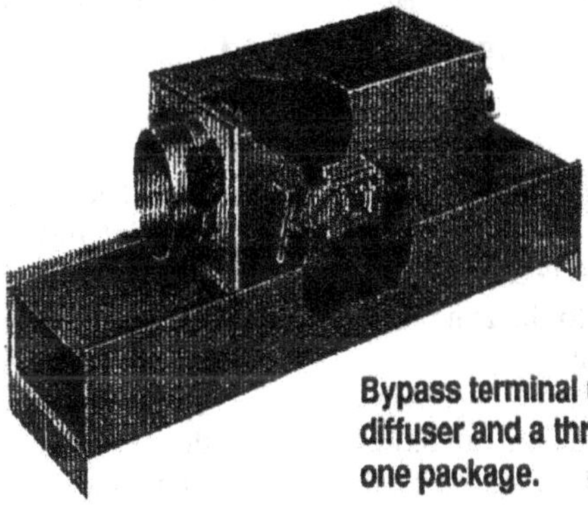

Bypass terminal units integrate a diffuser and a throttling valve into one package.

Bypass distribution boxes throttle air flow to one diffuser or to a group of diffusers.

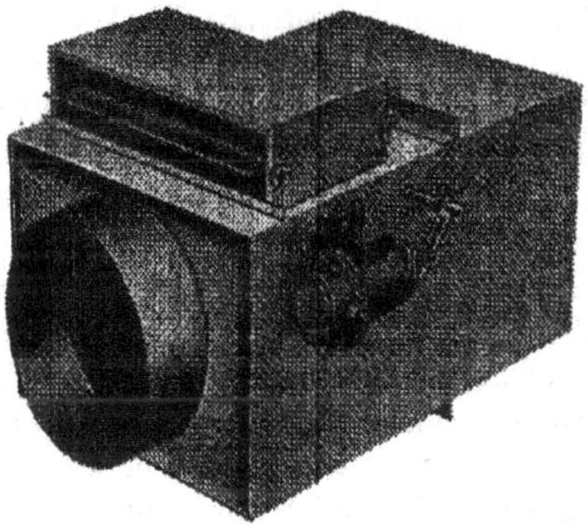

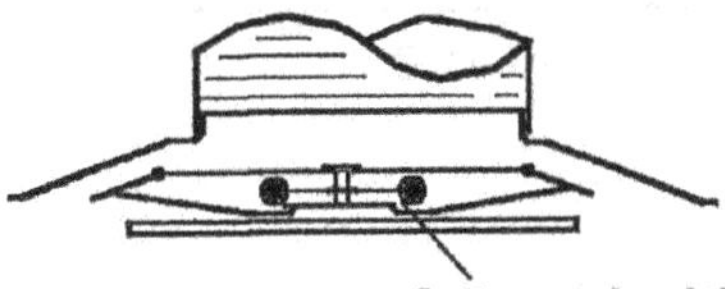

Self-contained throttling outlets do not require a remote thermostat.

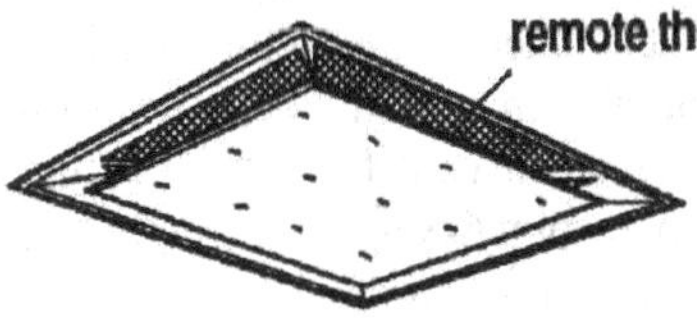

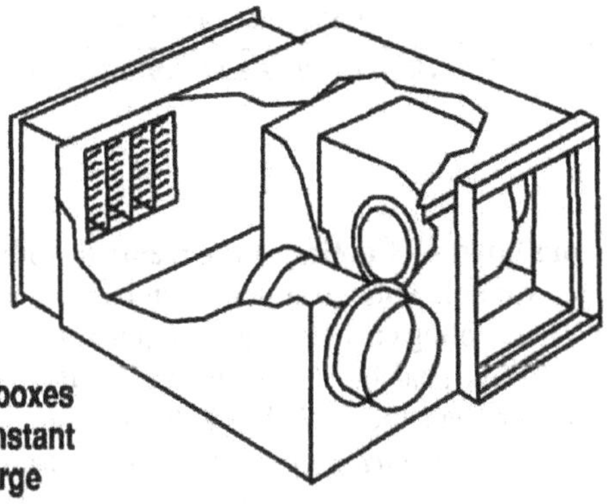

Fan-powered boxes maintain a constant flow of discharge

outlets are not acceptable. These "variable volume" diffusers may look like slot diffusers or they may resemble square or round ceiling diffusers, but they always discharge the supply air in a concentrated jet that is directed parallel to the ceiling. This discharge pattern allows the diffuser to take full advantage of the Cononda effect.

Some manufacturers market a VAV terminal that has a self-contained thermostat and a bellows that powers a modulating flow control device. These units do not require an external control system and they are self-powered. They are available in a throttling or a bypass configuration. These devices provide a simple and convenient way to convert a single-zone, constant volume system into a multiple-zone, variable volume system. Refer to ACCA **Manual Q**, Section 13 for more information about these devices.

Sizing Loads — Zone Cooling

The cooling capacity that is required for each zone is equal to the corresponding peak zone load. One **Manual N**, Form N-1 must be filled out for each zone. The sensible and latent loads that are associated with a given zone will correspond to the values that are found on line 8 of Form N-1 (lines 9 through 12 are not required). Note that if there is a supply duct gain (Form N-1, line 7), it is added to the zone load. This is done so that the duct gain is offset by an increase in the cooling CFM that is supplied to the zone.

Diversity

Each zone will experience a load peak during certain days of the year and during certain hours of the day. However, in most buildings the various zone loads do not peak simultaneously. Therefore, there will be a difference between the peak load that is associated with the whole building and the load that is calculated by summing the peak zone loads. "Diversity" is a term that is used to describe this difference. The concept of diversity is important because if a throttling system is used, it allows the designer to reduce the size of the central equipment. The following equation can be used to calculate a diversity factor. This factor varies from building to building, but values between 0.70 and 0.80 are common.

$$\textit{Diversity} = \frac{\textit{Peak Block Space Load}}{\textit{Sum of the Peak Zone Loads}}$$

- If reheat is not required in any zone during the summer design condition, the cooling loads at the central unit are equal to the peak block load. One **Manual N**, Form N-1 must be filled out for the whole building. The sensible and latent loads at the coil will correspond to the values that are found on line 12 of Form N-1.

- If reheat is required in one or more zones during the summer design condition, the sensible cooling load at the central unit is equal to the peak block load plus the sum of the zone reheat loads. Figure 11-8, shows how to estimate the reheat load that is associated with a critical zone that does not peak when the block load peaks.

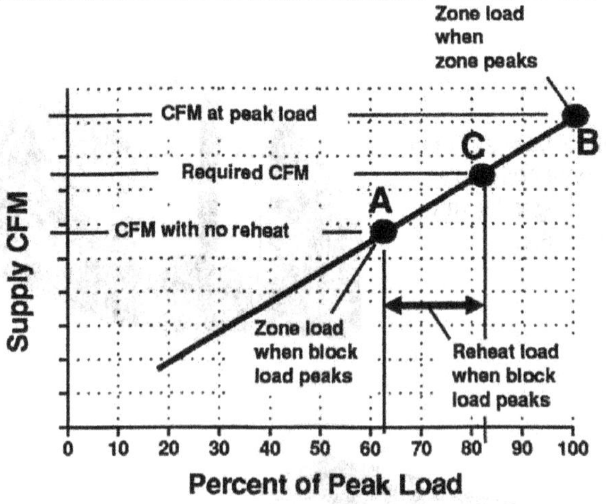

Point A indicates how much supply air the zone would recieve if reheat is not available when the block load peaks. Point C indicates the minimum acceptable supply air CFM. The differance between the load at point C and the load at point A is equal to the zone reheat that is required when the block load peaks.

Note that point C could fall anywhere along the load line. No reheat would be required if point C is to the left of point A. It is also possible for point A to be coincident with point B.

Figure 11-8

Sizing Load — Cooling at the Central Coil

Since variable volume systems can take advantage of diversity, the cooling capacity that is required at the central coil is based on the peak block load. However, the design load could be larger than the peak block load if one or more zones requires reheat *during summer design conditions.* (Zones that do not experience a peak load when the block load is at its peak could require reheat to maintain a minimum air flow; which might be required to satisfy special temperature, humidity or fresh air requirements.)

Fan heat is a system load that is not included on Form N-1. However, the sensible heat that is added by the supply fan and by the return fan should be accounted for. Sometimes this is accomplished (as per the manufacturer's equipment selection instructions) by subtracting the fan heat from the cooling capacity of the unit. (Refer to Section 6-8 for an example that illustrates this procedure.) Any portion of the fan heat that is not accounted for by a capacity reduction should be added to the other system loads that are imposed on the cooling coil.

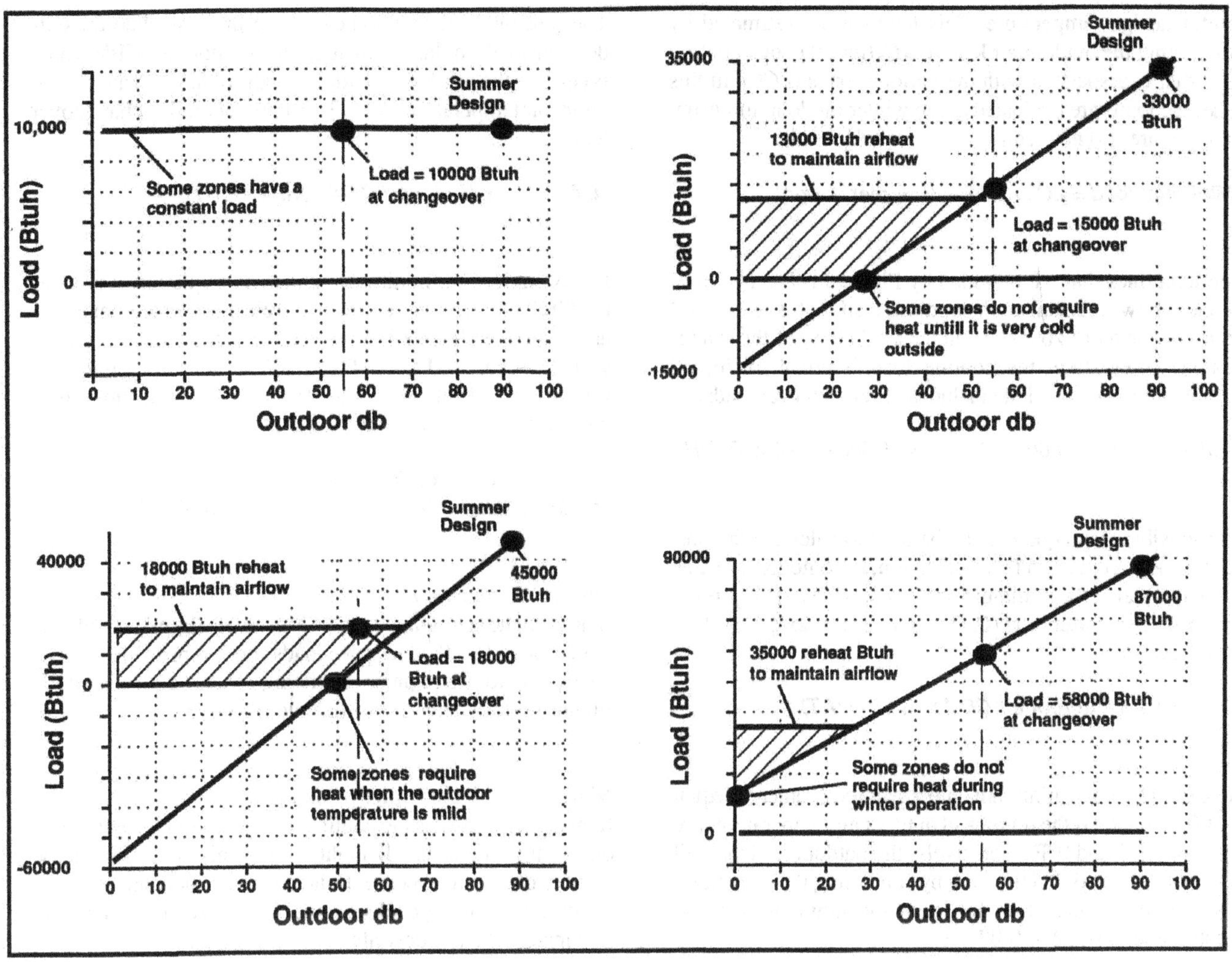

Figure 11-9

Pumping heat and piping gains are loads that are associated with chilled water systems. These sensible loads are not accounted for on form N-1, but if these loads exist, they should be added to the other loads that are imposed on the chiller.

Minimum Cooling Loads — Zones

The CFM that is supplied to a zone will be reduced in direct proportion to the reduction in the zone cooling load. Therefore, the air circulation rate within the zone, the dehumidifying capacity of the supply air and the CFM of fresh air that is delivered to the zone also will be reduced in direct proportion to the reduction in the zone cooling load. Consequently, minimum zone loads are important because they identify the zones that must be equipped with reheat coils and devices that prevent excessive throttling of the supply air. Load lines (refer to Section 4-10) can be used to evaluate the behavior of zone cooling loads. Figure 11-9, provides load-line diagrams that illustrate four possible scenarios. Figure 11-9 also shows how reheat can be used to maintain a minimum air flow.

Minimum Cooling Loads — Central Coil

During part-load operating conditions, the minimum load on the cooling coil will depend on the zone cooling loads, the zone reheat loads, the outdoor air load plus any other system loads that may apply. The minimum load on the cooling coil also will depend on whether the system is a throttling system or a bypass system. In either case, it is possible to make a reasonable estimate of the minimum cooling load at the central coil, but this work requires an analysis of the cooling loads and the reheat loads that are associated with each zone. As explained below, the minimum coil load also will depend on the strategy that is used to control the outdoor air damper. (Figure 11-1, on page 11-4 summarizes three basic outdoor air control strategies.)

No Economizer

If no economizer cycle is used, the minimum load on the coil will occur when the outdoor air temperature is equal to the winter design temperature. This load can be estimated by subtracting the outdoor air load (*OAL*) from the total cooling load that is associated with the various zones (*ZCL*). In this case, the loads are evaluated at the winter outdoor air design temperature and no credit is taken for solar loads.

Minimum load = ZCL - OAL *(at winter design)*

The load lines that are provided by Figure 11-9 can be used to show how to calculate the cooling load that is associated with two or more zones. As indicated below, at the winter design temperature, the cooling load is equal to 76,000 BTUH. Note that this load includes the zone reheat loads.

ZCL = 10,000 + 13,000 + 35,000 + 18,000 = 76,000 BTUH

The sensible heat equation can be used to calculate the outdoor air load (*OAL*). In this calculation, the outdoor air CFM value is equal to the minimum outdoor air CFM, *(RAT)* is the return air temperature and *(OAT) is the winter design, outdoor air temperature.*

OAL = 1.1 x Outdoor air CFM x (RAT - OAT)

For example, if the minimum outdoor air requirement is equal to 500 CFM and if the return and outdoor air temperatures are equal to 75°F and 0°F respectively, the outdoor air load will be equal to 41,250 BTUH. And, by combining this (negative) load with the Figure 11-9 zone loads, the minimum coil load is estimated to be 34,750 BTUH.

OAL = 1.1 x 500 x (75 - 0) = 41250 BTUH
Minimum Load = 76000 - 41250 = 34750 BTUH

Simple Economizer
If a simple dry-bulb economizer is installed, the minimum load on the coil will occur just before the system switches to the economizer cycle. This load can be estimated by subtracting the outdoor air load (*OAL*) from the total load that is associated with the various zones (*ZCL*). In this case, the loads are evaluated at the economizer changeover temperature and no credit is taken for solar loads.

Minimum load = ZCL - OAL *(at economizer changeover)*

The load lines that are provided by Figure 11-9 can be used to show how to calculate the cooling load that is associated with the various zones. As shown below, just before changeover, the cooling load is equal to 101,000 BTUH. Note that this load includes some, but not all of the zone reheat loads.

ZCL = 10000 + 15000 + 18000 + 58000 = 101000

The sensible heat equation can be used to calculate the outdoor air load. In this calculation, the outdoor air CFM value is equal to the minimum outdoor air CFM, *RAT* is equal to the return air temperature and *COT* is the economizer changeover temperature.

OAL = 1.1 x Minimum CFM x (RAT - COT)

For example, if the minimum outdoor air requirement is equal to 500 CFM and if the return air and changeover temperatures are equal to 75°F and 55°F respectively, the outdoor air load will be equal to 11,000 BTUH. And, by combining this (negative) load with the Figure 11-9 zone loads, the minimum coil load is estimated to be 90,000 BTUH.

OAL = 1.1 x 500 x (75 - 55) = 11000 BTUH
Minimum Load = 101000 - 11000 = 90000 BTUH

Ultimate Economizer
If an ultimate economizer cycle is installed, the minimum coil load is not related to the space loads and it is possible to have the refrigeration equipment operating with no load on the coil. Figure 11-1, on page 11-4 illustrates this point.

Night-Cycle (Unoccupied) Cooling
In many applications the cooling system is simply shut down during the (unoccupied) night cycle. This control strategy reduces operating expense. If the cooling system has to operate during the night cycle, a control strategy that reduces or eliminates reheat is desirable.

A minimum flow of outdoor air may not be required during the night cycle. If outdoor air is not required, the outdoor air damper should be closed whenever cooling is provided by mechanical refrigeration. However, the economizer cycle should still be used whenever the condition of the outdoor air is suitable for free cooling.

Internal loads are usually smaller during the night cycle, so the minimum load on the cooling coil could be less than the minimum day-cycle load. An analysis of both operating cycles is required to determine which cycle produces the smallest load on the cooling coil.

Zone Sensible Heat Ratios
A sensible heat ratio calculation should be made for each zone. These calculations should be based on the sensible and latent loads that are associated with the summer design conditions (the peak zone loads). There is no reason to expect all of the zones to have the same sensible heat ratio. As explained in the following paragraph, this information is required to establish an acceptable condition for the air leaving the central cooling coil.

Condition of the Air Leaving the Cooling Coil
The performance of the cooling coil must be compatible with the conditioning requirements of each zone. This means that the condition of the air leaving the cooling coil must be cold enough to offset the sensible load in any zone and dry (cold) enough to neutralize the latent load in any zone. Section 4-17 in this manual, which discusses this subject, indicates that the designer will have the following options:

- If humidity control is not critical, the required cooling coil performance can be based on a sensible heat ratio that falls near the midrange of the various zone sensible heat ratios.
- If humidity control in one or more zones is critical, the cooling coil performance must be based on the lowest sensible heat ratio that is associated with the critical zones.

Reheat Requirements
In theory, a flow control device would throttle the supply CFM to zero just before a (zone) load changes from cooling to heating. In practice, this "no-flow" condition is undesirable because supply air is required to provide air circulation, humidity control and fresh air. Therefore, reheat is used to create a "low-limit" cooling load that establishes a minimum value for the supply CFM. This heat may be supplied directly by a reheat coil or indirectly by baseboard radiation.

Sizing Loads — Zone Heating
There are two possible types of zone heating loads, a reheat load and a space load. Figure 11-10, shows these two loads. Note that reheat is not energized until the supply CFM falls below the minimum CFM requirement.

The load associated with neutralizing relatively cold supply air (raising the temperature of the supply air up to room temperature) is the reheat load. Call this load *L1* and use the sensible heat equation to calculate this load (*SAT* represents the supply air temperature and *RAT* represents the room temperature.)

L1 = 1.1 x Minimum CFM x (RAT - SAT)

The space load is the amount of heat that the zone would require at the winter design condition when the envelope losses are balanced against the internal gains. Call this load *L2*. Form N-1 can be used to calculate the envelope losses and the internal gains. (One form is required for each zone. The zone envelope loss is equal to the sum of lines 16 and 17 and the internal gain is equal to the line 4 value.)

L2 = (Envelope losses - Internal gains)

The amount of heat that a zone will require during any operating condition is equal to the reheat load plus the space load (*L1* + *L2*). However, this load will be different for each operating condition. Figure 11-11, provides a diagram that illustrates this point. This diagram shows how the net heating load is calculated for three typical operating conditions:

Condition 1
H1 represents the amount of heat that would be required if the central system was in the cooling mode and the zone was occupied (internal loads help heat the zone).

Condition 2
H2 represents the amount of heat that would be required if the central system was in the cooling mode and the zone was unoccupied (no internal loads).

Condition 3
H3 represents the amount of heat that would be required if the central system was in the night-cycle mode (no internal loads, heating-only).

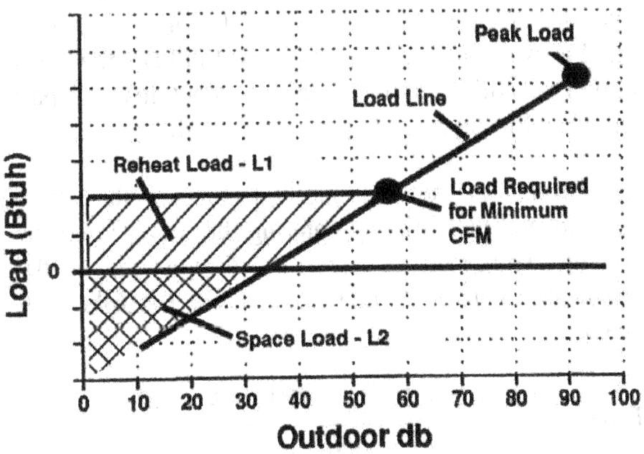

Figure 11-10

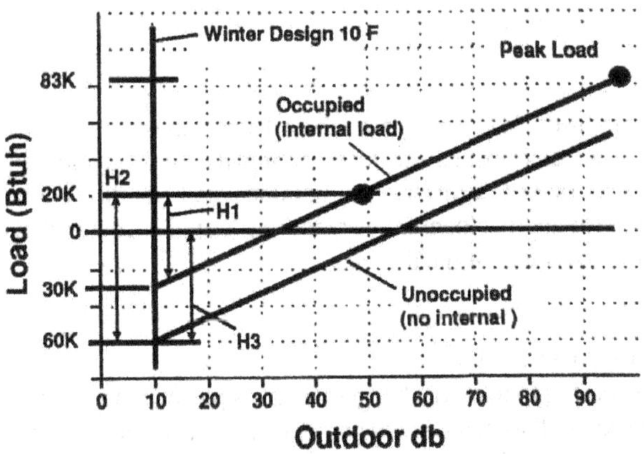

Figure 11-11

For this particular diagram, Condition 2 produces the largest heating load and *H2*, which is equal to 80,000 BTUH, represents the design load. However, the operating conditions that were used in this example do not necessarily apply to every terminal reheat system. In practice, the analysis must conform to the operating conditions that are actually expected to occur. For example, if Condition 2 could not occur, the design heating load would be determined by Condition 3 and *H3*, which is equal to 60,000 BTUH, would represent the design heating load.

Heat Required at the Central Unit
The amount of heat required at the central unit will depend on how the system satisfies the zone loads and the outdoor air load. No central heat will be required, providing that:

- The mixed air temperature (due to the minimum ventilation requirement) does not fall below the cooling cycle supply air temperature setpoint.
- Adequate heating capacity is installed in each zone.

Night-Cycle Heating Loads
The night-cycle reheat load (*L1*) will depend on how the system is operated. Consider the following possibilities:

- If cooling or dehumidification is not required when the space is unoccupied, the cooling system can be shut down during the night cycle. And, if fresh air or make up air is not required when the space is unoccupied, the outdoor air dampers can be closed. Therefore the reheat load (*L1*) will be almost eliminated because the air leaving the air handler will be close to the temperature of the return air. (The supply air will be a little colder than the return air because of damper leakage).
- If cooling or dehumidification is required when the space is unoccupied, the supply air temperature will be colder than the room air temperature. In this case, the night-cycle reheat load (L1) will depend on the strategy that is used to control the VAV devices and the reheat coils.

The night-cycle space load (*L2*) will depend on whether or not the zone thermostats are reset during the night cycle and whether or not there are any uninterrupted internal loads. Credit for internal loads may or may not be applied, depending on the judgment of the designer.

Changeover to Heating
The cooling to heating changeover takes place on a zone-by-zone basis. This changeover automatically occurs as the zone thermostats call for additional heat. The central equipment does not changeover to a heating cycle, but in some designs the central unit is switched to a night cycle.

Psychrometrics
An elementary psychrometric analysis of the cooling cycle should be made because it allows the designer to predict the condition of the air entering the cooling coil. This analysis also defines the range of acceptable conditions for the air that will be supplied to the zones during the cooling cycle. This information is directly related to the procedures that are used to select the cooling equipment. Refer to Sections 4-17 and 5-1 in this manual for more information about elementary psychrometrics.

Figure 11-12, shows a comprehensive psychrometric analysis of the variable volume cooling cycle. This diagram also shows how to use the psychrometric chart to estimate the design load on the cooling coil.

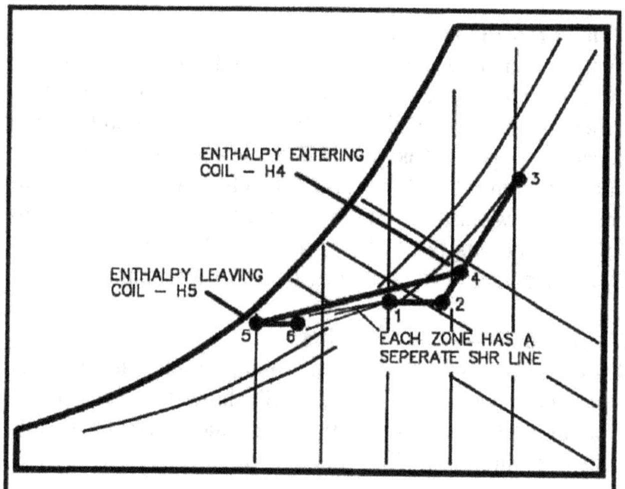

Air leaves the room [1] and picks up return-side heat [2], (the return-side gain could be fan heat, a return air plenum gain or a duct gain). Return air [2] is mixed with outdoor air [3] before the mixed air [4] enters the cooling coil. Cold, dry air off of the cooling coil [5] picks up supply-side heat [6], (the supply-side heat could be fan heat or a duct gain). The tempered, dehumidified air [6] enters the room and absorbs the sensible and latent loads and is routed back through the return [1]. At full load, the CFM through the coil is based on the peak block load. At part load, the CFM through the coil is throttled.

The total load on the cooling coil depends on the difference between the enthalpy entering the coil [4] and the enthalpy leaving the coil [5]. The design load on the cooling coil can be estimated by using the following equation.

$$Tons = \frac{4.45 \times \text{Block Load CFM} \times (H4 - H5)}{12000}$$

Figure 11-12

Equipment Selection — Primary Equipment
The capacity of the primary cooling equipment should be based on the peak block load (as explained above in the paragraph titled *Sizing Load — Cooling at the Central Coil*) and comprehensive engineering data should be used to select equipment. This performance data should correspond to the actual operating conditions that are expected to occur at the site. Part-load performance also must be evaluated and an appropriate method of capacity control should be provided. The procedures that are used to select various types of cooling equipment are presented in Section 6 of this manual.

Equipment Selection — Secondary Equipment
Special air outlets must be used with VAV systems so that proper air distribution is maintained at reduced air flow. The terminals may be slot diffusers or ceiling diffusers, but they must be designed to provide an adequate throw as the supply CFM is throttled. It also is important to make sure that the VAV supply air outlets and the associated flow control boxes are sized for acceptable performance at maximum and minimum flow rates. In some cases, low limit stops will be required to maintain minimum air flow.

Reheat coils must have enough capacity to maintain the setpoint temperature during any operating condition. Reheat coil capacities should be equal to or greater than the corresponding zone heating cycle, design loads. The procedure that is used for calculating the zone heating loads is provided above, refer to the paragraph titled *Sizing Loads — Zone Heating*. The procedures that are used to select heating coils are presented in Section 7 of this manual.

Equipment Selection — Humidifiers
If a humidifier is required, its capacity should be equal to or greater than the largest humidification load that is expected to occur during any operating condition. Refer to Section 10 for more information about selecting humidifiers.

Supply CFM Design Value — Zones
The zone CFM values are based on the peak sensible loads that are associated with the various zones. These loads correspond to the value that is found on Form N-1, line 8. One Form N-1 is required for each zone.

$$\text{Cooling CFM} = \frac{\text{Peak sensible zone load (BTUH)}}{1.1 \times (\text{Room design DB - Supply air DB})}$$

Minimum Zone CFM
Acceptable space conditions cannot be maintained if the supply air is severely throttled. If close control over space conditions is not required, the minimum CFM value can be arbitrarily set at 40 percent of the design CFM value. However, if close control over space conditions is required, the minimum flow requirement can exceed this 40 percent value. In this case, the value for the minimum CFM that can be delivered to any zone could depend on the air circulation requirement, the humidification requirement or the fresh air requirement. These relationships are explained below. Also, note that if none of these requirements apply and if a zone is unoccupied, the supply CFM can be throttled to zero.

- Air quality is related to air circulation. Therefore, the air circulation rate in a zone should not be allowed to fall below 0.5 CFM per square foot of floor area.
- The ability of the supply air to absorb a constant latent load will decrease in direct proportion to the reduction in the supply CFM. This means that the relative humidity will rise from 50 to (about) 58 percent if the supply CFM is cut in half — and the humidity will rise to 72 percent if the supply CFM is reduced to one quarter of the design value.
- The introduction of fresh air affects air quality. When the outdoor air damper is set at its minimum position, the CFM of outdoor air that is delivered to a zone will decrease in direct proportion to the reduction in the supply CFM. This means that if 20 CFM of fresh air is provided at full flow, only 10 CFM will be provided when the supply CFM is halved — and less than 5 CFM will be provided when the supply CFM is reduced to one quarter of the design value.

Supply CFM at the Central Unit — Throttling Systems
If no reheat is required during the summer design condition, the supply CFM at the central unit is based on the peak block sensible cooling load. If reheat is required during the summer design condition, the supply CFM at the central unit is based on a load that is equal to the peak block sensible load plus the reheat load. Refer above in the paragraph titled *Sizing Load — Cooling at the Central Coil* for more information about these loads.

$$\text{CFM} = \frac{\text{Sensible Design Load}}{1.1 \times (\text{Room design DB - Supply air DB})}$$

> Oversizing a VAV fan on a throttling system can cause poor fan performance, reduced fan efficiency and in some cases, poor control over space conditions. Therefore, it is important to make an accurate estimate of the supply CFM requirement.

Supply CFM at the Central Unit — Bypass Systems
In a bypass system, each dumping terminal always receives full air flow, so there is no air-side diversity associated with the fan. Therefore, the supply CFM at the central unit is equal to the sum of the zone, peak load, CFM values. Refer above to the paragraph titled *Supply CFM Design Value— Zones* for more information about the zone CFM calculations.

Minimum CFM at the Central Unit—Throttling Systems
For a throttling system, the minimum CFM that is required at the central unit is equal to the sum of the minimum CFM values that are associated with the various zones. These minimum values are based on the sensible cooling loads that are associated with the zones when the outdoor temperature is equal to the winter design temperature. Load lines (refer to Section 4-10) can be used to evaluate these sensible cooling loads. Figure 11-9 provides load-line diagrams that illustrate four possible scenarios. Figure 11-9 also shows how reheat can be used to maintain a minimum air flow.

Minimum CFM at the Central Unit — Bypass Systems
In a bypass system, each dumping terminal always receives full air flow, so there is no air-side diversity associated with the fan. Therefore, the minimum supply CFM at the central unit is equal to the design CFM value. Refer above to the paragraph titled *Supply CFM at the Central Unit — Bypass Systems* for more information about this calculation.

Duct Sizing — Throttling System
For a throttling system, the runout and branch duct sizes are determined by the air flow that is required during peak load conditions. Therefore, these sizes should be based on the peak load CFM values that are associated with the zones that are served by these ducts. Refer to the paragraph above titled *Supply CFM Design Value — Zones* for more information about zone air flow requirements.

Air-side diversity can be applied to the main trunk ducts. The size of these ducts should be based on the same air flow values that would be used to estimate the supply CFM requirement at the central unit. Refer to the paragraph above titled *Supply CFM at the Central Unit — Throttling Systems* for more information about this air flow requirement. Also refer to ACCA **Manual Q**, Sections 13, 14 and 16 for detailed information about duct sizing and fan selection.

Duct Sizing — Bypass Systems
Since excess air is bypassed rather than throttled, the CFM flowing through each section of duct remains constant during all modes of operation. There is no air-side diversity. Therefore, the size of all the supply and return ducts (trunks and branches), should be based on CFM values that correspond to the peak loads that are associated with the zones that are served by given section of duct. Refer to ACCA **Manual Q**, Sections 13, 14 and 16 for detailed information about duct sizing and fan selection.

Temperature and Humidity Controls
The temperature in each conditioned space can be maintained by a zone thermostat that controls a VAV device and in some cases, a heating device. At peak cooling load conditions, the thermostat should call for full air flow. At part-load conditions, the thermostat should throttle the supply CFM until thermostat setpoint is satisfied. This throttling action should be allowed to continue until a (predetermined) minimum flow condition occurs and then further throttling should be prevented. After the VAV device has been throttled to its minimum position, the thermostat should activate and control the heating device to maintain the thermostat setpoint.

As explained below, the method that is used to control the cooling coil at the central air handler will depend on whether the system is a throttling system or a bypass system and on whether the cooling coil is a chilled water coil or a DX coil. In all cases, it is necessary to make sure that the supply air has adequate dehumidification capacity during any operating condition.

- When cooling is provided by a chilled water coil, the same control strategy can be applied to both types of variable volume systems (throttling or modulating). In either case, the supply air leaving the central air handler — which could be operating in a mechanical cooling mode or an economizer mode — can be maintained at a constant (dry-bulb) temperature.

- A constant supply temperature control strategy also is appropriate when a DX coil is used with a throttling system.

- The capacity of a DX coil that is used with a bypass system should be controlled by a temperature sensor that is located in the main return air duct.

> It is possible to reset the temperature of the supply air as the cooling load decreases. This strategy reduces the amount of reheat that is required to maintain a minimum flow of supply air, but it also reduces the dehumidification capacity of the supply air.

It also is necessary to make sure that the capacity of the refrigeration equipment (chiller or DX condensing unit) tracks the coil load during any operating condition. There are no special refrigerant-side control problems when chilled water is used for cooling. However, DX systems require a direct acting (electric or pneumatic), step or modulating control. In either case, (chilled water or DX system) a minimum load analysis should be made to determine if hot gas bypass is required.

- When DX coils are used with a throttling system, the compressor and the coil circuits can be controlled by the temperature of the supply air.

- When DX coils are used with a bypass VAV system, the compressor and coil capacity should be controlled by the return air temperature. Return air temperature control is recommended because a supply-side temperature sensor could cause the compressor capacity controls to short cycle. (Since the system CFM is constant, the return air temperature is directly related to the space load.)

Control of a Separate Heating System
It is possible to use a separate heating system, such as a baseboard radiation system, to supplement or to replace the reheat coil. In this case, it is necessary to make sure that the various cooling and heating devices work in sequence. This can be accomplished by using one thermostat to control all the devices. Heat should be locked out until the VAV cooling device is throttled to its minimum position and the baseboard heat (if used) should be off if a reheat coil can maintain the desired space temperature.

Fan and Damper Control
The fan (or fans if a return air fan is used) should operate continuously when any zone is occupied. If outdoor air is required for make up air, the fan(s) should operate continuously whenever make up air is required, even if all the zones are unoccupied.

Fan capacity control devices are commonly installed on throttling systems, but they are not normally used with bypass systems. Fans that are equipped with these devices can match CFM output to fluctuating CFM requirements, while maintaining adequate flow and inlet pressure at every air flow throttling device. Fan capacity-control hardware can be operated by a signal from a static pressure sensor located in a supply air duct. (For pressure independent terminals, the static pressure sensor should be located about two thirds of the way downstream in the supply duct, and the controller must be set so that the required minimum static pressure is maintained at the most remote VAV device. For pressure dependent terminals, the sensor should be located about one half of the way downstream in the supply duct.) When fan control devices are used, it is important to make sure that the supply and return fans work in unison so that a uniform pressure is maintained in the conditioned space. Refer to ACCA **Manual Q**, Sections 16 and 17 for more information about VAV fans, fan control and space pressure.

Outdoor air, return air and exhaust air dampers should be controlled so that the minimum ventilation requirements are met whenever the building is occupied. Special controls may be required to maintain a constant flow of outdoor air through the (minimum) outdoor air damper as the supply CFM is modulated. If outdoor air is required for make up air, the various dampers should be controlled so that an adequate flow of make up air is always available. (The outdoor air requirement will be dictated by the ventilation requirement or by the make up air requirement, whichever is larger.)

Night-Cycle Control
Providing that the day cycle space conditions do not have to be maintained, control over the entire conditioning system can be transferred to a *heating-only* cycle when the building is unoccupied. In this case, the heat may be supplied entirely from the central unit, by reheat coils or by some combination of heat from the central unit and the reheat coils. (It is also possible to supply all or part of the heat with baseboard radiation.)

When control is transferred to the night cycle, the outdoor and exhaust air dampers (usually) can be closed and the fan(s) (usually) can be de-energized. When the night-cycle thermostat — which should be located in the zone that has the most rapid rate of temperature drop — calls for heat, the fan(s) should operate, but the outdoor air and the exhaust air dampers can remain closed. If desired, the night-cycle thermostat(s) can be set back. It also may be necessary to reverse the control action of some or all of the VAV flow control devices.

Precautions and Suggestions
Variable volume systems must be designed with greater care than other types of all-air systems. A list of important design considerations is provided below.

- A constant volume cooling system would be better suited for conditioning an internal zone — or any other type of space that has a cooling load that does not fall below 80 percent of the design cooling load.

- Basic, cooling-only, variable volume systems are best suited for spaces that have a minimum cooling load that is less than 80 percent of the design load, but not less than 40 percent of the design load.

- Variable volume with reheat is required for spaces that experience a minimum cooling load that is less than 40 percent of the design load — or for spaces that require a substantial flow of supply air to meet specific air circulation, humidity control or fresh air requirements.

- Variable volume with reheat or an independent heating system, is required for spaces that experience a cooling to heating changeover during the daytime (occupied), operating cycle.

- Some designers use two variable volume systems when a building has a large interior space. One for the interior zones, which is a "cooling-only" system and one for the perimeter zones, which is a cooling and heating system. However, a single system is usually more cost effective for buildings that do not have a large interior space.

- Throttling, variable volume systems are more appropriate for large buildings or large areas that have many zones (more than 10,000 to 15,000 CFM).

- Bypass, variable volume systems are good choices for small commercial buildings or areas that only have a few zones (less than 10,000 CFM).

- If one variable volume system serves the perimeter and the interior spaces, perimeter heat will (usually) be required to maintain a minimum flow of supply air to the perimeter zones. In addition to providing the required amount of reheat, the perimeter heating capacity must be capable of neutralizing the zone transmission and infiltration losses.

- VAV reheat boxes, which combine the reheat function with a maximum CFM control, simplify the design, installation and balancing work. Refer to air terminal manufacturer's catalogs and engineering manuals for additional information about these devices.

- Equipment sizing and selection procedures must be based on load calculations that determine the maximum and minimum zone cooling and heating loads, as well as the maximum and minimum block loads.

- The condition (dry-bulb and wet-bulb temperatures) of the air leaving the central equipment must be capable of neutralizing the sensible and latent loads that occur in all the zones during any operating condition.

- The percentage of outside air that is introduced through the central equipment must satisfy the minimum ventilation requirements in all the zones during every possible operating condition.

- Since the quantity of outside air introduced into a zone is reduced as the zone supply CFM is throttled, reheat may be required to maintain adequate air circulation rates or ventilation rates within a zone.

- If latent loads remain relatively constant while sensible loads vary, it may be necessary to install reheat to maintain humidity control within a zone.

- In general, the air flow to an occupied zone should not drop below 40 percent of the design CFM value. To maintain proper air circulation, the air flow to an occupied zone should not drop below 0.50 CFM per square foot of floor area. Zones with special ventilation, exhaust or humidity requirements may have a minimum air flow requirement that falls anywhere between 40 percent and 100 percent of the design CFM.

- Ceiling supply outlets must be selected so that an acceptable air pattern is achieved at minimum CFM as well as at the design CFM. Constant volume ceiling diffusers, sidewall outlets and floor outlets are not compatible with variable volume systems.

- Fan-powered VAV boxes can be used to combine constant volume supply outlets with variable volume capacity control devices.

- A VAV supply fan should never be oversized. Excessive fan capacity reduces the effective operating range of the flow control devices, reduces efficiency and increases operating cost.

- An appropriate method of fan control must be selected for throttling systems. Historically, inlet vanes have been used on larger fans and no control, or discharge damper control, has been used on smaller fans. But, since the cost of variable-speed drives has been reduced, speed control has become a viable option.

- When a throttling system is used, the supply fan should be controlled by a static pressure sensor that is located downstream in the supply duct.

- The return CFM should be less than the supply CFM by an amount that is equal to the total zone exhaust CFM plus the excess supply CFM that is used to pressurize the conditioned space.

- For throttling systems, the supply fan and the return fan must work in unison to maintain the desired air flows and to control the pressure in the conditioned space. Refer to ACCA **Manual Q**, Section 17 for more information about building pressure.

- When the system is started in the cooling mode after being shut down for a number of hours, all the air terminals will modulate to the full-open position. If a forward curved fan is used, this full-flow operating condition can produce an overload on the fan motor. Therefore, the motor must be specifically selected for this operating condition.

- Zone thermostats should be staged so that the air outlets are modulated to their minimum position setting before any zone heat is activated. It is good practice to select a thermostat that has a deadband between the minimum position and the heating setpoints.

- Heating, without cooling, may be required when the building is unoccupied. Systems equipped with a heating coil in the supply air system must changeover from the cooling mode to the heating mode. Changeover systems must have the ability to reverse the action of the air distribution devices so that supply CFM increases as the zone temperature drops. This feature also is desirable for fast morning warm-up following a night set-back.

- Extreme care must be exercised to prevent excessive operating costs when VAV systems are used with perimeter radiation that is scheduled by outdoor air temperature. The perimeter radiation should not provide heat until there is at least one zone that requires heat; and, the heating capacity should not be excessive during any operating condition. A careful part-load analysis is required to determine the heating requirements. A separate analysis is required for the day cycle and the night cycle.

- There are no special refrigerant-side control problems when chilled water coils are used with variable volume systems. However, certain refrigerant-side capacity control strategies are required when DX coils are used.

- The supply air temperature off a VAV cooling coil can be optimized by using a discriminator control. This will keep the coil temperature as high as possible without allowing the warmest zone to overheat. (Discriminator control will improve the room air circulation but, it will degrade humidity control.)

- VAV systems may not require air-side balancing. (Balancing is not usually required upstream from the VAV devices, but balancing could be required downstream from the devices.) However, a field test is required to insure that the VAV devices have adequate static pressure at the inlet.

- Inlet static pressures that are too high when the system operates at the design CFM indicate that the fan speed should be reduced.

- The minimum position setting of the outside air damper should be checked to insure that the minimum outside air quantity is available during any operating condition. In some cases, a special modulating control will be required to maintain the required flow of outdoor air.

11-5 Multizone and Dual-Duct Systems

This section discusses the performance characteristics, hardware, design procedures and control cycles that are associated with packaged multizone systems. This section also mentions the dual-duct system, but most of the discussion is devoted to multizone packaged equipment because it is specifically designed for small commercial applications. The information that is provided by Section 11-1 also applies to these systems.

Function
Multizone and dual-duct systems can either provide sensible and latent cooling or they can provide heat. They also can provide outdoor air for ventilation or for make up air, filtration and humidification.

Applications
Packaged multizone equipment is primarily designed for applications that have a limited number of zones, but multiple units can be used for applications that have a large number of zones. (The actual number of zones that can be accommodated by one unit depends on the product. Zoning capabilities typically range from 4 to 14 zones.) Because dual-duct systems can provide an unlimited number of zones, they are used to condition large buildings.

Multizone equipment and dual-duct systems that are designed as blow-through systems are suitable for applications that do not require precise control over humidity. Draw-through systems are less common, but they provide better humidity control. (More information about blow-through and draw-through configurations is provided below.)

Because it is packaged, multizone equipment is often selected for projects that have budgets that emphasize the importance of zoning and competitive installation costs. Higher installation costs are associated with dual-duct systems. Neither of these systems would be a primary candidate for projects that emphasize energy efficiency and low operating costs.

Configuration
Multizone and dual-duct systems are very similar. They both adjust to changes in the zone loads by mixing cold air with warm air. The fundamental differences between the two types of systems is associated with the location of the mixing devices and the arrangement of the duct system.

Multizone systems are parallel coil, single-duct cooling and heating systems. Cooling is provided by routing air through the cooling coil (cold-deck) and heat is provided by routing air through the heating device (hot-deck). Mixing dampers, which are located immediately downstream from the cold-deck and the hot-deck, control the supply air temperature by regulating the air flow through the decks. One set of dampers and one supply air duct is required for each zone. A diagram of a multizone system is provided at the top of the next page.

Dual-duct systems are also parallel coil systems. Cooling is provided by air that flows through the cold-deck and heat is provided by air that flows through the hot-deck. Two separate ducts are used to deliver warm air and cold air to mixing boxes, which are usually located in the ceiling space above the corresponding zones. Cooling and heating capacity is controlled at the mixing box by blending hot and cold air. (Normally the mixing box is equipped with a constant volume flow regulator.) At least one mixing box and two supply air ducts are required for each zone. A diagram of a dual-duct system is provided at the top of page 11-29.

Multizone and dual-duct system configurations also can be classified by the location of the supply air fan. On blow-through units, the fan is located upstream from both of the coils. On draw-through units, the fan is located downstream from the cooling coil. A blow-through unit will use less energy than the draw-through unit because the air that flows through the hot-deck is not precooled. (The draw-through design is essentially the same as a terminal reheat system.) The disadvantage of the blow-through design is that the air that bypasses the cooling coil is not dehumidified. For most applications, precise control over humidity is not as important as the operating cost. Therefore, the blow-through design is by far the most common arrangement. A diagram of both configurations is provided on the next page.

Characteristics
The characteristics that are generic to all types of all-air systems are listed in Section 11-1. The characteristics that are peculiar to multizone systems are listed below.

- Packaged multizone systems provide a limited number of zones, (typically 14 or less).
- Multizone systems provide good temperature control.
- Blow-through multizone systems do not provide exact humidity control because the air that passes through the hot-deck is not dehumidified.

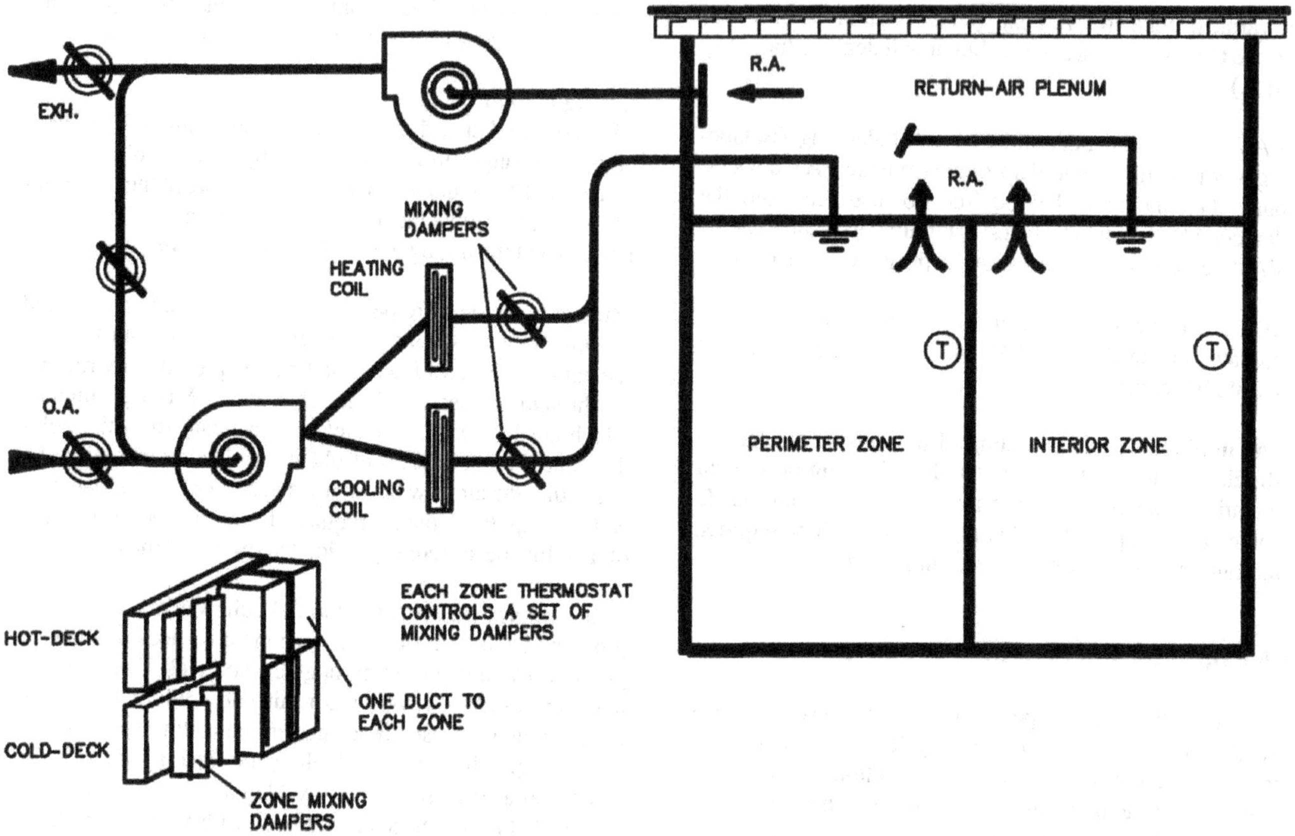

Multizone System

- Draw-through multizone systems provide full dehumidification capacity at all operating conditions because all the supply air is routed through the cooling coil.
- Multizone systems are classified as constant volume systems, but air circulation rates fluctuate, depending on how the air flow is split between the hot-deck and the cold-deck.
- Blow-through multizone systems can take advantage of the diversity that is associated with the zone loads.
- Draw-through multizone systems cannot take advantage of the diversity that is associated with the zone loads.
- Control over the capacity of the cooling coil is not directly related to the sensible space loads.
- Control over the capacity of the cooling coil is not directly related to the latent loads.
- Because they are packaged, multizone systems are relatively easy to design, install and operate. However, there could be some problems associated with routing ten or twelve separate supply ducts to the various zones.
- Seasonal changeover (heating or cooling) consist of starting and stopping the heating and cooling plants.
- Every zone has simultaneous heating and cooling.
- Since the fresh air requirements may vary by zone, the percentage of outdoor air introduced at the air handler may not match the outdoor air percentage that is required for each zone.
- Since the sensible heat ratios may vary by zone, the condition of the air leaving the cooling coil may not be perfectly compatible with all the zone sensible heat ratios.

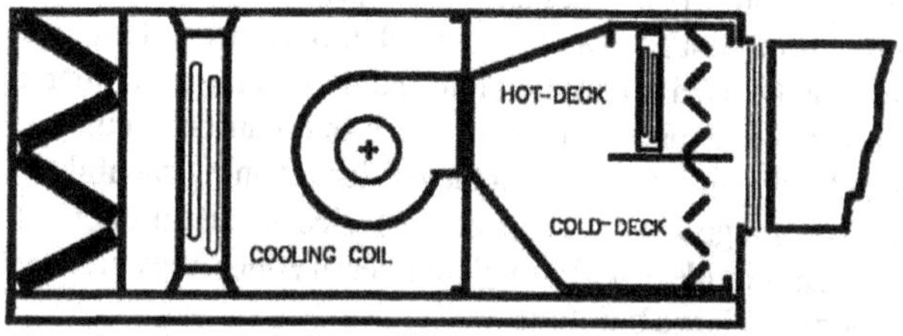

Draw-through — all air dehumidified, hot-deck air reheated

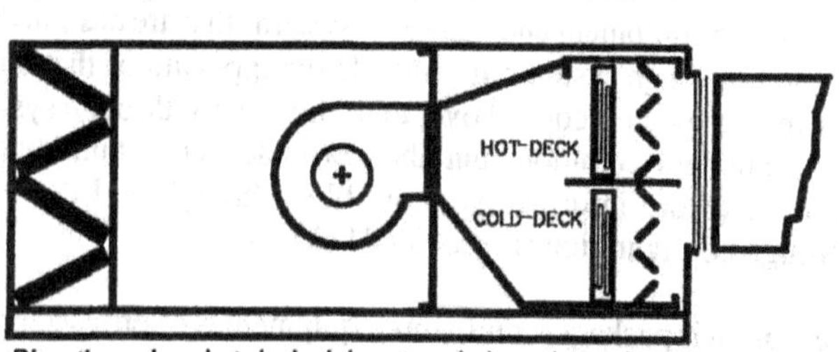

Blow-through — hot-deck air is not cooled or reheated

Blow-through and Draw-through Configurations

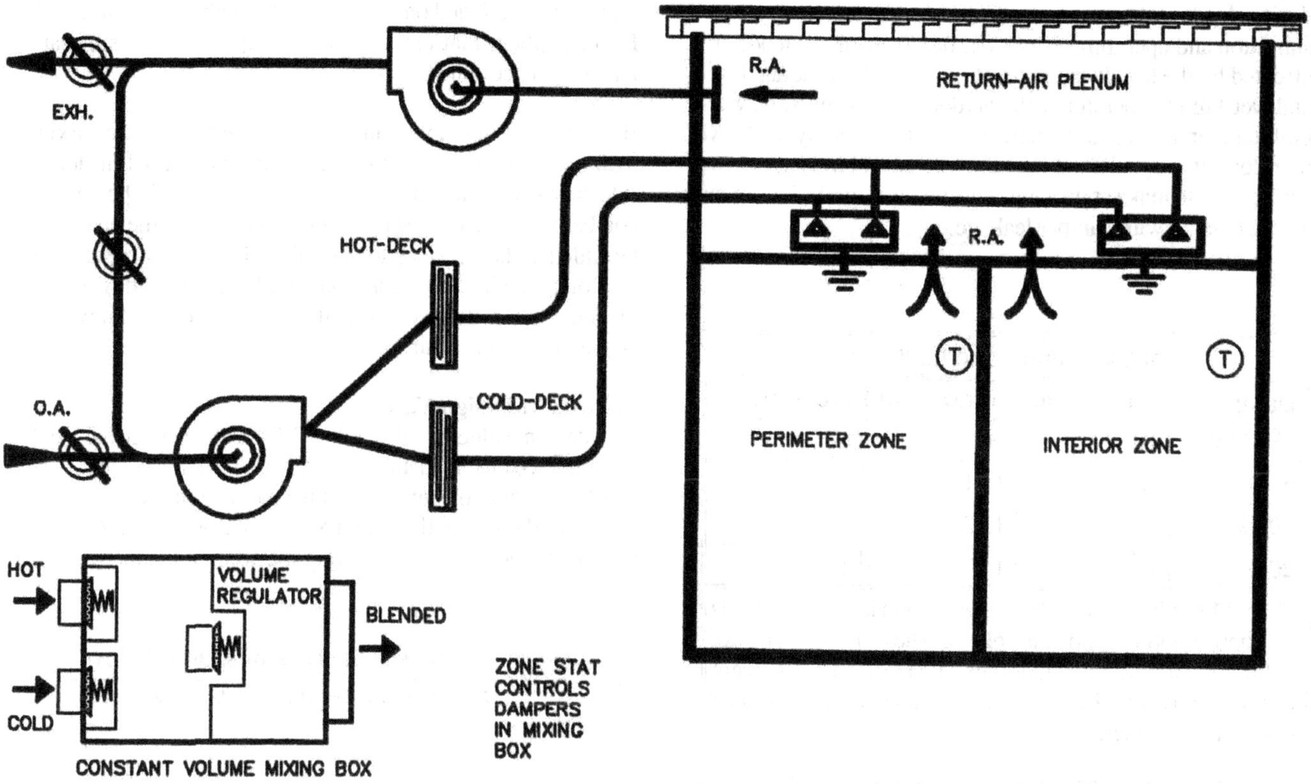

Dual-Duct System

- The central equipment must be used to condition a single occupied zone even though the rest of the building is vacant.
- Night-cycle control cannot be applied to the central equipment unless all the zones are vacant.
- Night-cycle control can be used on a zone-by-zone basis.
- It is not easy to account for and to meter energy consumption on a zone-by-zone basis.
- Multizone systems do not accommodate extensive floor plan changes.
- Hot-deck and cold-deck setpoints are critical for efficient operation. Optimizer controls can save energy.
- Because mixing hot and cold fluids wastes energy, a blow-through multizone system has higher operating costs than some other types of all-air systems.
- A draw-through multizone system uses about the same amount of energy as a terminal reheat system.

Primary Equipment

The primary cooling equipment may consist of a DX cooling system or an air handler that is equipped with a chilled water coil. The primary source of heat could be supplied by a hot water coil, an electric coil or by gas-fired equipment. (Some manufacturer's use the heat that is rejected from the DX cooling equipment to provide some of the heat that is required at the hot-deck.) Economizers are commonly used for free cooling and return fans are required to control the flow of return air and exhaust air. Return fans also provide control over the pressure in the conditioned space. Filters, dampers, and in some cases, humidifiers are installed in the air handler.

Secondary Equipment

No special secondary equipment or supply air diffusers are required for multizone systems. Mixing boxes are required for dual-duct systems. If winter humidification is required, either system can be fitted with a single central humidifier or individual humidifiers can be provided for each zone.

Mixing Damper Leakage — Multizone Units

Ideally, at peak load conditions, all the air that is supplied to a zone should be routed through the cold-deck (cooling) or the hot-deck (heating). However, leakage at the mixing dampers will cause some of the supply air to pass through the hot-deck (cooling) or through the cold-deck (heating). With standard dampers, 10 to 20 percent of the supply air could leak through a closed damper. If the unit is equipped with low-leakage dampers, about 3 to 5 percent of the supply air could leak through a closed damper.

Damper leakage is important because it produces secondary heating and cooling loads that must be added to the zone design loads. (Damper leakage also increases energy consumption and operating costs.) The damper leakage loads are affected by the hot-deck temperature, the cold-deck dry-bulb and wet-bulb temperatures, the outdoor air dry-bulb and wet-bulb temperatures and the percentage of the supply air CFM that flows through the outdoor air damper. Table 11-2 below, can be used to make approximate corrections for the loads that are associated with damper leakage.

Zone Cooling Load Multipliers			
Damper Leakage	Minimum Outdoor Air Percentage		
	10	20	30
5 %	1.005	1.015	1.025
10 %	1.010	1.035	1.050
20 %	1.025	1.070	1.100

Note: Multiply the Form N-1 sensible and latent zone loads by these cooling load multipliers. These factors are not valid if the minimum outdoor air CFM exceeds 35 percent of the supply air CFM or if heat is added at the hot-deck during hot weather.

Zone Heating Load Adjustment	
Damper Leakage	Additional Heating Load (BTUH) (CDTD = Room DB — Cold-Deck DB)
5%	0.055 x Zone CFM Design Value x TDCD
10%	0.110 x Zone CFM Design Value x TDCD
20%	0.220 x Zone CFM Design Value x TDCD

Note: Add the heating load adjustment to the Form N-1 zone heating load. (The TDCD value is equal to the temperature difference between the room and the cold-deck when the outdoor temperature is equal to the winter design temperature.)

Table 11-2

Sizing Loads — Zone Cooling
The cooling capacity that is required for each zone is equal to the corresponding peak zone load. One **Manual N**, Form N-1 must be filled out for each zone. The sensible and latent loads that are associated with a given zone will correspond to the values that are found on line 8 of Form N-1, (lines 9 through 12 are not required). Note that if there is a supply duct gain (Form N-1, line 7), it is added to the zone load. This is done so that the duct gain is offset by an increase in the cooling CFM that is supplied to the zone.

If a blow-through unit is installed, the leakage at the mixing dampers produces secondary (sensible and latent) zone loads that are not included on Form N-1. Refer above to Table 11-2 for multipliers that can be used to adjust the (sensible and latent) zone cooling loads.

If a draw-through unit is installed, the leakage at the mixing dampers should not produce secondary zone cooling loads. (On draw-through units, all the air passes through the cooling coil before it reaches the hot-deck. Therefore, some heat must be added at the hot-deck to neutralize the cooling capacity of the hot-deck air. This means that on a hot summer day, the air leaving the hot-deck will be fully dehumidified and it will be at a neutral temperature.)

Zone CFM Design Value
The design value for the zone CFM is based on the peak sensible cooling load that is associated with the zone. The following equation can be used to calculate this value. Refer above, to the subsection titled *Sizing Loads — Zone Cooling* for information about calculating the cooling load.

$$\textit{Zone CFM} = \frac{\textit{Peak sensible load for space (BTUH)}}{\textit{1.1 x (Room design DB - Supply air DB)}}$$

Sizing Load — Cooling at the Central Coil
Blow-through units push the supply air through parallel hot-deck and cold-deck coils. Only the air that passes through the cold-deck is cooled and dehumidified. This allows blow-through configurations to take advantage of diversity. (Diversity is defined on page 11-18.) Therefore, the cooling capacity that is required at the central coil is based on the peak block load. One **Manual N**, Form N-1 must be filled out for the whole building. The sensible and latent loads at the coil will correspond to the values that are found on Form N-1, line 12.

Reheat is not normally used with blow-through multizone systems. But, reheat could be used if humidity control in one or more zones is critical. If reheat is required in one or more zones, the sensible cooling load at the central coil is equal to the peak block load plus the sum of the zone reheat loads. In this case, the reheat load that is associated with a critical zone is equal to the difference between the peak zone load and the zone load that is associated with the peak block load estimate.

Draw-through units pull all the supply air through the cooling coil before it is distributed to the conditioned space. Therefore, the temperature of the air that passes through the hot-deck must be raised from the cooling coil leaving air temperature to the (neutral) room design temperature. This hot-deck heat is the same as a reheat load, so the concept of diversity does not apply to the draw-through configuration.

Therefore, the design load on the draw-through cooling coil is based on the sum of the peak zone loads.

Fan heat is a system load that is not included on Form N-1. However, the sensible heat that is added by the supply fan and by the return fan should be accounted for. Sometimes, this is accomplished (as per the manufacturer's equipment selection instructions) by subtracting the fan heat from the cooling capacity of the unit. (Refer to Section 6-8 for an example that illustrates this procedure.) Any portion of the fan heat that is not accounted for by a capacity reduction should be added to the other system loads that are imposed on the cooling coil.

Pumping heat and piping gains are loads that are associated with chilled water systems. These sensible loads are not accounted for on Form N-1, but if these loads exist, they should be added to the other loads that are imposed on the chiller.

Note that even though mixing damper leakage increases a zone load, it does not increase the load at the cooling coil. (As far as the zone is concerned, the damper leakage is an external load that is similar to an infiltration load. But since the leakage at the hot-deck damper occurs within the unit, it is not an external load on the unit.)

Air Flow (CFM) at the Central Unit
Multizone systems are classified as constant volume systems. Therefore, full air flow is delivered to every zone during all operating conditions. For design work, the value for the supply fan CFM is equal to the sum of the zone CFM values.

CFM at Central Unit = Sum of Zone CFM Values

Even though multizone systems are classified as constant volume systems, the air circulation rates can fluctuate during part-load conditions. These fluctuations are related to the air flow resistance (pressure drop) that is associated with the cold-deck coil and the hot-deck coil (or heat exchanger). The supply fan has to work against more resistance when most of the air is flowing through the cold-deck than when the air flow is split between the cold-deck and the hot-deck. (The pressure drop across a coil varies with the square of the flow rate through the coil.) These swings in the zone supply CFM should not affect comfort because the temperature of the supply air is automatically adjusted to provide the proper amount of cooling or heating capacity.

Minimum Cooling Loads — Zones
On a blow-through system, an evaluation of the minimum zone cooling loads is not required because the mixing dampers automatically compensate for load changes. However, the heating plant and the hot-deck control strategies should be based on knowledge of the zone cooling-heating changeover temperatures. Load lines (refer to Sections 4-9 and 4-10) can be used to evaluate the behavior of the zone loads.

On a draw-through system, all the supply air is cooled and dehumidified and the hot-deck heat performs the same function as a reheat coil. Therefore, the zone cooling loads are always equal to their peak load values. (Refer to Figure 11-3 on page 11-10.)

Minimum Cooling Load — Central Coil
On a blow-through system, the minimum load on the cooling coil depends on the CFM that is flowing through the coil when the system is operating at the minimum load condition. The coil load also depends on the condition of the air that enters and leaves the coil. And, as explained below, the minimum coil load depends on the strategy that is used to control the outdoor air CFM. (Figure 11-1, on page 11-4 summarizes three basic outdoor air control strategies.)

No Economizer
If no economizer cycle is used, the refrigeration equipment must operate in cold weather. In this case, the minimum coil load will be related to the block space load that is expected to occur when the outdoor temperature is equal to the winter design temperature. As indicated on the next page by Figure 11-13, the calculations that are used to estimate this load depend on whether the block space load is a net heating load or a net cooling load. (The block space load is equal to the sum of the transmission, infiltration and internal loads that are associated with the various zones. *These loads are evaluated at the winter design temperature.* No credit should be taken for solar gains. Form N-1, lines 4, 16 and 17 provide the required values.)

Simple Economizer
If a simple dry-bulb economizer is installed, the minimum cooling load on the coil will occur just before the system switches to the economizer cycle. In this case, the minimum load on the coil will be based on the block space load that occurs when the outdoor temperature is equal to changeover temperature. (Refer to page 4-15, Figure 4-12.) As indicated by Figure 11-13, the calculations that are used to estimate this load depends on whether the block space load is a net heating load or a net cooling load. (The block space load is equal to the sum of the transmission, infiltration and internal loads that are associated with the various zones. *These loads are evaluated at the economizer changeover temperature.* No credit should be taken for solar gains. A load-line analysis of the block space load will provide the required values. Refer to Section 4-10 of this manual for more information about load lines.)

Ultimate Economizer
If an ultimate economizer cycle is installed, it is possible to have the refrigeration equipment operating with no load on the coil. Figure 11-1 on page 11-4 illustrates this point.

On a draw-through system, all of the supply air is cooled and dehumidified and the hot-deck heat performs the same function as a reheat coil. Therefore, the zone cooling loads are

Minimum Cold-Deck Load and Hot-Deck Load — Blow-Through Units

Calculate the block space load that would be expected to occur for a given set of operating conditions. If this load is a net cooling load, use Procedure A. If this load is a net heating load, use the Procedure B.

Procedure A (Net Cooling Load)

Step 1) Calculate the CFM that is required to neutralize the block cooling load. (RAT is the room air dry-bulb temperate and CDT is the dry-bulb temperature of the air leaving the cooling coil.)

$$\text{Cooling CFM} = \frac{\text{Cooling Load}}{1.1 \times (RAT - CDT)}$$

Step 2) Calculate the difference between the system design CFM (the sum of the zone CFM values) and this cooling CFM value. This difference represents the excess CFM that must be introduced into the space as neutral air.

Excess CFM = Sum of zone CFM - Cooling CFM

Step 3) Use the following equation to calculate how much of the excess CFM must pass through the cold-deck in order to neutralize the portion of the excess CFM that will pass through the hot-deck. (TDHD refers to the dry-bulb temperature difference between the hot-deck and the room and TDCD refers to the temperature difference between the room and the cold-deck.)

$$\text{Cold-deck excess CFM} = \frac{\text{Excess CFM} \times TDHD}{(TDHD - TDCD)}$$

Step 4) Add the cold-deck excess CFM to the cooling CFM to find the total CFM that will flow through the cold-deck.

Cold-deck CFM = Cold-deck excess CFM + Cooling CFM

Subtract the cold-deck CFM from the system design CFM (the sum of the zone CFM values) to find the total CFM that will flow through the hot-deck.

Hot-deck CFM = Sum of zone CFM - Cold-deck CFM

Step 5) Use the sensible heat equation to estimate the cooling coil load. (MAT is the dry-bulb temperature of the mixture of outdoor air and return air that is entering the cooling coil and CDT is the dry-bulb temperature of the air leaving the cooling coil.)

Minimum Coil load = 1.1 x Cold-deck CFM x (MAT - CDT)

Use the sensible heat equation to estimate the heating load at the hot-deck. (MAT is the dry-bulb temperature of the air entering the hot-deck and HDT is the dry-bulb temperature of the air leaving the hot-deck.)

Hot-deck load = 1.1 x Hot-deck CFM x (HDT - MAT)

Procedure B (Net Heating Load)

Step 1) Calculate the CFM that is required to neutralize the block heating load. (RAT is the room air dry-bulb temperate and HDT is the dry-bulb temperature of the air leaving the hot-deck.)

$$\text{Heating CFM} = \frac{\text{Heating Load}}{1.1 \times (HDT - RAT)}$$

Step 2) Calculate the difference between the system design CFM (the sum of the zone CFM values) and this heating CFM value. This difference equals the excess CFM that must be introduced into the space as neutral air.

Excess CFM = Sum of zone CFM - Heating CFM

Step 3) Use the following equation to calculate how much of the excess CFM must pass through the cold-deck in order to neutralize the portion of the excess CFM that will pass through the hot-deck. (TDHD refers to the dry-bulb temperature difference between the hot-deck and the room and TDCD refers to the temperature difference between the room and the cold-deck.)

$$\text{Cold-deck excess CFM} = \frac{\text{Excess CFM} \times TDHD}{(TDHD - TDCD)}$$

Step 4) Calculate the cold-deck CFM. In this case, the cold deck CFM is equal to the cold-deck excess CFM value.

Cold-deck CFM = Cold-deck excess CFM

Subtract the cold-deck CFM from the system design CFM (the sum of the zone CFM values) to find the total CFM that will flow through the hot-deck.

Hot-deck CFM = Sum of zone CFM - Cold-deck CFM

Step 5) Use the sensible heat equation to estimate the cooling coil load. (MAT is the dry-bulb temperature of the mixture of outdoor air and return air that is entering the cooling coil and CDT is the dry-bulb temperature of the air leaving the cooling coil.)

Minimum Coil load = 1.1 x Cold-deck CFM x (MAT - CDT)

Use the sensible heat equation to estimate the heating load at the hot-deck. (MAT is the dry-bulb temperature of the air entering the hot-deck and HDT is the dry-bulb temperature of the air leaving the hot-deck.)

Hot-deck load = 1.1 x Hot-deck CFM x (HDT - MAT)

Figure 11-13

always equal to their peak load values. In this case, the procedure that is used to evaluate the minimum load on the central cooling coil is the same as the procedure that is applied to a terminal reheat system. Refer to page 11-10 for this information.

Night-Cycle Cooling
In many applications, the cooling system is simply shut down during the (unoccupied) night cycle. This control strategy reduces operating expense. On blow-through systems, a thermostat set-up control strategy is desirable if the cooling system has to operate during the night cycle. However, on draw-through systems, a thermostat set-up control strategy could waste energy.

A minimum flow of outdoor air may not be required during the night cycle. If outdoor air is not required, the outdoor air damper should be closed whenever cooling is provided by mechanical refrigeration. However, the economizer cycle should still be available whenever the condition of the outdoor air is suitable for free cooling.

The internal loads are usually smaller during the night cycle. If night-cycle cooling is required, the minimum load on the cooling coil may or may not be less than the minimum coil load during the day cycle. An analysis of both operating cycles is required to determine which cycle produces the smallest load on the cooling coil.

Zone Sensible Heat Ratios
A sensible heat ratio calculation should be made for each zone. These calculations should be based on the sensible and latent loads that are associated with the summer design condition (the peak zone loads). There is no reason to expect all the zones to have the same sensible heat ratio. As explained in the following paragraph, this information is required to establish an acceptable condition for the air leaving the central cooling coil.

Condition of the Air Leaving the Cooling Coil
The performance of the cooling coil must be compatible with the summer design cooling requirements that are associated with each zone. This means that the condition of the air leaving the cooling coil must be cold enough to offset the sensible load in any zone and dry (cold) enough to neutralize the latent load in any zone. Section 4-17 in this manual, which discusses this subject, indicates that the designer will have the following options:

- If humidity control is not critical, the required cooling coil performance can be based on a sensible heat ratio that falls near the midrange of the various zone sensible heat ratios.

- If humidity control in one or more zones is critical, the required cooling coil performance must be based on the lowest sensible heat ratio that is associated with the critical zones.

Cold-Deck Temperature at Part Load
If the building is occupied during mild or cold weather, some zones will usually require cooling. Therefore, at part-load conditions, the cold-deck temperature will be dictated by the zone that is the least sensitive to variations in the outdoor air temperature. For example, an interior zone could require a 55°F cold-deck temperature during the coldest weather.

Also, note that if an economizer cycle is used to maintain the cold-deck temperature during cold weather, the temperature of the air entering the hot-deck will be equal to the cold-deck temperature. This also will be the case if mechanical cooling is used with a draw-through unit during cold weather. However, if mechanical cooling is used with a blow-through unit during cold weather, the temperature of the air entering the hot-deck will be equal to the mixed air temperature downstream from the minimum outdoor air damper.

Reheat Requirements
Zone reheat is not normally used with blow-through multizone systems. However, reheat could be used to improve humidity control in one or more zones. In this case, the reheat load would increase the flow of dehumidified air (cold-deck CFM) and reduce the flow of humid air (hot-deck CFM). This heat may be supplied directly by a reheat coil or indirectly by baseboard radiation.

Zone Heating Load
The heating capacity that is required for a zone is equal to the sum of the values that are found on lines 16 and 17 of Form N-1. One Form N-1 must be filled out for each zone.

For either configuration (blow-through or draw-through), the leakage at the cold-deck mixing damper produces a secondary zone heating load that is not included on Form N-1. Refer to Table 11-2 on page 11-30 for a damper leakage load that can be added to the Form N-1 zone heating load.

Zone Supply Air Temperature — Heating
During the winter, the heating capacity of the air that is supplied to a zone depends on the zone supply air temperature (ZSAT). The following equation can be used to calculate the supply air temperature that will neutralize the zone heating, design load. One calculation is required for each zone. (The Zone CFM is determined by the cooling load — refer above to the sub-section titled *Zone CFM Design Value*).

$$ZSAT = Room\ Temperature + \frac{Zone\ Heating\ Load}{1.1 \times Zone\ CFM}$$

Hot-Deck Temperature During Cold Weather
One or more zones may require heat during cold weather and it is possible that each zone could require a different supply air temperature (ZSAT). Therefore the hot-deck temperature must be selected to satisfy the zone that requires the highest supply air temperature. Note that the zone heating loads and

the corresponding zone supply air temperatures could be reduced if credit is taken for the heat added by the internal loads that are associated with the zones. This means that different hot-deck setpoints could be used during the day cycle (internal loads) and the night cycle (no internal loads).

Hot-Deck Temperature at Part-Load

At part-load conditions, hot-deck heat will be required whenever one zone requires heat. However, the hot-deck setpoint can be reduced as the outdoor temperature rises and the heating loads decrease. And, hot-deck heat can be eliminated when all the zones have switched to the cooling mode. Load lines can be used to estimate when the various zones will changeover from heating to cooling. (Refer to Section 4-10 for more information about using load lines.)

Heat Capacity Required at the Hot-Deck

For a blow-through system, the calculation procedure provided by Figure 11-13 can be used to estimate the heating capacity that is required at the hot-deck. In this case, the hot-deck load is based on the block space load that is expected to occur when the outdoor temperature is equal to the winter design temperature. The hot-deck load also depends on the mixed air temperature, the hot-deck temperature, the cold-deck temperature and the room temperature. However, the values of these various parameters will vary, depending on the operating conditions. Some typical scenarios are described below, but other conditions are possible. It is the designer's responsibility to evaluate the various possible operating conditions and to determine the condition that produces the largest hot-deck load.

Scenario —1

(This is a worst case scenario.) The building is occupied, no credit is taken for the internal loads, cooling is required at the cold-deck and outdoor air is required for ventilation. In this case, the effect of the internal loads should not be included in the block space load; the mixed air temperature will depend on the minimum outdoor air requirement or the economizer setpoint; and the hot-deck, cold-deck and room temperatures correspond to the day-cycle setpoints.

Scenario —2

The building is occupied, the internal loads help heat the space, cooling is required at the cold-deck and outdoor air is required for ventilation. In this case, the effect of the internal loads should be included in the block space load; the mixed air temperature will depend on the minimum outdoor air requirement or the economizer setpoint; and the hot-deck, cold-deck and room temperatures correspond to the day-cycle setpoints.

Scenario —3

The building is unoccupied, the internal loads do not help heat the space, cooling is not required at the cold-deck and outdoor air damper is closed. In this case, the effect of the internal loads should not be included in the block space load; the mixed air temperature will depend on the damper leakage; and the hot-deck, cold-deck and room temperatures correspond to the night-cycle setpoints.

The procedure that is used to estimate the hot-deck load for a draw-through unit is the same as the procedure that is used for a blow-through unit (above). However, one small adjustment must be made to Figure 11-13. With a draw-through unit, the temperature of the air entering the hot-deck is always equal to the cold-deck temperature. Therefore, the mixed air temperature value (MAT) in Step 5 should be replaced with the cold-deck temperature value (CDT).

Hot-deck load = 1.1 x Hot-deck CFM x (HDT - CDT)

Night-Cycle Heating Loads

The night-cycle load at the hot-deck will depend on how the system is operated. Consider the following possibilities:

- If cooling or dehumidification is not required when the zones are unoccupied, the cooling system can be shut down during the night cycle. In this case, the cold-deck temperature will be the same as the mixed air temperature. And, if fresh air or make up air is not required when the zones are unoccupied, the outdoor air dampers can be closed. In this case, the mixed air temperature will depend on the damper leakage.

- If cooling or dehumidification is required when the zones are unoccupied, the cold-deck setpoint must be maintained during the night cycle. But, the outdoor air dampers can be closed if fresh air or make up air is not required when the zones are unoccupied. In this case, the mixed air temperature will depend on the damper leakage. (The economizer cycle should still be used to provide free cooling.)

The night cycle, zone heating loads will depend on whether the zone thermostats are reset during the night cycle and on whether there are any uninterrupted internal loads. Credit for internal loads may or may not be applied, depending on the judgment of the designer.

Changeover to Heating

The cooling to heating changeover takes place on a zone-by-zone basis. Zone changeover automatically occurs as the zone thermostats call for additional heat. Changeover at the central equipment takes place when the heating plant is activated. Load lines (refer to Section 4-10) can be used to determine the outdoor temperatures that are associated with the zone changeovers. The highest changeover temperature will determine when the heating plant should be activated.

Psychrometrics

An elementary psychrometric analysis of the cooling cycle should be made because it allows the designer to predict the condition of the air entering the cooling coil. This analysis also defines the range of acceptable conditions for the air that will be supplied to the zones during the cooling cycle. This information is directly related to the procedures that are used to select the cooling equipment. Refer to Sections 4-17 and 5-1 in this manual for more information about elementary psychrometrics.

Equipment Selection — Primary Equipment
For blow-through equipment, the capacity of the primary cooling equipment should be based on the peak block load. For draw-through equipment, the capacity of the primary cooling equipment should be based on the sum of the peak zone loads. (Refer above to the paragraph titled *Sizing Load — Cooling at the Central Coil.*) In either case, comprehensive performance data should be used to select equipment. This performance data should correspond to the actual operating conditions that are expected to occur at the site. Part-load performance also must be evaluated and an appropriate method of capacity control should be provided. The procedures that are used to select various types of cooling equipment are presented in Section 6 of this manual.

Equipment Selection — Secondary Equipment
Special secondary equipment is not usually required for multizone systems. Air distribution is provided by constant volume supply air outlets. Occasionally a reheat coil may be required to control the humidity in a critical zone.

Equipment Selection — Humidifiers
If a humidifier is required, its capacity should be equal to or greater than the largest humidification load that is expected to occur during any operating condition. Refer to Section 10 for more information about selecting humidifiers.

Duct Sizing
Multizone systems are classified as constant volume systems. There is no air-side diversity. Therefore, the size of all the supply and return ducts (trunks and branches), should be based on CFM values that correspond to the peak loads that are associated with a given section of duct (refer to ACCA **Manual Q**, page 15-3).

Temperature and Humidity Controls
The temperature in each zone can be maintained by a zone thermostat that controls the zone hot-deck and cold-deck dampers. The cold-deck and hot-deck temperatures should be carefully selected so that adequate cooling, dehumidification and heating is always available for every zone. Reset controls, which optimize the cold-deck and hot-deck setpoints, are often used to reduce energy consumption. It also is necessary to make sure that the capacity of the refrigeration equipment (chiller or DX unit) tracks the cooling load during any operating condition.

> Some packaged multizone units have individual heating and cooling coils for each zone. These units use less energy because they allow the cold-deck and hot-deck temperatures to be optimized for each zone.

Fan and Damper Control
The fan (or fans if a return air fan is used) should operate continuously when any zone is occupied. If outdoor air is required for make up air, the fan(s) should operate continuously whenever make up air is required, even if all the zones are unoccupied.

Outdoor air, return air and exhaust air dampers should be controlled so that the minimum ventilation requirements are met whenever the building is occupied. If outdoor air is required for make up air, the various dampers should be controlled so that adequate make up air is always available. (The outdoor air requirement will be dictated by the ventilation requirement or by the make up air requirement, whichever is larger.) Economizer cycles can be used to provide free cooling.

Night-Cycle Control
Providing that the day-cycle space conditions do not have to be maintained, control over the entire conditioning system can be transferred to a heating-only night cycle when the building is unoccupied. In this case, heat is supplied at the hot-deck, but no cooling is provided at the cold-deck.

Usually, when control is transferred to the night cycle, the outdoor and exhaust air dampers can be closed and the fan(s) can be de-energized. When the night-cycle thermostat — which should be located in the zone that has the most rapid rate of temperature drop — calls for heat, the fan(s) operate, but the outdoor air and exhaust air dampers can remain closed. If desired, the night-cycle thermostat(s) can be set-back.

11-6 Changeover Variable Volume Systems

This section provides an overview of the performance characteristics, hardware, design procedures and control cycles that are associated with changeover variable volume systems (CO-VAV). These systems are often referred to as Variable-Volume-Variable-Temperature (VVT) systems, but the VVT designation is registered trademark. The information provided by Section 11-1 also applies to these systems.

Function
Changeover variable volume systems can either provide sensible and latent cooling or they can provide heat, but they do not provide simultaneous heating and cooling. They also can provide filtration and outdoor air for ventilation or for make up air. Winter humidification is not normally provided by these systems.

The system will operate in the heating mode or the cooling mode, depending on which mode satisfies the greatest number of zone thermostats. The equipment automatically switches to the opposite operating mode when the temperature in one or more zones drifts to far from the desired setpoint. Zone dampers located in the zone supply air ducts control the zone CFM in response to signals from the zone thermostats. A

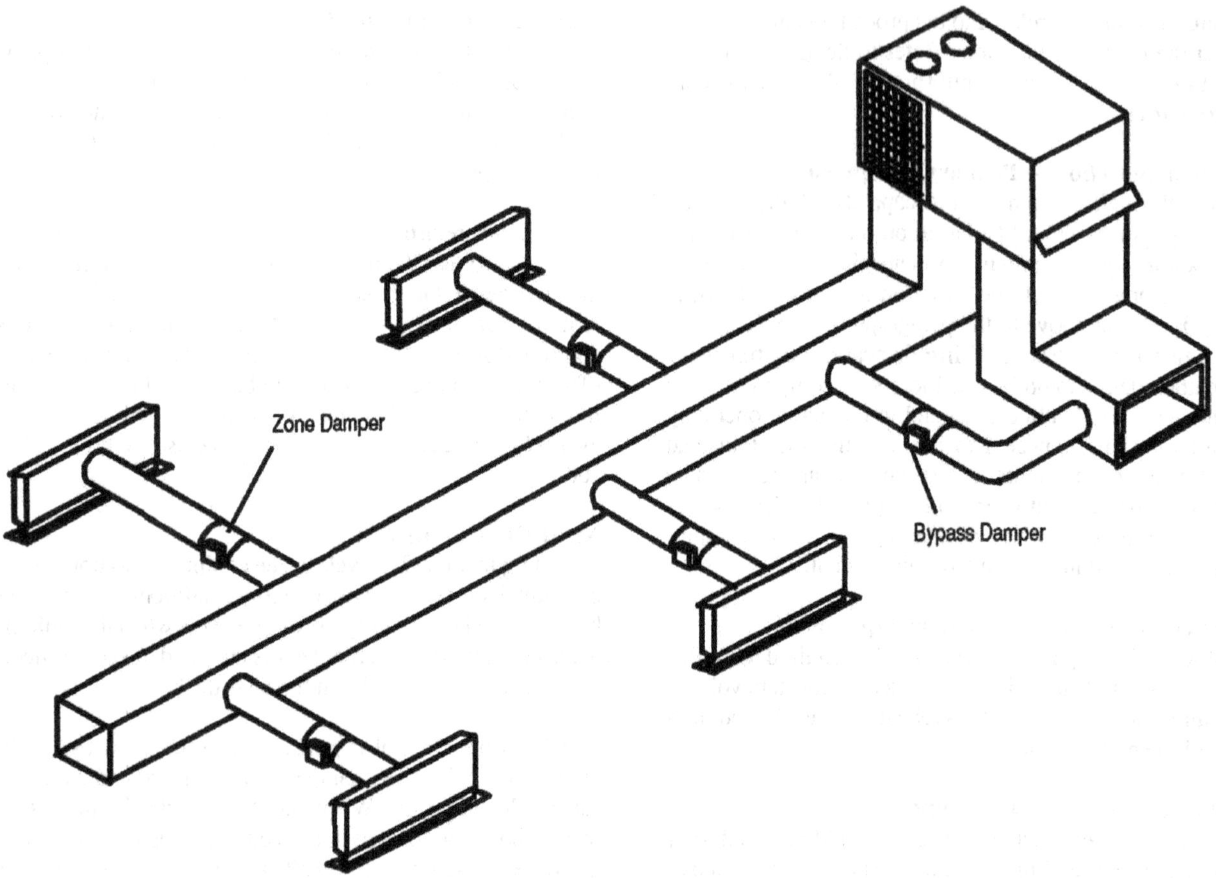

Changeover VAV System

bypass duct and a bypass damper is usually required to insure that adequate flow through the central equipment is maintained during any operating condition.

Applications
When precise control over space conditions is not required, basic varieties of changeover variable volume systems can be properly applied to small commercial buildings that have a few, compatible zones. The number of zones and/or the range of operating conditions that can be accommodated can be increased by using two or more units. (Compatibility means that, as a group, the zones do not require a considerable amount of simultaneous heating and cooling. Ideally, the zone loads should changeover from cooling to heating over a narrow range of operating conditions.)

Configuration
In their simplest form, changeover variable volume systems are single-zone, constant volume, heating and cooling systems that are "accessorized" with branch duct, flow-control dampers and microprocessor-based controls. Since the fans, coils, heat exchanges and air terminals are not designed to accommodate variable air flow conditions, it is important to make sure that adequate air flow is maintained through all the components during any operation condition.

Some of the newer heating and cooling equipment is more sophisticated. State of the art changeover variable volume systems could be equipped with a variable-speed compressor, variable-speed fans and a modulating heating plant. These components could automatically adjust their capacity to match the loads and air flow requirements that are associated with any operating condition.

Characteristics
The characteristics that are generic to all types of all-air systems are listed in Section 11-1. The characteristics that are peculiar to changeover variable volume systems are listed below.

- CO-VAV systems can condition a few compatible zones.
- CO-VAV systems provide fair temperature control.
- CO-VAV systems do not provide exact humidity control, full dehumidification capacity is not always available.
- The maximum humidity in any zone may not be limited to a specific level.
- Air circulation rates will vary, depending on the fluctuation of the space loads.
- CO-VAV systems can take advantage of the diversity that is associated with the zone cooling loads.

- Control over the capacity of the cooling coil is directly related to the sensible space loads.
- Control over the capacity of the cooling coil is not directly related to the latent loads.
- CO-VAV systems require an analysis of the zone load changeover patterns.
- CO-VAV systems are relatively easy to install and operate.
- No special provisions for seasonal changeover (heating or cooling) is required.
- The equipment switches between the cooling mode and the heating mode, but simultaneous heating and cooling is not provided.
- Since the fresh air requirements may vary by zone, the percentage of outdoor air introduced at the air handler may not match the outdoor air percentage that is required for each zone.
- Without special damper controls, the quantity of outdoor air delivered to a given zone will be throttled in proportion to the reduction in the supply CFM.
- Since the sensible heat ratios may vary by zone, the condition of the air leaving the cooling coil may not be perfectly compatible with all the zone sensible heat ratios.
- Without special controls, the latent capacity of air delivered to a given zone will be throttled in proportion to the reduction in the supply CFM.
- The central equipment must be used to condition a single zone even though the rest of the building may not require conditioning when the other zones are vacant.
- Night-cycle control cannot be applied to the central equipment unless all the zones are vacant.
- Night-cycle control can be applied to the terminal equipment on a zone-by-zone basis.
- CO-VAV systems provide zone control at a reasonable operating cost.
- CO-VAV systems provide flexibility to accommodate floor plan changes.
- Constant volume supply air outlets are often used with CO-VAV systems, but variable volume outlets would improve zone air circulation rates at part-load conditions.

Primary Equipment

On a basic CO-VAV system, the primary equipment consists of a central unit that contains a constant volume fan, a heating plant and a DX cooling system. The central equipment also could be a unitary or split-system heat pump. Economizers are commonly used for free cooling and return fans are required to control the flow of return air and exhaust air. Return fans also provide control over the pressure in the conditioned space. Filters and dampers are installed in the air handler.

Secondary Equipment

Secondary equipment consists of a packaged control system that includes variable volume dampers that are installed in the branch ducts, a bypass damper, local zone controls and an electronic control system. Room air distribution is commonly provided by constant volume supply outlets. However, variable volume supply outlets would improve room air circulation when the supply CFM is throttled by the zone dampers.

Sizing Loads — Zone Cooling

The cooling capacity that is required for each zone is equal to the corresponding peak zone load. One **Manual N**, Form N-1 must be filled out for each zone. The sensible and latent loads that are associated with a given zone will correspond to the values that are found on line 8 of Form N-1, (lines 9 through 12 are not required). Note that if there is a supply duct gain (Form N-1, line 7), it is added to the zone load. This is done so that the duct gain is offset by an increase in the cooling CFM that is supplied to the zone.

Sizing Load — Cooling at the Central Coil

Since changeover variable volume systems can take advantage of diversity, the cooling capacity that is required at the central coil is based on the peak block load. (Diversity is defined on page 11-18.) One **Manual N**, Form N-1 must be filled out for the whole building. The sensible and latent loads at the coil will correspond to the values that are found on Form N-1, line 12.

Fan heat is a system load that is not included on Form N-1. However, it is necessary to account for the sensible heat that is added by the supply and return fans. Sometimes this is accomplished (as per the manufacturer's equipment selection instructions) by subtracting the fan heat from the cooling capacity of the unit. (Refer to Section 6-8 for an example that illustrates this procedure.) Any portion of the fan heat that is not accounted for by a capacity reduction should be added to the other system loads that are imposed on the cooling coil.

Minimum Cooling Loads — Zones

Load lines (refer to Section 4-10) can be used to evaluate the behavior of the zone loads. Minimum zone cooling loads are important because they identify the zones that may require reheat coils and positioners that prevent excessive throttling of the supply air. (Refer below to the sub-section titled *Minimum Zone CFM* for more discussion on this subject.) Load lines also can be used to predict when the various zone changeover from cooling to heating. This changeover temperature information is useful for evaluating the compatibility of the various zones. (Refer below to the sub-section titled *Changeover to Heating* for more discussion on this subject.)

Minimum Cooling Load — Central Coil

The load on the cooling coil is always equal to the block space load, plus the outdoor air load, plus any other system loads that may apply. The behavior of the coil load can be analyzed by making a load-line analysis. (Refer to Section 4-11 of this manual, also refer to Section 4-16.) As explained below, the minimum coil load also will depend on the strategy that is used to control the outdoor air damper. (Figure 11-1 on page 11-4 summarizes three basic outdoor air control strategies.)

No Economizer
If no economizer cycle is used, the minimum load on the coil will depend on whether the block load load-line crosses over from cooling to heating:

- If a cooling-heating changeover occurs (with no credit for solar gains), the minimum load on the coil will be zero. (Refer to page 4-15, Figure 4-12.)

- If a cooling-heating changeover does not occur (with no credit for solar gains), the minimum load on the coil will equal the load that occurs when the outdoor temperature is equal to the winter design temperature.

Simple Economizer
If a simple dry-bulb economizer is installed, the minimum cooling load on the coil (with no credit for solar gains) will occur just before the system switches to the economizer cycle. In this case, the minimum load on the coil will equal the block load that occurs when the outdoor temperature is equal to changeover temperature. The block load load-line can be used to evaluate this load. (Refer to page 4-15, Figure 4-12.)

Ultimate Economizer
If an ultimate economizer cycle is installed, it is possible to have the refrigeration equipment operating with no load on the coil. (Figure 11-1 on page 11-4 illustrates this point.)

Night-Cycle Cooling
In many applications the cooling system is simply shut down during the (unoccupied) night cycle. This control strategy reduces operating expense. A set-up control strategy will reduce energy use if the cooling system has to operate during the night cycle.

A minimum flow of outdoor air may not be required during the night cycle. If outdoor air is not required, the outdoor air damper should be closed whenever cooling is provided by mechanical refrigeration. However, the economizer cycle should still be used whenever the condition of the outdoor air is suitable for free cooling.

Internal loads are usually smaller during the night cycle, so the minimum load on the cooling coil could be less than the minimum day-cycle load. If night-cycle cooling is required, an analysis of both operating cycles should be made to determine which cycle produces the smallest load on the coil.

Zone Sensible Heat Ratios
A sensible heat ratio calculation should be made for each zone. These calculations should be based on the sensible and latent loads that are associated with the summer design conditions (the peak zone loads). There is no reason to expect all the zones to have the same sensible heat ratio. As explained in the following paragraph, this information is required to establish an acceptable condition for the air leaving the central cooling coil.

Condition of the Air Leaving the Cooling Coil
The performance of the cooling coil must be compatible with the conditioning requirements of each zone. This means that the condition of the air leaving the cooling coil must be cold enough to offset the sensible load in any zone and dry (cold) enough to neutralize the latent load in any zone. Since changeover variable volume systems are used for applications that do not require precise humidity control, the cooling coil performance can be based on a sensible heat ratio that falls near the midrange of the various zone sensible heat ratios.

Reheat Requirements
Ideally, the zone loads should changeover from cooling to heating over a narrow range of operating conditions. Zones that have similar changeover patterns are compatible with each other. Incompatibility results when one or more zones create situations that require extended periods of simultaneous cooling and heating. In some cases, reheat can be used to reduce the simultaneous cooling and heating requirement.

During some operating conditions, the zone dampers will severely throttle the supply CFM. Usually, this low-flow condition is undesirable because some amount of supply air is required to provide air circulation, humidity control and fresh air. Therefore, reheat could be used to create a low-limit cooling load that establishes a minimum value for the supply CFM. Refer below to the sub-section titled *Minimum Zone CFM* for more discussion on this subject.

Sizing Loads — Zone Heating
The heat load that is associated with a zone corresponds to the sum of the values that are found on lines 16 and 17 of Form N-1. One form N-1 is required for each zone.

Heat Required at the Central Unit
The amount of heat required at the central unit will depend on the space heating load, the internal gains and the outdoor air load. Designers typically calculate a "worst case" load that includes the space load and the outdoor air load, but does not include a credit for the heat added by the internal gains. In this case, one **Manual N**, Form N-1 should be filled out for the whole building. The heat load at the unit will correspond to the value that is found on Form N-1, line 21.

Night-Cycle Heating Loads
The size of the night-cycle space load will depend on whether the space temperature is set back during the night cycle. And, the outdoor air load will depend on whether or not outdoor air is required during the night cycle. If outdoor air is not required, the outdoor air damper should be closed. (Remember to allow for some damper leakage.)

Changeover to Heating
The amount of (cooling-heating) mode cycling that is required at the central unit is related to the simultaneous heating and

cooling requirements that are associated with the zones. Mode cycling can be minimized and zone temperature control can be improved if the zones have compatible changeover temperatures. Load lines (refer to Section 4-10) can be used to evaluate the zone changeover temperatures. When a changeover is required, it takes place automatically when the control system logic switches the operating mode. The action of the zone dampers is automatically reversed when a changeover occurs.

Psychrometrics
An elementary psychrometric analysis of the cooling cycle should be made because it allows the designer to predict the condition of the air entering the cooling coil. This analysis also defines the range of acceptable conditions for the air that will be supplied to the zones during the cooling cycle. This information is directly related to the procedures that are used to select the cooling equipment. Refer to Sections 4-17 and 5-1 in this manual for more information about elementary psychrometrics.

Equipment Selection — Primary Equipment
The capacity of the primary cooling equipment should be based on the peak block load (as explained above in the paragraph titled *Sizing Load — Cooling at the Central Coil*) and comprehensive engineering data should be used to select equipment. This engineering data should correspond to the actual operating conditions that are expected to occur at the site. Part-load performance also must be evaluated and an appropriate method of capacity control should be provided. The procedures that are used to select various types of cooling equipment are presented in Section 6 of this manual.

Equipment Selection — Secondary Equipment
Follow the manufacturer's instructions regarding zone damper sizing, bypass damper sizing and control system hardware requirements. If constant volume supply outlets will be installed, they should be sized for the highest outlet velocity possible. (Noise is usually the limiting factor.) However, VAV supply outlets will provide better air circulation at part-load conditions.

Supply CFM Design Value — Zones
The zone cooling CFM estimate is based on the peak sensible zone load. This load corresponds to the value that is found on Form N-1, line 8. One Form N-1 is required for each zone.

$$\text{Cooling CFM} = \frac{\text{Peak sensible zone load (BTUH)}}{1.1 \times (\text{Room design DB} - \text{Supply air DB})}$$

The zone heating CFM estimate is based on the zone heating load. This load can be estimated by adding lines 16 and 17 on Form N-1. One Form N-1 is required for each zone.

$$\text{Heating CFM} = \frac{\text{Zone heating load for space (BTUH)}}{1.1 \times (\text{Supply air DB} - \text{Room design DB})}$$

Usually, the design value for the zone supply CFM is based on the cooling load, but in some cases, the heating load could establish the required air flow. If there is any doubt about which load to use, calculate the CFM required for each load and use the larger value. This design CFM value is used to size the zone air outlet(s) and the zone runout duct(s).

Minimum Zone CFM
When the system is operating in the cooling or heating mode, the CFM that is supplied to a zone will be reduced in direct proportion to the reduction in the zone load. Since the zone thermostat controls the zone damper, the zone temperature will always be maintained as the supply CFM is throttled. However, the air circulation rate within the zone, the CFM of fresh air that is delivered to the zone and the (cooling mode) dehumidifying capacity of the supply air will be reduced in direct proportion to the reduction in the supply CFM.

However, the cooling (or heating) plant usually does not operate continuously at part-load conditions. Therefore, neutral air can be circulated through open zone dampers when the cooling (or heating) demand (at the unit) is satisfied. This control strategy will cause the average zone air circulation — in terms of air changes per hour — to be substantially larger than the air circulation rates that are associated with a throttled damper position.

If, for any reason, the average air circulation rates are not adequate, minimum position "stops" can usually be programmed into the zone damper control sequences. These stops will insure that a minimum amount of supply air will be delivered to a zone during any operating condition. However, these stops could cause a zone to overcool (or to overheat) because the supply CFM would be excessive at low load conditions. But, these zone temperature excursions would eventually be corrected when the polling controls detect the problem and switch the unit to a different operating mode. However, it is possible that this "mode-cycling" could be undesirable and in extreme cases, the zone temperature excursions could be excessive. (This might occur if the stops in one or two zones are set to maintain a substantial air flow rate.) Therefore, zone reheat might be required to maintain control over the zone temperature and air flow.

Supply CFM at the Central Unit
Normally, cooling requires more air flow than heating, so the supply CFM at the central unit is usually based on the cooling load. Because the system can take advantage of diversity, the peak block sensible cooling load should be used to calculate the design value for the supply CFM. Refer above to the paragraph titled *Sizing Load — Cooling at the Central Coil* for more information about the coil load.

$$\text{CFM} = \frac{\text{Sensible Design Load}}{1.1 \times (\text{Room design DB} - \text{Supply air DB})}$$

Also, note that if heat is provided by a fossil fuel furnace, the heating CFM must be compatible with the manufacturer's temperature rise (TR) or discharge temperature (TS) limits. The following equations can be used to check for compliance with these requirements.

$$TR = \frac{\textit{Output heating capacity (BTUH)}}{\textit{Heating CFM}}$$

$$TS = \textit{Entering temperature} + TR$$

Minimum CFM at the Unit
Minimum air flow through the fan, DX coil, heating coil or heat exchanger must be maintained during any operating condition. A bypass duct equipped with an electronically controlled damper is recommended in situations where the part-load supply CFM (for the entire system) is expected to fall below 80 percent of the peak block load CFM.

Zone and Bypass Damper Sizing
Since the zone dampers and the bypass damper are acting as control valves they should not be oversized. Elliptical dampers are preferred because they provide a linear response to changes in the damper position. Refer to the manufacturer's engineering data for damper sizing recommendations and installation instructions.

Duct Sizing
The runout and branch duct sizes are determined by the air flow that is required to neutralize the peak zone load. Refer to the paragraph above titled *Supply CFM Design Value — Zones* for more information about zone air flow requirements.

Air-side diversity can be applied to the main trunk ducts. The size of these ducts should be based on the same air flow values that would be used to estimate the supply CFM requirement at the central unit. Refer to the paragraph above titled *Supply CFM at the Central Unit* for more information about this air flow requirement. Also refer to ACCA **Manual Q**, Section 15 for more information about duct sizing and fan selection.

Temperature and Humidity Controls
The temperature in each conditioned space is maintained by a zone thermostat that controls the zone damper. At peak-load conditions the thermostat will call for full air flow. At part-load conditions the thermostat initiates action that throttles the supply CFM until thermostat setpoint is satisfied. In some cases, this throttling action may only be allowed to continue until a (predetermined) minimum flow condition occurs, then further throttling is prevented. (If reheat is used, the thermostat should activate and control the heating device after the zone damper has been throttled to its minimum position.)

The central equipment is controlled by an electronic package that is provided by a vendor. (This package usually includes the zone dampers, the bypass damper and the associated devices.) This control package monitors the conditions in the various zones and the conditions at the unit. When a control action is required, the control package automatically initiates the required action and monitors the result.

It also is necessary to make sure that the capacity of the refrigeration equipment tracks the coil load at all points of operation. A minimum load analysis should be made to determine how much compressor unloading is required and to decide if hot gas bypass is required.

Fan and Damper Control
The fan (or fans if a return air fan is used) should operate continuously when any zone is occupied. If outdoor air is required for make up air, the fan(s) should operate continuously whenever make up air is required, even if all the zones are unoccupied.

Fan capacity control devices are commonly installed on large VAV systems, but they are not normally used with CO-VAV systems. (Variable-speed fans may be available with the newest types of equipment.) If fan control devices are used, it is important to make sure that the supply and return fans work in unison.

Outdoor air, return air and exhaust air dampers should be controlled so that the minimum ventilation requirements are met whenever the building is occupied. Special controls may be required to maintain a constant flow of outdoor air through the (minimum) outdoor air damper as the supply CFM is modulated. If outdoor air is required for make up air, the outdoor air, return air and exhaust air dampers should be controlled so that adequate make up air is always available. (The outdoor air requirement will be dictated by the ventilation requirement or by the make up air requirement, whichever is larger.)

Night-Cycle Control
Providing that the day-cycle space conditions do not have to be maintained, control over the entire conditioning system can be transferred to a heating-only night cycle when the building is unoccupied. When control is transferred to the night cycle, the outdoor and exhaust air dampers can (usually) be closed and the fan(s) can (usually) be de-energized. When the night-cycle controls call for heat, the fan(s) operate, but the outdoor air and the exhaust air dampers can remain closed. If desired, the night-cycle thermostat(s) can be set back.

11-7 Make Up Air Systems

This section provides an overview of the performance characteristics, hardware, design procedures and control cycles that are associated with make up air systems. The information that is provided by Section 11-1 also applies to these systems.

Function
The primary function of a make up air system is to replace air that is exhausted from the conditioned space with outdoor air. Usually this make up air must be introduced into the conditioned space at the same temperature as the room air — and in some cases, it must have the same moisture content as the room air. Therefore, make up air systems can be used to add heat and humidity to the outdoor air in the winter or they can be used to cool and dehumidify the outdoor air in the summer. Make up air systems also can be designed to partially or fully satisfy the heating and cooling loads that are associated with the building envelope.

Applications
Make up air systems are used to maintain an air balance in all types of small and large commercial, institutional and industrial buildings. Make up air systems are often used to heat storage areas, work areas and recreational areas.

Configuration
Make up air systems are usually single-zone, single-duct systems. (Direct-fired make up air units are often installed without a duct system.) The cooling coil (if required) and the heating coil (or heat exchanger) are usually arranged in series. In this configuration the central equipment can add to, or extract heat from, the flow of make up air.

Characteristics
The performance characteristics that are generic to all types of all-air systems are listed on page 11-2. The characteristics that apply to a single-zone constant volume systems also apply to make up air systems. These characteristics are listed on page 11-3.

Equipment
The make up air unit could be equipped with a hot water coil, an electric heating coil, a steam coil or fossil fuel heating equipment. If a fossil fuel heating plant is used, it could have a direct-fired burner or an indirect-fired heat exchanger. If cooling is required, it could be provided by a DX system or by a chilled water system. Filters, dampers and in some cases humidifiers, are installed in the air handler.

Sizing Loads — Cooling
When a make up air system is just used to condition outdoor air, the sensible and latent cooling loads are equal to the outdoor air loads. Form N-1, line 9 can be used to calculate these loads.

Fan heat is a system load that is not included on Form N-1. However, it is necessary to account for the sensible heat that is added by the supply fan. Sometimes, this is accomplished (as per the manufacturer's equipment selection instructions) by subtracting the fan heat from the cooling capacity of the unit. (Refer to Section 6-8 for an example that illustrates this procedure.) Any portion of the fan heat that is not accounted for by a capacity reduction should be added to the other system loads that are imposed on the cooling coil.

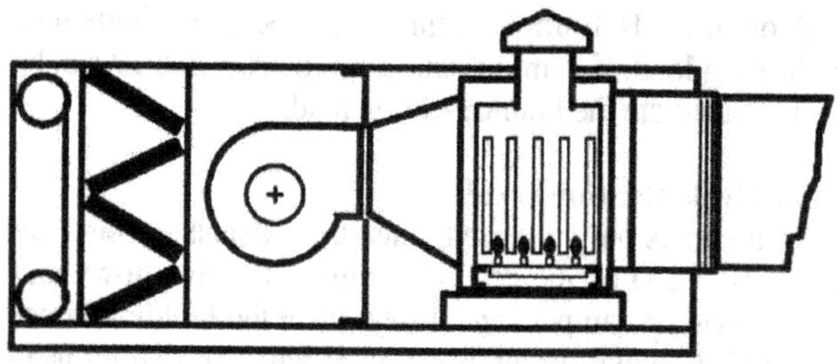

Indirect-fired unit

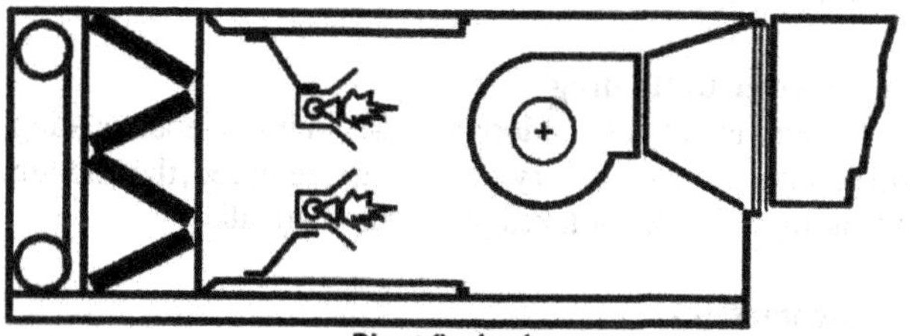

Direct-fired unit

Indirect- and Direct-Fired Make Up Air Units

Pumping heat and piping gains are system loads that are associated with chilled water systems. These sensible loads are not accounted for on the N-1 form, but if these loads exist, they should be added to the other loads that are imposed on the chiller.

Minimum Cooling Loads
When a basic make up air system is just used to condition outdoor air, the cooling load (during every operating condition) is equal to the outdoor air load. Since 100 percent outdoor air is always flowing through the unit, it is possible to have the refrigeration equipment operating with no load on the coil. This condition would exist whenever the dry-bulb temperature and the moisture content of the outdoor air is less than the temperature and moisture content of the room air.

Night-Cycle (Unoccupied) Cooling
Make up air is only required when the exhaust systems are operating. Usually, the cooling equipment is shut down and the outdoor air dampers are closed when the building is not occupied. If make up air is required when the building is unoccupied, a set-up control strategy would reduce energy consumption.

Condition of the Air Leaving the Cooling Coil
The performance of the cooling coil must be compatible with the conditioning requirements of the space. Therefore, the temperature and relative humidity of the air leaving the cooling coil must be equal to the temperature and humidity in the room (or a little colder to offset duct gains).

Sizing Load — Heating
When a basic make up air system is just used to heat outdoor air, the heating and humidification loads are equal to the

outdoor air loads. Form N-1, line 15 can be used to calculate the heating load. If humidification is required, line 19 can be used to calculate the humidification load.

Night-Cycle Heating Loads
Make up air is only required when the exhaust systems are operating. Usually, the heating equipment is shut down and the outdoor air dampers are closed when the building is not occupied. If make up air is required when the building is unoccupied, a set-back control strategy would reduce energy consumption.

Changeover to Heating
Since there is only an outdoor air load, simultaneous heating and cooling is not necessary. If cooling is required, the cooling to heating changeover takes place automatically.

Psychrometrics
The dry-bulb and wet-bulb temperatures of the air entering the cooling coil are equal to the summer outdoor design temperatures. The dry-bulb and wet-bulb temperatures of the air leaving the cooling coil are equal to the summer indoor design temperatures. (The leaving dry-bulb temperature might have to be a little colder to offset duct gains.)

Equipment Selection — Primary Equipment
The capacity of the equipment should be based on the design cooling and heating loads and comprehensive performance data should be used to select equipment. This performance data should be based on the actual operating conditions that are expected to occur at the site. Part-load performance also must be evaluated and an appropriate method of capacity control should be provided. Direct-fired heating equipment is preferred because it is equipped with modulating burners. If an indirect-fired furnace is used, it should be equipped with a burner that provides an adequate turn-down ratio. The procedures that are used to select various types of cooling and heating equipment are presented in Sections 6 and 7 of this manual.

Equipment Selection — Humidifiers
If a humidifier is required, its capacity should be equal to or greater than the largest humidification load that is expected to occur during any operating condition. Refer to Section 10 in this manual for more information about the selection of humidifiers.

Supply CFM
If the space is to be maintained at neutral pressure, the supply CFM should be equal to the CFM that is exhausted from the space. If the space is to be maintained at positive pressure, the supply CFM should be a little greater (5 to 10 percent) than the CFM that is exhausted from the space.

Duct Sizing
The CFM flowing through each section of duct and out of each supply outlet remains constant during all modes of operation. Therefore, the size of all the supply and return ducts (trunks and branches) should be based on the peak load CFM values.

Temperature and Humidity Controls
A modulating-type control, which maintains the supply air temperature setpoint during any operating condition, is preferred. On-off control or step control should be avoided. If this type of control is used, the equipment will not be able to adjust capacity at part-load conditions. If this happens, the supply air could be too hot or too cold — and raw unconditioned air will be introduced into the space when the equipment cycles off.

Fan and Damper Control
The fan (or fans if a return air fan is used) should operate continuously when make up air is required, even if the space is unoccupied. The outdoor air damper should be open when the fan is running and closed when the fan is off.

Section 12
Water Systems

Water systems are commonly used to heat and cool commercial, industrial and institutional buildings. With this type of system, water is the only medium (distribution fluid) that is used to transfer heat between a centrally located heating or cooling plant (boiler or chiller) and the secondary equipment that is located in the rooms or zones.

12-1 Overview of Hydronic Systems

This section provides a summary of the components and performance characteristics that are generic to all types of chilled water and low-temperature hot water systems. (Low-temperature hot water systems operate at temperatures that are below 250°F.) Additional information about the basic types of hydronic systems is provided in Sections 12-2 through 12-7.

Function
Hydronic systems can provide sensible and latent cooling by pumping chilled water through remote air handlers, fan-coils and unit ventilators. Hydronic systems also can provide heat by pumping hot water through fan-powered terminal devices, cabinet convectors or fin-tube radiation. Water-loop heat pump systems can provide sensible cooling, latent cooling and heat. If outdoor air is required, it must be introduced into the conditioned space by the secondary equipment (air handler or unit ventilator) or by a separate all-air system. If winter humidification is required, it must be introduced directly into the room (or zone), or it can be added to the air that is supplied to the room (or zone) by an air handler or fan-coil unit.

Applications
Hot water systems are commonly used to heat buildings of all types and sizes. Hydronic heating and cooling systems are normally used to condition large (commercial, institutional and industrial) buildings that have many zones, but they are occasionally used to condition small buildings. Heat pump, water-loop systems have been installed in a variety of buildings. Installed cost, energy efficiency and operating cost vary, depending on the type of system.

Possibly the most important feature that differentiates an all-water system from an all-air system is that the ducts and air handlers that are associated with an all-air system are much larger than the pumps and pipes that are associated with a water system. Therefore, water systems are often used when there is not enough space (within the building envelope) to accommodate an all-air system.

Water systems are suitable for applications that do not require precise humidity control. When cooling is provided by a fan-coil unit or unit ventilator, the humidity in a room or zone can be expected to drift as the room thermostat adjusts the sensible capacity of the secondary equipment. (The capacity can be controlled by modulating the water flow through the cooling coil or by varying the fan speed.)

Water systems are appropriate for applications that do not require a high degree of air filtration. Fan-coil units and unit ventilators are equipped with low efficiency filters, unit heaters and convectors do not have filters.

Basic Configurations
The configuration of a water system is related to the supply-side piping arrangement. Single-path systems only have one supply water circuit, which can be used to distribute hot water or cold water. Dual-path systems have two supply water circuits, one for hot water and one for chilled water. The configuration of a water system also is related to the return-side piping arrangement, which is either classified as *direct* or as *indirect*.

Single-Path Systems
Single-path systems can provide heat only, cooling only and they can changeover between heating and cooling. They cannot provide simultaneous heating and cooling. A single-path piping arrangement could be a series loop made up of one or more circuits (page 12-8), a one-pipe system (page 12-8) or a two-pipe system (page 12-10). As explained in Section 12-2, the series-loop and the one-pipe systems are only suitable for heating small areas. The two-pipe system can be used to heat or cool any size building or area.

Dual-Path Systems
Dual-path systems can provide heat only, cooling only and they can provide simultaneous heating and cooling. A dual-path piping arrangement could be a three-pipe system (page 12-12) or a four-pipe system (page 12-13). Three and four-pipe systems can be used to heat or cool any size building or any size area.

Direct and Reverse Return Piping
Figure 12-1, at the top of the next page, shows a direct return arrangement. Notice that the distance that the water must travel from the central equipment to a remote device and back to the central equipment is different for every device. This means that the pressure drop that is associated with piping runs is different for every device. The smallest piping loss is associated with the closest device and the largest piping loss is associated with the furthest device. Therefore, this arrangement produces piping system pressure losses that are not balanced.

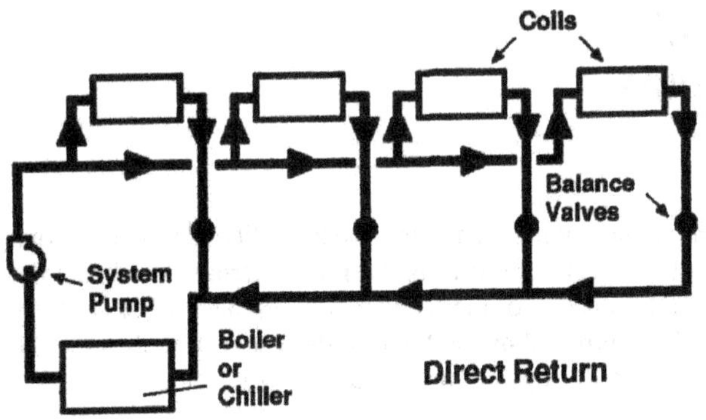

Figure 12-1

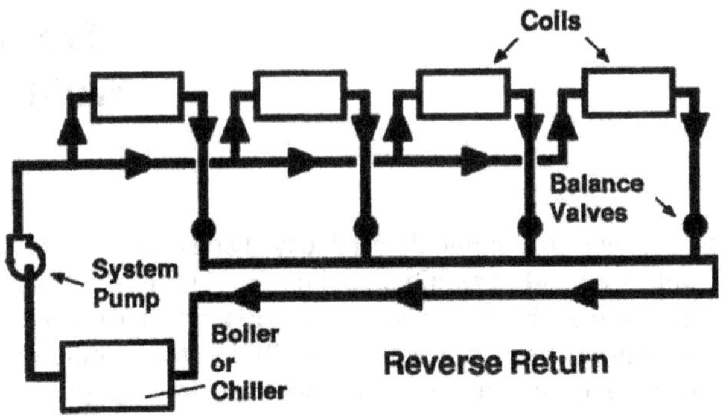

Figure 12-2

Figure 12-2 shows a reverse return arrangement. In this case, the distance that the water must travel from the central equipment to a remote device and back to the central equipment is the same for every device. This means that the pressure drop that is associated with the piping runs is the same for every device. Therefore, this arrangement produces piping system pressure losses that are balanced.

The advantage of direct return arrangement is that less pipe is required for the return run. The advantage of reverse return arrangement is that the piping losses are balanced. The disadvantage of direct return arrangement is that balancing valves must be installed in the branch risers. The disadvantage of reverse return arrangement is that more pipe is required for the return run. The (technical) reasons for selecting a direct or reverse return arrangement depend on the relationship between the piping losses and the pressure drops across the remote devices.

- If the pressure drops across the remote devices are not equal, the system will be inherently unbalanced, so there is no advantage to the reverse return layout. A direct return arrangement with balancing valves is appropriate in this situation.
- A direct return arrangement with balancing valves is appropriate when the pressure drops across the remote devices are large in comparison to the piping losses. (Fan-coils, unit heaters and unit ventilators have relatively large pressure drops.)
- If the pressure drops across the remote devices are equal, balancing valves are not required — providing that the piping losses are balanced by a reverse return layout.
- The pressure drops across the remote devices are not important if they are considerably smaller than the piping losses. (Fin-tube radiation and cabinet convectors may have relatively small water-side pressure drops.) In this case, balancing valves are not required, providing that the piping losses are balanced by a reverse return layout.

Combination Systems
Two or more of the basic piping configurations can be combined to create a hybrid system. Hybrid systems are used to solve the problems that are associated with zoning, controlling the water temperature that enters the primary or secondary equipment, controlling the flow rate through the primary or secondary equipment, limiting the pumping requirements (head, flow rate or power) and reducing the size of the distribution piping. Refer to hydronic system design manuals for more information about hybrid systems.

Primary-Secondary Pumping Systems
Primary-secondary pumping systems use a primary piping circuit and a primary pump to supply water to secondary circuits that are equipped with secondary pumps. The Figure at the top of the next page provides an example of a primary-secondary pumping system. These systems are usually used to heat and cool buildings that have two or more large areas. (These areas may be subdivided into a number of separate rooms or smaller zones.)

One important advantage that is associated with these systems is that they allow independent control over the temperature and flow rate of the supply water in the secondary circuits. They also provide independent control over the temperature drops that are associated with the various piping circuits. And, they allow the designer to use smaller pipes and a smaller pump in the primary circuit. Refer to hydronics design manuals for more information about these systems.

Characteristics
The characteristics that are generic to all types of water systems are listed below. The characteristics that are peculiar to specific types of water systems are provided in the following sections.

- Water systems are relatively easy to design and to control.
- The primary heating and cooling plants are centrally located.
- Secondary heating and cooling equipment is distributed throughout the building.

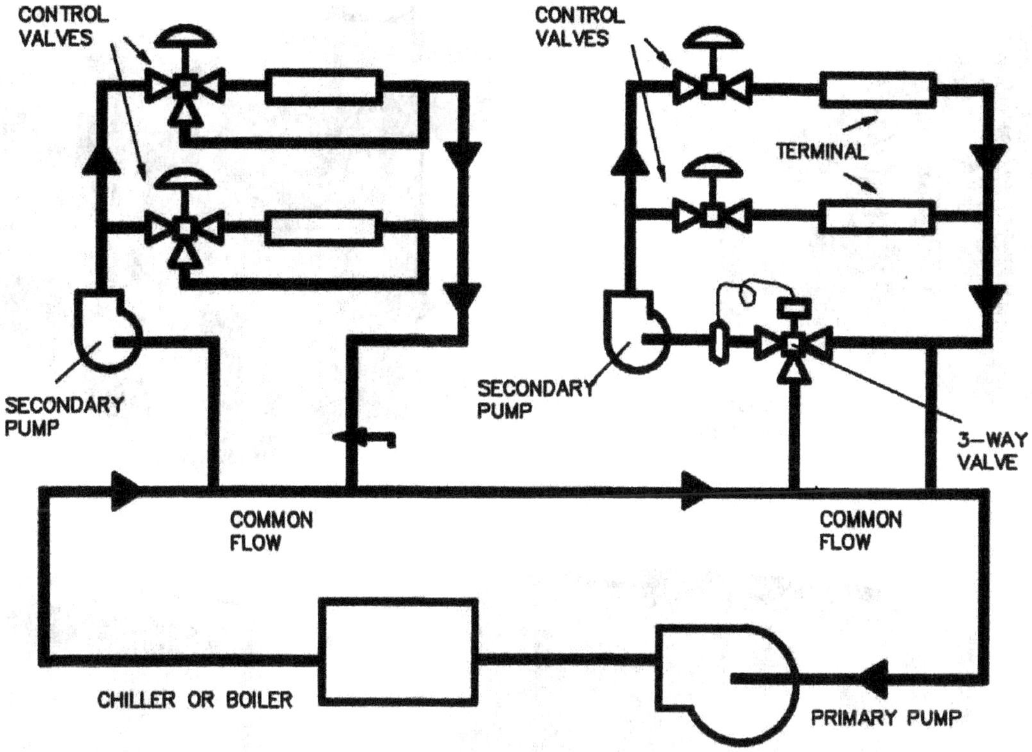

Primary - Secondary Pumping

- No ducts are required. A relatively small amount of interior space is required for pipe runs.
- Maintenance is not consolidated at one location. Many small units have coils, filters, valves, dampers and controls that require attention.
- A failure of the primary equipment affects a large area or the entire building.
- A failure of a piece of secondary equipment only affects a small area.
- These systems are not appropriate for applications that require close humidity control.
- Outdoor air must be introduced through secondary forced-air equipment or by a separate all-air system.
- These systems are not appropriate for zones that require a considerable amount of make up air.
- Multiple envelope penetrations increase infiltration loads when outdoor air is introduced through unit ventilators.
- Heat can be introduced at the perimeter of the building and under the windows.
- Fan-powered secondary equipment can provide positive room air circulation.
- Fan noise is produced at every secondary (forced-air) unit.
- Convectors do not provide positive room air circulation.
- The filters that are installed in the secondary equipment are not very efficient.
- Each piece of secondary cooling equipment requires a condensate drain.

Primary Equipment

The primary equipment is associated with cooling and heating water at a central location. The primary cooling equipment includes a water chiller and a heat rejection device. The primary heating equipment consists of a boiler. Central pumps are used to circulate water through the piping system.

Secondary Equipment

Secondary HVAC equipment (see next page) is used to condition individual rooms or zones. Cooling can be provided by small air handlers, fan-coils and unit ventilators that are equipped with a chilled water coil. Heat can be provided by small air handlers, fan-coils, unit ventilators and unit heaters that are equipped with a hot water coil. (On a water-loop heat pump system, heat and cooling is provided by water-source heat pumps.) Heat also can be provided by various types of convectors. Air handlers, unit ventilators and some types of fan-coil units can be used to supply outdoor air to the conditioned space. Some of these devices can provide economizer cycle cooling. Filters are usually installed in air handlers, unit ventilators, and fan-coil equipment. Filters usually are not provided with unit heaters or convection equipment.

Fittings and Hydronic Accessories

Water-side fittings and accessories are required to protect the equipment and to control the system. Piping hardware includes check valves, strainers, drains, shut-off valves, control

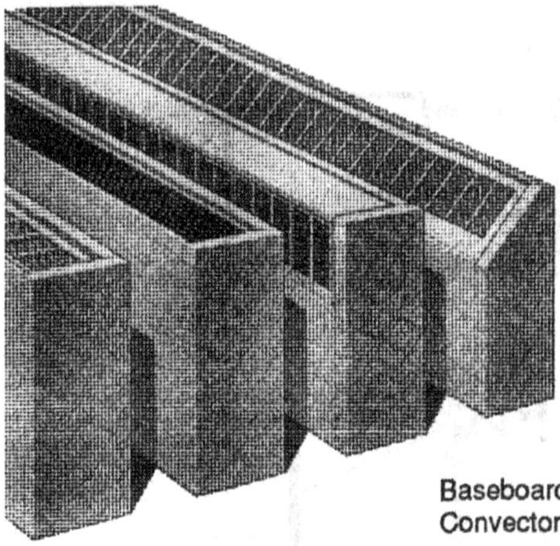

Baseboard Convectors

Boiler

Air-Cooled Chiller

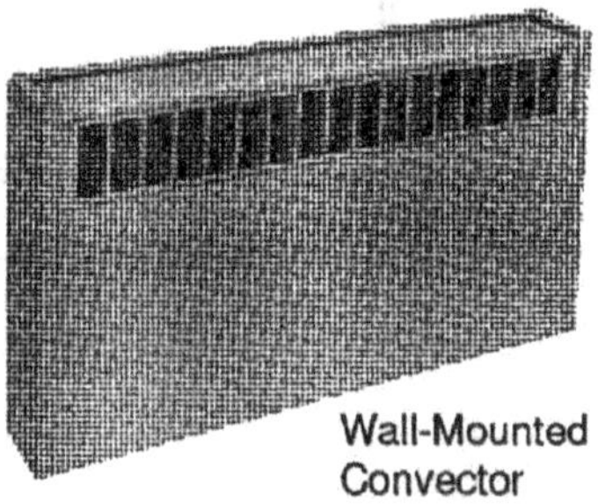

Wall-Mounted Convector

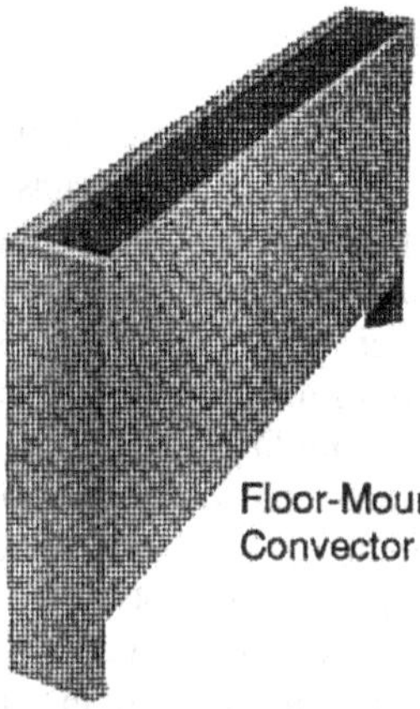

Floor-Mounted Convector

Ceiling-Mounted Unit Ventilator

Ceiling-Mounted Fan-Coil Unit

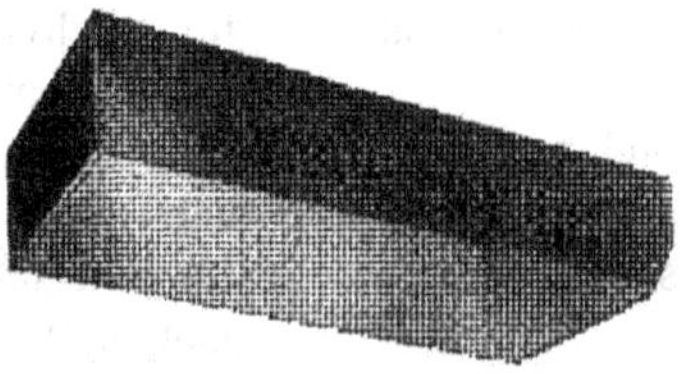

Unit Ventilator

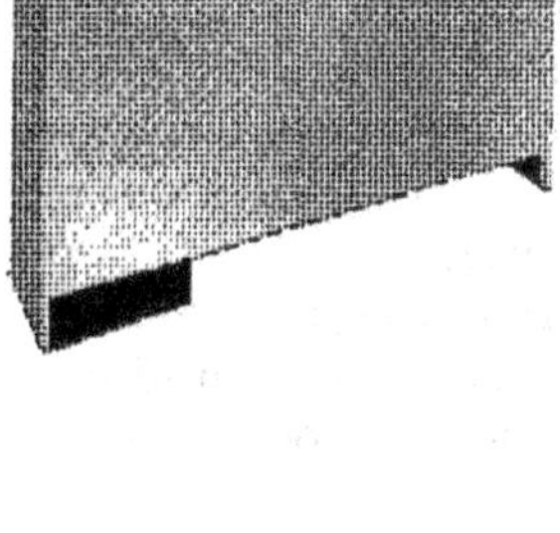

Wall-Mounted Fan-Coil Unit

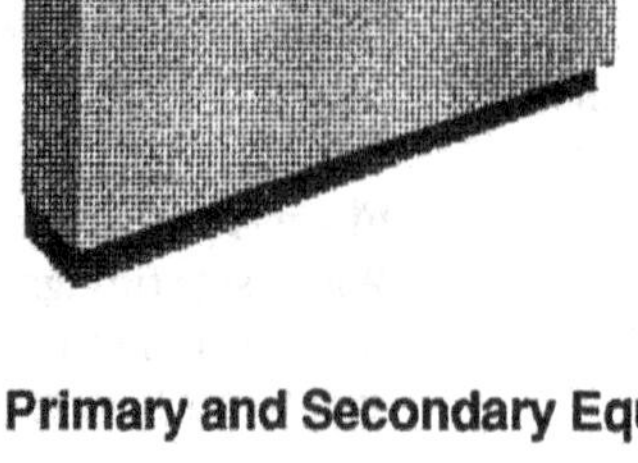

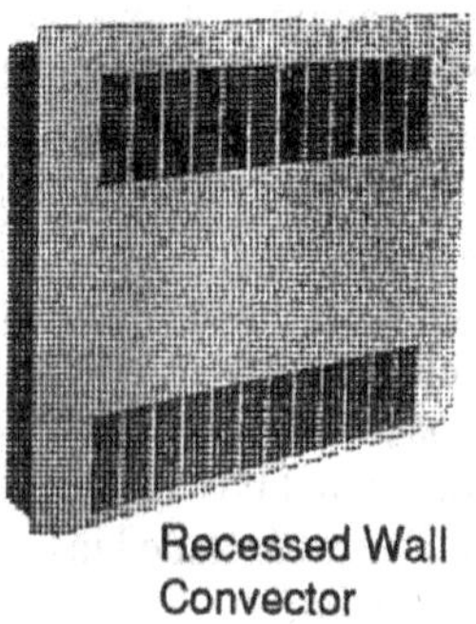

Recessed Wall Convector

Primary and Secondary Equipment

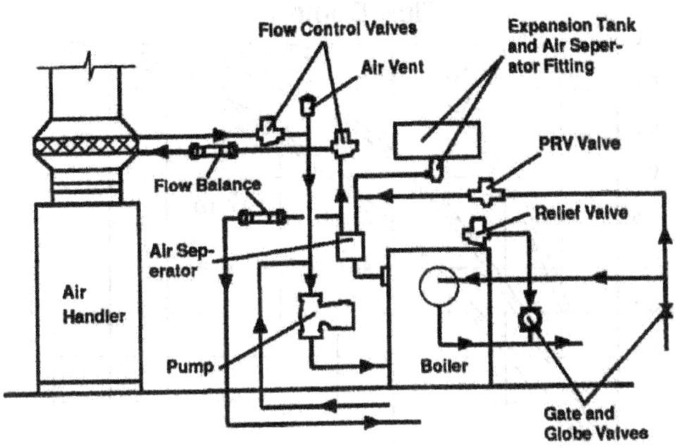

Accessories and Fittings at Central Equipment

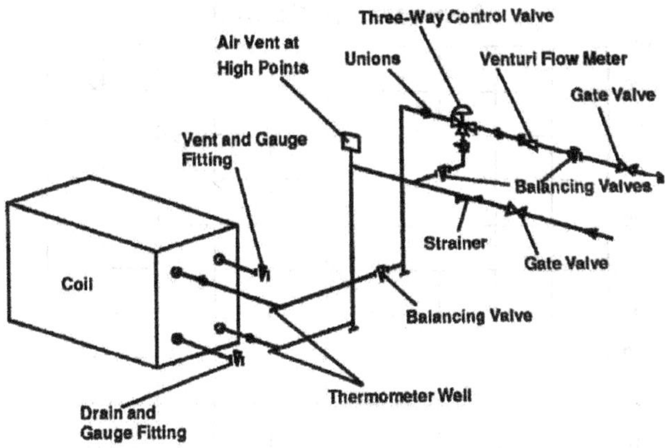

Accessories and Fittings at Remote Equipment

valves, relief valves, balancing valves, gages and flexible connections. Hydronic accessories include the expansion tank, pressure reducing valves, air-separators and air-vents.

Design Loads

The maximum heating and cooling loads must be estimated for each zone. Estimates of the minimum zone cooling loads and the heating-cooling changeover temperatures also are required. In addition, it is necessary to calculate the design loads that are associated with the central equipment. Estimates of the maximum and minimum cooling loads on the central chiller and an estimate of the maximum heating load on the boiler are necessary. The load calculations that apply to each type of water system will be discussed in the following sections.

Pump Performance

Pumps must be sized to deliver the design flow (GPM) against the maximum resistance (or head—which is measured in feet of water) that is associated with the piping circuit(s). This resistance is determined by the pressure drops that are associated with the primary and secondary equipment, the pipe runs, the control valves and the piping accessories. Appendix 4 in this manual provides a summary of the design work that is associated with pump selection and pipe sizing. Pump manuals, piping manuals and hydronics manuals provide detailed information about the procedures and calculations that are associated with designing water piping systems.

Control Valves Affect System Pressures

Designers usually try to maintain constant conditions (GPM and Head) at the pump and reasonably constant pressures at the control valves. Therefore, the type of control valve that is used to adjust the capacity of the secondary equipment is very important. Two-way throttling valves and shut-off valves are undesirable because they cause the head and flow rate (at the pump) to vary when they adjust the flow through the coil(s). The pressure and the flow rate at other points in the system also are affected by two-way valves. Three-way valves (mixing or diverting) adjust the flow through the coil by bypassing some or all the supply water around the coil. This bypass arrangement does not seriously affect the flow or head at the pump or at other points in the system.

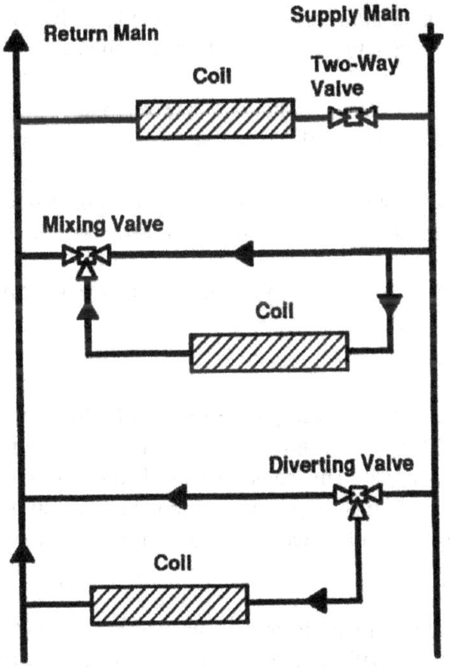

Two-way and Three-way Valves

Water-Side Control at the Secondary Equipment

The capacity of a forced-air coil or a convective device can be controlled from the water-side by adjusting the flow

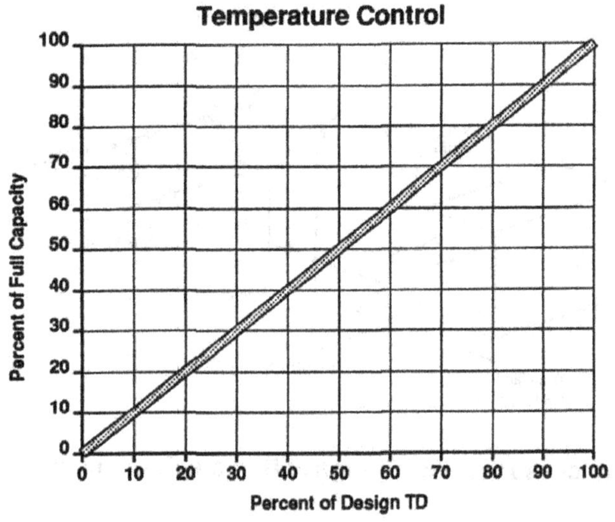

Figure 12-3

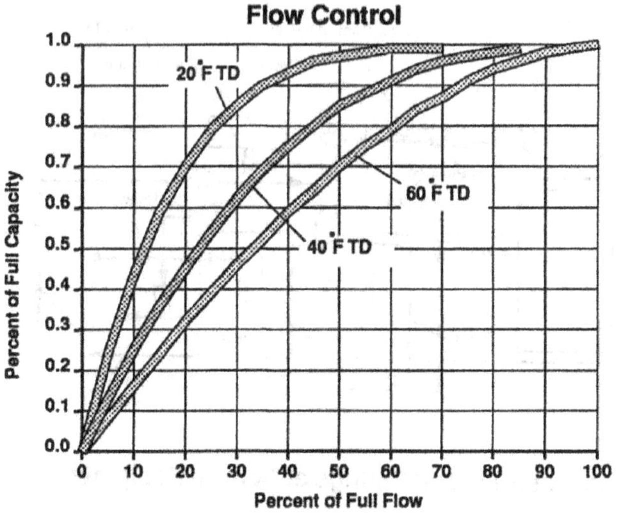

Figure 12-4

(GPM) through the coil or by changing the temperature of the water entering the coil. Temperature control (Figure 12-3) has an advantage over flow control (Figure 12-4) because temperature control is linear and flow control can be very nonlinear. However, flow control is far more common because:

- At any particular operating condition, a single (supply) water temperature may not satisfy the load that is associated with each piece of secondary equipment.
- Flow control devices (valves) are relatively simple and inexpensive to install.
- Equal percentage valves can be used to compensate for the nonlinearity that is associated with flow control.
- Flow control is reasonably linear when it is applied to chilled water coils.

Equipment Room Piping — Zoning and Changeover
Figures 12-5 and 12-6 show how control valves or pumps can be used to zone small hydronic systems. In this case, each zone is served by a series loop that is controlled by a pump or by a valve. These type of systems are usually limited to a few zones (loops).

Multiple pumps are preferred to valves because pumps insure that each loop will operate as an independent circuit. This means that the pumping requirements (flow rate and head) that are associated with any pump will not be affected when the other pumps are started and stopped.

If control valves are used, only one pump will be required to supply water to the parallel loops. In this case, the pumping requirements could be affected as the zone valves open and close. (If two-way valves are used for zone control, the pumping requirements will vary as the valves open and close. However, the pumping requirements remain nearly constant when supply water is bypassed to the return main through three-way valves.)

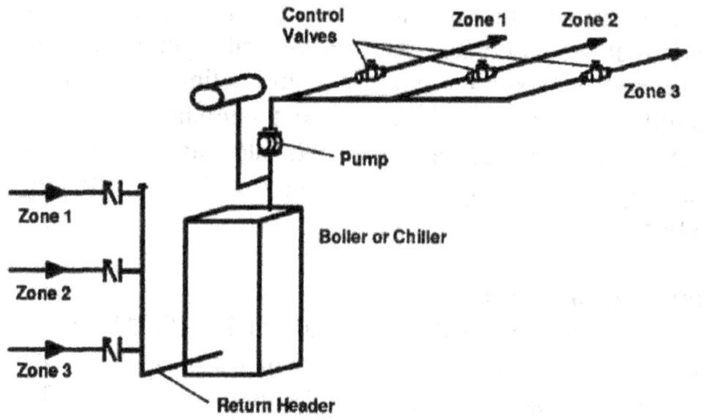

Figure 12-5

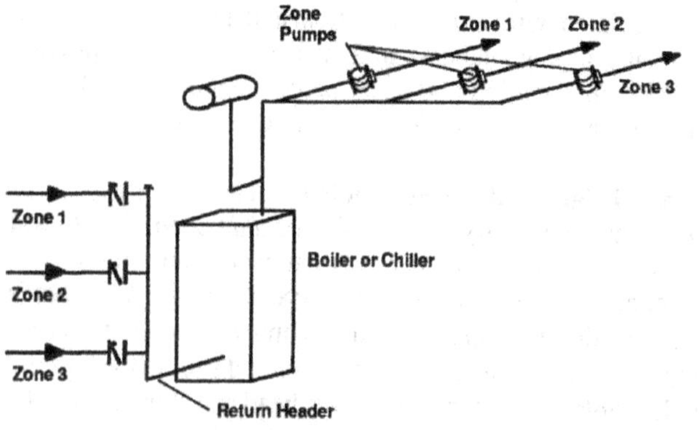

Figure 12-6

Equipment Room Piping — Changeover
The temperature of the water entering a chiller must never be too warm and the temperature of the water entering the boiler must never be too cold. This is usually not a problem with simple heating-only or cooling-only systems, but this can be a problem with changeover systems or with systems that have a large temperature drop across the primary circuit. Fortunately, special piping and pumping arrangements have been developed to control the temperature of the water that is returned to a boiler or a chiller. Figure 12-7 below provides a diagram that shows how equipment room piping circuits can be used to control the flow rate and the water temperature at the boiler and at the chiller. This method — which uses a special loop to mix system return water with boiler/chiller supply water — and a number of other methods are discussed in detail in hydronic system design manuals.

Control Systems
Controls are required for each piece of secondary equipment that is installed in a room or zone. The capacity of hydronic equipment can be controlled by water valves, fan speed changes or by supply air dampers. The capacity of water-source heat pump equipment can be controlled by starting and stopping the compressor or by adjusting the capacity of the compressor. Outdoor air, return air and exhaust air dampers must also be controlled when the local equipment is used to provide fresh air. Economizer controls also may be required on some types of secondary equipment.

Central controls are required to start and stop the chiller, heat rejection device, boiler and pump(s). Other controls may be required to initiate changeover and to supervise changeover. Central controls also are required to maintain supply water temperature setpoints at the primary equipment. Additional controls can be used to optimize the supply water temperature setpoints. Normally, the chiller and the boiler will require capacity controls. Various types of interlocks, safety and operating controls also are required. Additional comments about how specific types of all-water systems control temperature, humidity and air quality will be provided in the following sections. Refer to hydronics design manuals for detailed information about system controls.

12-2 Series-Loop and One-Pipe Systems

This section provides an overview of the performance characteristics, hardware, design procedures and control cycles that are associated with series-loop and one-pipe systems. Diagrams of these systems are provided on the next page. The information provided by Section 12-1 also applies to these systems

The important thing to notice about these systems is that the difference between the water temperature and the room temperature continuously decreases as the water flows through

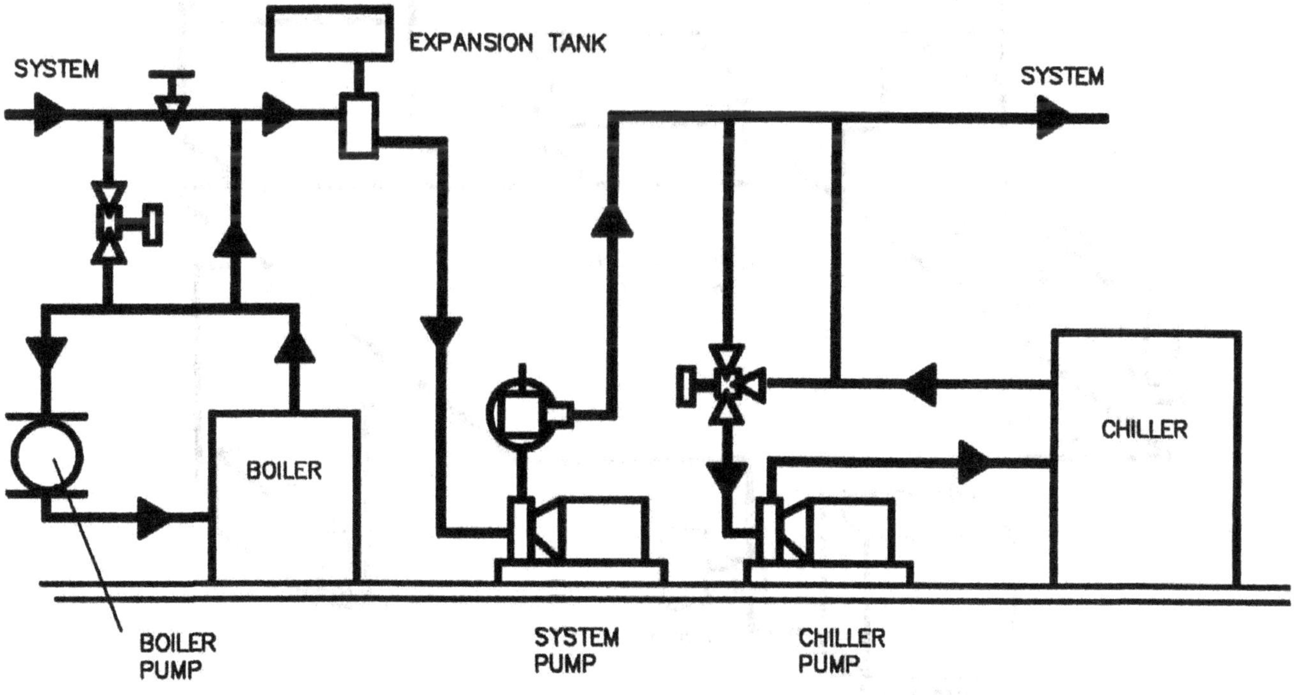

Figure 12-7

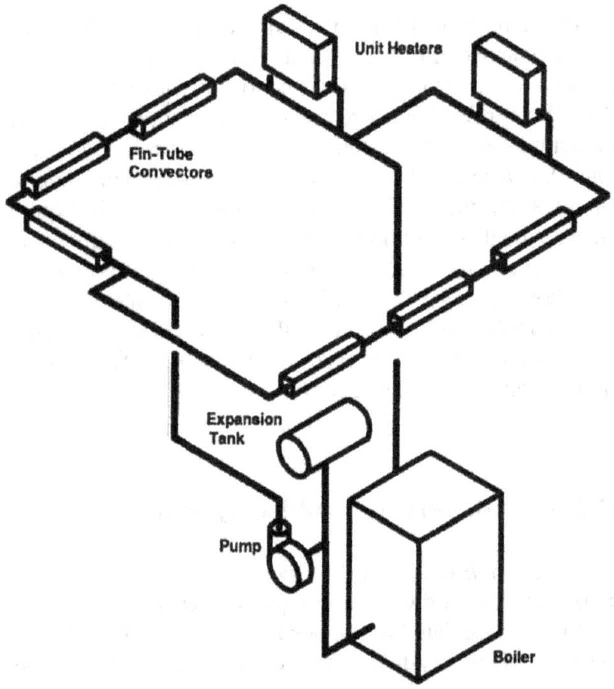

Series-Loop System (Two Circuits)

the loop. Therefore, the heating potential of the supply water and the output of the heating devices (usually baseboard radiation of cabinet convectors) continuously decrease in the direction of flow.

Also notice that, on series-loop systems, the output capacity of a particular device cannot be controlled by adjusting the flow rate or the supply water temperature because these changes would affect the performance of the other devices. (Air-side dampers can be used to provide some control over the capacity of a single device.) However, the one-pipe system does provide an opportunity to control individual devices. In this case, a two-position control valve can be installed in the branch riser.

Function
Series-loop and one-pipe systems are used to provide heat. If more than one zone is involved, a separate loop should be provided for each zone. Other types of systems should be considered if cooling or ventilation is required.

Application
Series-loop and one-pipe systems are used to heat small buildings or small areas within a large building. These systems are not recommended for heating large single-zone or multizone areas or for cooling applications.

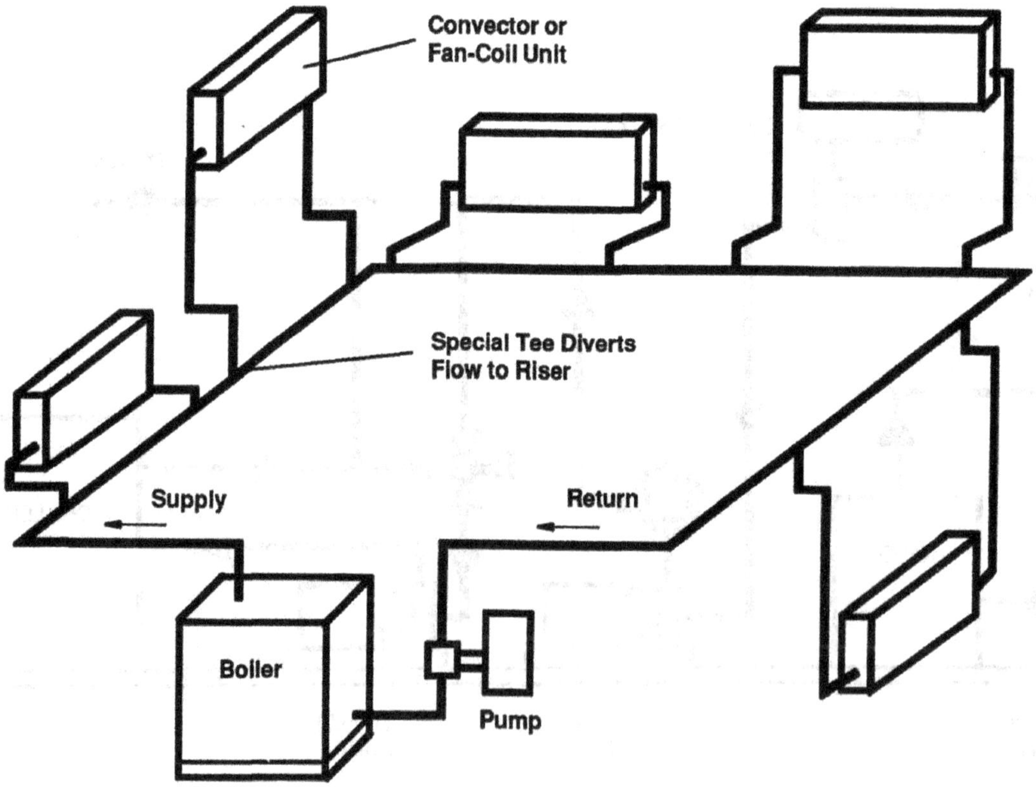

One-Pipe System

The total temperature drop around the loop provides a good indication of whether one of these systems is suitable for heating a particular space. Ideally, the system should be designed for a temperature drop of 20°F or less. This 20°F value will be compatible with the temperature rise across the boiler (refer to Section 5-4) and all the devices on the loop will have about the same heat output per unit of heat-transfer-surface-area.

Characteristics
The characteristics that are generic to all types of water systems are listed in Section 12-1 — refer to the paragraph titled *Characteristics*. The characteristics that are peculiar to series-loop and one-pipe systems are listed below.

- These systems are best suited for heating-only applications. (Cooling and ventilation has to be provided by other systems.)
- These systems are good for heating small areas.
- Multiple (parallel) loops can be used to heat more than one zone.
- Room air circulation rates are induced by convective forces.
- A constant flow (GPM) of water circulates through each loop and through each device.
- The temperature of the (boiler) supply water is not adjusted to track changes in the heating load.
- The controls that are used to adjust the capacity of individual heating devices are crude.
- Special tees are used on one-pipe systems to divert water into the risers.

Equipment
The primary equipment consists of a boiler and one or more pumps. The secondary equipment consists of fin-tube radiation, cabinet convectors and hydronic accessories.

Sizing Loads — Zone Heating
Usually there is one piping loop per zone and usually there is no outdoor air load imposed on the equipment that is connected to a loop. Therefore, the heating load that is associated with a zone (or loop) will correspond to the values that are found on line 16 of Form N-1. (Add the value from line 18 if outdoor air is heated by the equipment that is connected to the loop.) One Form N-1 is required for each zone (loop).

Boiler Sizing Load — Occupied
The design heating load at the boiler is equal to the sum of the zone (or loop) loads. If there is an outdoor air load associated with supplemental ventilating equipment, it should be added to the boiler load. An adjustment for piping losses is not included on Form N-1, but these losses should be estimated and added to the boiler load. Designers usually do not take credit for heat added by internal loads. Many designers do include an adjustment for what is known as a "pick-up load." (Some building codes specify assumptions and adjustments that must be incorporated into the boiler sizing calculations.)

Night-Cycle Heating Loads
The (unoccupied) night-cycle heating load will depend on whether the zone thermostats are set back. If there are any outdoor air loads associated with the system, they will be smaller (still considering damper leakage) if outdoor air is not required during the night cycle. Piping losses should be added to the other night-cycle loads.

Water Temperatures
At the boiler, the hot water supply temperature (*HWS*) usually ranges between 160°F and 210°F, but lower and higher temperatures can be used. The water temperature at the entrance to a heating device (*EWT*) will vary, depending on it's position on the loop. These entering water temperatures can be calculated, but it is acceptable to use the mean loop temperature (*MLT*) when the total temperature drop across the loop (*LTD*) is 20°F or less.

$$EWT = HWS - \frac{\text{Total load on upstream units (BTUH)}}{500 \times \text{Loop GPM}}$$

$$MLT = HWS - LTD / 2$$

The temperature rise across the boiler is usually 20°F or less. The temperature drop across a heating device or a zone loop usually ranges from a few degrees to 20°F, but it is possible to design for larger temperature drops. It is the designer's responsibility to select operating temperatures that are compatible with the heating loads, the equipment performance characteristics and the piping layout. Refer to Section 5-4 for additional information about water temperatures.

Equipment Selection
The designer must certify that the heating capacity of each device is adequate for any operating condition and that the various operating temperatures and water flow rates are within acceptable ranges. Comprehensive engineering data should be used to select the equipment. The equipment performance data should always correspond to the actual operating conditions that are expected to occur at the site. Part-load performance also must be evaluated and an appropriate method of capacity control should be provided. The procedures that are used to select hot water heating equipment are outlined in Sections 7-11, -12, -13 and -14 of this manual.

Control Strategy
Separate loops (which should be isolated from each other) provide the ability to supply heat on a zone-by-zone basis. The heating capacity of each loop can be controlled by switching

the zone pump on and off or by opening and closing a zone control valve. Refer to the paragraph that discusses equipment room piping on page 12-6.

The capacity of a convection device that is installed on a series-loop system can be controlled by an air-side damper. On a one-pipe system, the capacity of a heating device can be controlled by an air-side damper or by installing a two-position valve in the riser.

The temperature of the supply water leaving the boiler can either be maintained at a constant setpoint or it can be reset (scheduled) by the temperature of the outdoor air. All the water flow rates (GPM) are constant. The boiler should be on-line whenever one or more zones require heat.

12-3 Two-Pipe Systems

This section provides an overview of the performance characteristics and hardware that is associated with two-pipe systems. Section 12-6 provides information about calculation procedures, design temperatures and control cycles. The information provided by Section 12-1 also applies to these systems.

With the two-pipe system, the same water temperature (ignoring pipe losses) can be maintained at the entrance to every piece of secondary equipment. Therefore, the full heating or cooling potential of the supply water is available at the entrance to every heat transfer device.

Function

Two-pipe systems can provide heat or they can provide sensible and latent cooling, but they cannot provide simultaneous heating and cooling. Secondary equipment can be used to provide outdoor air for ventilation. Winter humidification and make up air must be provided by other systems.

Application

If simultaneous heating and cooling is not required, two-pipe systems can be used to heat and cool any number of zones in any size building. (Comfort will be compromised during the intermediate seasons if a two-pipe system is installed in a building that does require simultaneous heating and cooling.) Two-pipe systems are particularly useful for zoning a building that has many small rooms or zones. These systems do not provide close control over the indoor humidity. They usually have low efficiency air filters and they are not appropriate for applications that require make up air. Two-pipe heating systems are often used with an all-air cooling system.

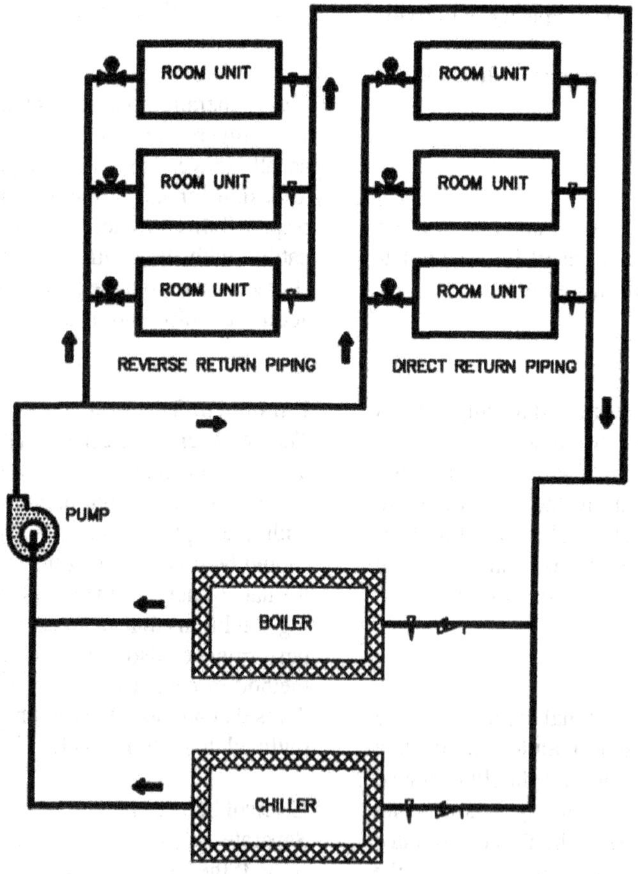

Two-Pipe System

Characteristics

The characteristics that are generic to all types of water systems are listed in Section 12-1 — refer to the paragraph titled *Characteristics*. The characteristics that are peculiar to two-pipe systems are listed below.

- Two-pipe systems can accommodate any number of zones.
- Two-pipe systems are not appropriate for applications that require simultaneous heating and cooling.
- Control valves can be used to adjust the (heating or cooling) capacity of the secondary equipment.
- Fan speed adjustments can be used to control the capacity of the secondary equipment.
- Outdoor air can be introduced through the secondary equipment.
- An economizer cycle is provided with some types of secondary equipment.
- Usually, the temperature of the boiler supply water is not adjusted to track the changes in the heating load.
- The expansion tank must be correctly sized for the volume changes that are associated with the changeover from chilled water to hot water.
- The equipment room piping details and controls must be designed to prevent the boiler and the chiller from being damaged during changeover.

Equipment

The primary equipment consists of a boiler, chiller and one or more pumps. The secondary equipment could consist of any of the devices mentioned in Section 12-1.

Changeover

The two-pipe system should not be used if simultaneous heating and chilled water cooling is required. (Simultaneous heating with economizer cooling is possible.) Changeover to the heating mode should occur when the critical rooms or zones require heat.

During changeover, there will be a large swing in the temperature of the water that flows through the distribution system. Special piping arrangements are required to implement a rapid changeover without loosing control over the temperature of the water entering the chiller or boiler. (The range of suitable return water temperatures at the chiller or boiler is usually specified by the equipment manufacturer.) Changeover piping and pumping layouts usually isolate the boiler circuit and the chiller circuit from the main distribution circuit. Figure 12-7 on page 12-7 provides an example of such an arrangement. Refer to hydronic system design manuals for comprehensive information about changeover piping arrangements.

12-4 Three-Pipe Systems

This section provides an overview of the performance characteristics, hardware and control cycles that are associated with the three-pipe system. Section 12-6 provides information about calculation procedures, design temperatures and control cycles. The information provided by Section 12-1 also applies to these systems.

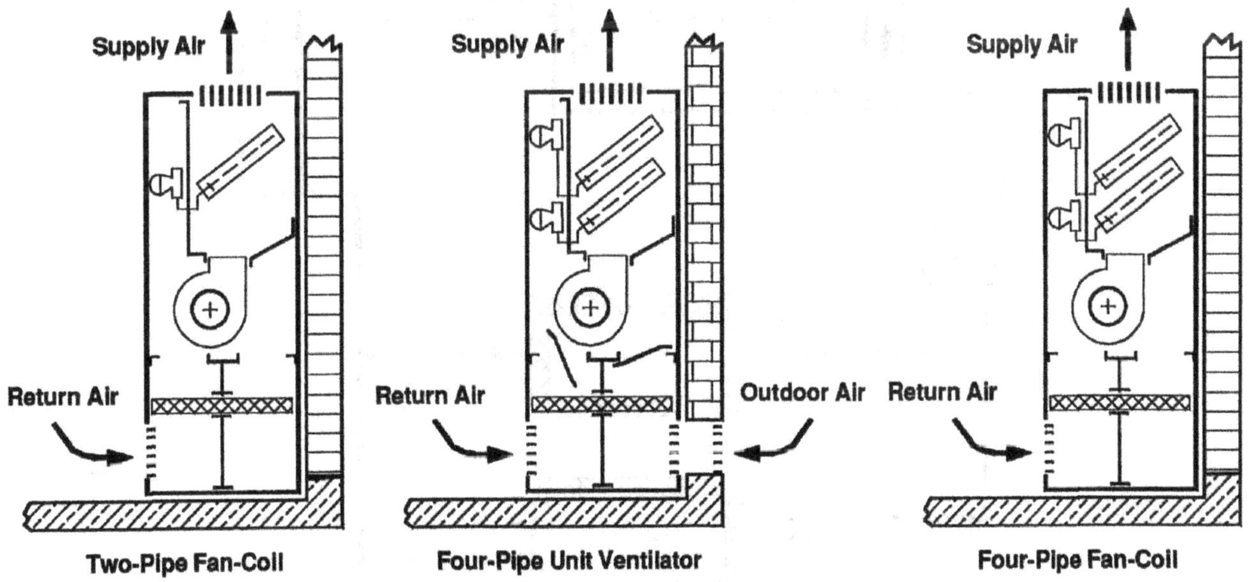

Unit Ventilator and Fan-Coil Units

The most important characteristic of the three-pipe design is that it has a common return. A common return reduces the amount of pipe that must be installed, but it wastes energy because hot water must be mixed with cold water. The common return also requires the designer to make sure that a suitable return water temperature at the chiller or boiler is maintained during any mode of operation.

Function

Three-pipe systems can provide heat or cooling and they can provide simultaneous heating and cooling. Secondary equipment can be used to provide outdoor air for ventilation. Winter humidification and make up air must be provided by other systems.

Application

Three-pipe system are not recommended because they waste energy. (Four-pipe systems provide a more efficient alternative.) Three-pipe systems have been used to heat and cool buildings that have many separate rooms and zones. They were created to solve the problems that are associated with simultaneous heating and cooling. These systems do not provide close control over the indoor humidity. They usually have low efficiency air filters and they are not appropriate for applications that require make up air.

Characteristics

The characteristics that are generic to all types of water systems are listed in Section 12-1. The characteristics that are peculiar to three-pipe systems are listed below.

- Three-pipe systems waste energy because they mix hot and cold water in the return pipe.
- These systems can accommodate any number of zones.
- These systems are appropriate for applications that require simultaneous heating and cooling.
- Control valves can be used to adjust the (heating or cooling) capacity of the secondary equipment.
- Usually, the terminal units have one coil and two supply water valves. The hot water valve closes before the chilled water valve opens.
- Fan speed adjustments can be used to control the capacity of the secondary equipment.
- Outdoor air can be introduced through the secondary equipment.

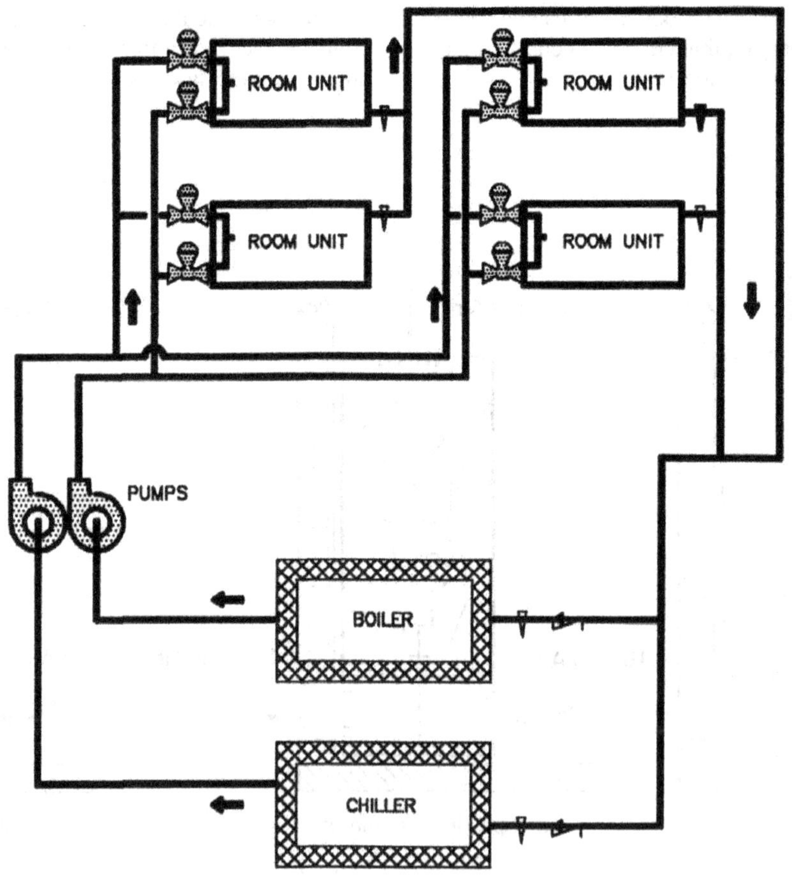

Three-Pipe System

- An economizer cycle is included with some types of secondary equipment.
- The temperature of the (boiler) supply water is (usually) not adjusted to track the changes in the heating load.
- The expansion tank must be correctly sized for the temperature changes that are associated with mixing hot and cold water.
- The equipment room piping details and controls must be designed to prevent the boiler and the chiller from being damaged by return water that is too cold or too hot.

Equipment

The primary equipment consists of a boiler, chiller and one or more pumps. The secondary equipment could consist of any of the devices mentioned in Section 12-1.

Changeover

Changeover at a terminal unit takes place automatically. The local control system modulates the heating (or cooling) valve from the full open position to the full closed position before the cooling (or heating) valve begins to open. And, there may be a deadband between the heating mode and the cooling mode. (Both valve ports may be contained in a single valve body.) The local control system also can incorporate economizer control cycles into the operating sequences.

There could be a large swing in the return water temperature when the boiler or chiller is put on-line or taken off-line. Special piping arrangements are required to implement a rapid changeover without loosing control over the temperature of the water entering the chiller or boiler. (The temperature of the water entering the chiller of boiler must be within the range that is recommended by the manufacturer during any possible operating condition.) These piping and pumping layouts usually isolate the boiler circuit and the chiller circuit from the main distribution circuit. Refer to hydronic system design manuals for more information about changeover piping arrangements.

12-5 Four-Pipe Systems

This section provides an overview of the performance characteristics, hardware and control cycles that are associated with a four-pipe system. Section 12-6 provides information about calculation procedures, design temperatures and control cycles. The information provided by Section 12-1 also applies to these systems.

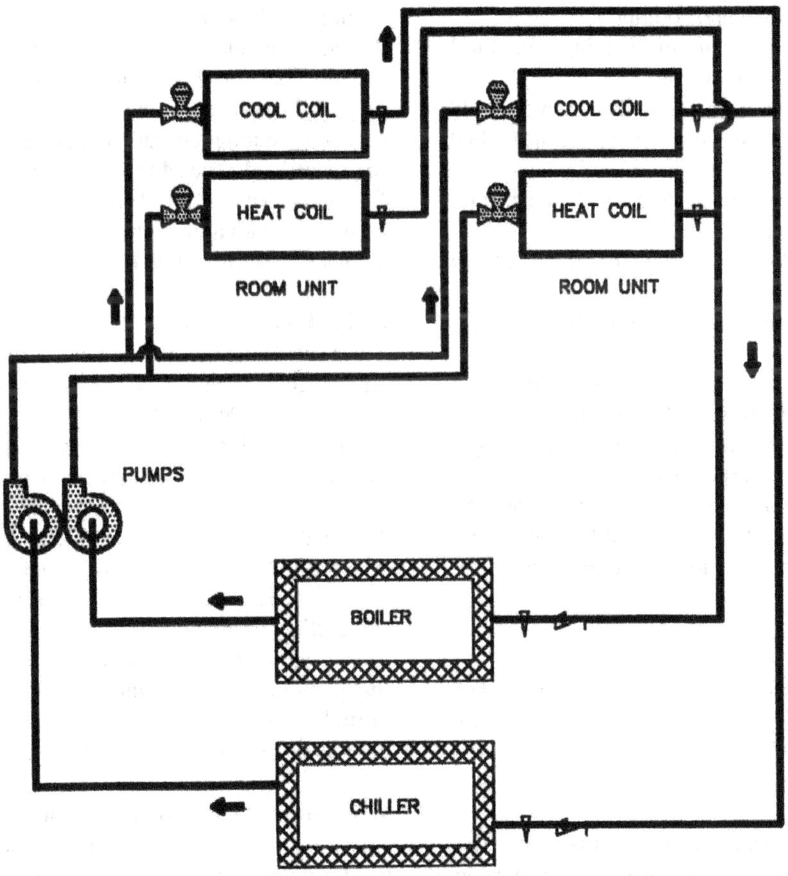

Four-Pipe System

The most important feature of the four-pipe design is that it can provide simultaneous heating and cooling without wasting energy. And, since the heating and cooling piping circuits are completely separate, there are no problems associated with wide swings in the return water temperature.

Function
Four-pipe systems can provide heat or cooling and they can provide simultaneous heating and cooling. Secondary equipment can be used to provide outdoor air for ventilation. Winter humidification and make up air must be provided by other systems.

Application
Four-pipe systems can be used to heat and cool buildings that have many separate rooms and zones. They are designed to provide simultaneous heating and cooling. These systems do not provide close control over the indoor humidity. They usually have low efficiency air filters and they are not appropriate for applications that require make up air.

Characteristics
The characteristics that are generic to all types of water systems are listed in Section 12-1. The characteristics that are peculiar to four-pipe systems are listed below.

- Four-pipe systems can accommodate any number of zones.
- Four-pipe systems are appropriate for applications that require simultaneous heating and cooling.
- Control valves can be used to adjust the (heating or cooling) capacity of the secondary equipment.
- Usually, the terminal units have two coils and two supply water valves. The hot water valve closes before the chilled water valve opens.
- Fan speed adjustments can be used to control the capacity of the secondary equipment.
- Outdoor air can be introduced through the secondary equipment.
- An economizer cycle is included on some types of secondary equipment.
- Usually, the temperature of the boiler supply water is not adjusted to track the changes in the heating load.
- The expansion tank is sized for the temperature changes that are associated with hot water system.
- The equipment room piping details and controls must be designed to prevent the boiler and the chiller from being damaged by return water that is too cold or too hot.

Equipment
The primary equipment consists of a boiler, chiller and one or more pumps. The secondary equipment could consist of any of the devices mentioned in Section 12-1.

Changeover
Changeover at a terminal unit takes place automatically. The local control system modulates the heating (or cooling) valve from the full open position to the full closed position before the cooling (or heating) valve begins to open. And, there may be a deadband between the heating mode and the cooling mode. The local control system also can incorporate economizer control cycles into the operating sequences.

The chiller and boiler can be put on-line or taken off-line as they are required. Special equipment room piping arrangements may be required to control the temperature of the water entering the chiller or boiler. (The temperature of the water entering the chiller of boiler must be within the range that is recommended by the manufacturer during any possible operating condition.) These piping and pumping layouts usually isolate the boiler circuit and the chiller circuit from the main distribution circuit. Refer to hydronic system design manuals for more information about equipment room piping arrangements.

12-6 Designing 2-, 3- and 4-Pipe Systems

This section provides information about the calculations and procedures that are used to design two-, three-, and four-pipe systems. This section also provides information about water-side operating temperatures and control strategies. Refer to Sections 12-3, -4 and -5 for additional information about a particular system.

Sizing Loads — Zone Cooling
The sensible and latent sizing loads that are associated with the secondary cooling equipment are based on the peak zone or peak room load. If outdoor air is introduced through the secondary equipment, the outdoor air load must be added to the space load. One **Manual N**, Form N-1, must be filled out for each room or zone. The sensible and latent loads that are associated with a given room or zone will correspond to the values that are found on line 12 of Form N-1.

Sizing Load — Chiller
Since distributed chilled water cooling equipment can take advantage of diversity, the cooling capacity that is required at the central chiller can be based on the peak block load. One **Manual N**, Form N-1, must be filled out for the whole building. The sensible and latent loads at the chiller will correspond to the values that are found on line 12, of Form N-1. In this case, the outdoor air load (line 9) should be based on the outdoor air CFM that is introduced through the secondary equipment. (If a separate all-air ventilation system is on the chiller circuit, the loads that are associated with this system should be added to the other cooling loads.) Also, note that pumping heat and piping gains are not accounted for on Form N-1. These loads should be estimated and added to the other loads that are imposed on the chiller.

Minimum Cooling Loads — Zones
An analysis of the behavior of the zone cooling loads allows the designer to determine if any zone requires cooling during cold weather. This analysis also allows the designer to predict

when the zones changeover from cooling to heating. This information is useful for formulating control strategies.

If outdoor air is introduced through the local equipment, the room or zone cooling load is equal to the space load plus the outdoor air load. If outdoor air is not introduced through the local equipment, the room or zone cooling load is just equal to the space load. In either case, the behavior of the zone load can be examined by making a load-line analysis. (Refer to Section 4-10 and 4-11 of this manual, also refer to Section 4-16.) As explained below, if outdoor air is involved, the minimum coil load also will depend on the strategy that is used to control the outdoor air damper. (Figure 11-1, on page 11-4, summarizes three basic outdoor air control strategies.)

No Economizer
If no economizer cycle is used, the minimum load at the secondary equipment will depend on whether the load line crosses over from cooling to heating.

- If a cooling-heating changeover occurs (with no credit for solar gains), the minimum load at the secondary equipment will be zero. (Refer to page 4-15, Figure 4-12.)

- If a cooling-heating changeover does not occur (with no credit for solar gains), the minimum load at the secondary equipment will be equal to the load that occurs when the outdoor temperature is equal to the winter design temperature.

Economizer
If an economizer is installed, the minimum cooling load at the secondary equipment (with no credit for solar gains) will occur just before the system switches to the economizer cycle. In this case, the minimum cooling load will equal to the load that occurs when the outdoor temperature is equal to economizer cycle changeover temperature (Refer to page 4-15, Figure 4-12).

Minimum Chiller Load — Two- and Four-Pipe Systems
The minimum load on the chiller will occur just before the outdoor temperature gets low enough to make chilled water unnecessary. This changeover temperature will depend on the outdoor air capabilities of the secondary cooling equipment and the behavior of the room or zone loads.

- If all the secondary cooling equipment is equipped with economizer cycles, the chiller can be taken off-line when the outdoor air is cold enough (around 55°F) to use economizer cycle cooling.

- If some or all of the secondary cooling equipment is not equipped with an economizer cycle, the chiller can be taken off-line when the outdoor temperature gets low enough to make cooling unnecessary. This changeover can occur when there are no critical rooms or zones that require cooling.

- It is possible that chilled water cooling will be necessary during the coldest weather. For example, chilled water cooling could be required for an internal zone. In this case, the chilled water system cannot be shut down. (It is possible to decommission the chiller and to use the outdoor air to produce the chilled water. This capability requires especially designed heat exchanges and a water tower.)

In either of the three situations described above, the minimum load on the chiller will occur just before the chiller is removed from duty. This load will be equal to the sum of the cooling loads that are associated with the various rooms or zones. These room or zone loads must be evaluated at the outdoor temperature that corresponds to the changeover temperature.

A block load analysis can be used to estimate the minimum chiller load when all of, or none of, the secondary equipment is equipped with economizers. A room-by-room analysis is required if some of the rooms are cooled by an economizer cycle while other rooms are cooled by chilled water. Load lines can be used to evaluate room, zone or block loads at any outdoor temperature.

Minimum Chiller Load — Three-Pipe Systems
The comments that were made above for the two- and four-pipe systems also apply to the three-pipe system. However, on a three-pipe system, the minimum chiller load also will be affected by the mixing that occurs in the return piping. This mixing load can be substantial and it must be added to the other chiller loads. An estimate of the flow rate (GPM) and temperature of the water that is leaving each piece of secondary equipment is required to calculate this mixing load.

Night-Cycle Cooling
In many applications, the cooling system can be shut down and the outdoor air dampers can be closed when the building is not occupied. This control strategy reduces operating expense.

If the cooling system has to operate when the building is unoccupied, a set-up control strategy would reduce energy consumption. Usually, outdoor air is not required during the night cycle, so the outdoor air dampers can be closed to obtain additional energy savings. However, economizer cycles should still be used whenever the condition of the outdoor air is suitable for economizer cooling. Internal loads are usually smaller during the night cycle, so the minimum load on the cooling coil could be less than the minimum day-cycle load. An analysis of both operating cycles is required to determine which cycle produces the smallest load on the chiller.

Sizing Loads — Zone Heating
The heating loads that are associated with the secondary equipment are based on the zone or room load. If outdoor air is introduced through the secondary equipment, the outdoor air load must be added to the space load. One **Manual N**, Form N-1, must be filled out for each room or zone. The load that

is associated with a given room or zone will correspond to the value that is found on line 21 of Form N-1.

Boiler Sizing Load — Two- and Four-Pipe Systems

The design heating load at the boiler is equal to the sum of the room or zone loads. If there is an outdoor air load associated with a separate ventilating system that is served by the boiler, it should be added to the boiler load. An adjustment for piping losses is not included on Form N-1, but these losses should be estimated and added to the boiler load. Designers usually do not take credit for heat added by internal loads. Many designers do include an adjustment for what is known as a "pick-up load." (Some building codes specify assumptions and adjustments that must be incorporated into the boiler sizing calculations.)

Boiler Sizing Load — Three-Pipe System

The comments that were made above for the two- and four-pipe systems also apply to the three-pipe system. However, on a three-pipe system, the boiler load also will be affected by the mixing that occurs in the return piping. This mixing load can be substantial and it must be added to the other boiler loads. An estimate of the flow rate (GPM) and temperature of the water that is leaving each piece of secondary equipment is required to calculate this mixing load.

Night-Cycle Heating Loads

The size of the night-cycle loads will depend on whether the space temperatures are set back during the night cycle. And, the outdoor air load will depend on whether or not outdoor air is required during the night cycle. If outdoor air is not required, the outdoor air damper should be closed. (Remember to allow for some damper leakage.) Piping losses should be added to the other night-cycle loads.

Supply Water Temperatures — Cooling

Chilled water supply temperatures usually range between 42°F and 46°F, but lower and higher temperatures can be used. The temperature drop across the chiller is usually 8°F to 10°F and the temperature rise across a cooling coil is usually 8°F to 12°F. It is the designer's responsibility to select operating temperatures that are compatible with the sensible and latent loads, the equipment performance characteristics and the piping layout. Refer to Section 5-3 for additional information about chilled water temperatures.

Supply Water Temperatures — Heating

At the winter design condition, the hot water supply temperatures usually range between 160°F and 210°F, but lower and higher temperatures can be used. Hot water supply temperatures can be reset (scheduled) by the outdoor air temperature. The temperature rise across the boiler is usually about 20°F and the temperature drop across a heating device could vary from a few degrees to 50°F or more. It is the designer's responsibility to select operating temperatures that are compatible with the heating loads, the equipment performance characteristics and the piping layout. Refer to Section 5-4 for additional information about hot water temperatures.

Equipment Selection

The designer must certify that the heating or cooling capacity of each device is adequate for any operating condition and that the various operating temperatures and water flow rates are within acceptable ranges. Comprehensive engineering data should be used to select the equipment. The equipment performance data should always correspond to the actual operating conditions that are expected to occur at the site. Part-load performance also must be evaluated and an appropriate method of capacity control should be provided. The procedures that are used to select hot water heating equipment are outlined in Sections 7-11, -12, -13 and -14 of this manual. The procedures that are used to select chilled water cooling equipment are outlined in Sections 6-13 and -14 of this manual.

Control Strategy

Room thermostats can be used to control the secondary equipment. The coil capacity (cooling or heating) of a fan-powered device is usually controlled by adjusting the water flow rate (GPM) and/or the fan speed. If the unit is equipped with an economizer cycle, the thermostat can be used to control the dampers and the water valves in sequence. The capacity of a convective heating device can be controlled by adjusting the water flow rate (GPM), the entering water temperature or by an air-side damper. Equipment manufacturers usually provide suitable controls as standard equipment or as an accessory item.

The temperature of the supply water leaving the chiller is normally maintained at a constant setpoint. The temperature of the water entering the chiller also must be under control. Hot water should not be allowed to enter the chiller during changeover or during any other operating condition. (Ask the chiller manufacturer for information about the maximum allowable entering water temperature.) The water flow rate (GPM) through the chiller could be constant or it could vary, depending on what type of valves are used to control the coils that are located in the secondary equipment. (The flow through the chiller should never be allowed to fall below the manufacturer's minimum GPM value.) The chiller should be maintained on a stand by status during the intermediate seasons and it should be taken out of service when cooling is no longer required.

The temperature of the supply water leaving the boiler can either be maintained at a constant temperature, reset (scheduled) by the temperature of the outdoor air or reset by other types of optimizer controls. The temperature of the water entering the boiler also must be under control. Cold water should not be allowed to enter the boiler during changeover or during any other operating condition. (Ask the boiler manufacturer for information about operating temperatures and thermal shock.) The water flow rate (GPM) through the boiler could be constant or it could vary, depending on what type of valves are used to control the coils in the secondary equipment. (The flow through the boiler should always be compatible with the temperature rise limits that are recommended by manufacturer.) The boiler should be maintained on a stand by status during the intermediate seasons and it should be taken out of service when heat is no longer required.

12-7 Water-Loop Heat Pump Systems

This section provides an overview of the performance characteristics, hardware and control cycles that are associated with a water-loop heat pump system. The information provided by Section 12-1 also applies to these systems.

The most important feature that is associated with water-loop heat pump systems is that they can reclaim energy that would otherwise be wasted. They accomplish this, during certain operating conditions, by transferring the heat that is removed from the rooms or zones that require cooling to the rooms or zones that require heat.

Function
Water-loop heat pump systems can provide heat or cooling and they can provide simultaneous heating and cooling. Some, but not all types of water-source heat pump equipment can provide outdoor air for ventilation. A separate all-air system will be required if the water-source heat pump equipment is not capable of supplying outdoor air for ventilation. Winter humidification and make up air must be provided by other systems.

Application
Water-loop heat pump systems can be used to heat and cool buildings that have many separate rooms and zones and they provide simultaneous heating and cooling. These systems can effectively reclaim energy if the building requires sizable amounts of heat and sizable amounts of simultaneous cooling for many hours per year. (These systems are often installed in buildings that do not have load characteristics that are favorable for reclaiming energy. This is done because these systems are simpler to design and to control than other types of simultaneous heating and cooling systems.) These systems do provide good humidity control, but they do not provide precise humidity control. Small, console heat pumps are usually equipped with low efficiency air filters.

Characteristics
The characteristics that are generic to all types of water systems are listed in Section 12-1. The characteristics that are peculiar to water-loop heat pump systems are listed below.

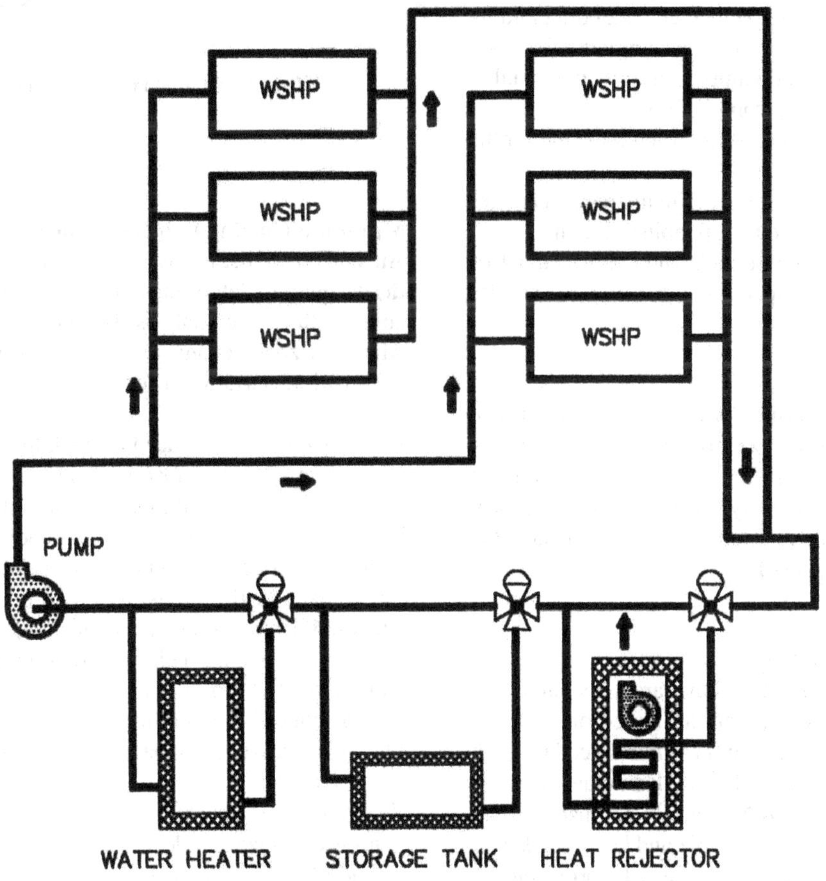

Water-Loop Heat Pump System

- Water-loop heat pump systems can accommodate any number of zones.
- Water-loop heat pump systems offer a simple way to condition buildings that require simultaneous heating and cooling.
- Outdoor air can be introduced through some types of console units (unit ventilator heat pumps) and rooftop units.
- An economizer cycle is included on some types of console units and rooftop units.
- Console units can be used to supply heat from a low, perimeter location. Other types of units supply heat from a ceiling location.
- These systems have the potential to save a considerable amount of energy in (certain types of) buildings that are located in colder climates.
- Maximum reclaim efficiency is obtained when (approximately) one third of the units are operating in the cooling mode and two thirds of the units are operating in the heating mode.
- A water storage tank will increase system efficiency if the heat that is associated with large internal loads during the day cycle can be used to heat the building during the night cycle.
- Economizer cycles increase system efficiency when all of the heat pump units are operating in the cooling mode.
- Economizer cycles can decrease system efficiency when simultaneous heating and cooling is required.
- The water loop has to be supplemented by a boiler and by heat rejection equipment because the energy that is added to the loop (by the heat pump units) is usually not equal to the energy that is extracted from the loop.
- The same piping can be used for the water loop and for the sprinkler system.
- The expansion tank is sized for the temperature changes that are associated with water-loop piping system.
- The equipment room piping details and controls must be designed to control the temperature rise across the boiler.

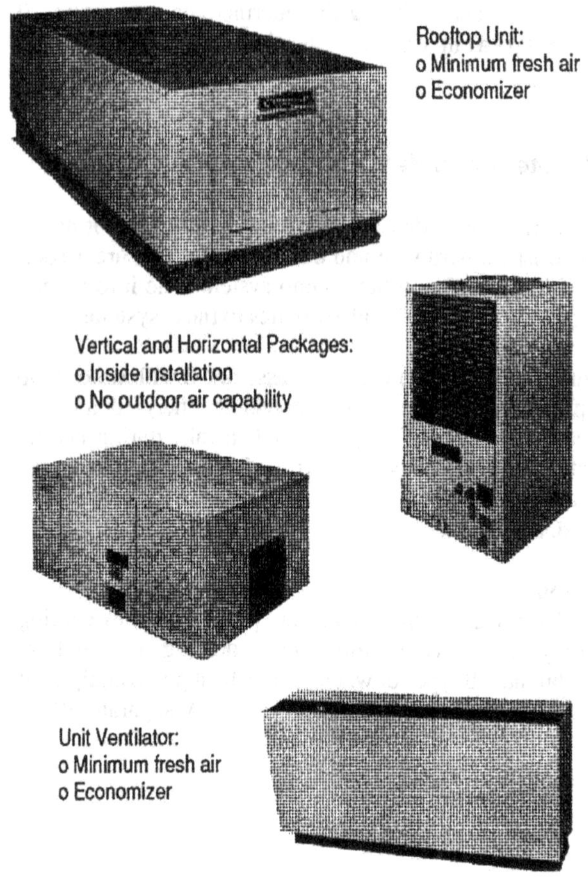

Water-Source Heat Pump Equipment

Equipment
The primary equipment consists of a boiler, a central heat rejection device (usually an evaporative cooler), system water pumps and possibly, a water storage tank. The secondary equipment consists of various types of water-source heat pumps. (Sometimes electric resistance heaters are installed in the secondary heat pump units.)

Sizing Loads — Zone Cooling
The sensible and latent sizing loads that are associated with the secondary heat pump equipment are based on the peak zone or peak room load. If outdoor air is introduced through the secondary equipment, the outdoor air load must be added to the space load. One **Manual N**, Form N-1, must be filled out for each room or zone. The sensible and latent loads that are associated with a given room or zone will correspond to the values that are found on line 12 of Form N-1.

Minimum Cooling Loads — Zones
An analysis of the behavior of the zone cooling loads allows the designer to determine if any zones require cooling during cold weather. This analysis also allows the designer to predict when the zones changeover from cooling to heating. This information is useful for formulating control strategies.

If outdoor air is introduced through the local equipment, the room or zone cooling load is equal to the space load plus the outdoor air load. If outdoor air is not introduced through the local equipment, the room or zone cooling load is just equal to the space load. In either case, the behavior of the zone load can be examined by making a load-line analysis. (Refer to Section 4-10 and 4-11 of this manual, also refer to Section 4-16.) As explained below, if outdoor air is involved, the minimum coil load also will depend on the strategy that is used to control the outdoor air damper. (Figure 11-1, page 11-4, summarizes three basic outdoor air control strategies.)

No Economizer
If no economizer cycle is used, the minimum load at the secondary equipment will depend on whether the load line crosses over from cooling to heating.

- If a cooling-heating changeover occurs, (with no credit for solar gains) the minimum load at the secondary equipment will be zero. (Refer to page 4-15, Figure 4-12.)
- If a cooling-heating changeover does not occur (with no credit for solar gains), the minimum load at the secondary equipment will be equal to the load that occurs when the outdoor temperature is equal to the winter design temperature.

Economizer
If an economizer is installed, the minimum cooling load at the secondary equipment (with no credit for solar gains) will occur just before the unit switches to the economizer cycle. In this case, the minimum cooling load will be equal to the load that occurs when the outdoor temperature is equal to changeover temperature (Refer to page 4-15, Figure 4-12).

Night-Cycle Cooling
During hot weather, the cooling system can be shut down and the outdoor air dampers can be closed when the building is not occupied. This control strategy reduces operating expense.

If the cooling system has to operate when the building is unoccupied, a set-up control strategy would reduce energy consumption. Usually, outdoor air is not required during the night cycle, so the outdoor air dampers can be closed to obtain additional energy savings.

Sizing Loads — Zone Heating
The heating loads that are associated with the secondary equipment are based on the zone or room load. If outdoor air is introduced through the secondary equipment, the outdoor air load must be added to the space load. One **Manual N**, Form N-1, must be filled out for each room or zone. The load that is associated with a given room or zone will correspond to the value that is found on line 21 of Form N-1.

Boiler Sizing Load
The largest heating load at the boiler will occur when the building is unoccupied (no internal gains) during coldest weather. Since there is little or no heat to reclaim, all the secondary units will be operating in the heating mode. Under these circumstances, the boiler should have a minimum capacity that is equal to 75 percent of the sum of the room or zone heating loads. (The remaining 25 percent of the required heat is added by the heat pump compressors.) Any outdoor air loads — unoccupied ventilation CFM or damper leakage — that is associated with this operating condition should be included in the zone loads. Any piping losses and any passive losses that are associated with the heat rejection equipment should be added to the boiler load.

Night-Cycle Heating Loads
The size of the night-cycle loads will depend on whether the space temperatures are set back during the night cycle. And, the outdoor air load will depend on whether or not outdoor air is required during the night cycle. If outdoor air is not required, the outdoor air dampers should be closed. (Remember to allow for some damper leakage.)

Equipment Selection
The designer must certify that the heating or cooling capacity of each device is adequate for any operating condition and that the various operating temperatures and water flow rates are within acceptable ranges. Comprehensive engineering data should be used to select the equipment. The equipment performance data should always correspond to the actual operating conditions that are expected to occur at the site. Part-load performance also must be evaluated and an appropriate method of capacity control should be provided. The procedures that are used to select hot water heating equipment are outlined in Sections 7-11, -12, -13 and -14 of this manual. The procedures that are used to select heat pump equipment are outlined in Section 8 of this manual.

Changeover
The cooling-heating changeover at a terminal unit takes place automatically when the local thermostat switches the operating mode of the heat pump. If economizer cycles are used, they should be locked out when simultaneous heating and cooling is required.

The boiler, the heat rejection device and the storage tank can be put on-line or taken off-line as they are required. Special piping arrangements may be required to control the temperature rise across the boiler.

Control Strategy
Room thermostats control the operating mode of the secondary heat pump equipment. The local control system also can incorporate economizer control cycles into the operating sequences. Equipment manufacturer's usually provide suitable controls as standard equipment or as an accessory item.

Water-Loop Temperature
The water-loop temperature is allowed to drift between 60°F and 90°F (approximately). The boiler controls maintain the low-limit setpoint and the heat rejection equipment controls (or the storage tank) maintain the high-limit setpoint. It is important to make sure that the temperature rise across the boiler is always under control. The boiler should be maintained on a stand by status during the intermediate seasons and it should be taken out of service when heat is no longer required.

Section 13
Unitary Refrigerant-Based Systems

Unitary, refrigerant-based systems are commonly used in all types of buildings. These systems could consist of one type, or a number of different types of packaged equipment. In many cases, the equipment is small and simple — a window-mounted cooling unit, for example — but it also could be large and sophisticated — a rooftop heat pump unit, for instance. Unitary systems are used for projects that have one, a few, or many small rooms or zones. They are desirable because they are simple to design and install, they are relatively inexpensive and they provide an efficient way to zone a building.

13-1 General Discussion

Since unitary systems are just a collection of self-contained packages, the system performance is defined by the characteristics of the individual pieces of equipment. This section provides a summary of the components and performance characteristics that are generic to all types of unitary equipment. Specific information about the most common types of unitary equipment is provided in Sections 11-2 through 11-7.

Function
All unitary, refrigerant-based, equipment can provide sensible and latent cooling. Some types of unitary equipment also can provide heat. Some packages can provide outdoor air for ventilation or make up air, but some units do not have this capability. Small packaged units cannot accommodate humidification devices, but humidifiers can be installed in large unitary equipment.

Applications
Unitary equipment has been installed in most types of small commercial buildings and it has been used for large commercial, institutional and industrial applications. Unitary equipment can be used when one or more rooms, or a few small zones, need to be isolated from a central system. Unitary equipment also provides an inexpensive way to condition an entire building or a portion of a building that has many rooms and zones.

This type of equipment is appropriate when precise temperature and humidity control is not required. Mechanically, small units are less efficient than large units. However, the overall efficiency of unitary systems can be quite good because there is an one-to-one relationship between the output of the unit and the conditioning requirements of the space. (There are no reheat or mixing loads and the unit can be shut down — or the thermostat can be set up or set back — when the space is not occupied.)

Outdoor air (ventilation) is always required for commercial applications. Some, but not all, types of unitary equipment have the ability to satisfy this requirement. But, even if the equipment does have outdoor air capability, it may not have an economizer cycle.

The economizer option is important if a room or zone requires cooling during cold weather. Mechanical, low ambient cooling provides an alternate way to neutralize the cooling loads during cold weather. However, some equipment may not offer either feature (low ambient cooling or an economizer); in which case, the equipment is not suitable for the application. Specific information about the outdoor air and low ambient cooling capabilities of the different types of unitary equipment is provided in the following sections.

Configuration
A variety of packaged equipment is available. Some equipment may require a duct system, but many units discharge air directly into the room. Examples of the type of equipment that could be found in a unitary system would include:

- Window air conditioners
- Packaged terminal air conditioners and heat pumps
- Ductless split-coil air conditioners and heat pumps
- Rooftop air conditioners and heat pumps
- Split-system air conditioners and heat pumps
- Indoor, unitary air conditioners
- Water-source heat pumps

Characteristics
The characteristics that are generic to unitary systems, regardless of the type of equipment, are listed below. The characteristics that are peculiar to specific types of unitary equipment are listed in the following sections.

- Zoning is simple — each space has its own equipment.
- Any space can be heated or cooled as required (simultaneous heating and cooling).
- Equipment can be shut down on a room-by-room basis.
- Thermostats can be set up or set back on a room-by-room basis.
- Controls are included with the equipment package.
- There is an one-to-one relationship between the coil sensible heat ratio and the room sensible heat ratio.
- Fans provide positive air distribution.
- Unitary systems are relatively simple to design.
- The equipment is relatively simple to install.
- The equipment is simple to operate.
- An equipment room is not required.
- The installation cost is relatively low.

- The overall system efficiency is good.
- Energy use can be metered on a tenant-by-tenant basis.
- Equipment failure only affects one room or zone.
- Many pieces of equipment must be maintained.
- Each unit requires a condensate drain.

Equipment Package

The equipment package always includes a DX cooling system and a low efficiency filter. Devices that provide heat also may be included, depending on the type of equipment. (Heat could be provided by reversing the refrigeration cycle, by an electric heating coil, a hot water coil, a steam coil; or in the case of a rooftop unit, by a fossil fuel furnace.) Some types of unitary equipment cannot provide outdoor air, some packages have minimum outdoor air dampers and some packages offer a full economizer cycle. Large packaged equipment may be available with filter options, damper options or a return air fan option. Controls are usually included with the package.

Design Loads

Equipment sizing loads are based on the peak zone cooling and heating loads. Calculations that estimate minimum loads are not necessary for small units that cycle on and off, but the minimum cooling load should be estimated when the equipment package offers capacity control options. See Section 4 in this manual for a general discussion of the load calculation and load analysis procedures that are used to determine equipment size and establish control strategies.

Control

Unitary equipment is usually constant volume, constant temperature equipment. Historically, small packaged units have controlled space temperature by cycling (on-off) the cooling or heating plant. However, improvements in motor and compressor technology provide the opportunity to use variable speed capacity control. Reheat coils, VAV devices, unloaders and hot gas bypass are still viable options when the application requires a large single-zone unit. Low ambient cooling and economizer cycle options are available on some, but not all, types of equipment. The details of how specific types of unitary equipment control temperature, humidity and air flow will be discussed in the following sections.

13-2 Window Air Conditioners and Heat Pumps

Window units are classified as appliances. They are used when low initial cost is significantly more important than good control over the conditions in the occupied space.

Window units can be used to cool or heat a room or zone. They can provide all the necessary cooling, or heating — or they can be used to supplement a central system. They are easy and inexpensive to install, and the newer models offer reasonably good efficiency. This type of equipment is usually installed without a duct, but some units have been fitted with a very short duct. Some units have an outdoor air damper and/or a vent damper. Window units do not include an economizer option or a low ambient cooling option. (Because there is no method for providing low ambient cooling, this type of equipment may not be appropriate when the room or zone has large internal loads.)

When the unit has a heating capability, the heat could be generated by an electric coil or it could be produced by reversing the refrigeration cycle. During cold weather, electric coils can be used to supplement the reversed cycle output. Capacities range from about 4,000 to 36,0000 BTUH — total cooling capacity.

Characteristics

The characteristics that are peculiar to window units are listed below. Also refer to page 13-1, which lists the characteristics that are generic to all types of unitary equipment.

- They can be installed at a low initial cost.
- Installation and operation is simple.
- Equipment selection and sizing is simple.
- They do not provide precise control over temperature and humidity.
- They do not provide precise control of ventilation air.
- They do not provide low ambient cooling.
- They do not offer an economizer cycle option.
- They have limited air filtration capability.
- Temperature is maintained by on-off cycling.
- The "free-delivery" air distribution is relatively crude.
- Equipment noise is generated near the room.
- Equipment durability and efficiency is based on appliance standards.
- Installation requires access to an exterior wall.
- Wall sleeves can be used in lieu of a window installation.

Equipment Selection

Window units do not require much design work, a simple load calculation is all that is needed. Performance data, which is minimal, provides information about total cooling capacity, heating capacity and efficiency. In some cases, the manufacturer may provide information about the coil sensible heat ratio (CSHR). This CSHR information is desirable because units that have a low CSHR are preferred for humid climates and units that have a high CSHR are better for dry climates.

Controls

All the comfort controls are integrated into the package. Cooling controls include a thermostat and switches that allow the user to select the operating mode — on-off, fan-only, cool-low and cool-high — for example. There also may be a knob or lever that allows the occupant to adjust a vent or an outdoor air damper. If the unit has heating capability, additional switches are provided to select the heating options. If safety and operating controls are provided, they will be included as standard equipment. Usually, there are no alternative control options.

13-3 Packaged Terminal Equipment

Packaged terminal equipment can be used to cool or heat, a room or zone. Packaged terminal air conditioners (PTAC) may just provide cooling or they may be equipped with a heating device, which could be an electric coil, a hot water coil or a steam coil. Packaged terminal heat pumps (PTHP) are usually equipped with an electric heating coil, but some units can be equipped with a hot water coil.

PTAC units do provide outdoor air capability, which is typically equal to 20 or 25 percent of the supply CFM. PTAC equipment does not include an economizer option or a low ambient cooling option. Mechanical cooling is usually not available when the outdoor temperature drops below 55°F. (Because there is no method for providing low ambient cooling, this type of equipment may not be appropriate when the room or zone has large internal loads.)

Packaged terminal equipment is manufactured to industry grade standards and it is relatively easy and inexpensive to install. The newer models have good efficiencies. These units are usually installed without a duct. If a ducted supply system is used, it must not produce more than 0.10 inches (water gage) of resistance (static pressure loss).

Characteristics
The characteristics that are peculiar to packaged terminal units are listed below. Also refer to page 13-1, which lists the characteristics that are generic to all types of unitary equipment.

- They are less expensive to install than a central system.
- Sizable wall penetrations are required for installation.
- They do not provide precise control over temperature and humidity.
- They do not provide precise control of ventilation air.
- They do not provide low ambient cooling.
- They do not offer an economizer cycle option.
- They have limited air filtration capability.
- Temperature is maintained by on-off cycling.
- The free delivery air distribution is relatively crude.
- Equipment noise is generated near the room.
- Installation requires access to an exterior wall.

Equipment Selection
Some, but not all, manufacturers publish comprehensive performance data. When it is available, this performance data should be used to select and to size the equipment. In any case, when the equipment is subjected to the indoor and outdoor design conditions, the sensible and latent cooling capacity and the heating capacity should be equal to, or slightly exceed, the design loads. A complete explanation of how applied performance data is used to select equipment is provided in Sections 6, 7 and 8 of this manual. Section 6 covers cooling equipment, Section 7 covers heating coils and Section 8 covers heat pump equipment. It also will be necessary to specify which accessories and options should be included with the basic equipment package.

Packaged Terminal Cooling Unit or Heat Pump

Controls
All the comfort controls are integrated into the package. Cooling controls include an integral or remote thermostat and switches that allow the user to select the operating mode — on-off, fan-only, cool-low and cool-high — for example. There also may be a knob or lever that allows the occupant to adjust a vent or an outdoor air damper. If the unit has heating capability, additional switches are provided to select one or more heating modes. If safety and operating controls are provided, they will usually be included as standard equipment. For example, some PTHP units may include defrost controls and some may not. Occasionally, it may be necessary to specify some control options. For example, there may be an optional interlock control that can be used to start and stop the equipment from a central location or a master-slave control that allows multiple units to be controlled by one thermostat.

13-4 Ductless Split-Coil Systems

Ductless split-coil equipment can be used to cool or heat, a room or zone. Multizone split-coil systems, which can cool or heat two or more rooms or zones, also are available. Split-coil air conditioners just provide cooling, but split-coil heat pumps provide cooling and heat. Usually, an electric heating coil is installed in the indoor unit of the heat pump equipment. In cold climates, it is important to make sure that this coil has enough capacity to satisfy the supplemental heating requirement.

Many different manufacturers produce ductless-split equipment. Depending on the product and model, the indoor unit may not have outdoor air capability. Even if the unit does have outdoor air capability, it is unlikely that it will have an economizer cycle option (but there is at least one manufacturer that offers this option). However, some manufactures do provide a low ambient cooling option. Therefore, certain makes and models are suitable for cooling rooms or zones that have large internal loads.

Ductless split-coil equipment is manufactured to industry grade standards, and it is easy to install. The mechanical efficiency of this type of equipment is good. Indoor units are available in a number of different configurations. Some take the form of a floor mounted console unit, others are wall mounted. Ceiling mounted units also are available. Ceiling mounted units provide the best type of air distribution for cooling applications.

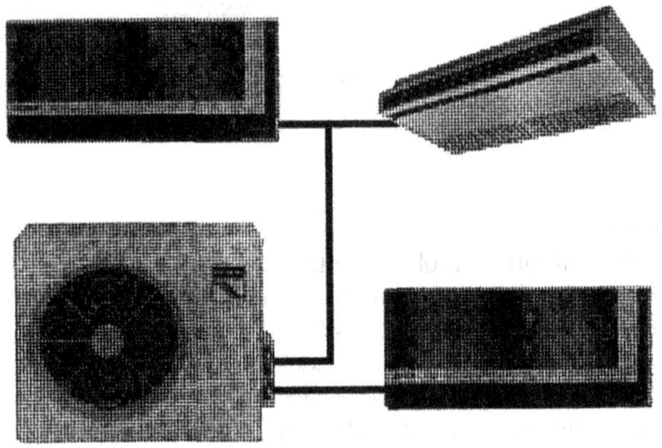

Ductless-Split Cooling Unit or Heat Pump

Characteristics
The characteristics that are peculiar to ductless split-coil units are listed below. Also refer to page 13-1, which lists the characteristics that are generic to all types of unitary equipment.

- They are less expensive to install than a central system.
- Envelope penetrations are required for refrigerant piping.
- They control temperature and humidity better than window units or PTAC units.
- Some indoor units provide minimum outdoor air.
- There is no economizer cycle option (with at least one exception).
- Many models have a low ambient cooling option, but some do not.
- They have limited air filtration capability.
- Temperature is maintained by on-off cycling — single-zone units.
- Temperature is maintained by cycling liquid line solenoid valves — multizone units.
- Some equipment has variable speed compressors and fans.
- Capacity control is available on some models.
- The free delivery air distribution of wall and floor units is relatively crude.
- Ceiling models provide air distribution that is similar to ceiling diffusers.
- Equipment noise is generated near the room.
- Installation requires access to an exterior wall or roof.

Equipment Selection
Some, but not all, manufacturers publish comprehensive performance data. When it is available, this performance data should be used to select and to size the equipment. In any case, when the equipment is subjected to the indoor and outdoor design conditions, the sensible and latent cooling capacity and the heating capacity should be equal to, or slightly exceed, the design loads. A complete explanation of how applied performance is used to select equipment is provided in Sections 6, 7 and 8 of this manual. Section 6 covers cooling equipment, Section 7 covers heating coils and Section 8 covers heat pump equipment. It also will be necessary to specify which accessories and options should be included with the basic equipment package.

Controls
The comfort controls are provided with the equipment. Cooling controls include an integral or remote thermostat and an integral or hand held control module that allows the occupant to select the operating mode, setpoint and fan speed. Many models include programmable controls that can be used to establish set-up and set-back schedules. If safety and operating controls are provided, they will usually be included as standard equipment. For example, most heat pump units include defrost controls. Occasionally, it may be necessary to specify control options — low ambient cooling, for example.

13-5 Outdoor Unitary Equipment

Outdoor unitary equipment is usually set on a roof, but it can be installed on-grade. The complete system consists of the unit, the supply and return ducts, the supply outlets, the return(s) and the controls. The unit might be a single-zone or a multizone unit. The comments made in this section will be limited to single-zone equipment. (A complete discussion of single-zone and multizone all-air systems is provided in Section 11 of this manual.)

A single-zone outdoor unit is completely self-contained. It could be a cooling-only unit, a cooling and heating unit, an air-source heat pump or a water-source heat pump. Cooling capacities range from a few tons to 50 tons or more. Heat could be provided by electric coils, a fossil fuel furnace, a hot water coil, a steam coil or by reversing the refrigeration cycle. These units usually have dampers or damper sections that can be used to control the percentage of outdoor air that is supplied to the conditioned space. And, most of the units can be equipped with an air-side economizer or accessorized with low ambient cooling controls. Therefore, this type of equipment is well suited for applications that require cooling during cold weather.

Outdoor unitary equipment is manufactured to industry grade standards and because it is packaged, it is relatively easy to install. The mechanical efficiency of this type of equipment is good, and the overall system efficiency that is associated with using multiple single-zone units is very good.

Rooftop Unit

Characteristics

The characteristics that are peculiar to outdoor unitary equipment are listed below. Also refer to page 13-1, which lists the characteristics that are generic to all types of unitary equipment.

- A ducted air distribution system is required.
- Duct runs occupy space within the building envelope.
- Roof penetrations are required for the duct runs.
- Temperature and humidity control can be fair to good.
- Dampers can be used to adjust the outdoor air percentage.
- Most manufacturers offer an economizer cycle option.
- Most units have a low ambient cooling option.
- The air filtration options may be somewhat limited.
- Some units control temperature by on-off cycling, others stage or modulate the cooling (or heating) plant.
- Some manufacturers may offer variable speed compressors and fans.
- Manufacturers usually offer some type of capacity control option.
- A return air fan or relief damper is required if the unit has an economizer.
- A return fan or relief damper is required if the resistance that is associated with the return duct system exceeds 0.10 inches (water gage) of pressure.
- The air distribution system can be designed to provide good air circulation.
- There is an opportunity to isolate equipment noise from the conditioned space.
- Weather conditions can be hard on rooftop units and on the people that service them.

Equipment Selection

Some, but not all, manufacturers publish comprehensive performance data. When it is available, this performance data should be used to select and to size the equipment. In any case, when the equipment is subjected to the indoor and outdoor design conditions, the sensible and latent cooling capacity and the heating capacity should be equal to, or slightly exceed, the design loads. A complete explanation of how applied performance is used to select equipment is provided in Sections 6, 7 and 8 of this manual. Section 6 covers cooling equipment, Section 7 covers heating coils and Section 8 covers heat pump equipment. It also will be necessary to specify which accessories and options should be included with the basic equipment package.

Controls

Usually the manufacturer provides one or more control package options. The basic controls include a remote thermostat or control panel that allows the occupant to select the operating mode and setpoint. A minimum outdoor air percentage control also may be included with the package. Many manufacturers offer programmable controls that can be used to establish set-up and set-back schedules. If safety and operating controls are provided, they may be included as standard equipment or provided as options. For example, most heat pump units include defrost controls, but an economizer or a low ambient control is usually an option.

13-6 Indoor and Split-System Unitary Equipment

Indoor equipment usually contains one or more compressors and the evaporator section. The unit may be air-cooled or water-cooled. A remote condenser will be required if it is air-cooled. The entire system consists of the indoor unit, the air distribution system, the controls and if required, a remote condenser.

Split-systems usually consist of an indoor unit that contains the evaporator section and an outdoor unit that contains one or more compressors and an air-cooled condenser. The entire system consists of the indoor unit, the outdoor unit, interconnecting refrigerant piping, the air distribution system and the controls.

Some, but not necessarily all, of the indoor units have dampers or damper sections that can be used to control the percentage of outdoor air that is supplied to the conditioned space. And, some units may have an air-side economizer option or a low ambient cooling option. Indoor equipment that has these features is suited for applications that require cooling during cold weather.

Unitary, indoor and split-system equipment is manufactured to industry grade standards and because it is packaged, it is relatively easy to install. The mechanical efficiency of this type of equipment is good and the overall system efficiency that is associated with using multiple single-zone units is very good. Note that, the information that is provided in Section 11-2 (*Single-Zone, Constant Volume Systems*) also applies to unitary indoor and split-system equipment.

Indoor Packaged Water-Cooled Unit

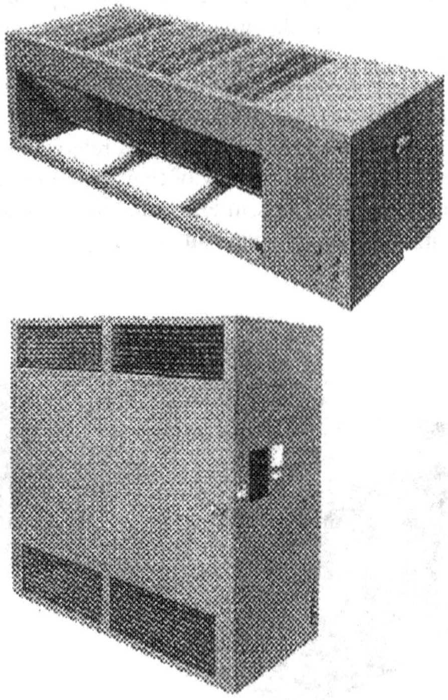

Split-System (Cooling or Heat Pump)

Characteristics

The characteristics that are peculiar to this type of equipment are listed below. Also refer to page 13-1, which lists the characteristics that are generic to all types of unitary equipment.

- Some indoor units supply conditioned air through free discharge nozzles.
- A ducted air distribution system may be required.
- The indoor unit and the duct runs occupy space within the building envelope.
- Envelope penetrations may be required for outdoor air and exhaust air ducts.
- Envelope penetrations may be required for the refrigerant piping runs.
- Temperature and humidity control can be fair to good.
- In some cases, dampers can be used to adjust the outdoor air percentage.
- Some units provide an economizer cycle option.
- Some units have a low ambient cooling option.
- The air filtration options may be somewhat limited.
- Some units control temperature by on-off cycling, others stage or modulate the cooling (or heating) plant.
- Most units have some type of capacity control option.
- A return air fan or relief damper is required if the unit has an economizer.
- A return fan or relief damper is required if the resistance that is associated with the return duct system exceeds 0.10 inches (water gage) of pressure.
- The air distribution system can be designed to provide good air circulation.
- Equipment noise may be generated near the room.

Equipment Selection

Some, but not all, manufacturers publish comprehensive performance data. When it is available, this performance data should be used to select and to size the equipment. In any case, when the equipment is subjected to the indoor and outdoor design conditions, the sensible and latent cooling capacity and the heating capacity should be equal to, or slightly exceed, the design loads. A complete explanation of how applied performance is used to select equipment is provided in Sections 6,

7 and 8 of this manual. Section 6 covers cooling equipment, Section 7 covers heating coils and Section 8 covers heat pump equipment. It also will be necessary to specify which accessories and options should be included with the basic equipment package.

Controls

Usually the manufacturer provides one or more control package options. The basic controls include a remote thermostat or control panel that allows the occupant to select the operating mode and setpoint. A minimum outdoor air percentage control also may be included with the package. Many manufacturers offer programmable controls that can be used to establish set-up and set-back schedules. If safety and operating controls are provided, they may be included as standard equipment or provided as options. For example, most heat pump units include defrost controls, but an economizer or a low ambient control is usually optional.

13-7 Water-Source Heat Pumps

Water-source heat pumps can be used to condition one or more rooms or zones. A complete system consists of a unit, the air distribution system, the controls and the water system. The supply water may be provided by (once-through) well water, or closed loop (reclaim) water. Some water-source heat pump units have dampers or damper sections that can be used to control the percentage of outdoor air that is supplied to the conditioned space — and, some units have an air-side economizer option. (There are many makes and models that have no outdoor air capability.) During cold weather, cooling can either be provided by an economizer — if it is available — or by the refrigeration cycle. (Low ambient cooling does not require any special controls because the water-cooled condenser is not affected by the outdoor temperature.) Therefore, water-source heat pump equipment is suitable for applications that require cooling during cold weather.

Water-source equipment is manufactured to industry grade standards. Because it is packaged, the equipment is relatively easy to install, but there is additional work associated with installing the water piping system. The mechanical efficiency of this type of equipment is excellent and the overall system efficiency that is associated with the reclaim configuration can range from good to outstanding. (A comprehensive discussion of the reclaim water-loop system is provided in Section 12-7.)

Characteristics

The characteristics that are peculiar to this type of equipment are listed below. Also refer to page 13-1, which lists the characteristics that are generic to all types of unitary equipment.

- Once-through systems use a lot of water.
- Reclaim systems require a boiler and a heat rejection device.
- Pumps, piping and hydronic accessories are required.
- The indoor units and duct runs occupy space within the building envelope.
- Envelope penetrations may be required for outdoor air and exhaust air ducts.
- Envelope penetrations may be required for water piping.
- Temperature and humidity control is good.
- In some cases, dampers can be used to adjust the outdoor air percentage.
- Some units provide an economizer cycle option.
- Low ambient cooling is intrinsic.
- The air filtration options may be somewhat limited.
- Some units control temperature by on-off cycling, others stage or modulate the cooling (or heating) plant.
- Larger units may have some type of capacity control option.
- A return air fan or relief damper is required if the unit has an economizer.
- A return fan or relief damper is required if the resistance that is associated with the return duct system exceeds 0.10 inches (water gage) of pressure.
- The air distribution system can be designed to provide good air circulation.
- Equipment noise may be generated near the room.

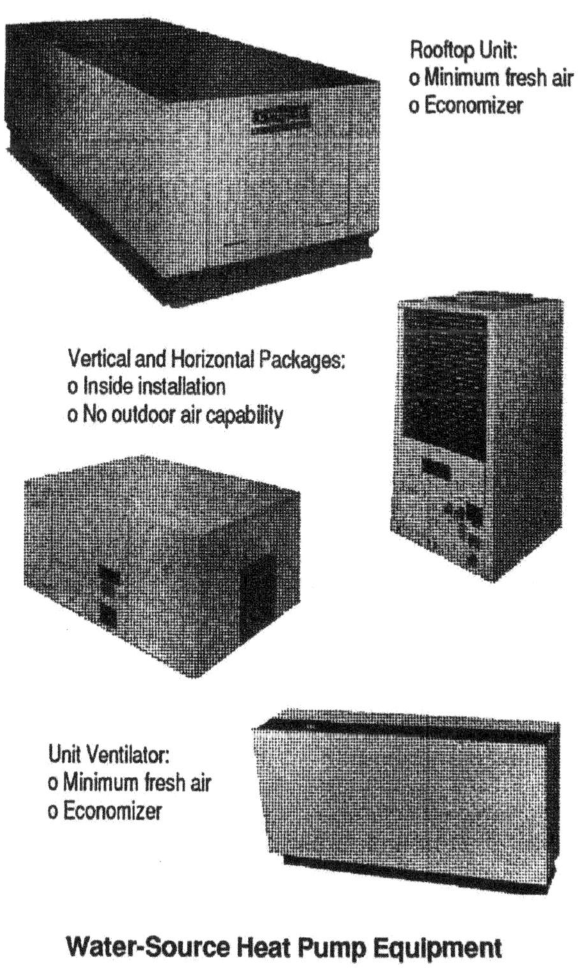

Water-Source Heat Pump Equipment

Equipment Selection

Some, but not all, manufacturers publish comprehensive performance data. When it is available, this performance data should be used to select and to size the equipment. In any case, when the equipment is subjected to the indoor and outdoor design conditions, the sensible and latent cooling capacity and the heating capacity should be equal to, or slightly exceed, the design loads. A complete explanation of how applied performance is used to select equipment is provided in Section 8. It also may be necessary to specify accessories and options.

Controls

Usually, the manufacturer provides one or more control package options. The basic controls include a remote thermostat or control panel that allows the occupant to select the operating mode and setpoint. A minimum outdoor air percentage control also may be included with the package. Many manufacturers offer programmable controls that can be used to establish set-up and set-back schedules. If safety and operating controls are required, they may be included as standard equipment or provided as options.

Appendix 1
Charts, Tables and Equations

This appendix presents a collection of the various charts, tables and equations that were used in this manual. For convenience, a brief description may be provided with each equation, chart or table. Refer to the appropriate section in the main body of this manual if there is any doubt about how to use a particular equation, chart or table.

A1-1 System Selection and Zoning

Some types of equipment only perform a single function — heating, for example. Some types of equipment perform many functions — heating, cooling dehumidification, filtration and ventilation, for example. Information about the capabilities of various types of equipment is provided below by Table 2-1. The "Zoning Evaluation Table" (see page A1-3) provides a

System Functions and Equipment		
Ventilation Only	Power Exhaust Systems	Propeller fans, exhaust fans, exhaust hoods
	Gravity Systems	Relief dampers, gravity vents
Heating-Only	Forced Air Systems	Electric or fossil fuel furnace with no outdoor air capability; electric, hot water or fossil fuel unit heaters, duct furnaces
	Convective Heating Systems	Electric or hot water baseboard and cabinet convectors
	Radiant Heating Systems	Electric radiant heating panels, hot water or electric floor radiation, high intensity electric or fossil fuel heating fixtures, low intensity gas-fired heating pipes
Heating and Ventilation	Forced Air Systems	Electric or fossil fuel make up air units, electric or fossil fuel furnace with outdoor air capability, central station air handlers with outdoor air capability
	Hot Water Systems	Distributed fan-coil units with outdoor air capability, unit ventilators
Cooling and Dehumidification, no Ventilation	All-Air Systems	Unitary or split DX equipment with no outdoor air capability
	Chilled Water Systems	Central station air handlers with no outdoor air capability, fan-coil units with no outdoor air capability
Cooling, Dehumidification and Ventilation	All-Air Systems	Unitary or split DX equipment with outdoor air capability, packaged terminal air conditioners (PTAC)
	Chilled Water Systems	Central station air handlers with outdoor air capability, fan-coil units with outdoor air capability, unit ventilators
Cooling, no Dehumidification and Ventilation	Evaporative Cooling Systems	Direct and indirect evaporative cooling equipment; direct/indirect multistaged evaporative cooling equipment. (In some situations indirect units may provide partial dehumidification.)
Heating, Cooling, Dehumidification; no Ventilation	All-Air Systems	Package heating and DX cooling equipment with no outdoor air capability, unitary or split heat pump equipment with no outdoor air capability
	Water Systems	Central station air handlers with no outdoor air capability, distributed fan-coil units, water-loop heat pump equipment with no outdoor air capability
Heating, Cooling, Dehumidification and Ventilation	All-Air Systems	Package heating and DX cooling equipment with outdoor air capability, unitary or split heat pump equipment with outdoor air capability, packaged terminal air conditioners (PTAC), packaged terminal heat pumps (PTHP).
	Water Systems	Central station air handlers with outdoor air capability, unit ventilators, water-loop heat pump equipment with outdoor air capability
Humidification	Ducted — Steam Systems	Steam Humidifiers
	Ducted — Water Systems	Pan, atomizing and wetted element humidifiers
	Nonducted Systems	Steam, pan, atomizing and wetted humidifiers

Table 2-1

Zoning Evaluation Table

Zoning Factors	Selection Schedule	Room Numbers①											
		1	2	3	4	5	6	7	8	9	10	11	12
Functions	Heat												
	Cool												
	Dehumidification												
	Ventilation												
	Humidification												
Space Conditions	Comfort Conditions												
	Special Temperature												
	Special Humidity												
Filter Req-uirements	Standard Filters												
	Special Filters												
Space Pressure	Neutral												
	Negative												
	Positive												
Exhaust From Room	Less Than 10%												
	More Than 10%												
	100 Percent												
Outdoor CFM	0% to 10%												
	11% to 20%												
	Above 20%												
	100%												
Occupancy Pattern	Schedule A												
	Schedule B												
	Schedule C												
Exposure②	Interior Zone												
	N, E, S, W												
	SE or SW (Corner)												
Peak Load③	Constant												
	9 AM, Noon, 3, 6 PM												
Heating-Cooling Change-over	Cool-Only												
	OAT Above 50 F												
	OAT 35 F to 50 F												
	OAT 20 F to 35 F												
	OAT Below 20 F												
Room SHR	Below 0.75												
	0.76 to 0.85												
	Above 0.85												
Utility Bills	Separate Billing												
	Central Billing												

Notes:
① Enter an ✗ or a ✔ in the room columns to indicate choice from selection schedule.
② Enter compass point in room columns to indicate exposure.
③ Enter a 9, N, 3 or 6 in room columns to indicate time of day peak.

System Performance Characteristics

General Characteristics and Performance Details		Single-Zone, Unitary & Self-Contained DX											All-Air Systems						Water Systems							
		Multiple 1-Zone Units	Rooftop H&C	Rooftop Heat Pump	Split-System- Cooling	Split-System Heat Pump	Indoor H&C Unit	PTAC Cooling Unit	PTAC Heat Pump	Furnaces and MUA Units	Unit /Duct Heaters	Radiant Heating	Terminal Reheat	Dumping VAV ②	Modulating VAV ②	Blow-Through Multizone	Draw-Through Multizone	VVT-Changeover	2-Pipe System	3-Pipe System	4-Pipe System	Water-Loop Heat Pumps	Fan-Coil Terminal Units	Unit Ventilators	Unit Heaters	Baseboard Radiation
Function	Heat	✗	✗	✗		✗	✗		✗	✗	✗	✗	✗	✗	✗	✗	✗	✗	✗	✗	✗	✗	✗	✗	✗	✗
	Cool / Dehumid	✗	✗	✗	✗	✗	✗	✗	✗				✗	✗	✗	✗	✗	✗	✗	✗	✗	✗	✗	✗		
	Ventilation	✗	✗	✗	①	①	①	✗	✗	✗			✗	✗	✗	✗	✗	✗	✗	✗	✗			✗		
Zones	One		✗	✗	✗	✗	✗	✗	✗	✗	✗	✗											✗	✗	✗	✗
	Many	F											B	G	B	F	F	F	P	B	B	B				
Mode (H&C)	Heat or cool		✗	✗		✗	✗		✗					✗	✗			✗	✗				✗	✗		
	Simultaneous	✗											✗			✗	✗			✗	✗	✗				
Comfort	Temp control	G	G	G	G	G	G	G	G	G	F	F	B	G	G	G	G	F	P	B	B	G	G	G	G	G
	RH control	G	G	G	G	G	G	G	G				B	F	F	G	B	F	P	B	B	G	G	F		
Air Quality	Circulation	B	B	B	B	B	B	F	F	G	F		B	G	P	G	G		G	F	F	G	F	F	F	
	Fresh Air	G	G	G	①	①	①	F	F	G			B	F	P	G	G	F	G	G	G			G		
	Filtration	G	G	G	G	G	G	P	P	G			B	G	G	G	G	G	P	P	P	F	P	P		
	Pressurization	✗	✗	✗	①	①	①			✗			B	P	P	F	F	P								
Efficiency Factors	Reheat												✗													
	Mixing															✗	✗			✗	✗					
	Economizer	✗	✗	✗	①	①	①						✗	✗	✗	✗	✗	✗						✗		
	Reclaim																					✗				
Spatial Requirement	Equip outdoors	✗	✗	✗	✗	✗		✗	✗	✗			✗	✗	✗	✗	✗	✗				✗				
	Equip inside				✗	✗	✗			✗	✗	✗	✗	✗	✗	✗	✗		✗	✗	✗	✗	✗	✗	✗	✗
	Ducts inside	✗	✗	✗	✗	✗	✗			✗	✗		✗	✗	✗	✗	✗	✗				✗				
Condenser	Air-Cooled	✗	✗	✗	✗	✗		✗	✗				✗	✗	✗	✗	✗	✗	✗	✗	✗					
	Water-Cooled						✗						✗	✗	✗	✗	✗		✗	✗	✗	✗				
Cooling Coil	DX	✗	✗	✗	✗	✗	✗	✗	✗				✗	✗	✗	✗	✗	✗				✗				
	Water												✗	✗	✗	✗	✗		✗	✗	✗		✗	✗		
Heating Coil	Electric	✗	✗	✗		✗	✗		✗	✗	✗	✗	✗	✗	✗	✗	✗	✗								
	Water												✗	✗	✗	✗	✗		✗	✗	✗		✗	✗	✗	✗
	Fossil	✗	✗				✗			✗	✗	✗	✗	✗	✗	✗	✗	✗								
Flexibility	Floor plan revs.												✗	✗	✗			✗								
Op-Cost	Energy Use	B	G	B	G	B	G	F	G	G	F	G	P	G	B	F	P	G	G	P	G	B	G	F	F	G

Note: ① Option not standard — Provide OA/EA ducts; field fabricated economizer section. ② No reheat P = poor, F = fair, G = good, B = best, Blank = not available form that can be used to analyze the zoning requirements that are associated with any application. The "System Performance Characteristics" information sheet (see page A1-4) summarizes the capabilities of the various types of systems. The "System Selection Form" (below) can be used to compare the performance characteristics of the various types of systems.

System Selection Form					
General Characteristics and Performance Details		Application	1-Zone H/C	VVT-CO	Multizone
Function	Heat				
	Cool / Dehumid				
	Ventilation				
Zones	One				
	Many				
Mode (H&C)	Heat or Cool				
	Simultaneous				
Comfort	Temp. Control				
	RH Control				
Air Quality	Circulation				
	Fresh Air				
	Filtration				
	Pressurization				
Efficiency Factors	Reheat				
	Mixing				
	Economizer				
	Reclaim				
Spatial Requirement	Equip. Outside				
	Equip. Inside				
	Ducts Inside				
Condenser	Air-Cooled				
	Water-Cooled				
Cooling Coil	DX				
	Water				
Heating Coil	Electric				
	Water				
	Fossil				
Noise	NC Rating				
Flexibility	Floor Plan Revs.				
Cost Analysis	Installed Cost				
	Energy Use				
	Maint. Costs				

P = Poor, F = Fair, G = Good, B = Best

X = Available or acceptable

Blank = Not available or not applicable

A2-2 Recommended Filter Efficiencies

High efficiency mechanical filters and electronic filters have the ability to remove small respiratory particles, so they can be used to improve air quality, providing that odors and gases are not a problem. Special filters, carbon activated filters for example, can be used to remove gaseous contaminates.

The filter efficiency that is used for a particular application is a compromise between the installed cost, operating cost and maintenance costs. High efficiency filters tend to increase the installed cost because they are more expensive, they take up more space and they affect the pressure difference across the duct walls. These filters also are more expensive to operate (higher fan pressures are required) and to maintain. (The space requirements and pressure drops that are associated with electronic filters are considerably less than the space requirements and pressure drops that are associated with media filters.)

The table on the next page recommends filter efficiencies (ASHRAE Dust Spot Efficiencies) for various applications. In most cases (efficiencies less than 90%), the filters will protect the equipment and remove the larger types of airborne particles. In some cases (efficiencies greater than 90%), the filters will remove the smaller types of airborne particles. If air quality problems are involved or if air quality problems are anticipated, the designer should disregard these recommendations and select a filter (or filtering system) that will provide the desired performance.

A1-3 Outdoor Air Quality Standard

If the proper amount of ventilation is used, the indoor air quality will depend on the outdoor air quality. If the quality of the outdoor air is substandard, special filters or scrubbing equipment will be required to clean the incoming air flow. Acceptable outdoor air quality, as defined by ASHRAE Standard 62-89, is defined by the following table.

Outdoor Air Quality Standard				
Pollutant	Long Term		Short Term	
	PPM	Average	PPM	Average
Sulfur dioxide	80	1 Year	365	24 Hours
Total particulate	75	1 Year	260	24 Hours
Carbon monoxide			40000	1 Hour
Carbon monoxide			10000	8 Hours
Oxidants (ozone)			235	1 Hour
Nitrogen dioxide	100	1 Year		
Lead	1.5	3 Months		

PPM = Parts Per Million

Suggested ASHRAE Dust Spot Efficiencies for NonSmoking[1] Applications

Apartments	20% to 30%
Arcades	20% to 30%
Arenas	
Public areas	30% to 35%
Floor	30% to 35%
Ticket Booth	30% to 35%
Stages & Work Rooms	30% to 35%
Bakeries	20% to 30%
Ball / Dance Rooms	30% to 35%
Bank	35% to 40%
Barber Shops	20% to 30%
Bars and Discos	20% to 30%
Beauty	20% to 30%
Bowling Centers	
Bowling & Spectator	20% to 30%
Cafeterias	35% to 40%
Church	
Assembly	30% to 35%
Classroom	30% to 35%
Meeting / Activity	30% to 35%
Coin Laundries	20% to 30%
Convenience Centers	
Stores	20% to 30%
Court Club/Fitness Center	
Tennis	20% to 30%
Racquetball / Squash	20% to 30%
Exercise Room	20% to 30%
Department Stores	
Shopping Areas	20% to 30%
Diners or Fast Food	30% to 35%
Doctor/Dentist Office	
Waiting	30% to 35%
Examination	30% to 35%
Medical Procedure	90% +
Dormitories	
Bedrooms	20% to 30%
Suites	20% to 30%
Lobbies	20% to 30%
Meeting / Game	20% to 30%
Dry Cleaners	
Commercial	20% to 30%
Coin	20% to 30%
Print / Duplicating	20% to 30%
Exhibition Hall	
Public areas	30% to 35%
Floor	30% to 35%
Ticket Booth	30% to 35%
Stages & Work Rooms	30% to 35%
Florist	20% to 30%
Funeral Parlor	30% to 35%
Garages	Below 20%

Government Buildings	
Court Room	30% to 35%
Legislative Chambers	30% to 35%
Committee-Conference	30% to 35%
Offices	30% to 35%
Press Room lounge	20% to 30%
Toilets	Below 20%
Jail Cells	20% to 30%
Guard Stations	20% to 30%
Police Station	20% to 30%
Fire House	20% to 30%
Gymnasiums	
Floor	20% to 30%
Spectator	20% to 30%
Locker	Below 20%
Hospital	
Patient Room	90% & PF
Medical Procedures	90% & PF
Operate Delivery	90% & PF
Recovery / Intens Care	90% & PF
Autopsy	90% & PF
Physical Therapy	90% & PF
Cafeteria	90% & PF
Amphitheater	30% to 35%
Industrial	10% to 99.%
Kitchens	Below 20%
Laboratories	80% to 99%
Lab Animal Rooms	Below 20%
Library or Museum	
General Areas	35% to 60%
Work Rooms	35% to 60%
Rare books & Exhibits	80% to 99%
Shopping Mall	
Mall	20% to 30%
Store	20% to 30%
Motels	
Bedrooms	20% to 30%
Suites	20% to 30%
Lobbies	20% to 30%
Conference	20% to 30%
Assembly Hall	20% to 30%
Nursing Homes	
Patient Room	80%
Medical Procedures	80%
Physical Therapy	80%
Office Building	
Office	35% to 60%
Bull Pen	35% to 60%
Conference	35% to 60%
Reception	35% to 60%
Pet Shops	20% to 30%

Pharmacy	
Work Room	30% to 35%
Sales Area	30% to 35%
Photo Studio	
Studio	30% to 35%
Darkroom	30% to 35%
Pool Hall	20% to 30%
Restaurant / Nite-club	35% to 40%
School	
Classroom	30% to 35%
Laboratories	30% to 60%
Shops	30% to 35%
Music Rooms	30% to 35%
Libraries	30% to 35%
Auditoriums	30% to 35%
Gyms	20% to 30%
Locker Rooms	Below 20%
Common Rooms	30% to 35%
Offices	30% to 35%
Lunch Room	30% to 35%
Shoe Repair	20% to 30%
Skating Rinks	20% to 30%
Small Stores	20% to 30%
Smoking Room	80%
Supermarket	35% to 40%
Swimming Pools	
Pool & Deck	20% to 30%
Spectator	20% to 30%
Theater	
Public areas	30% to 35%
Floor	30% to 35%
Ticket Booth	30% to 35%
Stages & Work Rooms	30% to 35%
Public Toilet	Below 20%
Transportation Terminal	
Public Areas	30% to 35%
TV-Radio	
Booths	35% to 40%
Studios	35% to 40%
Electronic Equipment	80% & PF
Audience	35% to 40%
Stages	35% to 40%
Variety Stores	20% to 30%
Veterinarian	
Kennel	10% to 15%
Operating Room	30% to 35%
Waiting Room	30% to 35%
Warehouse	10% to 35%

① Filter efficiency must be 80% or higher to remove smoke.

② PF = Prefilter

ASHRAE Standard 62-89 Ventilation Rates (CFM per Person)③

Apartments ① ②	30 / Room
Arcades	0.3 / SQ.FT.
Arenas	
Public areas	15
Floor	20
Ticket Booth	20
Stages & Work Rooms	15
Bakeries	15
Ball / Dance Rooms	25
Bank	20
Barber Shops	15
Bars and Discos	30
Beauty	25
Bowling Centers	
Bowling & Spectator	25
Cafeterias	20
Church	
Assembly	15
Classroom	15
Meeting / Activity	15
Coin Laundries	15
Convenience Centers	
Stores	0.30 / SQ.FT.
Court Club/Fitness Center	
Tennis	20
Racquetball / Squash	20
Exercise Room	20
Department Stores	
Shopping Areas	0.30 / SQ.FT.
Diners or Fast Food	20
Doctor/Dentist Office	
Waiting	15
Examination	15
Medical Procedure	30
Dormitories	
Sleeping Area ②	15
Suites	30 / Room
Lobbies	15
Meeting / Game	20
Dry Cleaners	
Commercial	30
Coin	15
Print / Duplicating	0.50 / SQ.FT.
Exhibition Hall	
Public Areas	15
Floor	20
Ticket Booth	20
Stages & Work Rooms	15
Florist	15
Funeral Parlor	15
Garages	1.5 / SQ.FT.

Government Buildings	
Court Room	15
Legislative Chambers	15
Committee-Conference	20
Offices ②	20
Press Room lounge	60
Toilets	50 / Fixture
Jail Cells	20
Guard Stations	15
Police Station	15
Fire House	15
Gymnasiums	
Floor	20
Spectator	15
Locker	0.5 /SQ.FT.
Hospital	
Patient Room	25
Medical Procedures	15
Operate Delivery	30
Recovery / Intens Care	15
Autopsy	100
Physical Therapy	15
Cafeteria	20
Amphitheater	15
Industrial	Survey
Kitchens	15
Laboratories	20
Lab Animal Rooms	0.50 / SQ.FT.
Library or Museum	
General Areas	15
Work Rooms	15
Rare books & Exhibits	15
Shopping Mall	
Mall	0.20 / SQ.FT.
Store	0.30 / SQ.FT.
Motels	
Bedrooms ②	30 / Room
Suites	30 / Room
Lobbies	15
Conference	20
Assembly Hall	15
Nursing Homes	
Patient Room ②	25
Medical Procedures	15
Physical Therapy	15
Office Building	
Office ②	20
Bull Pen	20
Conference	20
Reception	15
Pet Shops	1.0 / SQ.FT.

Pharmacy	
Work Room	15
Sales Area	0.3 / SQ.FT.
Photo Studio	
Studio	15
Darkroom	0.5 / SQ.FT.
Pool Hall	20
Restaurant / Nite-club	20
School	
Classroom	15
Laboratories	20
Shops	20
Music Rooms	15
Libraries	15
Auditoriums	15
Gyms	20
Locker Rooms	0.5 / SQ.FT.
Smoking Lounges	60
Offices ②	20
Lunch Room	20
Shoe Repair	20
Skating Rinks	0.5 / SQ.FT.
Small Stores	0.3 / SQ.FT.
Smoking Room	60
Supermarket	15
Swimming Pools	
Pool & Deck	0.5 / SQ.FT.
Spectator	15
Theater	
Public areas	20
Floor	15
Ticket Booth	20
Stages & Work Rooms	15
Public Toilet	50 / Fixture
Transportation Terminal	
Public Areas	15
TV-Radio	
Booths	20
Studios	20
Electronic Equipment	20
Audience	15
Stages	15
Variety Stores	0.30 / SQ.FT.
Veterinarian	
Kennel	1.0 / SQ.FT.
Operating Room	30
Waiting Room	15
Warehouse	0.15 / SQ.FT.

① 100 CFM kitchen Exhaust
② 50 CFM Private Bath Exhaust
③ Unless otherwise noted

A1-3 Ventilation Rates

The table on the previous page summarizes the ventilation rates that are recommended by ASHRAE Standard 62-89. These ventilation rates should be used unless the designer can show that lower ventilation rates are acceptable. (Lower ventilation rates can be used, providing that they are generated by the calculation procedures that are included in the 62-89 Standard. However, these procedures are not easy to use and they may require information that is not readily available.)

A1-4 Comfort Chart

The ASHRAE comfort chart summarizes the various combinations of dry-bulb temperature and relative humidity that will produce a comfortable environment (for most people). Two sets of conditions are used, one for cooling and one for heating. Note that the shaded areas, which represent the "applied" comfort zones, are considerably smaller than the ASHRAE comfort zones. (The "applied" comfort zones recognize the ability of the cooling equipment to remove water from the air and consider the potential for condensation to form on cold surfaces during the heating season.)

A1-5 Recommended Supply Outlet Noise Levels

The amount of noise that can be generated at the supply outlets depends on the application. The information sheet on the next page provides a list of acceptable NC values and supply outlet air velocity values. (The NC value is determine by plotting the noise level that is associated with each octave band on a special NC evaluation chart (see **Manual Q**, page 21-3).

Recommended Supply Outlet Noise Levels for Various Applications

Application	NC Level	Maximum Velocity
Private Residence		
Rural/Suburban	20 to 25	700
Urban	25 to 30	700
Apartments	30 to 35	700
Hotels/Motels		
Private Room or Suite	30 to 35	700
Meeting/Banquet Room	30 to 35	700
Halls, Corridors, Lobbies	35 to 40	800
Service and Support Areas	40 to 45	900
Offices		
Executive	25 to 30	700
Conference Rooms	25 to 30	700
Private	30 to 35	700
Open Bay Areas	35 to 40	800
Computer Room	35 to 40	900
Business Machine Room	40 to 45	900
Waiting Rooms	40 to 45	900
Retail Stores		
Clothing Department	35 to 40	800
General Sales Areas	40 to 50	900
Supermarkets	40 to 50	900
Hospitals/Nursing Homes/Clinics		
Private/Semi-private Rooms	25 to 30	700
Operating Rooms	25 to 30	700
Wards	30 to 35	700
Churches and Schools		
Sanctuaries	20 to 30	700
Libraries	30 to 35	700
Classrooms	30 to 35	700
Laboratories and Shops	35 to 40	800
Gyms and Rec. Areas	40 to 50	900
Kitchens	40 to 50	900

Application	NC Level	Maximum Velocity
Public Assembly		
Concert Halls	20 to 25	700
Legitimate Theaters	20 to 25	700
Auditoriums	25 to 30	700
Motion Picture Theaters	30 to 35	700
Libraries / Museums / Courts	30 to 35	700
Coliseums	30 to 35	700
Convention Halls	30 to 35	700
Exhibition Halls	35 to 40	800
Bowling Centers	40 to 50	1000
Bank / Post Office	35 to 40	800
Lobbies	35 to 40	800
Swimming Pools	40 to 55	1000
Transportation Terminals		
Ticketing Areas	30 to 35	700
General Waiting	35 to 45	800
Studios		
Recording Studios	20 to 25	700
Broadcasting Studios	20 to 25	700
TV audience Studios	30 to 35	700
Restaurants/Cafeterias/Lounges		
Restaurant Eating Area	35 to 40	800
Lounges and Bars	35 to 45	800
Night Clubs	35 to 45	800
Cafeterias	40 to 50	900
Kitchens	40 to 50	900
Industrial Production Areas	Background noise sets noise limits	

Note: *Maximum Velocity* refers to the face or neck velocity of the supply outlet (FPM).

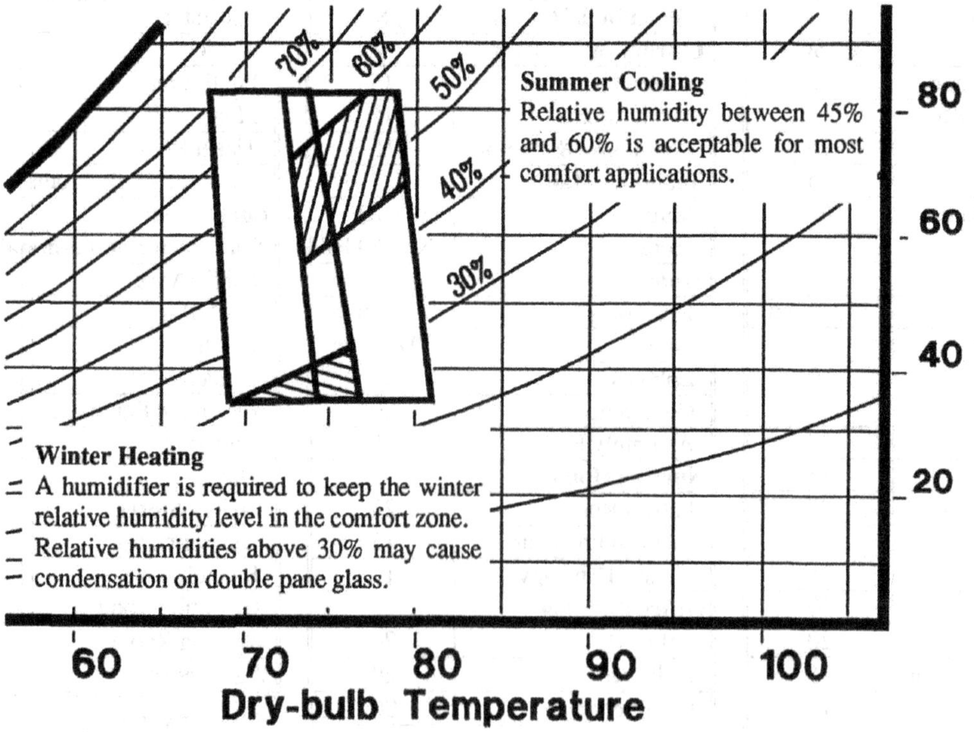

ASHRAE Comfort Chart

A1-6 Air-Side Equations (below 2000 feet elevation)

The CFM that must be supplied to a room, zone or space depends on the sensible load (cooling or heating), the room temperature and the temperature of the air that enters the room through the supply outlet.

$$\text{Cooling CFM} = \frac{\text{Room Sensible Cooling Load (BTUH)}}{1.1 \times (\text{Room DB} - \text{Supply Air DB})}$$

$$\text{Heating CFM} = \frac{\text{Room Heating Load (BTUH)}}{1.1 \times (\text{Supply Air DB} - \text{Room DB})}$$

The sensible heat equation relates the heat that is added to (or subtracted from) air that flows through a piece of heating (or cooling) equipment to the CFM and the air-side temperature change.

$$\text{BTUH} = 1.1 \times \text{CFM} \times (T_{out} - T_{in})$$

The latent heat equation relates the latent heat that is added to or extracted from an air flow (by latent loads, humidification equipment or dehumidification equipment) to the CFM and the change in the air-side moisture content.

$$\text{BTUH} = 0.68 \times \text{CFM} \times (\text{Grains Difference})$$

The total heat equation relates the total amount of heat (sensible plus latent) that is added to or extracted from an air flow (by any type of load or by any type of HVAC equipment) to the CFM and the change in the air-side enthalpy.

$$\text{BTUH} = 4.45 \times \text{CFM} \times (\text{Enthalpy Difference})$$

The temperature of the air that is exiting from a duct run may be warmer or cooler than the temperature or the air that enters the duct run, depending on the gains (during cooling) or the losses (during heating). The total gain is equal to the sum of the duct gain, the return air ceiling plenum gain and the fan heat (plus any other gain). The total loss is equal to the sum of the duct loss and the return air ceiling plenum loss (plus any other loss).

$$\text{Exiting DB} = \text{Entering DB} + \frac{\text{Duct Gains (BTUH)}}{1.1 \times \text{Return CFM}}$$

$$\text{Exiting DB} = \text{Entering DB} - \frac{\text{Duct Losses (BTUH)}}{1.1 \times \text{Return CFM}}$$

The temperature of the mixed air (as it leaves the mixing box) depends on the return air temperature, the outdoor temperature and the percentage of the supply air that flows through the outdoor air damper.

$$\text{MAT} = \text{RAT} - \text{Percent OA} \times (\text{RAT} - \text{OAT})$$

The temperature of the air that leaves a piece of equipment depends on the entering air temperature, the heat that is subtracted (cooling) or added (heating) by the equipment and the CFM that flows through the equipment.

$$\text{LAT} = \text{EAT} - \frac{\text{BTUH Removed}}{1.1 \times \text{CFM}} \quad \text{(cooling)}$$

$$\text{LAT} = \text{EAT} + \frac{\text{BTUH Added}}{1.1 \times \text{CFM}} \quad \text{(heating)}$$

A1-7 Water-Side Equations

The water-side equation relates the heat that is added to (or subtracted from) water that flows through a piece of heating (or cooling) equipment with the GPM and the water-side temperature change.

$$\text{BTUH} = 500 \times \text{GPM} \times \text{Temperature Change}$$

$$\text{GPM} = \frac{\text{BTUH (in or out)}}{500 \times \text{Temperature Change}}$$

$$\text{Temperature Change} = \frac{\text{BTUH (in or out)}}{500 \times \text{GPM}}$$

The temperature of the water that is exiting from a pipe run may be warmer or cooler than the temperature or the water that enters the same pipe run, depending on the heat gains (during cooling) or the heat losses (during heating). The total heat gain is equal to the sum of the pipe heat gains plus the heat that is added by the pump (plus any other heat gain). The total heat loss is equal to the sum of the pipe heat losses (plus any other heat loss).

$$\text{Rise } (^\circ F) = \frac{\text{Gain (BTUH)}}{500 \times \text{GPM}}$$

$$\text{Drop } (^\circ F) = \frac{\text{Loss (BTUH)}}{500 \times \text{GPM}}$$

A1-8 Cooling Equipment Information

The cooling equipment sizing load depends on whether the system can take advantage of diversity. The following table shows how the sizing load is determined for various types of systems.

Combining Zone Loads

System Type	Cooling Capacity	Coil CFM	Fan CFM	Duct Diversity
TRH	Sum Zones	Sum Zones	Sum Zones	No
M-VAV	Block	Block	Block	Yes
BP-VAV	Block	Sum Zones	Sum Zones	No
DT-DD	Sum Zones	Sum Zones	Sum Zones	Yes
BT-DD	Block+XS	Block+XS	Sum Zones	Yes
DT-Mult	Sum Zones	Sum Zones	Sum Zones	No
BT-Mult	Block+XS	Block+XS	Sum Zones	No

XS = An allowance to compensate for air that is diverted through the hot-deck — approximately 10 percent.

TRH — Terminal reheat
M-VAV — Full modulating variable air volume
BP-VAV — Bypass variable air volume
DT-DD — Draw-through dual-duct
BT-DD — Blow-through dual-duct
DT-Mult — Draw-through multizone
BT-Mult — Blow-through multizone

Cooling equipment should not be excessively oversized. The following table provides sizing guidelines for sizing cooling equipment and heat pumps.

Summary of Sizing Limits — Cooling

Cooling-Only or Heat Pump in Mild Climate

Sensible capacity exceeds sensible load
Latent capacity exceeds latent load
Total capacity less than 1.15 x total load

Heat Pump in Cold Climate

Sensible capacity exceeds sensible load
Latent capacity exceeds latent load
Total capacity less than 1.25 x total load

A1-9 Heating Equipment Information

Heating equipment should not be oversized. The following table provides some guidelines for sizing hot air furnaces, hot water boilers and electric heating coils. (Air- and water-source pump capacity is determined by the cooling load.)

Summary of Sizing Limits — Heating

Hot Air Furnaces

Heating capacity must exceed the design load
Capacity less than 1.15 x design load preferred
Capacity should not exceed 1.40 x design load

Hot Water Boilers

Heating capacity must exceed the design load
Capacity less than 1.15 x design load preferred
Capacity should not exceed 1.40 x design load

Electric Coils

Heating capacity must exceed the design load
Capacity less than 1.15 x design load preferred

Use a time clock or programmable stat for set-back recovery.

With an air-source heat pump, the capacity of the supplemental heaters is determined by drawing a balance point diagram. (The same procedure is used for water-source equipment.) An example of a balance point diagram for an air-source unit is shown below.

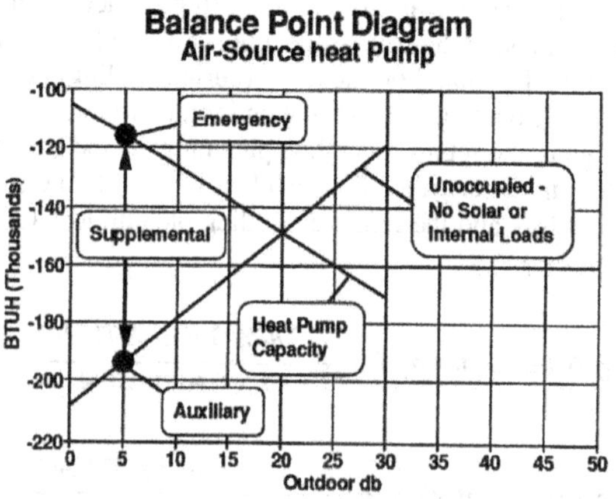

Appendix 2
Applications

This appendix provides a summary of the design requirements that are associated with specific applications. Since it is not practical to list every possible application, try to find one that is similar to what you are working on. In some cases, it may be necessary to use two or more different applications. Refer to Section 3 for a generic discussion of application design requirements. Try to apply the principles outlined in Section 3 to the application you are working on.

Refer to Section 3 or Appendix 1 for tables and charts that provide specific information about the indoor design conditions, filter requirements and outdoor air requirements for various applications. Note that these tables and charts do not take priority over local codes and regulations. Always refer to codes and regulations first. Only use the tables and charts that are provided in this manual when codes and regulations do not provide the necessary guidance.

Applications Included in this Section

Industrial and Factory Applications

Occupancy Multipurpose occupancy — offices, production areas, parts department, shop area, lunch room, locker rooms and storage areas (refer to other applications for more information).

Functions Heating and ventilation is normally required, cooling and dehumidification may be required. In some cases, humidification could be required. The functions depend on the environment that is required for people, processes and product.

Zoning Zone the production areas according to their functional requirements. Zone offices, parts departments, shop areas and lunch rooms. Separate interior zones from perimeter zones. Perimeter zones may be sensitive to the exposure. All zones must be compatible with the occupancy and production schedules.

Lighting Low to high intensity levels depend on the area — make a survey.

Occupancy Loads Population density is usually moderate in most areas. Expect large occupancy loads in the lunch room. Get occupancy information from the plant superintendent.

Equipment Loads Sensible and latent loads are associated with production equipment, manufacturing processes, computers, office equipment, minor appliances, lunch room appliances and vending machines. Make a survey (determine the actual power requirements and usage factors).

Cooling Load The cooling loads in the production areas are closely associated with the production equipment and the manufacturing processes. The envelope loads also may be important. The cooling load in non-production areas depend on the occupancy and the envelope loads. Relatively steady loads usually characterize the interior-zone areas, but the loads in the perimeter-zone areas will be affected by the exposure and the envelope construction details.

Operating Mode Operating modes vary from area to area, depending on the space requirements, outdoor temperature, time-of-year, time-of-day and production schedule. Make a thorough survey for each area.

Humidity Control The humidity loads and humidity control requirements vary from area to area. Make a thorough survey for each area. The humidity levels in non-production areas can be allowed to drift within a "comfort zone." Some processes may require close humidity control. Some production areas may not require any humidity control — ask the plant superintendent.

Air Quality Air quality is an important factor in all areas. Make a survey to evaluate all sources of pollution. Refer to codes and standards.

Exhaust CFM Exhaust systems may be required to control contaminated air. Make a survey to evaluate the exhaust requirements. Refer to codes and standards. Exhaust lunch rooms, locker rooms and toilets.

Outdoor Air Fresh air is required to control air quality. Make a survey to evaluate the fresh air requirements for each area. Refer to codes and standards. Outdoor air also is required to replace exhaust air.

Space Pressure The pressure in the production areas must be compatible with the manufacturing process and the production equipment. The pressure in other areas depends on the air quality and safety requirements that are associated with the area. Get information from the plant superintendent and refer to codes and standards.

Schedules The plant superintendent should be able to provide information about the operating schedules.

Noise HVAC system noise is not usually a problem in the production areas. Talk to plant superintendent to determine if there are any special noise level requirements.

Special Problems Controlling the temperature, humidity and air quality in an industrial environment can be difficult and expensive. Some applications present serious health, fire and explosion hazards.

Laboratories

Occupancy

Multiple purpose occupancy — laboratory rooms, animal rooms, special storage rooms, library, auditorium, lobby, offices, meeting rooms, lounges, computer rooms, cafeteria, kitchen, locker rooms, laundry, maintenance shops, service areas and more (refer to other applications for more information).

Functions

Any combination of heating, cooling, dehumidification, ventilation and humidification may be required.

Zoning

Each laboratory room is a separate zone. Zone the services areas, storage areas, animal rooms and the administrative areas. Zone libraries, laboratories, auditoriums, lobbies, lounges, cafeterias and kitchens. Some areas may be subdivided into more than one zone. Interior areas may require year-round cooling. Perimeter zones are sensitive to exposure — consider the solar loads and heat loss that is associated with glass. Large heating loads usually occur near doors (heated vestibules are desirable). Refer to other applications for more information.

Lighting Loads

Low to high, to very high lighting loads, depending on the area — make a survey.

Occupancy Loads

The population density in the various areas may be constant, variable by schedule or variable in a random way. Public areas may experience wide variations in population density. Maximum population density is based on the expected "full-room" occupancy. Large populations increase humidity loads, reduce room sensible heat ratios and reduce the heating changeover points. Areas that are associated with steady, low-density populations tend to have lower latent loads and higher sensible heat ratios. Get schedules and occupancy information from a staff authority. Refer to other applications for more information.

Equipment Loads

Laboratories have a wide variety of electrical and fossil fuel equipment that produces sensible and latent loads. Sensible and latent loads also are associated with laboratory animals. Many of these loads are space loads, but some loads may be neutralized by exhaust hoods or laboratory equipment cooling systems. Televisions, sound equipment, vending machines, audio-visual equipment and small appliances may be found in public, staff and administrative areas. A rigorous load analysis is required for the kitchen and cafeteria areas. Offices have internal loads, which are generated by computers, typewriters, duplicating equipment and small appliances. Service areas have loads that are associated with laundry and shop equipment. Make a thorough survey of every room in the building to determine the sensible and latent loads that are associated with equipment and appliances. Use load factors and the usage factors to de-rate "name-plate" inputs. Consider the effects of diversity and exhaust hoods. Refer to other applications for more information.

Laboratory Animal Metabolisim and Outdoor Air Requirements

	Weight (LB)	Sensible	Latent	ACH
Mouse	0.05	1.11	1.65	15
Hamster	0.25	4.02	6.0	15
Rat	0.62	7.77	11.6	15
Guinea Pig	0.90	10.2	15.2	15
Rabbit	5.41	39.2	58.5	10
Cat	6.61	45.6	68.1	10
Primate	12.0	71.3	106.0	15
Dog (Small)	22.7	105.0	161.0	10
Dog (Medium)	50.0	231.0	355.0	10

Sensible and latent heat gains = BTUH; ACH = Air changes per hour (Outdoor air)

Laboratories — Continued

Cooling Load

The peak cooling load that is associated with any perimeter area occurs when the occupancy loads, lighting loads, equipment loads and envelope loads combine to produce the peak-load condition. This peak may not coincide with the peak envelope load because the occupancy, lighting and equipment loads may not be synchronized with the envelope load. The peak cooling load that is associated with any interior area occurs when the occupancy loads, lighting loads and equipment loads combine to produce a peak-load condition. The sensible and latent cooling loads in some areas may be quite steady. Other areas may experience wide load-swings (depending on occupancy schedules, equipment operating schedules and solar gains). A rigorous analysis of full-load and part-load conditions must be made for every room. Refer to other applications for more information.

Operating Mode

Perimeter rooms typically changeover to heating as the outdoor temperature drops. However, perimeter rooms with large internal loads may not changeover until it gets very cold outside — or they may not changeover at all. Interior zones require year-round cooling. Simultaneous heating and cooling is required for incompatible zones. Reheat may be required to control humidity in mild weather. Humidification might be required in cold weather. Analyze the part-load requirements of every room to determine the operating requirements (mode and schedules). Refer to other applications for more information.

Humidity Control

Humidity control is critical in some areas, in which case, reheat, humidification and sensitive controls will be required. Laboratory humidity levels can range from less than 30 percent to more than 95 percent, depending on the experiment or process. The relative humidity can be allowed to drift within the "comfort zone" in the other areas. Humidity control in some areas will be affected by internal loads, exhaust systems and the make up air requirements — make a thorough survey of the factors that affect each room. Low occupancy rates and large sensible (lighting and equipment) loads will produce small latent loads and high room sensible heat ratios. Large populations and moderate sensible (lighting and equipment) loads will produce large latent loads and low room sensible heat ratios. Refer to other applications for more information.

Air Quality

Air quality may be a critical factor throughout the facility. Most laboratory and supporting service areas have specific filter, ventilation, exhaust and space pressure requirements. Biological contaminants, odors, fumes and moisture must not be allowed to migrate from room to room. Large quantities of outdoor air are required for some rooms. Special exhaust hoods may be required. In some cases, contaminated exhaust flows must be sterilized before they are expelled. Some rooms and duct systems have to be made of special materials. Some ducts must be completely free of dirt and dust. Some ducts must be air tight. Cross-contamination between outdoor intakes and exhaust openings is totally unacceptable. Make a thorough survey of each room to determine the exact nature of the problems and requirements. Use general ventilation and standard filters for administrative, staff and public areas. Contaminated air that is associated with the kitchen, dedicated smoking areas, duplicating equipment, laundries, showers, locker rooms, toilets and maintenance shops must not be allowed to enter other areas. Refer to other applications for more information.

Exhaust CFM

Many laboratory areas have specific exhaust requirements. Special exhaust hoods may be required. In some cases, contaminated exhaust flows must be filtered or sterilized before they are expelled. Exhaust systems may consist of multiple "single-apparatus" exhaust systems or a central "collective" exhaust system. Use field-measured exhaust CFM values and account for the effects of diversity. Some exhaust systems have to be air tight and/or made of special materials. Make a thorough survey of the exhaust requirements associated with each room. Exhaust systems also are required for food preparation areas, maintenance shops, laundries, dedicated smoking areas, locker rooms, showers and toilets. Refer to other applications for more information.

Outdoor Air

Many laboratory areas have specific ventilation requirements ranging from 100 percent recirculated air to 100 percent outdoor air. Animal rooms require 10 to 15 air changes per hour (see previous page). Make a thorough survey of the ventilation requirements that are associated with each room or area. In the public, staff and administrative areas, the outdoor air CFM depends on population density and exhaust CFM. Refer to other applications for more information.

Laboratories — Continued

Space Pressure

Many laboratory areas have specific pressure requirements ranging from negative to positive. Make a thorough survey of the pressure requirements that are associated with each room or area. Food preparation areas, duplicating departments, maintenance shops, laundries, dedicated smoking areas, locker rooms, showers and toilets should be negative. All other public, staff and administrative areas can be neutral or a little positive. Refer to other applications for more information.

Schedules

Some laboratories, administrative and public areas may be occupied or operated on a 24-hour basis. Other areas may be occupied or operated on a schedule. Get schedule information from a staff authority.

Noise

Noise levels may be important or even critical for some types of laboratory rooms. Get the noise level requirements from a staff authority. Meeting rooms, offices, the library and the auditorium should be quiet. Noise is important, but not critical, in some public areas, and it may not be important in the unoccupied areas.

Special Problems

Almost every room is a zone. A large number of areas may require simultaneous heating and cooling. Many rooms are not compatible because of conflicting temperature, humidity, filter, pressure, exhaust and ventilation requirements. Temperature, humidity, air quality and pressure may have to be held to close tolerances. Through-the-wall equipment may not be suitable for some rooms or zones. Large sensible and latent loads and load swings are associated with population density, equipment, lighting and envelope construction details. Year-round cooling and dehumidification may be required in some areas. Reheat may be required to control humidity. Humidification may be required in cold weather. The operating modes, space conditions and operating hours that are associated with various occupancies and zones may not compatible with each other. Large heating loads occur near entrances. Special exhaust and ventilation equipment may be required in some areas. Special fire and safety requirements may apply. Installation, operating and maintenance costs may be high.

Radio, TV and Communications Buildings

Occupancy

Multipurpose occupancy — general and private offices, conference and meeting rooms, broadcast booths, control rooms, studios, stages, editing rooms, audience areas, reception area, printing and duplicating departments, computer rooms, audio-visual rooms, dressing rooms, library, communications equipment rooms, workshops, mail room, lunch room, storage rooms. Refer to other applications for more information.

Functions

Any combination of heating, cooling, dehumidification, ventilation may be required.

Zoning

Zones are required for offices, meeting rooms, bull pen areas, broadcast booths, control rooms, studios, stages, audience areas, editing rooms, audio-visual rooms and electronic equipment rooms. Zone reception areas, lunch rooms, computer rooms, printing and duplicating departments, workshops, mail rooms and storage rooms. All zones must be compatible with occupancy schedules and equipment operating schedules. Also consider the heating loads near the outside doors and the cooling loads that are associated with exposed glass and electrical equipment.

Lighting Loads

Medium, high or very high lighting levels, depending on the area — talk to manager and make a field survey.

Occupancy Loads

Population density varies by area. Within a particular area, the population density may be steady or variable. High density and large variations in density can be expected to occur in audience areas and lunch rooms. Plan for low population densities in private offices. Make provisions for wide occupancy swings in conference and meeting rooms. Check the population densities that are associated with support and services areas. Get occupancy information from the manager.

Equipment Loads

Large sensible loads are associated with cameras, electronic equipment, broadcasting and communications equipment, computers and audio-visual equipment. Sensible and latent loads are associated with coffee brewers, lunch room appliances, vending machines and duplicating equipment. Get information about the equipment loads from the manager. Use load factors and the usage factors to de-rate "name-plate" inputs. Consider the effects of diversity

Cooling Load

The large sensible loads that are associated with lights and equipment may consist of a steady "base load" plus a variable "task load." Large (but variable), sensible and latent loads are associated with audiences. The peak loads in the broadcasting areas are usually dictated by the production schedule. Relatively steady loads are associated with interior offices and supporting services areas. Anticipate variable sensible and latent loads in meeting rooms and lunch rooms. The load swings in perimeter zones will depend on the internal loads, exposure and envelope construction details.

Operating Mode

Zones with large internal loads require year around cooling. Audience areas may require cooling and dehumidification in cold weather. Perimeter zones may or may not changeover to heating — depend on internal loads, exposure and envelope construction details. Simultaneous heating and cooling is required for zones with incompatible operating modes.

Humidity Control

Some areas may have high sensible heat ratios because the lighting and equipment loads are large and latent loads are (relatively) small. However, the latent loads in the audience area can be relatively large and the corresponding room sensible heat ratios can be relatively low. Audience areas also may experience wide load and sensible heat ratio swings, depending on occupancy. Reheat may be required to control humidity in the audience areas. The humidity levels in most other spaces can be allowed to drift within a "comfort zone." Ask the manager if there are any zones or pieces of equipment that require precise humidity control.

Air Quality

General ventilation is required for most spaces. Special filters are not required for most spaces. Some electronic equipment may require high efficiency filters. Isolate and exhaust the fumes that are associated with food services, work shops and duplicating equipment.

Radio, TV and Communications Buildings — Continued

Exhaust CFM

Exhaust equipment is required for lunch rooms, duplicating departments, workshops, (some) equipment rooms and the toilets. An exhaust system may be required for the stage area.

Outdoor Air

The outdoor air CFM depends on the population density and exhaust CFM requirements that are associated with each area — make a thorough survey.

Space Pressure

The pressure in lunch rooms, duplicating departments, workshops, some equipment rooms and toilets should be slightly negative. A negative pressure may be required in the stage area. All other areas can be neutral or a little positive.

Schedules

Some areas may require 24-hour conditioning. Talk with the manager to determine the occupancy and equipment operating schedules. Take advantage of diversity.

Noise

Noise levels are important, especially in the broadcasting and production areas. Private offices, meeting rooms and most of the supporting services rooms should be quiet.

Special Problems

Large sensible loads are generated by lights and equipment. Large sensible and latent loads are associated with audiences. Operating and occupancy schedules cause large load (sensible and latent) swings. Year-round cooling and dehumidification is required. The operating hours that are associated with the various areas may not be compatible with each other. Noise control is critical in some areas. Some areas may require a very large amount of supply CFM per-square-foot of floor area. Large heating loads occur near the audience entrance doors.

Warehouses

Occupancy

Multipurpose occupancy — offices, computer room, shipping rooms, lunch room, locker rooms and high-bay storage areas.

Functions

Heating and ventilation is required. Cooling and dehumidification is usually required for the office areas. In some cases, the warehouse area may require cooling or humidification.

Zoning

Zone the offices, data processing rooms and lunch rooms. Separate the interior office areas from the perimeter office areas. Perimeter areas may have to be zoned by exposure. The warehouse area usually requires a separate heating system. All zones must be compatible with occupancy schedules.

Lighting

Lighting levels are usually low in the warehouse area and medium to high in office areas — make a survey.

Occupancy Loads

The population density is usually light and steady in the office areas. Large occupancy loads are associated with the lunch room. Get occupancy information from the manager.

Equipment Loads

In the office areas, sensible loads are generated by computers, typewriters and communications equipment. Sensible and latent loads are associated with coffee brewers, lunch room appliances, vending machines and duplicating equipment.

Cooling Load

Loads are relatively steady in the interior office areas, but there are wide, occupancy related, load swings in the lunch room. Load swings can be expected to occur in the perimeter office areas. Peak perimeter loads depend on the exposure and the envelope construction details. Cooling is not usually required in the warehouse.

Operating Mode

The interior office areas require year-round cooling. The perimeter zone heating changeover points depend on the exposure and the envelope construction details. Simultaneous heating and cooling is required for incompatible zones. Cooling may not be required in the warehouse area.

Humidity Control

In the office areas, the latent loads are relatively low and steady, so the room sensible heat ratios are relatively high. Humidity levels can be allowed to drift within a "comfort zone."

Air Quality

General ventilation is required for the offices and the lunch room. Special filters are not usually required. Isolate the offices and lunch room from fumes generated by forklift engines or battery charging equipment. Dilute any fumes that are released into warehouse area.

Exhaust CFM

Exhaust equipment is required in the lunch room, locker rooms, battery charging area and toilets. Large exhaust fans can be used to cool the warehouse in the summer.

Outdoor Air

In the offices and the lunch room, the outdoor air CFM depends on occupancy and exhaust requirements. In the warehouse area, the outdoor air CFM depends on the air quality and the exhaust requirements.

Space Pressure

The pressure in the lunch room, locker room and toilets should be negative — the office areas can be neutral or a little positive. The pressure in the warehouse should be neutral or a little negative.

Schedules

Ask the manager for the operating and occupancy schedules.

Noise

Noise levels must be controlled in the offices areas.

Special Problems

Economics may dictate less than desirable temperatures in the warehouse. Hot air will stratify in high-bay areas. Large heating loads are caused by infiltration at the shipping doors (special infiltration barriers and door heaters may be required). Fumes are generated by lift trucks engines and there is an explosion risk in the battery charging area.

Bars

Occupancy Multipurpose occupancy — bar area, table area, dancing area, grill or kitchen, office and storage areas.

Functions Heating, cooling, dehumidification and ventilation are usually required.

Zoning The bar, table and dancing areas create one or more zones. The grill or kitchen area must be isolated from the patron areas. Zone the office areas and the storage areas. Consider the large solar loads and heat losses that are associated with exposed glass. Large heating loads occur near entrances (heated vestibules are desirable).

Lighting Loads The lighting loads are usually low, especially in the patron areas.

Occupancy Loads The population density can vary widely. Periods of high densities are normal. Additional sensible and latent loads are associated with food service. Patrons increase the humidity loads, reduce the room sensible heat ratio and reduce the heating changeover point. Dancing causes even higher sensible and latent loads. Get occupancy information from the manager.

Equipment Loads Sensible and latent loads are generated by small appliances, motors, video games and food serving counters that are located in patron area. A separate load analysis is required for the kitchen.

Cooling Load Interior area cooling loads depend on the internal loads. Perimeter areas are affected by the internal loads, the exposure and the envelope construction details. Wide load swings are caused by occupancy variations. The peak load condition is usually dictated by the occupancy schedule.

Operating Mode The interior zones require year-round cooling. The perimeter zone, heating changeover points depend on the exposure, the envelope construction details and the occupancy loads. Simultaneous heating and cooling is required for incompatible zones.

Humidity Control Wide swings in latent loads and room sensible heat ratios are caused by changes in the occupancy level. If occupancy is high, the latent loads can be very high and the room sensible heat ratios can be low. Humidity levels can be allowed to drift within a "comfort zone."

Air Quality The grill or kitchen area has special ventilation and filter requirements. General ventilation and standard filters are required for the patron and the auxiliary areas. Smoking may be heavy in the patron areas. Capture and exhaust the cooking odors. Do not recirculate air from the kitchen, locker rooms and toilets.

Exhaust CFM The grill or kitchen exhaust CFM depends on the cooking and the exhaust hood equipment (see ACCA **Manual Q**). Exhaust equipment is requird for the toilets and locker rooms.

Outdoor Air Outdoor air CFM depends on the occupancy and the exhaust CFM requirement. Introduce the outdoor air in patron areas. Dancing areas have larger ventilation requirements than eating areas. Make up air may be required in kitchen.

Space Pressure The patron areas should be under a slight positive pressure. The grill area or kitchen, locker rooms and toilets should be under a negative pressure.

Schedules Talk to manager to determine the occupancy schedule, employee hours and kitchen operating schedules.

Noise There are usually no special noise level requirements.

Special Problems Large variations in cooling loads are related to occupancy levels. Incompatible spaces create ventilation, pressure and air balance problems. Large heating and cooling loads occur near exposed glass. Large heating loads occur near the main entrance.

Diners and Fast Food Stores

Occupancy Basically, a single purpose occupancy characterized by a perimeter zone eating area, cooking facilities, a storage area and possibly, an office.

Functions Heating, cooling, dehumidification and ventilation are usually required.

Zoning The counter and table areas create one or more zones. The grill or kitchen areas must be isolated from the patron areas. Zone the office areas and storage areas. Consider the large solar loads and the heat loss that is associated with exposed glass. Large heating loads occur near the entrances (heated vestibules are desirable).

Lighting Loads Lighting loads are medium to high.

Occupancy Loads The population density can vary widely. Periods of high densities are normal. Consider the additional sensible and latent loads associated with food service. The occupants increase the humidity load, reduce the room sensible heat ratio and reduce the heating changeover point. Get occupancy information from the manager.

Equipment Loads Sensible and latent loads are associated with the small appliances, motors, video games and food serving counters that are located in patron area. A separate load analysis is required for grill or kitchen area.

Cooling Load Usually, the entire seating area is in a perimeter zone, which is affected by the internal loads, the exposure and the envelope construction details. Wide load variations are associated with solar gains, glass transmission loads and changes in population density. The peak loads is usually determined by the combined affect of the solar gains and the occupancy load.

Operating Mode The perimeter zone, heating changeover points depend on the exposure, the envelope construction details and the occupancy loads. Simultaneous heating and cooling will not be required if there is no interior zone. An office might require its own source of heat.

Humidity Control Wide swings in the latent load and the room sensible heat ratio are associated with the solar gains, the transmission loads, the infiltration loads and the occupancy loads. Humidity levels can be allowed to drift within a "comfort zone."

Air Quality The grill or kitchen area has special ventilation and filter requirements. General ventilation and standard filters are required for the other areas. Smoking may be heavy in the patron areas. Capture and exhaust the cooking odors. Do not recirculate air from cooking area.

Exhaust CFM Exhaust the cooking area. The exhaust CFM depends on the cooking and exhaust hood equipment (see ACCA **Manual Q**). Exhaust the toilets.

Outdoor Air Outdoor air CFM depends on th occupancy and the exhaust CFM requirements. Introduce outdoor air in the patron areas. Make up air may be required in the cooking area.

Space Pressure The patron areas should be under a slight positive pressure. The cooking area and the toilets should be under a negative pressure.

Schedules Talk to manager to determine occupancy schedule and employee hours.

Noise There are no special noise level requirements.

Special Problems Large variations in the cooling loads are associated with the occupants and the solar gains. Incompatible spaces create ventilation, pressure and air balance problems. Large heating and cooling loads occur near exposed glass. Large heating loads occur near the entrances.

Kitchens and Bakeries

Occupancy
Basically, a single purpose occupancy, but one or more small office areas could be involved. Bakeries usually have a sales area. Most kitchens and some bakeries are part of a larger facility.

Functions
Heating and ventilation is required in the food preparation areas (cooling may be requested). Heating, ventilation, cooling and dehumidification is usually required in an office or sales area.

Zoning
Zone the food preparation areas, offices and sales areas. In the sales area, large cooling loads may occur near glass and large heating loads may occur near the doors (a heated vestibule is desirable).

Lighting Loads
Lighting loads may be medium or high — make a survey.

Occupancy Loads
The population density is low to medium in a bakery sales area or a kitchen. Normally, the occupancy loads only have a small effect on the room sensible heat ratio or the heating changeover point.

Equipment Loads
Determine the sensible and latent loads that are associated with kitchen equipment. (Refer to ACCA **Manual** N for information about appliance loads. Contact the cooking equipment manufacturer. Talk to the chief.) Use load factors and the usage factors to de-rate "name-plate" inputs. Consider the effects of diversity. Account for load reductions that are associated with the exhaust hoods.

Cooling Load
Small kitchen offices may be interior zones. The peak cooling load in a bakery sales area might be caused by the combined effect of the solar gain and the occupancy schedule. (Changes in the solar gain and the population density can cause wide load swings in the sales area.) Kitchen cooling loads are primarily associated with the equipment loads and the make up air loads. (Try to capture and exhaust the heat and moisture generated by cooking equipment.)

Operating Mode
Year-round cooling may be required in an "interior-zone" office. The heating changeover in a sales area will depend on the relative size of the internal loads and the envelope loads. If there is a lot of glass, this changeover may occur during mild weather. In a kitchen, the heating changeover point will usually depend on the relative size of the equipment loads and the outdoor air load.

Humidity control
Office and sales area population densities are usually not high enough to cause humidity control problems. Allow the humidity levels to drift within the "comfort zone." Effective exhaust systems are required to control the humidity in the kitchen.

Air Quality
The heat, mists, vapors and odors that are associated with cooking equipment should not exfiltrate into adjoining areas. Special filters are required for exhaust hoods.

Exhaust CFM
General exhaust systems or special exhaust hoods are required in the kitchen. Refer to ACCA **Manual** Q for more information about exhaust systems. Contact exhaust system equipment manufacturers.

Outdoor Air
Outdoor air is required to replace the air that is removed by exhaust equipment. Some, or perhaps all, of this make up air can be introduced (as ventilation) into the public areas and then allowed to exfiltrate to the kitchen. In some cases, a make up air unit will be required for the cooking area.

Space Pressure
Negative pressure is required in the cooking area. The adjoining areas should be maintained at a slightly positive pressure.

Schedules
Operating schedules vary — ask the chef or manager.

Noise
Normally, there are no unusual limitations on noise levels.

Special Problems
Large loads are associated with make up air. It is important to maintain the pressure differential between the cooking area and the public areas. During hot weather, the cooking areas will not be comfortable unless cooling is installed. The grease that is associated with cooking equipment presents a fire and safety hazard.

Restaurants and Night Clubs

Occupancy Multipurpose occupancy — general dining areas, banquet rooms, bar, dance floor, stage, waiting room, dressing rooms, offices, kitchen, locker rooms and store rooms. Refer to other applications.

Functions Any combination of heating, cooling, dehumidification and ventilation may be required.

Zoning The general dining area may require more than one zone. Banquet rooms, bars, dancing areas, stage areas, dressing rooms, offices, locker rooms and store rooms should be separate zones. The kitchen must be isolated for the patron areas. Consider the large solar gains and heat loss that is associated with glass and the heat loss near entrances (heated vestibules are desirable).

Lighting Loads Low lighting levels in the patron areas. Make a survey of the lighting loads in the other areas.

Occupancy Loads The population density varies widely. Periods of high densities are normal. Consider the additional sensible and latent loads that are associated with food service. The occupants increase the humidity load, reduce the room sensible heat ratio and reduce the heating changeover point. Dancing causes even higher loads. Get occupancy information from the manager.

Equipment Loads Sensible and latent loads are associated with small appliances, motors and food serving counters that are located in patron area. A load analysis is required for kitchen. The loads in other areas depend on the application — make a survey.

Cooling Load Interior-zone loads depend on the internal loads. Perimeter areas are effected by the internal loads and the envelope loads. Wide load swings are associated with the occupancy schedule. Peak loads are always related to the occupancy schedule.

Operating Mode Interior zones require year-round cooling. Perimeter zone heating changeover points depend on exposure, envelope construction details and occupancy loads. Simultaneous heating and cooling is required for incompatible zones.

Humidity Control Expect wide swings in the latent loads and the room sensible heat ratios. Latent loads can be very high and room sensible heat ratios can be low. Allow humidity levels to drift within the "comfort zone."

Air Quality Kitchens have special ventilation and filter requirements. General ventilation and standard filters are required for the other areas. Smoking may be heavy in some areas. Isolate the odors that are associated with the food and liquor service. Do not recirculate air from kitchen, locker rooms and toilets.

Exhaust CFM Exhaust the kitchen. (The kitchen exhaust CFM depends on the kitchen equipment — see ACCA **Manual Q**.) Exhaust the toilets and locker rooms.

Outdoor Air The outdoor air CFM depends on occupancy and the exhaust requirements. Introduce the outdoor air in patron areas. Dancing areas have larger ventilation requirements than eating areas. Make up air may be required in kitchen.

Space Pressure The eating areas should be under a positive pressure. The kitchen, locker rooms and toilets should be under negative pressure. Bars, dancing areas, stage areas and auxiliary areas should be at a neutral or negative pressure.

Schedules Ask the manager about the occupancy schedule, working hours and the kitchen operating schedule. Take advantage of diversity.

Noise Noise levels may or may not be important depending on the desired setting and mood.

Special Problems Large load variations are related to occupancy. Incompatible applications create ventilation, pressure and air balance problems. Frequent floor plan changes and tenant turn-over. Large heating and cooling loads near glass. Large heating loads near the entrance. Zones not isolated by walls.

Hospitals

Occupancy

Multiple purpose occupancy — patient rooms, treatment, therapy, operating, recovery, delivery, x-ray, radiology, cardiology, intensive care, trauma, isolation, pathology, autopsy, medication and anesthesia storage rooms, darkrooms, nursery, waiting rooms, morgue, pharmacy, library, laboratories, auditorium, lobby, offices, meeting rooms, lounges, computer rooms, cafeteria, kitchen, game rooms, locker rooms, laundry, maintenance shops, service areas and more. Also refer to other applications.

Functions

Any combination of heating, cooling, dehumidification, ventilation and humidification may be required.

Zoning

Each patient room or ward is a separate zone. (Some areas or wards may require more than one zone.) Zone each and every "surgery and care" room, "nursing" room, "diagnostics and treatment" room and "sterilizing and supply" room. Zone the hospital services areas and the administrative areas. Zone the library, laboratories, auditorium, lobby, lounges, cafeteria and kitchen. Interior spaces may require year-round cooling. Perimeter zones are sensitive to exposure — analyze the solar loads and the heat loss that is associated with exposed glass. Large heating loads occur near doors (heated vestibules are desirable). Refer to other applications for more information.

Lighting Loads

Lighting loads can vary from low to high, depending on the area — make a survey.

Occupancy Loads

The population density in the various areas may be constant; or variable, according to a schedule; or variable in a random way. Some areas may experience wide variations in population density. Maximum population density is based on the expected "full room" occupancies. Large populations increase humidity loads, reduce room sensible heat ratios and reduce the heating changeover points. Areas that are associated with steady, low density populations tend to have lower latent loads and higher sensible heat ratios. Get occupancy schedules and occupancy information from a staff authority. Refer to other applications for more information.

Equipment Loads

In the patient rooms, the equipment loads are generated by medical equipment, radios and televisions. A wide variety of appliances and equipment can be found in the rooms that are used to provide or to support medical procedures, treatments and services. Laboratories have a wide variety of equipment that generates sensible and latent loads. Televisions, sound equipment, vending machines, video games and small appliances may be found in public, staff and administrative areas. A rigorous load analysis is required for the kitchen and cafeteria serving area. Offices have equipment loads that are produced by computers, typewriters, duplicating equipment and small appliances. Service areas have laundry and maintenance shop equipment. Make a thorough survey of every room in the building; determine the sensible and latent loads that are associated with the equipment and appliances. Use load factors and the usage factors to de-rate "name-plate" inputs. Consider the effects of diversity and exhaust hoods. Also refer to other applications.

Cooling Load

The peak cooling load that is associated with any perimeter area occurs when the occupancy loads, lighting loads, equipment loads and envelope loads combine to produce the peak-load condition. This peak may not coincide with the peak envelope load because the occupancy, lighting and equipment loads may not be synchronized with the envelope load. The peak cooling load that is associated with any interior area occurs when the occupancy loads, lighting loads and equipment loads combine to produce a peak-load condition. The sensible and latent cooling loads in some areas may be quite steady. Other areas can experience wide load-swings (depending on occupancy schedules, equipment operating schedules and solar gains). A rigorous analysis of full-load and part-load conditions must be made for every room. Refer to other applications for more information.

Operating Mode

Perimeter rooms typically changeover to heating as the outdoor temperature drops. However, perimeter rooms with large internal loads may not changeover until it gets very cold outside — or they may not changeover at all. Interior zones require year-round cooling. Simultaneous heating and cooling is required for incompatible zones. Reheat may be required to control humidity in mild weather. Humidification might be required in cold weather. Analyze the part-load requirements of every room to determine the operating requirements (mode and schedules). Also refer to other applications.

Hospitals — Continued

Heat Gain from Hospital Laboratory Equipment			
Appliance Type	Size	Maximum Input Rating, BTU/HR	Recommended Rate, BTU/HR①
Autoclave (bench)	0.7 FT^3	4270	480
Bath, hot or cold circulation	1.0 to 9.7 gallons, -22 to 212°F	2560 to 6140	440 to 1060 sensible
			850 to 2010 latent
Blood analyzer	120 samples/hour	2510	2510
Blood analyzer with CRT screen	115 samples/hour	5120	5120
Centrifuge (large)	8 to 24 places	3750	3580
Centrifuge (small)	4 to 12 places	510	480
Chromatograph		6820	6820
Cytometer (cell sorter/analyzer)	1000 cells/second	73230	73230
Electrophoresis power supply		1360	850
Freezer, blood plasma, medium	13 FT^3, down to -40°F	340 per FT^3	136 per FT^3
Hot plate, concentric ring	4 holes, 212°F	3750	2970
Incubator, CO_2	5 to 10 FT^3, up to 130°F	9660	4810
Incubator, forced draft	10 FT^3, 80 to 140°F	2460	1230
Incubator, general application	1.4 to 11 FT^3, up to 160°F	160 to 220 per FT^3	80 to 110 per FT^3
Magnetic stirrer		2050	2050
Microcomputer	256 kbytes to 15 mbytes	341 to 2047	300 to 1800
Minicomputer		7500 to 15000	7500 to 15000
Oven, general purpose, small	1.4 to 2.8 FT^3, 460°F	2120 per FT^3	290 per FT^3
Refrigerator, laboratory	22 to 106 FT^3, 39°F	80 for 10 FT^3	34 for 10 FT^3
Refrigerator, blood, small	7 to 20 FT^3, 39°F	260 per FT^3	102 per FT^3
Spectrophotometer		1710	1710
Sterilizer, freestanding	3.9 FT^3, 212 to 270°F	71400	8100
Ultrasonic cleaner, small	1.4 FT^3	410	410
Washer, glassware	7.8 FT^3 load area	15220	10000
Water still	5 to 15 gallons	14500 per gallon	320 per gallon

① Assume heat gain is zero if equipment is installed under a exhaust hood.

Hospitals — Continued

Humidity Control

Humidity control is critical in some areas; in which case, reheat, humidification and sensitive controls will be required. The desired humidity levels can range from 30 percent to 95 percent. The relative humidity can be allowed to drift within the "comfort zone" in the other areas. Humidity control in some areas will be effected by the internal loads, the exhaust equipment and the make up air requirements—make a thorough survey of the factors that affect each room. Low occupancy rates and large sensible (lighting and equipment) loads will produce small latent loads and high room sensible heat ratios. Large populations with moderate sensible (lighting and equipment) loads will produce large latent loads and low room sensible heat ratios. Refer to other applications for more information.

Air Quality

Air quality is a critical factor throughout the facility. Most hospital and supporting service areas have specific filter, ventilation, exhaust and space pressure requirements. Biological contaminants, odors, fumes and moisture must not be allowed to migrate from room to room. Large quantities of outdoor air are required for some rooms. Special exhaust hoods may be required. In some cases, contaminated exhaust flows must be sterilized before being expelled to the outdoors. Some rooms and duct systems have to be made of special materials. Some duct systems must be completely free of dirt and dust. Some ducts must be air tight. Cross contamination between outdoor intakes and exhaust openings is totally unacceptable. Make a thorough survey of each room to determine the exact nature of the air quality requirements. Refer to codes and standards. Use general ventilation and standard filters for administrative, staff and public areas. Contaminated air that is associated with the kitchen, dedicated smoking areas, duplicating equipment, laundry, showers, locker rooms, toilets, maintenance shops must not be allowed to enter other areas. Refer to other applications for more information.

Filter Recommendations

Minimum Number Filter Beds	Area or Room	Filter Efficiency %	
		Bed 1	Bed 2
2	Operating, Delivery Nurseries, Intensive Care	25	90
	Bone Marrow or Organ Transplant	25	90
	Patient Care, Treatment Diagonistic	25	90
1	Food Preparation, Laundry	80	na
	Adminstrative, Clean/Soiled Storage	25	na

Exhaust CFM

Many hospital areas have specific exhaust requirements. Special exhaust hoods may be required. In some cases, contaminated exhaust must be filtered or sterilized before they are expelled. Some exhaust systems may have to be made of special materials. Make a thorough survey of the exhaust requirements that are associated with each room or area. Refer to codes. Exhaust systems also are required for food preparation areas, maintenance shops, laundry, dedicated smoking areas, locker rooms, showers and toilets. Refer to other applications for more information.

Outdoor Air

Many hospital areas have specific ventilation requirements, which can range from 100 percent recirculated air to 100 percent outdoor air. Make a thorough survey of the ventilation requirements that are associated with each room or area. Refer to codes and standards. In the public, staff and administrative areas, the outdoor air CFM depends on the population density and the exhaust CFM requirements. Refer to other applications for additional information.

Hospital Pressure, Ventilation, Circulation and Exhaust Requirements	Space Pressure	Outdoor ACH	Circulation ACH	Exhaust to Outdoors	Room Unit Recirculate
Surgery and Critical Care					
Operating Room — 100% Outdoor Air	P	15	15	Yes	No
Operating Room — Recirculating	P	5	25	Optional	No
Recovery Room	P	2	6	Optional	No
X-Ray Room	P	3	15	Optional	No
Delivery and Birthing	P	5	12	Optional	No
Nursery	P	5	12	Optional	No
Trauma Room	P	5	12	Optional	No
Anesthesia Storage (see codes)	Neutral	Optional	8	Yes	No
Nursing					
Patient Room	Neutral	2	2	Optional	Optional
Intensive Care	N	Optional	10	Yes	No
Isolation Room	P	2	6	Optional	No
Isolation Ward	Neutral	2	6	Yes	No
Patient Corridor	Neutral	2	10	Yes	No
Toilet Room	Neutral	2	4	Optional	Optional
Diagnostic and Treatment					
Examination Room	Neutral	2	6	Optional	Optional
Medication Room	P	2	4	Optional	Optional
Pharmacy	P	2	4	Optional	Optional
Treatment Room	Neutral	2	6	Optional	Optional
X-Ray Room	Neutral	2	6	Optional	Optional
Physical Therapy and Hydrotherapy	N	2	6	Optional	Optional
Autopsy Room	N	2	12	Yes	No
Darkroom	N	2	10	Optional	No
Nonrefrigerated Body Holding Room	N	Optional	10	Yes	No
Lab — Bacteriology, Cytology, Histology	N	2	6	Yes	No
Lab — Biochemistry	P	2	6	Optional	No
Lab — Glasswashing or Sterilizing	N	2	10	Yes	No
Lab — Histology, Pathology	N	2	6	Yes	No
Lab — Nuclear Medicine	N	2	6	Yes	No
Lab — Serology	P	2	6	Optional	No
Lab — Media Transfer	P	2	4	Optional	No
Sterilizing, Supply and Service					
Sterilizer Equipment Room	N	Optional	10	Yes	No
Linen and Trash Chute Room	N	Optional	10	Yes	No
Laundry, General	Neutral	2	10	Yes	No
Soiled Linen Sorting and Storage	N	Optional	10	Yes	No
Clean Linen Storage	P	2	2	Optional	Optional
Soiled or Decontamination Room	N	2	6	Yes	No
Clean Workroom and Sterile Storage	P	2	4	Optional	Optional
Equipment Storage	Neutral	2	2	Optional	Optional
Food Preparation Room	Neutral	2	10	Yes	No
Warewashing	N	Optional	10	Yes	No
Dietary Day Storage	Neutral	Optional	2	Optional	No
Bedpan Room	N	Optional	10	Yes	No
Bathroom	N	Optional	10	Optional	No
Janitors Closet	N	Optional	10	Optional	No

P = positive; N = negative — This table does not replace code requirements. — Extracted from 1987 ASHRAE Handbook

Hospitals — Continued

Space Pressure

Many hospital areas have specific space pressure requirements, which can range from negative to positive. Make a thorough survey of the pressure requirements that are associated with each room or area. Refer to codes and standards. Food preparation areas, duplicating department, maintenance shops, laundry areas, dedicated smoking areas, locker rooms, showers and toilets should be under a negative pressure. All other public, staff and administrative areas can be maintained at a neutral or slightly positive pressure. Refer to other applications for more information.

Schedules

Patient rooms, nursing rooms, care rooms, trauma rooms, some hospital services areas, some administrative areas, some laboratories and some lounges are occupied on a 24-hour basis. Surgery, diagnostic and treatment rooms are occupied on a scheduled basis. Some hospital services areas, some administrative areas, some laboratories and some lounges are operated on a scheduled basis. Get schedule information from a staff authority. Also make a survey to determine the operating schedules for various public, staff and administrative areas.

Noise

Patient rooms, some "medical services" rooms, meeting rooms, offices, staff sleeping rooms, the library and the auditorium should be quiet. Noise levels are important, but not critical in most of the other areas. Noise levels may not be critical in the maintenance, shop and laundry areas.

Special Problems

Almost every room is a zone. A large number of rooms and areas require simultaneous heating and cooling. Many rooms and zones are incompatible because of conflicting temperature, humidity, filter, pressure, exhaust and ventilation requirements. Through-the-wall equipment can be noisy and a source of drafts — and it may not be able to meet all of the requirements that are associated with some zones. Large sensible and latent loads and load swings are associated with population density, equipment loads, lighting loads and envelope loads. Year-round cooling and dehumidification is required in some areas. Reheat may be required to control humidity in some areas. Humidification may be required in cold weather. The operating modes, space conditions and operating hours that are associated with various occupancies and zones are not compatible with each other. Large heating loads occur near entrances. There are special exhaust and ventilation requirements in many areas. This application is characterized by high capital, operating and maintenance costs.

Nursing Homes

Occupancy

Multiple purpose occupancy — dormitory rooms, apartments, lobby, lounges, television rooms, game rooms, dining room, kitchen, administrative offices, examination rooms, therapy rooms, laundry rooms, and building maintenance areas. Refer to other applications for additional information.

Functions

Any combination of heating, cooling, dehumidification, ventilation and humidification may be required.

Zoning

Each dormitory room is a separate zone; suites may require two or more zones. Zone the lobby, lounges, television rooms, game rooms, dining room, kitchen, administrative offices, examination rooms, therapy rooms, laundry rooms, and building maintenance areas. Some public areas may require more than one zone. Interior spaces may require year around cooling. Perimeter zones are sensitive to exposure — analyze the solar loads and the heat losses that are associated with glass. Large heating loads occur near doors (heated vestibules are desirable).

Lighting Loads

Low lighting levels in the guest rooms and corridors do not create large cooling loads. Low to medium lighting loads in the public areas, medium to high lighting loads in the examination and therapy rooms — make a survey.

Occupancy Loads

Dormitory rooms are occupied on a 24-hour basis. Wide variations in population density can occur in the public areas. Maximum population densities are based on the expected "full-room" occupancy. The occupancy in dining room is usually scheduled. Consider the additional sensible and latent loads that are associated with food service. Large populations increase the humidity load, reduce the room sensible heat ratio and reduce the heating changeover point. The administrative areas are characterized by a steady, scheduled occupancy. Low occupancy levels are associated with examination and therapy rooms. Get occupancy information from the manager.

Equipment Loads

In the dormitory rooms, the internal loads are produced by televisions, radios, small refrigerators and electric space heaters. Televisions, sound equipment, vending machines, and small appliances are normally found in the public areas. Small appliances are usually found in the dining room. A rigorous load analysis is required for kitchen. Office loads are generated by computers, typewriters, duplicating equipment and small appliances. Examination and therapy rooms may contain medical equipment that produces sensible and latent loads. Service areas contain laundry and shop equipment. Make a thorough survey.

Cooling Load

In the dormitory rooms, the peak loads usually coincide with peak solar gains. The public areas and the dining room are characterized by large sensible and latent loads and wide load swings. In these areas, the maximum loads occur when the occupancy loads, envelope loads and lighting loads combine to produce the peak-load condition. In the public areas, large sensible and latent load swings are caused by variations in population density. Sensible load swings also are caused by changes in the lighting level and the envelope loads. Solar gains are important, but the peak design loads may not be coincident with peak solar loads. Interior areas are not sensitive to solar gains, so the cooling load depends on the occupancy, the lighting loads and the equipment loads. Wide load swings in the interior public areas are caused by variations in the population density. Relatively steady loads characterize the interior administrative areas and the (supporting) services areas. Variable loads occur in the perimeter administrative areas. Load swings in perimeter administrative areas depend on the internal loads, the exposure and the construction details.

Operating Mode

Dormitory rooms typically changeover to heating at moderate outdoor temperatures, but the change-over temperature is sensitive to the solar gain. The public areas and the dining room may require year-round cooling. Interior zones require year-round cooling. The public area and the office perim-eter-zone heating changeover points depend on the population density, the lighting loads, the exposure and the envelope construction details. Simultaneous heating and cooling is required for all incompatible zones. Reheat may be required to control humidity in mild weather. Humidification may be required to control humidity in cold weather.

Nursing Homes — Continued

Humidity Control

Large variations in the dormitory room sensible loads cause the humidity to drift, but close control over humidity is not required. In the summer, the relative humidity can range between 50 to 60 percent (75 °F dry-bulb). In the winter, the relative humidity can range between 25 to 30 percent (72 °F dry-bulb). In the public areas, variations in occupancy, lighting loads and equipment loads cause swings in the sensible loads, latent loads and room sensible heat ratios. Small populations combined with substantial lighting and equipment loads will cause small latent loads and high room sensible heat ratios. Large populations combined with moderate lighting and equipment loads will cause large latent loads and low room sensible heat ratios. Allow the humidity levels to drift within the "comfort zone." In the administrative offices, the latent loads will be relatively low and steady and the room sensible heat ratios will be relatively high — the humidity levels can be allowed to drift within the "comfort zone." Do not recirculate air from the therapy rooms because the whirlpools and other therapy equipment produce large latent loads.

Air Quality

Some ASHRAE filter and ventilation recommendations are summarized below. Also refer to codes and standards. Use general ventilation and standard filters for public areas. Air that is associated with the kitchen, dedicated smoking areas, duplicating equipment areas, laundry areas and workshops must not be allowed to infiltrate into other areas.

Filter Recommendations

Minimum Number Filter Beds	Area or Room	Filter Efficiency
1	Patient Care, Treatment, Therapy Diagonistic and Related Areas	80
	Food Preparation, Laundry Areas	80
	Adminstrative, Bulk/Soiled Storage	30

Nursing Home Pressure, Ventilation, Circulation and Exhaust Requirements

	Space Pressure	Outdoor ACH	Circulation ACH	Exhaust to Outdoors	Room Unit Recirculate
Patient Care					
Patient Room	Neutral	2	2	Optional	Optional
Patient Corridor	Neutral	Optional	2	Optional	Optional
Toilet Room	N	Optional	10	Yes	No
Diagnostic and Treatment					
Examination Room	Neutral	2	6	Optional	Optional
Physical Therapy and Hydrotherapy	N	2	6	Optional	Optional
Sterilizing, Supply and Service					
Sterilizer Equipment Room	N	Optional	10	Yes	No
Linen and Trash Chute Room	N	Optional	10	Yes	No
Laundry, General	Neutral	2	10	Yes	No
Soiled Linen Sorting and Storage	N	Optional	10	Yes	No
Clean Linen Storage	P	Optional	2	Yes	No
Food Preparation Room	Neutral	2	10	Yes	Yes
Warewashing	N	Optional	10	Yes	Yes
Dietary Day Storage	Neutral	Optional	2	Yes	No
Bathroom or Janitors Closet	N	Optional	10	Yes	No

P = positive; N = negative — This table does not replace code requirements. — Extracted from 1987 ASHRAE Handbook

Nursing Homes — Continued

Exhaust CFM

Patient rooms should be equipped with bathroom exhausts that operate continuously. Exhaust systems also may be required for the therapy rooms — refer to local codes. Exhaust food preparation areas, workshops, laundry areas, dedicated smoking areas and toilets.

Outdoor Air

Ventilation is required for patient rooms, examination rooms, therapy rooms, food preparation areas and laundry rooms — refer to local codes. In the public areas, administrative areas and ancillary areas, the outdoor air CFM depends on population density and exhaust CFM requirements — make a survey.

Space Pressure

The pressure in the therapy rooms, some of the supply and service rooms, the food preparation areas, the duplicating department, the maintenance and service areas, the laundry areas, the dedicated smoking areas, the showers and the toilets should be negative. All other areas can be neutral or a little positive — depending on the air quality requirements. Refer to codes.

Schedules

Space conditioning is required in the dormitory rooms on a 24-hour basis. If a central plant is used, it can take advantage of the diversity that is associated with the various exposures. Make a survey to determine the operating schedules for the examination and therapy rooms, the various public areas and the administrative areas.

Noise

Dormitory rooms, examination rooms, meeting rooms and offices should be quiet. Noise levels are important but not critical in the other public areas.

Special Problems

Through-the-wall equipment can be noisy and a source of drafts. In the public areas, large sensible and latent loads and load swings are associated with changes in the population density, the lighting loads and the envelope loads. Significant load swings are associated with large glass areas. Year-round cooling and dehumidification is required in some areas. Reheat and humidification may be required to control humidity during any season. Excessive moisture is released in therapy rooms. The operating modes and operating hours that are associated with the various areas may not be compatible with each other. Large heating loads occur near entrances.

Doctor or Dentist Offices

Occupancy Multipurpose occupancy — waiting room, general and private offices, conference rooms, examination rooms, medical procedure rooms, audio-visual rooms and storage rooms.

Functions Heating, cooling, dehumidification, ventilation is required. In some cases, humidification is required.

Zoning Zone the waiting room, the general and private offices, the conference rooms, the examination rooms, the medical procedure rooms and the audio-visual rooms. Zone the interior areas. Multiple perimeter zones may be required, depending on the exposure.

Lighting Loads Lighting loads can vary from medium to high, depending on the area — make a survey.

Occupancy Loads Design for a high population density in the waiting room. The population density that is associated with a conference room can vary from low to high. A low population density is typical for the other areas. Get occupancy information from the staff.

Equipment Loads Sensible and latent loads are produced by computers, office equipment, medical equipment (see Hospitals), audio-visual equipment, coffee brewers, and duplicating equipment.

Cooling Load Relatively steady loads are associated with the interior zones. In the perimeter zones, the peak loads and the load swings depend on the exposure and the envelope construction details.

Operating Mode Interior zones require year-round cooling. The perimeter-zone heating changeover points depend on the exposure and the envelope construction details. Simultaneous heating and cooling is required for incompatible zones. Winter humidification may be required.

Humidity Control Latent loads are relatively low and steady, room sensible heat ratios are relatively high. Humidity levels can be allowed to drift within the "comfort zone."

Air Quality General ventilation is required for the offices. Special filters, cxhaust systems, ventilation and pressure levels may be required for medical treatment rooms (see Hospitals). Discuss the air quality requirements with the staff and refer to codes. Isolate the fumes that are associated with duplicating equipment and toilets.

Exhaust CFM Specific types of exhaust systems may be required for the medical treatment rooms (refer to Hospitals). Discuss the exhaust system requirements with the staff and refer to codes. Exhaust the toilet areas.

Outdoor Air The outdoor air CFM depends on the fresh air and exhaust CFM requirements that are associated with each zone.

Space Pressure The pressure in most of the areas can be neutral or a little positive. The medical treatment rooms may have specific pressure requirements (refer to Hospitals). Discuss the pressure requirements with the staff and refer to codes. Toilets should be under a negative pressure.

Schedules Talk with the staff to determine occupancy and medical equipment operating schedules.

Noise Low noise levels are required throughout the facility.

Special Problems The filter, outdoor air CFM and exhaust CFM requirments, and the space pressure relationships, may depend on air quality requirements that are associated with the treatment rooms.

Offices

Occupancy Multipurpose occupancy — general and private offices, conference and meeting rooms, reception area, printing and duplicating departments, computer rooms, audio-visual rooms, library, communications equipment room, mail, lunch and storage rooms. Refer to other applications.

Functions Heating, cooling, dehumidification and ventilation are required. Humidification may be required.

Zoning Each different type of occupancy is a separate zone. Zone the interior areas. Multiple, perimeter zones are associated with the exposure. Perimeter zones that are not defined by a inside wall can be considered to extend 12 to 18 feet in from the outside wall. Zone each private office and meeting room. Zone the bull pen areas, reception areas, lunch rooms, computer rooms, printing department, audio-visual rooms, mail rooms and storage rooms. Consider the heating loads near entrances (heated vestibules are desirable) and the sensible loads that are generated by electrical equipment. All zones must be compatible with occupancy schedules.

Lighting Loads Lighting loads vary from medium to high, depending on the area — talk to manager and make a survey.

Occupancy Loads The population density varies by area, but it usually remains steady within an area. Medium or high density characterizes bull pen and lunch room areas. Low density is associated with private offices. Wide variations in population density occur in conference and meeting rooms. Low population density is associated with supporting services areas. Get occupancy information from the manager.

Equipment Loads Sensible loads are generated by computers, typewriters, communications and audio-visual equipment. Sensible and latent loads are generated by coffee brewers, lunch room appliances, vending machines and duplicating equipment.

Cooling Load In the interior zones, the loads are relatively steady. Load swings are typical in the perimeter zones. In the perimeter zones, the peak loads depend on the exposure and the envelope construction details.

Operating Mode The interior zones require year-round cooling. The perimeter zone heating changeover points depend on the exposure and the envelope construction details. Simultaneous heating and cooling is required for incompatible zones.

Humidity Control Latent loads are relatively low and steady, room sensible heat ratios are relatively high. Humidity levels can be allowed to drift within the "comfort zone."

Air Quality General ventilation is required for offices. Special filters are not usually required. Isolate the fumes that are associated with duplicating equipment.

Exhaust CFM Exhaust systems are typically required for the lunch room, duplicating department, electrical equipment rooms and toilets.

Outdoor Air The outdoor air CFM depends on the occupancy and exhaust CFM requirements. The outdoor air requirements vary by zone — make a survey.

Space Pressure The pressure in the lunch room, duplicating department and toilets should be negative. All other areas can be neutral or a little positive.

Schedules Talk to manager to determine occupancy and equipment operating schedules. Take advantage of diversity. Data processing areas may require 24-hour conditioning.

Noise Most areas should be quiet, especially the private offices, meeting rooms and audio-visual rooms.

Special Problems Large heating and cooling loads are associated with exposed glass. Large heating loads occur near the main entrance. Zones are not defined by walls. The system must accommodate zones that are occupied after hours. Frequent floor plan changes. Multiple tenant utility bills.

Veterinarian Offices

Occupancy Multipurpose occupancy — waiting room, offices, examination rooms, medical procedure rooms, audio-visual rooms, animal rooms and storage rooms.

Functions Heating, cooling, dehumidification and ventilation are required.

Zoning Zone the waiting room, offices, conference rooms, examination rooms, medical procedure rooms and audio-visual rooms. Zone the interior areas. The perimeter zones are determined by the exposure.

Lighting Loads Lighting loads vary from medium to high, depending on the area — make a survey.

Occupancy Loads Most areas are characterized by low, steady population densities. Expect large occupancy variations in waiting room. Get occupancy information from the staff. (Refer to Pet Shops and Laboratories for information about animals.)

Equipment Loads Sensible and latent loads are generated by computers, office equipment, medical equipment, audio-visual equipment, coffee brewers and duplicating equipment.

Cooling Load Relatively steady loads are associated with the interior zones. In the perimeter zones, the peak loads and the load swings depend on the exposure and the envelope construction details.

Operating Mode Interior zones require year-round cooling. The perimeter zone heating changeover points depend on the exposure and the envelope construction details. Simultaneous heating and cooling is required for incompatible zones.

Humidity Control The latent loads are relatively low and steady and the room sensible heat ratios are relatively high. The humidity levels can be allowed to drift within the "comfort zone."

Air Quality General ventilation is required in the office areas. Special filters, ventilation, exhaust systems and pressure levels may be required for medical treatment rooms. Large amounts of outdoor air are required for the animal rooms. Discuss the air quality requirements with the staff and refer to codes. Isolate the fumes that are associated with the animal rooms, duplicating equipment and toilets.

Exhaust CFM Special exhaust may be required for medical treatment rooms. Discuss exhaust system requirements with the staff and refer to codes. Exhaust the animal rooms and toilet areas.

Outdoor Air The outdoor air CFM depends on the fresh air and exhaust requirement that is associated with each zone. Large amounts of outdoor air are required for animal rooms. (Refer to Pet Shops or Laboratories for more information.)

Space Pressure The pressure in most of the areas can be neutral or a little positive. The medical treatment rooms may have specific pressure requirements. Discuss the pressure requirements with the staff and refer to codes. The animal rooms and toilets should be under negative pressure.

Schedules Ask the staff about the occupancy and medical equipment operating schedules.

Noise Except for the animal rooms, low noise levels are required throughout the facility.

Special Problems The filter requirements, outdoor air CFM, exhaust CFM and space pressure relationships depend on treatment room air quality requirements. The animal rooms must not cause air quality problems.

Apartments and Multifamily Structures

Occupancy A single purpose occupancy is defined by the residential living areas. Offices, game rooms, pool and hot-tub areas, laundry rooms and service areas should be treated as separate applications.

Functions Heating, cooling, dehumidification, ventilation and in some cases, winter humidification, is required.

Zoning Each apartment is a zone. Sometimes, the rooms within an apartment are zoned. The corridors and entrances could be zones (heated vestibules are desirable). The offices, game rooms, pool and hot-tub areas, laundry rooms and service areas have nothing in common with the apartments, so they should be treated as separate applications.

Lighting Loads Low lighting levels are associated with the apartments and corridors. The lighting loads in the other areas depend on the application.

Occupancy Loads Low, but steady population densities are associated with the apartments (assume two persons per bedroom). The occupancy levels in the other areas depend on the application.

Equipment Loads In the apartment areas, use 1500 to 3000 BTUH for the appliance load. The equipment loads in the other areas depend on the application.

Cooling Load In the apartments, the internal loads are fairly constant. Apartment areas are sensitive to solar gains, which cause load swings that depend on the exposure, month-of-year and time-of-day. The cooling loads in the other areas depend on the application.

Operating Mode Apartments typically changeover to heating at a moderate outdoor temperature, but the changeover temperature depends on the size of the solar gain. Simultaneous heating and cooling is required for apartments that have incompatible exposures. The operating mode in the other areas depends on the application.

Humidity Control Large variations in sensible loads cause humidity to drift, but close control over humidity is not required. The relative humidity should be allowed to drift within the "comfort zone."

Air Quality The air quality requirements are the same as those that are associated with single family housing applications. The air quality requirements in the other areas depend on the application.

Exhaust CFM Local toilet and kitchen exhausts are required in the apartment areas, but they can operate intermittently. The exhaust requirements in the other areas depend on the application.

Outdoor Air In the apartments areas, the outdoor air CFM may be provided by infiltration or by mechanical ventilation, refer to local codes. The ventilation quality requirements in the other areas depend on the application.

Space Pressure To isolate odors and fumes, the apartments should be negative with respect to the corridors. The pressure in the other areas depends on the application.

Schedules Space conditioning is required on a 24-hour basis. If a central plant is used, there is some diversity associated with the various exposures and the common spaces.

Noise Residential systems must be very quiet.

Special Problems The standard **Manual J** load calculation procedures do not apply to apartments that only have one or two exposed walls. Central systems must provide simultaneous heating and cooling. Utility bills must be provided for each apartment. Mechanical ventilation may be required. Through-the-wall equipment can be noisy and a source of drafts. The partition loads that are associated with adjacent spaces depend on whether the adjacent space is occupied.

Dormitories

Occupancy

Multiple purpose occupancy — private rooms, lobby, offices, lounges, TV rooms, dining room, kitchen, game rooms, laundry and service areas. Refer to other applications for more information.

Functions

Heating, cooling, dehumidification and ventilation are required.

Zoning

Each private room is a separate zone. Zone the lobby, offices, lounges, TV rooms, dining room, kitchen, game rooms, laundry and service areas. Some public areas may require more than one zone. Interior spaces may require year-round cooling. Perimeter zones are sensitive to exposure, consider the solar loads and the heat loss that is associated with glass. Large heating loads near entrance doors (heated vestibules are desirable).

Lighting Loads

The low lighting levels in the private rooms and corridors do not produce large cooling loads. Low to medium lighting loads are associated with the other areas — make a survey.

Occupancy Loads

The private rooms are occupied on a 24-hour basis. Wide variations in population density are associated with some of the public areas. The maximum population densities are based on the expected "full room" occupancy. The occupancy in dining room is usually scheduled (consider the additional sensible and latent loads that are associated with food service). Large occupancy loads increase the humidity loads, reduce the room sensible heat ratios and reduce the heating changeover points. The areas that are associated with administrative activities are characterized by steady, scheduled occupancy loads. Ask the manager for occupancy information.

Equipment Loads

In the private rooms, internal loads are produced by computers, televisions, sound equipment or (possibly) a small refrigerator. Televisions, sound equipment, vending machines, video games and small appliances are normally found in public areas. Small appliances are usually found in the dining room. A rigorous load analysis is required for kitchen. Offices have equipment loads that are generated by computers, typewriters, duplicating equipment and small appliances. Service areas have laundry and shop equipment. Make a thorough survey of each area.

Cooling Load

In the private rooms, the peak loads usually coincide with the peak solar gains. The design loads in the perimeter-zone dining room or public areas — which might be characterized by wide load swings — occur when the occupancy loads, envelope loads and lighting loads combine to produce a peak load condition. (The solar gains are important in the perimeter areas, but the peak design loads may not be coincident with the peak solar loads.) In the interior-zone public areas, the sensible and latent design loads depend on the population density, the lighting levels and the equipment loads. Since the interior areas are not sensitive to solar gains, the load swings depend on the occupancy level, the lighting level and the equipment loads. Wide load swings in interior public areas are usually caused by large changes in the population density. Relatively steady loads characterize the interior administrative areas and the supporting services areas. Variable loads characterize the perimeter-zone administrative areas. The load swings in the perimeter administrative zones depend on the internal loads, the exposure and the envelope construction details.

Operating Mode

The private rooms typically changeover to heating at a moderate outdoor temperature, but the changeover temperatures depend on the size of the solar gains. Some of the public areas and the dining room may require year-round cooling. The interior zones require year-round cooling. The heating changeover points in the perimeter-zone public and administrative areas depend on the population density, lighting levels, exposure and envelope construction details. Simultaneous heating and cooling is required for all incompatible zones. In some areas, reheat may be required to control the humidity during mild weather.

Humidity Control

In the private rooms, large variations in the sensible loads can cause the humidity to drift, but close control over humidity is not required. In these rooms, the relative humidity should be allowed to drift within the "comfort zone." Wide swings in occupancy levels, lighting levels and equipment loads affect the sensible loads, latent loads and room sensible heat ratios in the public areas. Small popula-

Dormitories — Continued

tions with considerable lighting and equipment loads will cause small latent loads and high room sensible heat ratios. Large populations with moderate lighting and equipment loads will cause large latent loads and low room sensible heat ratios. Allow the humidity levels to drift within the "comfort zone." In the administrative offices, the latent loads will be relatively low and steady and the room sensible heat ratios will be relatively high. In these areas, the humidity levels can be allowed to drift within the "comfort zone."

Air Quality

Standard filters, infiltration or a small amount of ventilation is required for the private rooms (refer to codes). Use general ventilation and standard filters for the public areas. Air that is associated with the kitchen, dedicated smoking areas, duplicating equipment, laundry and workshops must not be allowed to contaminate other areas.

Exhaust CFM

Private rooms with toilet facilities should be equipped with bathroom exhausts that operate intermittently. Exhaust systems also are required for food preparation areas, some types of office equipment, workshops, laundry, dedicated smoking areas, community showers and toilets.

Outdoor Air

Infiltration or a small amount of ventilation is required for the private rooms (refer to codes). In the public areas, the outdoor air CFM depends on the population density and the exhaust CFM requirements. The outdoor air requirements for each different area can vary — make a survey.

Space Pressure

The pressure in the food preparation areas, duplicating department, workshop, laundry, dedicated smoking areas, showers and toilets should be negative. The pressure in all the other areas can be neutral or a little positive (depending on the air quality requirements).

Schedules

Space conditioning is required in the private rooms on a 24-hour basis. If a central plant is used, there is some diversity associated with the various exposures and the occupancy schedule that is associated with the administrative areas and the public areas. Make a survey to determine the operating schedules for various public areas and the administrative areas.

Noise

The private rooms, meeting rooms and offices should be quiet. Noise levels are important, but not critical in the public areas.

Special Problems

In the private rooms, through-the-wall equipment can be noisy and a source of drafts. In the public areas, large sensible and latent loads and large load swings are associated with population density, lights and envelope loads. Large load swings are associated with large glass areas. Year-round cooling and dehumidification may be required in some areas. In some areas, reheat may be required to control humidity during mild weather. The operating modes and operating hours that are associated with various occupancies and zones may not be compatible with each other. Large heating loads occur near entrances.

Motels

Occupancy Multiple purpose occupancy — sleeping rooms, lobby, offices, meeting rooms, ballrooms, bar, restaurant, game rooms, pool and hot-tub areas, laundry rooms, and service areas. Refer to other applications for additional information.

Functions Heating, cooling, dehumidification and ventilation are required.

Zoning Each guest room is a separate zone. Zone the lobby, offices, meeting rooms, ballrooms, bar, restaurant, game rooms, laundry rooms, service areas, pool and hot-tub areas. Some areas may require more than one zone. The interior spaces may require year-round cooling. The zoning strategies in the perimeter areas will depend on the exposure (analyze the solar loads and heat loss that is associated with glass). Perimeter zones that are not defined by an interior wall can be considered to extend 12 to 18 feet inward from the outer wall. Large heating loads are associated with entrances (heated vestibules are desirable).

Lighting Loads Low lighting loads are associated with the guest rooms and corridors. The lighting loads in the other areas can vary from low to high, depending on the area — make a survey.

Occupancy Loads The guest rooms are usually occupied in the morning, late afternoon and evening. Some guest rooms are used for small parties and receptions. Wide swings in population density are associated with the conference rooms, meeting rooms and other public areas (daily events usually run from morning into the late evening.) The maximum population density that is associated with any public area is based on the event that produces the largest number of occupants (also consider the additional sensible and latent loads associated with food service). Large populations increase the humidity load, reduce the room sensible heat ratio and reduce the heating changeover point. Dancing causes even higher sensible and latent loads. The areas that are associated with the administrative activities are characterized by steady, scheduled occupancy rates. Get occupancy information from the manager.

Equipment Loads In the guest rooms, sensible and latent loads are generated by televisions and small refrigerators, coffee makers and occasionally, a kitchenette. Audio-visual equipment and small appliances are normally found in conference rooms, ballrooms and meeting rooms. Bars and dining areas are subjected to sensible and latent loads that are generated by appliances, motors and food serving counters. A rigorous load analysis is required for the kitchen. Offices have internal loads that are generated by computers, typewriters, duplicating equipment and small appliances. In the lobby, corridors, ancillary areas and service areas, internal loads are generated by vending machines, video games, laundry equipment and shop equipment. Make a thorough survey of each area.

Cooling Load The peak loads in the guest rooms usually coincide with the peak solar gains. The loads in the meeting rooms, conference rooms, ballrooms and the other public areas are characterized by large sensible and latent loads and wide load swings. In these areas, the occupancy loads, envelope loads and lighting loads combine to produce a peak load condition. In the interior areas, large sensible and latent load swings are related to changes in occupancy and lighting levels. In the perimeter areas, large sensible and latent load swings are associated with changes in the envelope loads, lighting loads and the occupancy levels. (Solar gains are important in the perimeter areas, but the peak design loads may not be coincident with the peak solar loads.) The interior administrative and service areas are characterized by relatively steady loads. The loads in the perimeter administrative and service areas will vary, depending on the envelope loads (which depend on the exposure and the construction details). Special load calculation procedures are required for the pool and hot-tub areas (refer to recreational facilities).

Operating Mode The guest rooms typically changeover to heating at moderate outdoor temperatures, but these changeover temperatures are affected by the solar gains. The meeting rooms, conference rooms, ballrooms and other highly populated areas may require year-round cooling. The interior zones require year-round cooling. The heating changeover points in the public areas and the perimeter administrative areas depend on the population density, lighting, exposure and envelope construction details. Simultaneous heating and cooling is required for all incompatible areas. In some areas, reheat may be required to control the humidity in mild weather. Special operating cycles are required for the pool and hot-tub areas (refer to recreational facilities).

Motels — Continued

Humidity Control

Large variations in the sensible loads that are associated with the guest rooms cause humidity to drift, but close control over humidity is not required. In these areas, the relative humidity can be allowed to drift within the comfort zone. Wide swings in the sensible loads, latent loads and room sensible heat ratios are associated with the public areas. These swings are caused by changes in the occupancy levels, lighting loads and envelope loads. Small populations combined with large lighting and envelope loads will produce small latent loads and high room sensible heat ratios. Large populations combined with moderate lighting and envelope loads will produce large latent loads and low room sensible heat ratios. Allow the humidity levels to drift within a "comfort zone." In the administrative offices, the latent loads will be relatively low and steady and the room sensible heat ratios will be relatively high. In these areas, the humidity levels can be allowed to drift within a "comfort zone." Special problems are associated with the pool and hot-tub areas (refer to recreational facilities).

Air Quality

Standard filters, infiltration or a small amount of ventilation is required for the guest rooms (refer to codes). Use general ventilation and standard filters for the meeting and assembly areas. The air that is associated with bars, food preparation areas, dedicated smoking areas, duplicating equipment, laundry areas and workshops must not be allowed to contaminate other areas. Special problems are associated with the pool and hot-tub area (refer to recreational facilities).

Exhaust CFM

The guest rooms should be equipped with bathroom exhausts that operate intermittently. Exhaust systems also are required for bars, food preparation areas, workshops, laundry areas, dedicated smoking areas and toilets. Special requirements are associated with the pool and hot tub area (refer to recreational facilities).

Outdoor Air

Infiltration or a small amount of ventilation is required for the guest rooms (refer to codes). In the assembly areas and the other public areas, the outdoor air CFM requirement will depend on population density and exhaust system make up air requirement. The outdoor air requirements will vary by area — make a survey. Special requirements are associated with the pool and hot-tub area (refer to recreational facilities).

Space Pressure

The pressure in the bars, food preparation areas, duplicating department, workshop, laundry areas, dedicated smoking areas and toilets should be negative. The pressure in all the other areas can be neutral or a little positive — depending on the air quality requirements. Special requirements are associated with the pool and hot-tub areas (refer to recreational facilities).

Schedules

Space conditioning is required in the guest rooms on a 24-hour basis. If a central plant is used, there is some diversity associated with the various exposures and the occupancy schedules that relate to the public areas and the administrative areas — make a survey to determine the operating schedules that are associated with each area.

Noise

The guest rooms, meeting rooms and offices should be quiet. The noise level is important in the conference rooms and in the meeting rooms. The noise level is important, but not critical in the other public areas.

Special Problems

In the guest rooms, through-the-wall equipment can be noisy and a source of drafts. In the public areas, large sensible and latent loads and load swings are associated with changes in the population density, lighting loads and envelope loads. Large load swings and zoning problems are associated with large glass areas. Year-round cooling and dehumidification is required in some areas. In some areas, reheat may be required to control the humidity during mild weather. The operating modes and operating hours that are associated with various areas may not be compatible with each other. Some zones may not be defined by interior walls. Large heating loads occur near entrances. It is difficult to control the pressure in areas that have a lot of door traffic. Equipment noise is unacceptable in many areas.

Arenas

Occupancy

Multipurpose occupancy — the main arena consists of the playing floor, private spectator booths, general spectator seating and entrance areas; the rest of the facility might include meeting rooms, locker rooms, training rooms, audio-visual rooms, ticket office, general and private administrative offices, radio and TV broadcasting booths, dressing rooms, restaurants, bars, concession stands, storage rooms, work shops and more. Refer to other applications for additional information.

Functions

Heating, cooling, dehumidification and ventilation are required.

Zoning

Each different type of occupancy defines a separate zone. The arena area is usually divided into more than one zone. Zone the private spectator booths, all the auxiliary areas, the audio-visual rooms, broadcasting booths, locker rooms and the ticket booths. Zone the administrative areas, food service areas, bars and maintenance areas. Usually, the main arena is an interior zone (unless there are substantial glass areas at the outside walls). Zone the interior and perimeter areas that are associated with the administrative offices. Zoning decisions should consider the effect of the solar loads and transmission loads that are associated with glass (in any area). Zone the lobby and the other entrance areas. All zones must be compatible with the operating modes and operating hours that are associated with the zone.

Lighting Loads

In the arena seating areas, the lighting loads range from low to medium. The lighting loads are large in the competition area. Medium lighting loads are associated with most of the auxiliary and administrative areas. Low to medium lighting levels are associated with the bars and food service areas. Large lighting loads are associated with stage production areas. Make a survey of each area.

Occupancy Loads

The arena areas and the auxiliary areas are characterized by events that can run from late morning into the late evening. The maximum population density for a particular area is based on the event that generates a "full house" occupancy. Large variations in the population density can be expected, depending on the nature of the event. Small groups may occupy the meeting rooms or auxiliary areas before or after the event. The areas that are associated with administrative activities are characterized by steady, scheduled occupancy loads. Large occupancy loads increase the humidity loads, reduce the room sensible heat ratios and reduce the heating changeover points (also consider that the additional sensible and latent loads are associated with food service). Get occupancy information from the manager.

Equipment Loads

Sensible loads are associated with sound equipment, electronic equipment, broadcasting and communications equipment, office equipment, computers and audio-visual equipment. Other sensible and latent loads are associated with the equipment that is found in the bars, food service areas and administrative areas — small appliances, vending machines and office equipment, for example. Make a survey and get information about the equipment loads from the manager. Use load factors and the usage factors to de-rate the "name-plate" inputs. Consider the effects of diversity and exhaust hoods.

Cooling Load

The arena and auxiliary areas are characterized by large sensible and latent loads and wide load swings. The design loads occur when the occupancy level produces a peak load condition. Large sensible and latent load swings are caused by variations in the population density. Wide load swings in the interior auxiliary areas are caused by changes in the population density. Relatively steady loads characterize the interior administrative areas and the supporting service areas. Load swings in the perimeter administrative areas are associated with the exposure and the envelope construction details.

Operating Mode

The arena's area and the interior auxiliary areas require year-round cooling. The perimeter zone heating changeover points depend on lighting and equipment loads, the population density, the exposure and the envelope construction details. Simultaneous heating and cooling is required for incompatible zones. In some areas, reheat may be required to control humidity in mild weather.

Humidity Control

In the public areas, wide swings in the sensible loads, latent loads and the room sensible heat ratios are caused by changes in the occupancy loads and the lighting loads. Small populations and large lighting loads will produce small latent loads and high room sensible heat ratios. Large populations

Arenas—Continued

and low lighting loads will produce large latent loads and low room sensible heat ratios. Exact control over humidity is not required — allow the relative humidity to drift within the "comfort zone." In the administrative areas, the latent loads will be relatively low and steady and the room sensible heat ratios will be relatively high. The relative humidity can be allowed to drift within the "comfort zone" in these areas.

Air Quality

Ventilation (outdoor air) is required for all the occupied areas. Standard filters are appropriate for most of the areas. Some broadcasting, communications or audio-visual equipment may require high efficiency filters. The air that is associated with the locker rooms, bars, food preparation areas, dedicated smoking areas and workshops must not be allowed to contaminate the other areas. Use outdoor air to dilute the odors and fumes that are associated with events that feature animals or internal combustion engines.

Exhaust CFM

Exhaust systems are required for locker rooms, projection booths, bars, food preparation areas, workshops, dedicated smoking areas and toilets.

Outdoor Air

The outdoor air requirements depend on the population density, the dilution ventilation requirements and the exhaust system make up air requirements. The outdoor air requirements vary by zone — make a survey.

Space Pressure

The pressure in the locker rooms, projection booth, bars, food preparation areas, workshops, dedicated smoking areas and toilets should be negative. The pressure in all the other areas can be neutral or a little positive (depending on the air quality requirements).

Schedules

Make a survey to determine the schedules that are associated with the arena, the ancillary areas and the administrative areas. The operating hours that are associated with the arena and the ancillary areas are usually not compatible with the operating hours that are associated with the ticket booth and administrative areas. Take advantage of the diversity that is associated with the various operating schedules.

Noise

Noise levels are important, but not critical, in the arena. Low noise levels are required in the audio-visual rooms and the broadcast areas. The meeting rooms and the offices should be quiet.

Special Problems

Large sensible and latent loads and load swings are associated with changes in the population density, lighting levels, equipment loads and envelope loads. Large load swings are associated with glass areas. Year-round cooling and dehumidification may be required in some areas. Reheat may be required to control humidity during mild weather in some areas. This type of application is characterized by very large energy requirements. The operating modes and the operating hours that are associated with the various occupancies and zones may not be compatible with each other. Some zones may not be defined by walls and partitions. Large heating loads occur near the entrances. It is hard to pressurize a zone that has a lot of door traffic. Equipment noise is unacceptable in many areas. The HVAC system must be able to accommodate events that feature animals or internal combustion engines.

Churches

Occupancy

Multipurpose occupancy — assembly hall, ancillary areas, meeting rooms, main vestibule, general and private offices, conference rooms, audio-visual rooms, class rooms, library, kitchen, printing and duplicating department, storage room, maintenance shop, residence and more. Refer to other applications for additional information.

Functions

Heating, cooling, dehumidification and ventilation are required.

Zoning

Zone the interior and perimeter areas of the assembly hall, the balcony and the choir areas. Zone the vestibules and the other ancillary areas. The perimeter zones are sensitive to the exposure and the construction details. Consider the effect of the solar loads and the transmission loads that are associated with glass. Perimeter zones that are not defined by an interior wall can be considered to extend 12 to 18 feet inward from the exterior wall. In the other areas of the complex, each different room or area should be treated as a separate zone. All zones must be compatible with the application, the equipment operating modes and the equipment operating hours.

Lighting Loads

The lighting loads will vary from low to high, depending on the area — make survey.

Occupancy Loads

The assembly areas are characterized by a weekly schedule that consists of a few short periods of peak occupancy, a few short periods of moderate occupancy and long periods of low (zero) occupancy. The ancillary areas that are associated with religious services are also characterized by scheduled occupancy, but the population densities may be more steady. The areas that are associated with the staff activities are characterized by steady, scheduled occupancy loads. Large occupancy loads increase the humidity loads, reduce the room sensible heat ratios and reduce the heating changeover points. In the dining areas, additional sensible and latent loads are associated with food service. Get occupancy information from the staff.

Equipment Loads

Sensible loads are associated with office equipment, minor appliances, audio-visual equipment, vending machines and duplicating equipment. Sensible and latent loads are generated by kitchen equipment and food service appliances.

Cooling Load

The assembly areas, meeting rooms and school rooms are characterized by large, occupancy related, peak loads that occur for short periods of time. (Precooling will help reduce the peak loads.) The solar gains may be an important factor in some of these areas, but the peak solar loads may not be coincident with the peak occupancy loads. The load swings in the perimeter areas are caused by variations in the occupancy level, the solar gain and the transmission load. The interior assembly areas are not subject to solar gains, so the cooling loads in these areas depend on the occupancy loads and the lighting loads. Wide load swings in the interior areas are caused by changes in the population density. Cooling is not required when assembly areas are vacant. (Thermal storage systems often prove to be practical because when the assembly areas are vacant for long periods of time.) Relatively steady loads characterize the interior administrative areas. Load swings in the perimeter administrative areas are associated with the exposure and the envelope construction details. Refer to other applications for more information about the other types of occupancies that are associated with the complex.

Operating Mode

The assembly areas and the interior zones require year-round cooling. The perimeter zone heating changeover points depend on the population density, internal loads, exposure and envelope construction details. Simultaneous heating and cooling is required for incompatible zones. Reheat may be required to control humidity in mild weather.

Humidity Control

Wide, occupancy related, swings in the latent loads and the room sensible heat ratios are associated with the assembly areas. The latent loads will be high and the room sensible heat ratios will be low when the population density is high. Conversely, the latent loads will be low and the room sensible heat ratios will be high when the population density is low. Allow the relative humidity levels to drift within the "comfort zone." In the staff offices, the latent loads will be relatively low and steady and the room sensible heat ratios will be relatively high. In these areas, the relative humidity levels can be allowed to drift within the "comfort zone."

Churches — Continued

Air Quality

Ventilation (outdoor air) is required for all the occupied areas. Standard filters are appropriate for most of the areas. Air that is associated with the duplicating equipment, work shop and food service areas must not be allowed to contaminate the other areas.

Exhaust CFM

Exhaust systems are required for the food service areas, duplicating department, workshop and toilets.

Outdoor Air

The outdoor air requirements depend on the population density and on the exhaust system make up air requirements. The outdoor air requirements will vary by zone — make a survey.

Space Pressure

The pressure in the kitchen, workshop, duplicating department and toilets should be negative. The pressure in all the other areas can be neutral or a little positive (depending on the air quality requirements).

Schedules

Make a survey to determine the schedules that are associated with the assembly areas and the staff areas. Also inquire about the unscheduled events that are associated with the church activities. The operating hours that are associated with the assembly areas are usually not compatible with the operating hours that are associated with the staff areas. The schedules that are associated with the offices, school rooms and meeting rooms may not be coincident with the religious services schedule. Take advantage of the diversity that is associated with the various operating schedules. Continuous conditioning is required for a residence that is part of the complex.

Noise

Noise levels are important. The assembly areas, school rooms, meeting rooms, private offices, library and audio-visual rooms should all be quiet.

Special Problems

Large sensible and latent loads and large load swings are associated with changes in the population density. Large load swings are associated with glass and entrance doors. Year-round cooling and dehumidification is required in some areas. Reheat may be required to control humidity during mild weather. The assembly areas are characterized by a short peak load condition that is much larger than average load condition. The operating modes and operating hours that are associated with the various occupancies and zones may not be compatible with each other. Some zones may not be defined by interior walls. Large heating loads occur near the entrance doors. It is hard to pressurize a zone that has a lot of door traffic. Equipment noise is unacceptable in many areas.

Convention Centers and Exhibition Halls

Occupancy

Multipurpose occupancy — one or more large exhibition halls, ancillary areas, main lobby, secondary entrances, ticket office, meeting rooms, general and private offices, audio-visual rooms, duplicating department, radio and TV broadcasting booths, projection booth, dressing rooms, restaurants, bars, concession stands, storage room, work shops and more. Refer to other applications for additional information.

Functions

Heating, cooling, dehumidification and ventilation are required.

Zoning

Each different type of occupancy or application is a separate zone. The exhibition areas are usually divided into more than one zone. Zone the main lobby and the other entrances. Zone the ancillary areas, audio visual rooms, communications booths and ticket booths. Zone the administrative areas, food service areas, bars and maintenance areas. Separate zones are required for the interior areas and the perimeter areas that are associated with large rooms. The loads that define the perimeter zones depend on the exposure and the envelope construction details (large solar loads and transmission loads that are associated with glass). Perimeter zones that are not defined by a wall can be considered to extend 12 to 18 feet inward from the exterior wall. All zones must be compatible with equipment operating modes and operating hours.

Lighting Loads

The lighting loads vary from low to high in most of the exhibition areas, depending on the event. The lighting levels in ancillary areas and the administrative areas range from low to medium. Large lighting loads are associated with the stage productions. Make a survey or each area and design for the largest expected load.

Occupancy Loads

The exhibition halls and the auxiliary areas are characterized by daily events that can run from morning into the late evening. The maximum population density that is associated with any particular area is based on the event that will generate the largest occupancy. The daily occupancy levels will vary widely, depending on the nature of the event. Small parties and receptions may take place in the meeting rooms or in the ancillary areas. The areas that are associated with administrative activities are characterized by steady, scheduled occupancy loads. Large occupancy loads increase the humidity loads, reduce the room sensible heat ratios and reduce the heating changeover points. In some areas, additional sensible and latent loads are associated with food service. Make a survey and ask the manager for occupancy information.

Equipment Loads

In the exhibition areas, the appliance and equipment loads will vary widely, depending on the nature of the event. In the other parts of the facility, sensible loads are generated by projectors, sound equipment, electronic equipment, broadcasting, communications equipment, office equipment, computers and audio-visual equipment. Additional sensible and latent loads are associated with the equipment (appliances, serving counters and vending machines) that is found in the bars and the food service areas. In the administrative areas, office equipment, computers and duplicating equipment generate sensible loads. Make a thorough survey of every room in the building to determine the sensible and latent loads that are associated with the equipment and appliances. Use load factors and the usage factors to de-rate "name-plate" inputs. Consider the effects of diversity and exhaust hoods.

Cooling Load

The exhibition halls and the ancillary areas are characterized by large sensible and latent loads and wide load swings. In the interior areas, the design loads occur when occupancy loads, lighting loads and equipment loads combine to produce a peak load condition. In these areas, large sensible and latent load swings are caused by changes in the population density, lighting levels and equipment loads. The design loads and the load swings in the perimeter areas are produced by various combinations of occupancy loads, lighting loads, equipment loads and envelope loads. (The solar gains are always important in the perimeter zones, but the peak load condition may not be synchronized with the maximum solar gain because the peak internal loads may not be coincident with peak solar loads.) Relatively steady loads characterize the interior administrative and supporting services areas. Variable loads characterize the perimeter administrative areas. The loads and load swings that are associated with the perimeter administrative zones depend on the internal loads, the exposure and the construction details.

Convention Centers and Exhibition Halls — Continued

Operating Mode

The exhibition areas and the interior zones require year-round cooling. The perimeter zone heating changeover points depend on lighting loads, equipment loads, population density, exposure and the envelope construction details. Simultaneous heating and cooling is required for incompatible zones. In some areas, reheat may be required to control humidity in mild weather.

Humidity Control

In the public areas, wide swings in the sensible loads, latent loads and room sensible heat ratios are caused by changes in the occupancy loads, lighting loads and equipment loads. Small populations with sizable lighting and equipment loads will generate small latent loads and high room sensible heat ratios. Large populations with moderate lighting and equipment loads will generate large latent loads and low room sensible heat ratios. The humidity levels can be allowed to drift within the "comfort zone" in these areas. In the administrative offices, the latent loads will be relatively low and steady and the room sensible heat ratios will be relatively high. the humidity levels can be allowed to drift within the a "comfort zone" in these areas.

Air Quality

Ventilation (outdoor air) is required for all the occupied areas. Standard filters are appropriate for most of the areas. Some broadcasting, communications or audio-visual equipment may require high efficiency filters. The air that is associated with bars, food preparation areas, dedicated smoking areas, duplicating equipment and workshops must not be allowed to contaminate the other areas. Use outdoor air to dilute the odors and fumes that are associated with events that feature animals or internal combustion engines.

Exhaust CFM

Exhaust systems are required in the projection booth, bars, food preparation areas, workshops, dedicated smoking areas and toilets.

Outdoor Air

Outdoor air requirements depend on population density, the dilution ventilation requirements and exhaust system make up air requirements. The outdoor air requirements vary by zone — make a survey.

Space Pressure

The pressure in the projection booth, bars, food preparation areas, duplicating department, workshop, dedicated smoking areas and toilets should be negative. The pressure in all other the areas can be neutral or a little positive (depending on the air quality requirements).

Schedules

Make a survey to determine the schedules that are associated with the exhibition hall and the ancillary areas. The operating hours that are associated with the exhibition areas are usually not compatible with the operating hours associated with the ticket booth and the administrative areas.

Noise

Noise levels are important, but not critical in the exhibition areas. The meeting rooms and private offices should be quiet.

Special Problems

Large sensible and latent loads and large load swings are associated with changes in the population density, lighting levels, equipment loads and envelope loads. Large load swings are associated with large glass areas. Year-round cooling and dehumidification is required in some areas. Reheat may be required to control humidity during mild weather in some areas. This type of application is characterized by very large energy requirements. The operating modes and the operating hours that are associated with the various occupancies and zones may not be compatible with each other. Zones may not be defined by walls and partitions. Large heating loads occur near the entrances. It is hard to pressurize a zone that has a lot of door traffic. Equipment noise is unacceptable in many areas. Air quality problems are associated with events that feature animals and internal combustion engines.

Theaters, Concert Halls and Auditoriums

Occupancy

Multipurpose occupancy — assembly hall, ancillary areas, main lobby, ticket office, meeting rooms, general and private offices, conference rooms, projection booth, stage area, dressing rooms, orchestra pit, concession stand, storage room, work shops and more. Refer to other applications for additional information.

Functions

Heating, cooling, dehumidification and ventilation are required.

Zoning

Each different type of occupancy is a separate zone. Zone the assembly hall seating areas (more than one zone may be required for the main floor and more than one zone may be required for the balcony); also zone the stage area. Zone the main lobby and the other entrances. Zone the ancillary areas, audio-visual rooms, communication areas and ticket booths. Zone the administrative areas, food service areas and maintenance areas. Separate zones are required for the interior areas and the perimeter areas that are associated with large rooms. The loads that define the perimeter zones depend on the exposure and the envelope construction details (large solar loads and transmission loads are associated with glass). Perimeter zones that are not defined by a wall can be considered to extend 12 to 18 feet inward from the exterior wall. All zones must be compatible with equipment operating modes and operating hours.

Lighting Loads

The auditorium lighting loads vary from negligible (for a movie house) to low or medium for live performances. The auditorium lights will be on during intermissions, before and after a performance, and when the cleaning crew is working. Large lighting loads are associated with the overhead stage lighting and the spot lights. The stage and auditorium lighting levels vary depending on the nature of the performance. The lighting levels in the rest of the facility depend on the type of occupancy. Make a survey.

Occupancy Loads

The occupancy loads in the auditorium areas are characterized by weekly or daily performance schedules and long unoccupied periods. The maximum population density is based on expected "full house" occupancy, but the day-to-day occupancy level will vary widely, depending on the nature of the event, day-of-week and time-of-day. The ancillary areas have schedules that are similar to the performance schedules, but the occupied periods are short (usually occurring just before or after the performance and during intermissions). Small parties and receptions may take place in the lobby or ancillary areas after the performance. The stage areas are used for rehearsals when seating area is vacant. The areas that are associated with the staff activities are characterized by steady, scheduled occupancy loads. Large occupancy loads increase the humidity loads, reduce the room sensible heat ratios and reduce the heating changeover points. (Additional sensible and latent loads are associated with food service.) Get occupancy information from the staff.

Equipment Loads

Large sensible loads are generated by projectors, sound equipment, TV cameras, electronic equipment, broadcasting equipment, communications equipment, computers and audio-visual equipment. Additional sensible and latent loads are associated with the equipment (appliances, serving counters and vending machines) that is found in the ancillary areas and the food service areas. Office equipment, computers and duplicating equipment generate sensible loads in the administrative areas. Make a thorough survey of every room in the building to determine the sensible and latent loads that are associated with the equipment and appliances. Use load factors and the usage factors to de-rate "name-plate" inputs. Consider the effects of diversity and exhaust hoods.

Cooling Load

The assembly areas and the ancillary areas are characterized by large peak loads that occur for a few hours a day or for a few hours per week. Large, variable, sensible and latent loads and loads swings are associated with these areas. The design loads in the interior zones occur when the occupancy loads and the lighting loads combine to produce a peak load condition. In these areas, large sensible and latent load swings are caused by changes in the population density and the lighting levels. (The sensible loads that are generated by the lights may be negligible, they may be produce a steady load or they may be characterized by a steady base load that is combined with a variable performance related load.) The design loads and the load swings in the perimeter zones are produced by various combinations of

Theaters, Concert Halls and Auditoriums — Continued

occupancy loads, lighting loads, equipment loads and envelope loads. (The solar gains are always important in the perimeter zones, but the peak load condition may not be synchronized with the maximum solar gain because the peak internal loads may not be coincident with peak solar loads.) Cooling is not required when assembly and ancillary areas are vacant. (Thermal storage systems often prove to be practical because when the assembly areas vacant for long periods of time.) Relatively steady loads characterize the interior administrative and supporting services areas. Variable loads characterize the perimeter administrative areas. The loads and load swings that are associated with the perimeter administrative zones depend on the internal loads, the exposure and the construction details.

Operating Mode

The assembly area and any interior zone requires year-round cooling. In the perimeter zones, the heating changeover points depend on the population density, the internal loads, the exposure and the envelope construction details. Simultaneous heating and cooling is required for incompatible zones. In some areas, reheat may be required to control humidity in mild weather.

Humidity Control

Wide, occupancy related, swings in the sensible loads, latent loads and room sensible heat ratios are associated with the assembly and ancillary areas. When the occupancy level is high, the latent loads will be high and room sensible heat ratios will be low. The room sensible heat ratios will vary, depending on the occupancy level. Allow the relative humidity level to drift within the "comfort zone." In the administrative areas, the latent loads will be steady and relatively low, and the room sensible heat ratios will be relatively high. Allow the relative humidity levels to drift within the "comfort zone."

Air Quality

Ventilation (outdoor air) is required for all the occupied areas. Standard filters are appropriate for most of the areas. Some broadcasting, communications or audio-visual equipment may require high efficiency filters. The air that is associated with the locker rooms, food service areas, dedicated smoking areas and workshops must not be allowed to contaminate the other areas.

Exhaust CFM

Exhaust systems are required for projection booths, bars, food service areas, workshops, dedicated smoking areas and toilets.

Outdoor Air

The outdoor air CFM depends on the population density and the exhaust CFM requirements. The outdoor air requirements vary by zone — make a survey.

Space Pressure

The pressure in the projection booth, food service areas, workshops, dedicated smoking areas and toilets should be negative. The pressure in all the other areas can be neutral or a little positive (depending on the air quality requirements).

Schedules

Make a survey to determine the theater, rehearsal and administrative schedules. Also inquire about the unscheduled events that are associated with the facility. The operating hours that are associated with the assembly areas are usually not compatible with the operating hours that are associated with the ticket booth and the administrative areas. Take advantage of the diversity that is associated with the various operating schedules.

Noise

Noise levels are important for legitimate theaters, concert halls and auditoriums. Noise levels are still important, but less critical for movie theaters. Assembly areas, meeting rooms and private offices should be quiet.

Special Problems

Large sensible and latent loads and load swings are associated with occupancy levels. Large load swings are associated with glass and entrance doors. Year-round cooling and dehumidification may be required in some areas. Reheat may be required to control humidity during mild weather in some areas. This type of application is characterized by large energy requirements. The operating modes and the operating schedules that are associated with the various areas may not be compatible with each other. Some zones may not be defined by walls and partitions. Large heating loads occur near the entrance. It is hard to pressurize a zone that has a lot of door traffic. Equipment noise is unacceptable in many areas.

Libraries and Museums

Occupancy

Multipurpose occupancy — public reading rooms, exhibit areas, lecture rooms, audio-visual rooms, auditorium, general offices, private offices, conference rooms, meeting rooms, book-stack areas, vaults, gift shop, large entrance area, printing and duplicating departments, computer rooms, communications equipment, mail room, cafeteria, lunch room, work shops, storage rooms. Refer to other applications for additional information.

Functions

Heating, cooling, dehumidification, ventilation and humidification are required.

Zoning

Each different type of occupancy is a separate zone. Some of the areas that are associated with a given type of occupancy may require more than one zone. Zone the interior areas and the perimeter areas. The perimeter zones depend on the exposure the envelope construction details. Perimeter zones that are not defined by an interior wall can be considered to extend 12 to 18 feet inward from the exterior wall. Zone all the storage, exhibit and work areas that have special temperature and humidity requirements. Some areas can be zoned in accordance with the occupancy schedule, but some areas require space conditioning on a 24-hour basis. Also consider the heating loads near entrances (heated vestibules are desirable). Refer to other applications for more information about the zoning requirements that are associated with the facility.

Lighting Loads

The lighting loads can vary from low to high, depending on the area. Consider the effect of the general lighting and the spot lighting. Get lighting information from the staff and make a survey.

Occupancy Loads

The population density depends on the area. Large swings will occur in the public areas, meeting rooms and conference rooms, but it usually remains steady in the administrative and work areas. Refer to other applications for more information about the occupancy loads that are associated with the various kinds of areas. Get occupancy information from the staff and make a survey.

Equipment Loads

Sensible and latent loads may be generated by the equipment that is associated with exhibits. Sensible loads are generated by computers, typewriters, communications and audio-visual equipment. Sensible and latent loads are associated with coffee brewers, lunch room appliances, vending machines, duplicating equipment and work shop equipment. Refer to other applications for more information about the equipment loads that are associated with the various kinds of areas. Get information about the equipment loads from the staff and make a survey.

Cooling Load

The loads in the interior public areas depend on the occupancy levels, the lighting loads and the internal loads. The peak loads that are associated with these areas occur during periods of peak occupancy and load swings are caused by changes in the population density. The loads in the perimeter public areas depend on the occupancy levels, the lighting loads and the internal loads. The peak loads that are associated with these areas occur when the envelope loads, occupancy loads and internal loads combine to generate the peak load condition (the peak solar loads may not be coincident with the peak occupancy loads). In these areas, loads swings are caused by changes in the envelope loads and the population density. Relatively steady loads characterize the interior administrative and storage areas. Load swings in the perimeter administrative areas are associated with the exposure and the envelope construction details. Refer to other applications for more information about the other types of occupancies that are associated with the facility.

Operating Mode

The interior zones require year-round cooling. The perimeter zone heating changeover points depend on the occupancy, the exposure and the envelope construction details. Dehumidification with reheat and heating with humidification may be necessary during any season, depending on the space humidity requirements. Simultaneous heating and cooling is required for incompatible zones.

Humidity Control

In the public areas, large variations in population density cause large swings in the latent load. The latent loads are normally small and steady in the private work areas and in the storage areas. Exact humidity control may be required in some public and private areas, but in the other areas, the relative humidity can be allowed to drift within the "comfort zone". Book stacks, book vaults, art storage and display areas, stuffed animal exhibits, clothing and tapestry exhibits, fossil and bone exhibits, scientific

Libraries and Museums — Continued

exhibits and work rooms are just a few examples of areas that are sensitive to the relative humidity. Some exhibits must be enclosed in special, environmentally controlled, cases. Reheat or special dehumidifying equipment will usually be required. The conditions that are ideal for the exhibits may not be comfortable for the patrons. Obtain detailed humidity requirements from the staff.

Air Quality

General ventilation and standard filters will suffice for some public and private areas, but some vaults, storage and display areas require specific ventilation rates and special filters. Ozone created by electronic filters may expedite the deterioration of some articles. Raw, unfiltered outdoor air may expedite the deterioration of some articles. Air that is associated with the workshop and food service areas must not be allowed to contaminate other areas. Obtain the air quality requirements from the staff.

Exhaust CFM

Exhaust systems are required for the workshops, the food service areas, the duplicating department and the toilets.

Outdoor Air

The outdoor air requirements depend on occupancy level, the exhibit fresh air requirements and the make up air requirements that are associated with the exhaust systems. The outdoor air requirements vary by zone — make a survey.

Space Pressure

The pressure in the workshops, lunch room, food service areas, duplicating department and toilets should be negative. The pressure in all other areas can be neutral or a little positive, depending on the air quality requirements.

Schedules

Ask the staff for information about the occupancy schedules that are associated with the visitors and the employees. Ask about the equipment operating schedules that are associated with the exhibits. One or more areas may require 24-hour conditioning. The schedules that are associated with the various areas may not be compatible with each other. Take advantage of the diversity that is associated with the various operating schedules.

Noise

Noise levels are important. Reading areas, exhibit areas, lecture rooms, private offices, meeting rooms and audio-visual rooms should be quiet.

Special Problems

There may be special temperature and humidity requirements for the books or the exhibits. The temperature and humidity requirements that are associated with various areas may not compatible with each other or with conditions that are associated with occupant comfort. Special temperature and humidity requirements increase capital cost and operating costs. The interior-zone load patterns are not compatible with perimeter-zone load patterns. Large heating and cooling loads are associated with large glass areas. Large heating loads occur near the entrance doors. Zones may not be defined by walls or partitions. The HVAC system(s) must accommodate zones that are occupied after hours. Incompatible applications create ventilation, pressure and air balance problems. Special air filters may be required for some areas. Frequent floor plans changes are associated with museum exhibits. The HVAC systems must accommodate wide variations in the occupancy levels. Noise levels must be low in most areas.

Schools

Occupancy

Multiple purpose occupancy — class rooms, library, shops, laboratories, gym, auditorium, swimming pool, lobby, offices, meeting rooms, lounges, cafeteria, kitchen, game rooms, laundry, and service areas. Refer to other applications for more information about the various areas.

Functions

Heating, ventilation, cooling and dehumidification are required.

Zoning

Each class room is a separate zone. Zone the library, shops, laboratories, gym, auditorium, swimming pool, lobby, offices, meeting rooms, lounges, cafeteria, kitchen, game rooms, laundry, and service areas. Some areas may require more than one zone. Interior spaces may require year-round cooling. Perimeter zones are sensitive to the exposure and the envelope construction details (especially the solar load and heat loss that is associated with glass). Large heating loads occur near entrance doors (heated vestibules are desirable).

Lighting Loads

The medium to high lighting levels that are associated with most areas produce significant internal loads — make a survey.

Occupancy Loads

The class rooms, laboratories and shops are occupied on a scheduled basis, but they may experience wide swings in population density. Wide variations in population density will occur in the library, in the assembly areas and in the other "activities" areas. The occupancy levels that are associated with the cafeteria are usually scheduled, but there may be large changes in the population density, depending on the schedule. (Additional sensible and latent loads are associated with food service.) The maximum population density that is associated with any zone is based on the expected "full room" occupancy. Large populations increase the humidity loads, reduce the room sensible heat ratios and reduce the heating changeover points. The areas that are associated with the administrative offices are characterized by steady, scheduled occupancy loads. Get occupancy information from the staff authority.

Equipment Loads

In the class rooms, the equipment loads are generated by computers and audio-visual equipment. Laboratories and shops have a wide variety of equipment that produce sensible and latent loads — make a survey. Televisions, sound equipment, vending machines, video games and small appliances may be found in the casual student assembly areas and in the administrative lounges. A rigorous load analysis is required for the kitchen and cafeteria serving area. Offices have equipment loads that are generated by computers, typewriters, duplicating equipment and small appliances. Service areas have laundry equipment and maintenance areas have shop equipment.

Cooling Load

The class room, shop and laboratory design loads occur when the occupancy loads, lighting loads, internal loads and solar loads combine to produce the peak load condition.(The solar gains are important, but in some areas, the peak occupancy loads may not be coincident with peak solar loads.) The perimeter areas experience wide load swings, depending on the population density, the lighting levels and the envelope loads. The lounges, assembly areas and the cafeteria are characterized by large sensible and latent loads and by wide load swings that are related to the population density. Since the interior areas are not sensitive to solar gains, the cooling loads depend on the occupancy, lighting and equipment loads. Wide load swings in the interior areas are caused by changes in the population density. Relatively steady loads characterize the interior administrative and supporting services areas. Variable loads characterize the perimeter administrative areas. The load swings in perimeter administrative zones depend on the internal loads, exposure and construction details. Special load calculation procedures are required for the pool and hot-tub areas (refer to recreational facilities).

Operating Mode

Class rooms, shops and laboratories typically changeover to heating at moderate outdoor temperatures, but the changeover point varies, depending on the envelope load and the internal load that is associated with each zone. Some assembly areas and the cafeteria may require year-round cooling. Interior zones require year-round cooling. The perimeter office heating changeover points depend on the population density, the lighting loads, the exposure and the envelope construction details. Simultaneous heating and cooling is required for all incompatible zones. Reheat may be required to control humidity in mild weather in some highly populated zones. Special operating cycles are required for the pool and hot-tub areas (refer to recreational facilities).

Schools — Continued

Humidity Control

In the classrooms, large variations in the sensible envelope loads affect the room sensible heat ratios. The sensible heat ratios in the shops and laboratories will be effected by the envelope loads, the internal loads, the exhaust systems and the make up air requirements — make a thorough survey. In the interior areas, wide swings in the sensible loads, latent loads and room sensible heat ratios are caused by changes in the population density, lighting loads and equipment loads. Small populations and sizable sensible loads will produce small latent loads and high room sensible heat ratios. Large populations and moderate sensible loads will produce large latent loads and low room sensible heat ratios. Allow the relative humidity levels to drift within the "comfort zone." In the administrative offices, the latent loads will be relatively low and steady and the room sensible heat ratios will be relatively high. The humidity levels can be allowed to drift within the "comfort zone." Special problems are associated with the pool and hot-tub areas (refer to recreational facilities).

Air Quality

Ventilation (outdoor air) is required for all the occupied areas. Standard filters are suitable for most of the areas. The contaminated air that is associated with the laboratories, student shops, kitchen, dedicated smoking areas, duplicating equipment, locker rooms, toilets, laundry and maintenance shops must not be allowed to contaminate other areas. Special problems are associated with the pool and hot-tub area (refer to recreational facilities).

Exhaust CFM

Exhaust systems are required in some shops and laboratories — make a thorough survey. Exhaust systems also are required for the food preparation areas, maintenance shops, janitors closets, laundry areas, dedicated smoking areas, locker rooms and toilets. Special requirements are associated with the pool and hot tub area (refer to recreational facilities).

Outdoor Air

The ventilation (outdoor air) requirements vary by zone, depending on the population density and the amount of make up air that is required for the exhaust system(s). The make up air requirements in shops and laboratories will depend on the type of equipment that is used in these areas. Make a thorough survey of each area. Special requirements are associated with the pool and hot-tub area (refer to recreational facilities).

Space Pressure

The pressure in the laboratories, classroom shops, food preparation areas, duplicating department, janitors closets, maintenance shops, laundry, dedicated smoking areas, locker rooms and toilets should be negative. The pressure in all other areas can be neutral or a little positive, depending on the air quality requirements. Special requirements are associated with the pool and hot-tub areas (refer to recreational facilities).

Schedules

Space conditioning is required in the classrooms, laboratories, assembly areas, shops, libraries, auditoriums, gyms, swimming pools and administrative areas on a scheduled basis. If a central plant is used, there may be some diversity associated with the various exposures and occupancy schedules. Some of the recreational areas may be used when classes are not in session. Ask the staff to provide information about the operating schedules that are associated with each area.

Noise

The library, auditorium, classrooms, meeting rooms and offices should be quiet. Noise levels are not as important in the recreational and casual assembly areas.

Special Problems

In the classrooms, through-the-wall equipment can be noisy and a source of drafts. Special exhaust and ventilation requirements are associated with the laboratories and shops. In many areas, large sensible and latent loads and swings are associated with changes in the population density, lighting loads and envelope loads. Large load swings are associated with large glass areas. Year-round cooling and dehumidification is required in some zones. The operating modes and the operating hours that are associated with the various areas may not be compatible with each other. Large heating loads occur near the entrance doors.

Transportation Terminals

Occupancy	Multipurpose occupancy — large open ticketing and waiting areas, executive clubs, offices, hard and soft goods shops, bars, lunch counters, restaurants, bakery, video games, barber shop, in-house repair shops, locker rooms, store rooms and more. Refer to other application sheets for information about the areas that support and supplement the ticketing and waiting areas.
Functions	Heating, cooling, dehumidification and ventilation are required.
Zoning	Each different occupancy is a separate zone. Some areas may have to be divided into two or more zones. All zones must be compatible with the HVAC equipment operating mode and operating hours. Some zones are created by the large solar loads and transmission loads that are associated with glass. Large heating loads occur near doors (heated vestibules are desirable).
Lighting Loads	Lighting loads vary from low to high, depending on the area — make a survey.
Occupancy Loads	Large variations in population density are associated with most of the areas. Extended periods of high densities are normal. Additional sensible and latent loads are associated with food service areas. Large population densities increase the humidity loads, reduce the room sensible heat ratios and reduce the heating changeover point. Make a thorough survey.
Equipment Loads	In the ticketing and waiting areas, sensible and latent loads are generated by computer, video terminals and food service counters. Other equipment is found in the subsidiary areas. Survey each area.
Cooling Load	Since the interior areas are not sensitive to solar gains, the design cooling loads depend on the internal loads. In these areas, wide load swings are cause by changes in the population density. The design cooling loads in the perimeter areas occur when the internal and the envelope loads combine to produce a peak load condition. Wide load swings in the perimeter areas are caused by variations in the occupancy level, the solar gains and the transmission loads.
Operating Mode	The interior zones will require year-round cooling. The perimeter areas will changeover to heating, but the changeover temperature depends on the relative size of the internal loads and the envelope loads. Simultaneous heating and cooling is required for incompatible zones.
Humidity Control	Most of the ticketing and waiting areas are characterized by wide, occupancy related, swings in the latent loads and the room sensible heat ratios. High population densities cause large latent loads and low room sensible heat ratios. Low population densities produce low latent loads and high room sensible heat ratios. Allow the humidity levels to drift within the "comfort zone."
Air Quality	Ventilation (outdoor air) is required for every occupied area. Standard filters will suffice for the ticketing and waiting areas. Use ventilation and exhaust systems to control smoke and odors. Locate the outdoor air intakes away from, and down wind from, engine exhaust fumes (charcoal filters may be required to control noxious fumes). Kitchen areas have special ventilation and filter requirements. Isolate cooking odors. Do not recirculate air from kitchens, locker rooms and toilets.
Exhaust CFM	Exhaust systems are required in the food service areas, repair shops, locker rooms, and toilets (survey each area).
Outdoor Air	The ventilation requirement (outdoor air CFM) depends on the population density and the make up air requirement — survey each area.
Space Pressure	Block noxious fumes by maintaining a positive pressure in the ticketing and waiting areas. The pressure in the cooking areas, workshops, locker rooms and toilets should be negative.
Schedules	Make a survey to determine operating schedules that are associated with each area. Take advantage of the diversity that is associated with the various exposures and operating schedules. Many areas require 24-hour conditioning.

Transportation Terminals — Continued

Noise

Noise levels are not critical in most areas. The administration offices should be quiet.

Special Problems

Large sensible and latent loads and swings are associated with the occupancy levels. Large load swings are associated with glass. Year-round cooling and dehumidification may be required in some areas. The operating modes and operating hours that are associated with the various areas may not be compatible with each other. Zones may not be defined by walls and partitions. Large heating loads occur near the entrance doors. Special infiltration barriers and heating equipment may be required in the boarding areas and baggage handling areas. It is hard to pressurize a zone that has a lot of door traffic. Noxious fumes are associated with engine exhaust and petroleum products. Each tenant may have to be billed for energy use.

Ballrooms and Roller Skating Rinks

Occupancy Multipurpose occupancy—offices, seating or dancing area, food and drink service, bar, lunch counter, sales areas, child care, kitchen, locker rooms and more. Refer to other applications.

Functions Heating, cooling, dehumidification and ventilation are required.

Zoning Each different type of occupancy is a separate zone. Some areas may require more than one zone. Some areas may have to divided into interior zones and perimeter zones. Perimeter zones are defined by the exposure and the envelope construction details (the solar and transmission loads that are associated with glass are especially important). Perimeter zones that are not defined by an interior wall can be considered to extend 12 to 18 feet inward from the exterior wall. Large heating loads are associated with the entrance doors (heated vestibules are desirable). All zones must be compatible with the corresponding occupancy schedules.

Lighting Loads Lighting loads normally vary from low to medium, depending on the area — make a survey. Special lighting effects may be associated with a dance floor.

Occupancy Loads A steady, low density occupancy characterizes the administrative areas. Large changes in the population density occur in the various patron areas. A large number of occupants increase the humidity loads, reduce the room sensible heat ratio and reduce the heating changeover points. (Food service, skating and dancing causes even higher, occupancy related, humidity loads.) Obtain occupancy information from the manager.

Equipment Loads Depending on the area, sensible loads are generated by office equipment, duplicating equipment, audio-visual equipment, video games and vending machines. Sensible and latent loads are associated with coffee brewers, cooking equipment, food service appliances and vending machines. Make a survey of the equipment that is associated with each area.

Cooling Load The interior-zone cooling loads depend on the internal loads (people, lights and equipment). The perimeter areas are effected by the internal loads and the envelope loads. When exterior glass is involved, the peak loads and the load swings depend on the solar gains and the population density. In some perimeter areas, the peak solar gain may not be coincident with the peak occupancy load.

Operating Mode The interior zones require year-round cooling. The perimeter-zone heating changeover points depend on the exposure, envelope construction details (especially the glass area) and the occupancy loads. Simultaneous heating and cooling is required for incompatible zones.

Humidity Control The patron areas may experience high latent loads, low sensible heat ratios and wide swings in the latent loads and the sensible heat ratios. The latent loads in the administrative areas are relatively low and steady. The relative humidity levels in all the areas can be allowed to drift within the "comfort zone."

Air Quality Food preparation areas have special ventilation and filter requirements. Ventilation (outdoor air) and standard filters are required for all the patron and administrative areas. Isolate the odors that are associated with smoking, food service, liquor service and skate rental. Do not recirculate air from the bar, kitchen or grill, skate rental area, locker rooms and toilets.

Exhaust CFM Exhaust systems are required for the kitchen or grill, skate rental area, locker rooms and toilets.

Outdoor Air The ventilation requirements depend on the population density and on the amount of make up air that is required for the exhaust systems. Introduce the outdoor air in patron and administrative areas. Make up air may be required in the kitchen or grill area.

Space Pressure The eating areas should be under a positive pressure. The bar, kitchen or grill, skate rental area, locker rooms and toilets should be under a negative pressure. Other areas can be maintained at a positive or a neutral pressure.

Ballrooms and Roller Skating Rinks — Continued

Schedules

Ask the manager about the operating hours that are associated with each area, the employee work schedules and the kitchen or grill operating schedule.

Noise

The office area should be quiet. The noise levels in the other areas are less critical.

Special Problems

The various areas are characterized by large variations in the population density and the cooling loads. Incompatible applications create ventilation, pressure and air balance problems. Large heating and cooling loads are associated with large glass areas. Large heating loads occur near the entrance doors. Zones may not be defined by walls and partitions. The HVAC system must accommodate zones that are occupied at different hours of the day.

Bowling Centers

Occupancy Multipurpose occupancy — bowling alleys, spectator areas, bar, restaurant, kitchen or grill, child care area, locker rooms and offices. Refer to other applications.

Functions Heating, cooling, dehumidification and ventilation are required.

Zoning Each different type of occupancy is a separate zone. Some areas may require more than one zone. Some areas may have to be divided into interior zones and perimeter zones. Perimeter zones are defined by the exposure and the envelope construction details (the solar and transmission loads that are associated with glass are especially important). Perimeter zones that are not defined by an interior wall can be considered to extend 12 to 18 feet inward from the exterior wall. Large heating loads are associated with the entrance doors (heated vestibules are desirable). In the bowling area, conditioning is not required for the area between the foul lines and the pen setters. All zones must be compatible with the corresponding occupancy schedules.

Lighting Loads Lighting loads normally vary from low to medium, depending on the area — make a survey of all the areas.

Occupancy Loads A low, steady occupancy level characterizes the office areas. Large changes in the population density occur in the various patron areas. Extended periods of high population density are normal. A large number of occupants increase the humidity loads, reduce the room sensible heat ratios and reduce the heating changeover points. (Food service and bowling causes even higher, occupancy related, humidity loads.) Obtain occupancy information from the manager.

Equipment Loads Depending on the area, sensible loads are generated by office equipment, duplicating equipment, audio-visual equipment, video games, vending machines, motors and pin setting equipment. Sensible and latent loads are associated with coffee brewers, food preparation equipment, food service equipment, appliances and vending machines. Make a survey of the equipment that is associated with each area.

Cooling Load The design cooling load for the bowling and spectator areas depends on the size of the internal loads (lights and people). Cooling is not required beyond the foul line. The other interior-zone cooling loads depend on the loads that are associated with people, lights and equipment. The perimeter areas are affected by the internal loads and the envelope loads. When exterior glass is involved, the peak loads and the load swings depend on the solar gains and the population density. In some perimeter areas, the peak solar gain may not be coincident with the peak occupancy load.

Operating Mode The interior zones require year-round cooling. The perimeter-zone heating changeover points depend on the exposure, envelope construction details (especially the glass area) and the occupancy loads. Simultaneous heating and cooling is required for incompatible zones.

Humidity Control The patron areas may experience high latent loads, low room sensible heat ratios and wide swings in the latent loads and the room sensible heat ratios. (High latent loads are associated with the people that are bowling.) The latent loads in the office areas are relatively low and steady. The relative humidity levels in all the areas can be allowed to drift within the "comfort zone."

Air Quality Food preparation areas have special ventilation and filter requirements. Ventilation (outdoor air) and standard filters are required for all the patron and office areas. Control the odors that are associated with the physical activity of the bowlers, the shoe rental area, tobacco smoke, the food service areas and the bar. Do not recirculate air from the bar, the kitchen or grill, the shoe rental area, locker rooms and toilets.

Exhaust CFM Exhaust the area between the foul lines and the pin setters. (A vertical baffle that extends down from the ceiling and runs down the foul lines will trap the stale air.) Exhaust systems also are required for the kitchen or grill, the shoe rental area, locker rooms and toilets.

Bowling Centers — Continued

Outdoor Air

The ventilation requirements depend on the population density and on the amount of make up air that is required for the exhaust systems. Introduce the outdoor air in patron and administrative areas. Make up air may be required in kitchen.

Space Pressure

The spectator areas and eating areas should be under a positive pressure. The bowling lanes, pin setting areas, bar, kitchen or grill, shoe rental area, locker rooms and toilets should be under negative pressure. Other areas can be maintained at a positive or a neutral pressure.

Schedules

Ask the manager about the operating hours that are associated with each area, the employee work schedules and the kitchen or grill operating schedule.

Noise

The office area should be quiet. The noise levels in the other areas are less critical.

Special Problems

Fresh air should flow through bowling area from the spectator area toward exhausts that are located near pin setters. Large variations in population density and cooling loads are associated with many of the areas. Incompatible applications create ventilation, pressure and air balance problems. Large heating and cooling loads are associated with large glass areas. Large heating loads occur near the entrance doors. Zones may not be defined by walls and partitions. The HVAC system must accommodate zones that are occupied at different hours of the day.

Country Clubs

Occupancy

Multipurpose occupancy — offices, meeting rooms, reception area, dining room, bar, lunch counter, sales areas, child care, kitchen, locker rooms, indoor pool, saunas, hot tubs and more. Refer to other applications for more information about particular areas.

Functions

Heating, cooling, dehumidification and ventilation are required.

Zoning

Each different type of occupancy is a separate zone. Special problems are associated with pools and hot tubs. Some areas may require more than one zone. Some areas may have to divided into interior zones and perimeter zones. Perimeter zones are defined by the exposure and the envelope construction details (the solar and transmission loads that are associated with glass are especially important). Perimeter zones that are not defined by an interior wall can be considered to extend 12 to 18 feet inward from the exterior wall. Large heating loads are associated with the entrance doors (heated vestibules are desirable). All zones must be compatible with the corresponding occupancy schedules.

Lighting Loads

Lighting loads normally vary from low to medium, depending on the area — make a survey.

Occupancy Loads

A low, steady occupancy level characterizes the office areas. Large changes in the population density occur in the various patron areas. Extended periods of high population density are normal in some areas. A large number of occupants increase the humidity loads, reduce the room sensible heat ratios and reduce the heating changeover points. (Food service and dancing causes even larger, occupancy related, humidity loads.) Obtain occupancy information from the manager.

Equipment Loads

Depending on the area, sensible loads are generated by office equipment, duplicating equipment, audio-visual equipment, video games and vending machines. Sensible and latent loads are associated with coffee brewers, food preparation equipment, food service equipment, appliances and vending machines. Make a survey of the equipment that is associated with each area.

Cooling Load

The design cooling loads in the interior-zones depend on the loads that are associated with people, lights and equipment. The perimeter areas are affected by the internal loads and the envelope loads. When exterior glass is involved, the peak loads and the load swings depend on the solar gains and the population density. In some perimeter areas, the peak solar gain may not be coincident with the peak occupancy load.

Operating Mode

The interior zones require year-round cooling. The perimeter-zone heating changeover points depend on the exposure, envelope construction details (especially the glass area) and the occupancy loads. Simultaneous heating and cooling is required for incompatible zones.

Humidity Control

The patron areas may experience high latent loads, low room sensible heat ratios and wide swings in the latent loads and the room sensible heat ratios. The latent loads in the office areas are relatively low and steady. The relative humidity levels in all the areas can be allowed to drift within the "comfort zone." Indoor swimming pool and hot tub areas require special systems to control humidity.

Air Quality

Food preparation areas have special ventilation and filter requirements. Ventilation (outdoor air) and standard filters are required for all the patron and office areas. Control the odors that are associated with tobacco smoke, food preparation, food service, and liquor service. Do not recirculate air from the bar, the kitchen or grill, locker rooms and toilets. (The areas that are associated with an indoor pool or hot tub require special attention.)

Exhaust CFM

Exhaust systems are required for the kitchen or grill, the locker rooms and the toilets. (The areas that are associated with an indoor pool or hot tub require special attention.)

Outdoor Air

The ventilation requirements depend on the population density and on the amount of make up air that is required for the exhaust systems. Introduce the outdoor air in patron and administrative areas. Make up air may be required in kitchen. (The areas that are associated with an indoor pool or hot tub require special attention.)

Country Clubs — Continued

Space Pressure

The eating areas should be under a positive pressure. The kitchen or grill area, the bar the locker rooms and the toilets should be under negative pressure. Other areas can be maintained at a positive or a neutral pressure. (The areas that are associated with an indoor pool or hot tub require special attention.)

Schedules

Ask the manager about the operating hours that are associated with each area, the employee work schedules and the kitchen or grill operating schedule.

Noise

Offices and meeting rooms should be quiet. The noise levels in the other areas are important but not critical.

Special Problems

Large variations in population density and cooling loads are associated with many of the areas. Incompatible applications create ventilation, pressure and air balance problems. Large heating and cooling loads are associated with large glass areas. Large heating loads occur near the entrance doors. Zones may not be defined by walls and partitions. The HVAC system must accommodate zones that are occupied at different hours of the day.

Gyms

Occupancy

Multipurpose occupancy — the competition area includes a playing floor, spectator seating and entrance areas; the gym facilities also include meeting rooms, locker rooms, training rooms, weight rooms, audio-visual rooms, general and private administrative offices. Refer to other applications for additional information.

Functions

Heating, cooling, dehumidification and ventilation are required.

Zoning

Each different type of occupancy is a separate zone. The gym seating and playing areas might have to be divided into more than one zone. The main gym area is essentially an interior zone unless there are substantial glass areas at the outside walls. Zone the ancillary areas, audio-visual rooms, training rooms, weight rooms and locker rooms. Zone the administrative offices. Some areas may have to be divided into interior zones and perimeter zones. Perimeter zones are defined by the exposure and the envelope construction details (the solar and transmission loads that are associated with glass are especially important). Perimeter zones that are not defined by an interior wall can be considered to extend 12 to 18 feet inward from the exterior wall. Large heating loads are associated with the entrance doors (heated vestibules are desirable). All zones must be compatible with the corresponding occupancy schedules.

Lighting Loads

Medium to high lighting loads are associated with the playing area. The lighting levels in the seating area may be the same as, or less than, those in the playing area (depending on the size of the gym). Medium lighting levels are associated with most of the auxiliary and administrative areas. Make a survey.

Occupancy Loads

Activities might be scheduled from the morning into the late evening in the gym and ancillary areas. The maximum population density in these areas is based on the activities that will generate a "full house" occupancy. But, the occupancy levels in these areas can vary widely, depending on the nature of the activity. (Small groups may use these areas when no activities are scheduled.) The administrative areas are characterized by steady, scheduled occupancy loads. In either case, large occupancy loads increase the humidity loads, reduce the room sensible heat ratios and reduce the heating changeover points. Get occupancy information from the staff.

Equipment Loads

Sensible loads are generated by sound equipment, office equipment, computers and audio-visual equipment. Other sensible and latent loads may be generated by training equipment, therapy equipment, small appliances, and vending machines. Make a survey of each area.

Cooling Load

The gym and the ancillary areas are characterized by large sensible and latent loads and large load swings. In these areas, the design cooling loads usually occur when the occupants produce a peak load condition. Relatively steady cooling loads (which depend on people, lights and equipment) characterize the interior administrative areas. Load swings in perimeter administrative zones depend on internal loads, exposure and construction details. When exterior glass is involved (for any area), the peak loads and the load swings depend on the solar gains and the population density. In some perimeter areas, the peak solar gain may not be coincident with the peak occupancy load.

Operating Mode

The gym, the interior ancillary areas and the interior administrative areas require year-round cooling. In the perimeter zones, the heating changeover points depend on the lighting loads, equipment loads, population density, exposure and the envelope construction details (especially the glass area). Simultaneous heating and cooling is required for incompatible zones.

Humidity Control

In the gym areas, wide swings in the sensible loads, latent loads and room sensible heat ratios are caused by changes in the occupancy loads and lighting loads. Small populations combined with large lighting loads will produce small latent loads and high room sensible heat ratios. Large populations combined with low lighting levels will cause large latent loads and low room sensible heat ratios. The relative humidity in these areas can be allowed to drift within the "comfort zone." (Reheat may be required to control humidity during mild weather.) In the administrative offices, the latent loads will

Gyms — Continued

be relatively low and steady and the room sensible heat ratios will be relatively high. In these areas, the relative humidity levels can be allowed to drift within the "comfort zone."

Air Quality

Ventilation (outdoor air) and standard filters are required for the gym and the administrative areas. Air that is associated with locker rooms, showers and toilets must not be allowed to contaminate the other areas.

Exhaust CFM

Exhaust systems are required for the locker rooms, therapy rooms, showers and the toilets.

Outdoor Air

The ventilation requirements that are associated with each area depend on the population density and on the amount of make up air that is required for the exhaust system — make a survey.

Space Pressure

The pressure in the locker rooms, showers and toilets should be negative. The pressure in all the other areas can be neutral or a little positive (depending on the air quality requirements).

Schedules

Question the staff about the occupancy and operating schedules that are associated with the gym, ancillary areas and the administrative areas.

Noise

Noise levels are not critical in the gym, but the audio-visual rooms, meeting rooms and offices should be quiet.

Special Problems

Large sensible and latent loads and large load swings are caused by changes in the population density and lighting levels. Year-round cooling and dehumidification is required in the gym, and reheat may be required to control humidity during mild weather. The operating modes and operating hours that are associated with various areas may not be compatible with each other. Some zones may not be defined by interior walls or partitions. Large heating loads occur near the entrance doors. It is hard to pressurize a zone that has a lot of door traffic.

Racquet Clubs

Occupancy

Multipurpose occupancy — the playing areas may include racquetball courts, squash courts or tennis courts, spectator seating areas and entrance areas. The club facilities also may include lounges, exercise rooms, weight training rooms, food service areas, bars, locker rooms and private administrative offices. Refer to other applications for additional information.

Functions

Heating, cooling, dehumidification and ventilation are required.

Zoning

The competition areas may be divided into one or more zones; depending on the activity that is associated with the area, the number of similar areas and the size of a given area or group of areas. The spectator areas and lounge areas should be separate zones. Zone the ancillary areas, exercise rooms, weight rooms and locker rooms. Zone the administrative offices. Some areas may have to divided into interior zones and perimeter zones. Perimeter zones are defined by the exposure and the envelope construction details (the solar and transmission loads that are associated with glass are especially important). Perimeter zones that are not defined by an interior wall can be considered to extend 12 to 18 feet inward from the exterior wall. Large heating loads are associated with the entrance doors (heated vestibules are desirable). All zones must be compatible with the corresponding occupancy schedules.

Lighting Loads

High lighting loads are associated with the competition areas. Low to medium lighting loads characterize the other areas — make a survey.

Occupancy Loads

Activities usually are scheduled from morning into the late evening. The occupancy levels in the various areas can vary widely, depending on the nature of the activity, the day of the week and the time of the day. The maximum population density in the various areas is based on the normal operating conditions that will generate a "full room" occupancy. (Extremely high population densities may be experienced during tournaments, but these events are sporadic, so they do not form the basis for estimating the occupancy loads.) The administrative areas are characterized by steady, scheduled occupancies. Large occupancy loads increase the humidity loads, reduce the room sensible heat ratios and reduce the heating changeover points. Get occupancy information from the staff.

Equipment Loads

Sensible loads are generated by exercise equipment, office equipment, computers, audio-visual equipment and video games. Sensible and latent loads are generated by cooking equipment, food service equipment, bar equipment, coffee brewers, small appliances and vending machines. Make a survey of the equipment that is associated with each area.

Cooling Load

The various areas are characterized by large sensible and latent loads and wide load swings. In these areas, the design cooling loads usually occur when the occupants produce a peak load condition. Relatively steady cooling loads (which depend on people, lights and equipment) characterize the interior administrative areas. Load swings in perimeter administrative zones depend on internal loads, exposure and construction details. When exterior glass is involved (for any area), the peak loads and the load swings depend on the solar gains and the population density. In some perimeter areas, the peak solar gain may not be coincident with the peak occupancy load.

Operating Mode

Racquetball courts, squash courts and exercise areas require year-round dehumidification — even when heat is required. (Reheat may be required to control the humidity in mild weather or even in cold weather.) Tennis courts require heat during cold weather. Tennis courts may be air conditioned during hot weather, but if they are, humidity control is not a major problem. The interior ancillary areas and the interior administrative areas require year-round cooling. In the perimeter zones, the heating changeover points depend on the lighting loads, equipment loads, population density, exposure and the envelope construction details (especially the glass area). Simultaneous heating and cooling is required for incompatible zones.

Humidity Control

The competition areas, exercise areas, spectator areas and lounge areas are subject to wide swings in the sensible loads, latent loads and room sensible heat ratios that are caused by changes in the occupancy loads and lighting loads. (Large latent loads — up to 7000 BTUH per hour per court —

Racquet Clubs — Continued

are associated with racquetball and squash courts. Even higher latent loads may be associated with exercise rooms.) Small populations combined with large lighting loads will produce small latent loads and high room sensible heat ratios. Large populations combined with low lighting levels will cause large latent loads and low room sensible heat ratios (especially if heavy exercise is involved). The relative humidity in the various patron areas can be allowed to drift within the "comfort zone." (In some areas, reheat may be required to control humidity during mild weather.) In the administrative areas, the latent loads will be relatively low and steady and the room sensible heat ratios will be relatively high and the relative humidity levels can be allowed to drift within the "comfort zone."

Air Quality

Ventilation (outdoor air) and standard filters are required for the various patron areas and the administrative areas. Air that is associated with the food service areas, locker rooms, showers and toilets must not be allowed to contaminate the other areas.

Exhaust CFM

Exhaust systems are for the bar areas, food service areas, locker rooms, showers and the toilets.

Outdoor Air

The ventilation requirements that are associated with each area depend on the population density and on the amount of make up air that is required for the exhaust system — make a survey.

Space Pressure

The pressure in the competition areas, exercise areas, bar areas, food service areas, locker rooms, showers and toilets should be negative. The pressure in all the other areas can be neutral or a little positive (depending on the air quality requirements).

Schedules

Question the staff about the occupancy and operating schedules that are associated with the competition areas, the ancillary areas and the administrative areas.

Noise

Noise levels are not critical in the competition and exercise areas, but the lounge and administrative areas should be quiet.

Special Problems

Large sensible and latent loads and large load swings are caused by changes in population density and activity levels. Year-round cooling and dehumidification may be required for the racquetball courts, squash courts and exercise areas. In some areas, reheat may be required to control humidity during mild or cold weather. The operating modes and operating hours that are associated with various areas may not be compatible with each other. Some zones may not be defined by walls and partitions. Large heating loads occur near the entrance. It is hard to pressurize a zone that has a lot of door traffic.

Swimming Pools

Occupancy

Multipurpose occupancy — the main natatorium area consists of the pool and the deck areas, spectator seating areas and entrance areas. The pool facilities also may include meeting rooms, locker rooms, training rooms and administrative offices. Refer to other applications for additional information.

Functions

In the pool areas, heating, dehumidification and ventilation are required — cooling is optional. Heating, cooling, dehumidification and ventilation are usually required in the ancillary and administrative areas.

Zoning

The pool and spectator areas should be divided into separate zones. The seating areas might have to be divided into more than one zone. Zone the ancillary areas, training rooms, and locker rooms. Zone the administrative offices. (The latent loads in the administrative areas and in some of the ancillary areas will depend on how much moisture migrates from the pool area. Ideally these areas should be completely isolated from the pool area.) Some areas may have to divided into interior zones and perimeter zones. Perimeter zones are defined by the exposure and the envelope construction details (the solar and transmission loads that are associated with glass are especially important). Perimeter zones that are not defined by an interior wall can be considered to extend 12 to 18 feet inward from the exterior wall. Large heating loads are associated with the entrance doors (heated vestibules are desirable). All zones must be compatible with the corresponding occupancy schedules.

Lighting Loads

Medium to high lighting loads are associated with the pool area. The lighting levels in the seating area may be the same as, or less than, those in the playing area. Medium lighting levels are associated with most of the ancillary and administrative areas. Make a survey.

Occupancy Loads

Activities might be scheduled from the morning into the late evening in the pool and ancillary areas. The maximum population density in these areas is based on the activities that will generate a "full room" occupancy. But, the occupancy levels in these areas can vary widely, depending on the nature of the activity. (Small groups may use these areas when no activities are scheduled.) The administrative areas are characterized by steady, scheduled occupancy loads. In either case, large occupancy loads increase the humidity loads, reduce the room sensible heat ratios and reduce the heating changeover points. Get occupancy information from the staff.

Equipment Loads

Sensible loads are generated by sound equipment, office equipment, computers and audio-visual equipment. Other sensible and latent loads may be generated by training equipment, therapy equipment, small appliances, and vending machines. Make a survey of each area.

Cooling Load

The pool and spectator areas are dominated by large latent loads. (The peak latent loads will occur at peak occupancy during wet weather.) A special type of HVAC system must be designed to control the humidity load that is associated with these areas. The ancillary areas may be characterized by large sensible and latent loads and large load swings. In these areas, the design cooling loads usually occur when the occupants produce a peak load condition. Relatively steady cooling loads (which depend on people, lights and equipment) characterize the interior administrative areas. Load swings in perimeter administrative zones depend on internal loads, exposure and construction details. When exterior glass is involved (for any area), the peak loads and the load swings depend on the solar gains and the population density. In some perimeter areas, the peak solar gain may not be coincident with the peak occupancy load.

Operating Mode

The pool and its associated areas are subject to humidity conditions that require year-round dehumidification. The interior, ancillary areas and administrative areas that are (should be) isolated from the pool areas require year-round cooling. In the perimeter ancillary and administrative areas (that should be isolated from the pool areas), the heating changeover points depend on the lighting loads, equipment loads, population density, exposure and the envelope construction details (especially the glass area). Simultaneous heating and cooling is required for incompatible zones.

Humidity Control

Humidity control is the primary consideration in the pool area and in the areas that are common to the pool area. Large amounts of outdoor air or special HVAC equipment is required to control the humidity

Swimming Pools — Continued

in these areas. (The ventilation CFM or the refrigeration equipment capacity is based on the latent loads that are associated with these areas and reheat may be required to control the dry-bulb temperature.) The ancillary areas that are isolated from the pool areas will have variable latent loads and variable room sensible heat ratios, depending on the population density and the sensible loads. In the administrative offices, the latent loads will be relatively low and steady and the room sensible heat ratios will be relatively high. Small populations combined with large sensible loads will produce small latent loads and high room sensible heat ratios. Large populations combined with low sensible loads will cause large latent loads and low room sensible heat ratios. The relative humidity in the various areas can be allowed to drift within the "comfort zone."

Air Quality

The filters that are used for the pool area must not react with chlorine. (In this type of environment, medium efficiency filters will reduce streaking near the supply air outlets.) Ventilation (outdoor air) and standard filters are required for the ancillary areas and the administrative areas. Air that is associated with pool areas, locker rooms, showers and toilets must not be allowed to contaminate the other areas.

Exhaust CFM

Specially designed exhaust systems can be used to control the humidity in the pool areas. Exhaust systems are required for the locker rooms, therapy rooms, showers and toilets.

Outdoor Air

Large amounts of outdoor air will be required if the humidity levels in the pool areas are controlled by an exhaust systems. The ventilation requirements that are associated with the other areas depend on the population density and on the amount of make up air that is required for the exhaust system — make a survey.

Space Pressure

The pressure in the pool area should be negative with respect to all the other areas, including the locker rooms. The pressure in the locker rooms, showers and toilets should be negative. The pressure in the administrative areas and in some of the ancillary areas should be positive.

Schedules

Question the staff about the occupancy and operating schedules that are associated with the pool, the ancillary areas and the administrative areas.

Noise

Noise levels are not critical in the pool areas, but some of the ancillary areas and the administrative areas should be quiet.

Special Problems

Major problems are associated with controlling the large latent loads that are generated in the pool areas. And, these latent loads must be controlled during any type of weather. In most climates, special mechanical refrigeration systems are required for year-round control. (Ventilation systems will not do the job if the outdoor air is to humid.) In any case, the capital costs and operating costs that are associated with humidity control are substantial. In some ancillary areas, large sensible and latent loads and large load swings may be caused by changes in the population density and sensible gains. The operating modes and operating hours that are associated with various areas may not be compatible with each other. Some zones may not be defined by interior walls or partitions. Large heating loads occur near the entrance doors. It is hard to pressurize a zone that has a lot of door traffic.

Convenience Centers and Mini-Malls

Occupancy Multipurpose occupancy — multiple single occupancy stores and some multipurpose occupancy stores; supermarkets, theaters, professional offices, etc. Refer to other applications for detailed information about each type of occupancy.

Functions Heating, cooling, dehumidification and ventilation are required. (Some applications may require winter humidification.)

Zoning Usually each different store has a separate system. Within a given store, each different type of area establishes a separate zone. Some areas may have to divided into interior zones and perimeter zones. Interior spaces require year-round cooling. The perimeter zones are defined by the exposure and the envelope construction details (the solar and transmission loads that are associated with glass are especially important). Perimeter zones that are not defined by an interior wall can be considered to extend 12 to 18 feet inward from the exterior wall. The area near a checkout counter is usually a separate zone. Large heating loads are associated with the entrance doors (heated vestibules are desirable). All zones must be compatible with the corresponding occupancy schedules. Special infiltration barriers and heating equipment may be required for a loading dock.

Lighting Loads The lighting loads usually range between medium and high, depending on the store or the department — make a survey of each area.

Occupancy Loads The population density can vary widely, depending on the store or the department. Get occupancy information from the store managers.

Equipment Loads The equipment loads vary, depending on the store or the department — make a survey of each area.

Cooling Load Interior areas are not sensitive to solar gains, so the cooling loads and load swings in these areas depend on the size of the internal loads. The perimeter areas are affected by the internal loads and the envelope loads. In these areas, the design loads occur when the internal loads and the envelope loads combine to produce a peak load condition; and large load swings are caused by changes in the population density and the envelope loads (particularly the loads that are associated with glass).

Operating Mode The interior zones require year-round cooling. The perimeter areas will changeover to heating, but the changeover temperature will depend on the relative size of the internal loads and the envelope loads. (The loads that are associated with the glass areas and the entrance doors are especially important.) Simultaneous heating and cooling is required for incompatible zones.

Humidity Control The humidification loads and the humidification requirements vary, depending on the store or department (make a survey). Changes in the population density will cause the relative humidity to drift, which is acceptable, providing that it stays in the "comfort zone."

Air Quality The air quality requirements depend on how the store is used. Ventilation (outdoor air) and standard filters are required for most of the stores (talk to the managers to determine if there are any special air quality requirements). Control the odors that are associated with tobacco smoke, food preparation, food service and liquor service. Do not recirculate air from cooking areas and toilets. (Refer to other applications.)

Exhaust CFM The exhaust system requirements depend on the application — make a survey and talk to the managers. Exhaust all odors and fumes, and exhaust the toilets.

Outdoor Air The ventilation requirements depend on the population density and on the amount of make up air that is required for the exhaust systems — make a survey.

Space Pressure The space pressure depends on the application. Spaces that are associated with activities that produce odors should be negative with respect to other spaces — make a survey.

Convenience Centers and Mini-Malls

Schedules Talk to the managers to determine the occupancy and operating schedules. Take advantage of diversity.

Noise Normally, there are no unusual limitations on the noise level — talk to the managers.

Special Problems Multiple applications must be accommodated. Some applications may be incompatible if they occur in the same store. Large heating and cooling loads are associated with glass. Large heating loads occur near the entrance doors. Wide load swings are caused by changes in the population density and envelope loads. Frequent floor plan changes and tenant turnover. Lease agreements may require or exclude certain HVAC system features.

Department Stores

Occupancy Multipurpose occupancy — offices, hard and soft goods sales areas, lunch counter, restaurant, bakery, flower shop, beauty shop, optometrist, dentist, insurance sales, pharmacy, auto service, in-house repair shops, locker room, warehouse and more. Refer to other applications additional information.

Functions Heating, cooling, dehumidification and ventilation are required. (Humidification may be required for some areas.)

Zoning Each different type of occupancy (department) is a separate zone. Some occupancies (departments) may require two or more zones. Some areas may have to divided into interior zones and perimeter zones. Interior spaces require year-round cooling. The perimeter zones are defined by the exposure and the envelope construction details (the solar and transmission loads that are associated with glass are especially important). Perimeter zones that are not defined by an interior wall can be considered to extend 12 to 18 feet inward from the exterior wall. Large heating loads are associated with the entrance doors (heated vestibules are desirable). All zones must be compatible with the corresponding occupancy schedules. Special infiltration barriers and heating equipment may be required for a loading dock.

Lighting Loads In the sales areas and office areas, the lighting loads usually range from medium to high. Special lighting may be required in some departments — make a survey.

Occupancy Loads The population density can vary widely for the building as a whole and from department to department. (Food service causes even larger, occupancy related, humidity loads.) Get occupancy information from the department managers.

Equipment Loads The equipment loads vary, depending on the area (department). Some of the equipment loads are associated with running the business; office equipment, for example. Some of the equipment loads are associated with merchandise displays; the loads that are generated in the television and lighting fixture display areas, for example. Some equipment loads are associated with food service and entertainment; cooking equipment, food warmers, video games and vending machines, for example. Refer to other applications for more information. Make a thorough survey of each area.

Cooling Load Interior areas are not sensitive to solar gains, so the cooling loads and load swings in these areas depend on the size of the internal loads. The perimeter areas are affected by the internal loads and the envelope loads. In these areas, the design loads occur when the internal loads and the envelope loads combine to produce a peak load condition; and large load swings are caused by changes in the population density and the envelope loads (particularly the loads that are associated with glass).

Operating Mode The interior zones require year-round cooling. The perimeter areas will changeover to heating, but the changeover temperature will depend on the relative size of the internal loads and the envelope loads. (The loads that are associated with the glass areas and the entrance doors are especially important.) Simultaneous heating and cooling is required for incompatible zones.

Humidity Control The humidification loads and the humidification requirements vary, depending on the department (make a survey). Changes in the population density will cause the relative humidity to drift, which is acceptable, providing that it stays in the "comfort zone." (Humidity control could be important in some departments.) Look for unusual sources of humidity. Refer to information about other applications.

Air Quality The air quality requirements depend on how the area is used. Ventilation (outdoor air) and standard filters are required for most of the departments (ask the staff if there are any special air quality requirements). Control the odors that are associated with tobacco smoke, food preparation and food service. Do not recirculate air from cooking areas and toilets. (Refer to other applications.)

Exhaust CFM The exhaust system requirements depend on the application — make a survey and talk to the staff. Exhaust all odors and fumes, and exhaust the toilets.

Department Stores — Continued

Outdoor Air

The ventilation requirements, which vary by department, depend on the population density and on the amount of make up air that is required for the exhaust systems — make a survey.

Space Pressure

The space pressure depends on the application. Spaces that are associated with activities that produce odors should be negative with respect to other spaces — make a survey.

Schedules

Talk to the department managers to determine the occupancy and operating schedules. Take advantage of diversity.

Noise

Normally, there are no unusual limitations on the noise levels that are associated with the sales areas. There may be some specific noise level requirements associated with the offices and the professional services areas — talk to the managers.

Special Problems

Multiple applications must be accommodated. Some applications may be incompatible with each other. Some zones are not defined by partitions and walls. Large heating and cooling loads are associated with glass. Large heating loads occur near the entrance doors. Wide load swings are caused by changes in the population density and envelope loads. Frequent floor plan changes.

Funeral Homes

Occupancy

Multipurpose occupancy — assembly rooms and ancillary areas, main vestibule, offices, conference rooms, work rooms, storage room and more.

Functions

Heating, cooling, dehumidification and ventilation are required.

Zoning

Zone the assembly areas, ancillary areas, offices, conference rooms and work rooms. Distinguish between the interior zones and the perimeter zones. The perimeter zones are defined by the exposure and the construction details (particularly the glass area). All zones must be compatible with the intended use and the occupancy schedule.

Lighting Loads

The lighting loads may vary from low to high, depending on the area — make a survey.

Occupancy Loads

The assembly and ancillary areas are characterized by relatively short periods of high population density. The areas that are associated with the staff activities are characterized by steady, scheduled occupancy loads. Large occupancy loads increase the humidity loads, reduce the room sensible heat ratios and reduce the heating changeover points. Get occupancy information from the manager.

Equipment Loads

Sensible and latent loads are associated with office equipment, minor appliances and the equipment that is used in the preparation rooms.

Cooling Load

The assembly areas are characterized by large, occupancy related, loads that occur for short periods of time. In the interior areas, the loads and the load swings depend on the occupancy and lighting loads. In the perimeter areas, the loads and the load swings depend on the population density, the lighting loads, the solar gains and the transmission loads. (The solar gains may be an important factor in some of these areas, but the peak solar loads may not be coincident with the peak occupancy loads.) In the offices and the work areas, the internal loads are fairly constant, so the peak loads and the load swings depend on the envelope loads.

Operating Mode

The assembly areas and the interior zones require year-round cooling. The perimeter-zone heating changeover points depend on the population density, exposure and envelope construction details (especially the glass area). Simultaneous heating and cooling is required for incompatible zones. Reheat may be required to control the humidity during mild weather.

Humidity Control

Wide, occupancy related, swings in the latent loads and the room sensible heat ratios are associated with the assembly areas. The latent loads will be high and the room sensible heat ratios will be low when the population density is high. Conversely, the latent loads will be low and the room sensible heat ratios will be high when the population density is low. Allow the relative humidity levels to drift within the "comfort zone." In the staff offices, the latent loads will be relatively low and steady and the room sensible heat ratios will be relatively high. In these areas, the relative humidity levels can be allowed to drift within the "comfort zone."

Air Quality

Ventilation (outdoor air) and standard filters are required for the assembly and office areas. Air that is associated with the preparation rooms and toilets must not be allowed to contaminate the other areas. Try to contain the odors that are associated with flowers. Refer to codes.

Exhaust CFM

Exhaust systems are required for the preparation rooms and the toilets (and possibly for the floral display areas). Refer to codes.

Outdoor Air

The outdoor air requirements depend on the population density and on the exhaust system make up air requirements that are associated with each zone — make a survey and refer to codes.

Space Pressure

The pressure in the preparation rooms, floral display areas and toilets should be negative. The pressure in all the other areas can be neutral or a little positive.

Schedules

Ask the staff about the assembly schedules and the (staff) work schedules.

Funeral Homes — Continued

Noise

Low noise levels are required throughout the facility.

Special Problems

Large sensible and latent loads and load swings are associated with population density. Year-round cooling and dehumidification is required in some areas. Reheat may be required to control humidity during mild weather. Large cooling loads occur for short periods of time. The operating modes and operating hours that are associated with various occupancies and zones may not be compatible with each other. Large heating loads occur near the entrance doors. Some areas have special exhaust and ventilation requirements. Equipment noise is unacceptable.

Garages and Auto Dealerships

Occupancy Multipurpose occupancy — offices, showroom, parts department, waiting rooms, lunch room, shop area, locker rooms and storage areas.

Functions Heating, cooling, dehumidification and ventilation are required (accept in the shop).

Zoning Zone the offices, showroom, waiting rooms, parts department and the lunch room. Isolate the interior zones from the perimeter zones. The perimeter zones are defined by the exposure and the construction details (particularly the glass area). The shop area should have a separate heating system. All zones must be compatible with the intended use and the occupancy schedule.

Lighting Medium to high lighting loads are associated with the various areas — make a survey.

Occupancy Loads The population density is usually moderate in the showroom and office areas. Large occupancy loads are associated with the lunch room. Get occupancy information from the staff.

Equipment Loads Sensible loads are generated by computers, office equipment and audio-visual equipment. Sensible and latent loads are generated by coffee brewers, lunch room appliances, vending machines and duplicating equipment.

Cooling Load The design cooling loads in the interior-zones depend on the loads that are associated with people, lights and equipment. The perimeter areas are affected by the internal loads and the envelope loads. When exterior glass is involved, the peak loads and the load swings depend on the solar gains and the population density. (The solar gains are especially important in the show room area.) Cooling is not usually provided in the shop.

Operating Mode The interior zones require year-round cooling. The perimeter-zone heating changeover points depend on the exposure, envelope construction details (especially the glass area) and the internal loads. Simultaneous heating and cooling is required for incompatible zones. Heating is required in the shop.

Humidity Control In the offices and the customer areas, the latent loads are relatively low and steady and the room sensible heat ratios are relatively high. In these areas, the humidity levels can be allowed to drift within the "comfort zone."

Air Quality Ventilation and standard filters are required for the showroom, waiting areas, offices, parts department and lunch room. Fumes that are generated in the shop must not be allowed to contaminate the other areas. Capture and exhaust the toxic fumes that are associated with the shop area.

Exhaust CFM Exhaust systems are required for the lunch room, locker rooms, shop area, battery charging area and toilets. Special exhaust equipment is required for the shop. General exhaust systems can be used to cool the shop in the summer.

Outdoor Air In each area, the ventilation requirements depend on the population density and the amount of make up air that is required for the exhaust systems.

Space Pressure The pressure in the shop, lunch room, locker rooms and toilets should be negative. The pressure in the other areas can be neutral or a little positive.

Schedules Ask the manager about the customer hours and the work schedules.

Noise The sales and offices areas should be quiet.

Special Problems Large heating and cooling loads are associated with large glass areas. Large heating loads occur near the entrance doors. Large heating loads are associated with the garage doors. In the garage areas, the infiltration loads must not produce adverse working conditions. Hot air stratifies in high-bay shop areas. Fumes are generated by engines. There is an explosion risk in battery charging area.

Small Print Shops, Paint Stores or Wallpaper Stores

Occupancy

Multipurpose occupancy — sales area, offices, conference room, reception area, work rooms and storage rooms.

Functions

Heating, cooling, dehumidification and ventilation are required — humidification may be required in a print shop.

Zoning

Zone the sales areas, office areas, work rooms and storage areas. Isolate the interior zones from the perimeter zones. The perimeter zones are defined by the exposure and the construction details (particularly the glass area). Perimeter zones that do not have an interior wall can be considered to extend 12 to 18 feet inward from the exterior wall. The work areas and store rooms may have special exhaust requirements. Large heating loads occur near the enterance doors (heated vestibules are desirable). All zones must be compatible with the intended use and the occupancy schedule.

Lighting Loads

Medium to high lighting loads are associated with most areas — make a survey.

Occupancy Loads

Population densities are usually light and they only have a small affect on the room sensible heat ratio and the heating changeover point.

Equipment Loads

Most stores have some small appliances and vending machines. Specialized equipment is associated with printing, duplicating and paint mixing operations. Talk to manager and make a survey.

Cooling Load

The peak cooling load in any zone is determined by the combined effect of the solar gains, transmission loads, occupancy loads and internal loads. Changing solar loads can cause wide swings in the cooling load (in some perimeter zones).

Operating Mode

The interior zones require year-round cooling. The perimeter areas will changeover to heating, but the changeover temperature (which is usually well below 70°F) will depend on the relative size of the internal loads and the envelope loads. (The loads that are associated with the glass areas and the entrance doors are especially important.) Simultaneous heating and cooling is required for incompatible zones.

Humidity control

Normally, the latent loads are not high enough to cause humidity control problems — allow the humidity levels to drift within the "comfort zone." (Year-round humidity control may be required for printing operations — ask the manager.)

Air Quality

Ventilation (outdoor air) and standard filters are required for the office and sales areas. Special ventilation and exhaust systems may be required for the work areas and the storage areas. Make a survey and talk with the manager.

Exhaust CFM

Exhaust all odors and fumes that are associated with the work areas and the storage areas. Exhaust the toilets.

Outdoor Air

The ventilation requirements depend on the population density and on the amount of make up air that is required for the exhaust systems — make a survey.

Space Pressure

Negative pressure is required in the areas that are could be contaminated by odors and fumes. The other areas can be maintained at a neutral or a slightly positive pressure.

Schedules

Ask the manager about the customer hours and the operating schedules.

Noise

Normally, there are no unusual limitations on noise levels.

Special Problems

Control odors, fumes and (for printing operations) humidity. The various areas may not be compatible with each other. Large heating loads, cooling loads and load swings are associated with large windows and entrance doors.

Shopping Malls

Occupancy

Multipurpose occupancy — single occupancy stores and multi-occupancy stores, supermarkets, theaters, professional offices, etc. Refer to other applications for more information about particular types of occupancies.

Functions

Heating, cooling, dehumidification and ventilation are required. (Winter humidification may be required for some areas.)

Zoning

There may be a central system or each different store may have a separate system. Within a given store, each different type of area establishes a separate zone. The enclosed mall area requires many zones; also zone the maintenance areas and locker rooms. Some areas may have to be divided into interior zones and perimeter zones. The perimeter zones are defined by the exposure and the envelope construction details (the solar and transmission loads that are associated with glass are especially important). Perimeter zones that are not defined by an interior wall can be considered to extend 12 to 18 feet inward from the exterior wall. Large heating loads are associated with the entrance doors (heated vestibules are desirable). All zones must be compatible with the corresponding occupancy schedules. Special infiltration barriers and heating equipment may be required for a loading dock.

Lighting Loads

Lighting loads vary, depending on the store and the area within the store — make a survey.

Occupancy Loads

Peak occupancy loads and occupancy related load swings vary, depending on the store and the area within the store — make a survey.

Equipment Loads

Equipment loads vary, depending on the store and the area within the store — make a survey.

Cooling Load

Interior areas are not sensitive to solar gains, so the cooling loads and load swings in these areas depend on the size of the internal loads. The perimeter areas are affected by the internal loads and the envelope loads. In these areas, the design loads occur when the internal loads and the envelope loads combine to produce a peak load condition; and large load swings are caused by changes in the population density and the envelope loads (particularly the loads that are associated with glass).

Operating Mode

The interior zones require year-round cooling. The perimeter areas will changeover to heating, but the changeover temperature will depend on the relative size of the internal loads and the envelope loads. (The loads that are associated with the glass areas and the entrance doors are especially important.) Simultaneous heating and cooling is required for incompatible zones.

Humidity Control

The humidification loads and the humidification requirements vary, depending on the store or department (make a survey). In any particular area, changes in the population density will cause the relative humidity to drift, which is acceptable, providing that it stays in the "comfort zone."

Air Quality

The air quality requirements depend on how the store or the area within the store is used. Ventilation (outdoor air) and standard filters are required for most of the areas (make a survey and talk to the managers to determine if there are any special air quality requirements). Control the odors that are associated with tobacco smoke, food preparation, food service and liquor service. Do not recirculate air from smoking areas, cooking areas and toilets. (Refer to other applications.)

Exhaust CFM

The exhaust system requirements depend on the application — make a survey and talk to the managers. Exhaust all the areas that are associated with activities that produce odors and fumes, and exhaust the toilets.

Outdoor Air

The ventilation requirements depend on the population density and on the amount of make up air that is required for the exhaust systems — make a survey.

Space Pressure

The pressure in the stores should be a little positive with respect to the mall. Spaces that are associated with activities that produce odors should be negative with respect to other spaces — make a survey.

Shopping Malls — Continued

Schedules

Ask the managers about the occupancy and equipment operating schedules. Take advantage of diversity.

Noise

Normally, there are no unusual limitations on the noise level — talk to the managers.

Special Problems

Multiple applications must be accommodated (some may be incompatible if they occur in the same store). Large heating and cooling loads are associated with glass and entrance doors. Wide load swings are caused by changes in the population density and envelope loads. Stores do not have a well defined envelope when there front is open to the mall area. Frequent floor plan changes and tenant turnover. Lease agreements may require or exclude certain HVAC system features.

Small Stores and Shops

Occupancy

Single purpose occupancy with no food service. Hard-goods stores, soft-goods stores, candy store, TV and appliance store, barber shop, beauty shop, dry cleaner, coin laundry, pet shop, pharmacy, photo studio, pool hall, shoe store and more. Refer to other applications if the store does not fit this category.

Functions

Heating, cooling, dehumidification and ventilation are required — humidification may be required for some stores.

Zoning

Within a store, each different type of area establishes a separate zone. Some areas may have to divided into interior zones and perimeter zones. Interior spaces require year-round cooling. The perimeter zones are defined by the exposure and the envelope construction details (the solar and transmission loads that are associated with glass are especially important). Perimeter zones that are not defined by an interior wall can be considered to extend 12 to 18 feet inward from the exterior wall. The area near a checkout counter is usually a separate zone. Large heating loads are associated with the entrance doors (heated vestibules are desirable). Some stores may have rooms that have special air quality and exhaust requirements. All zones must be compatible with the corresponding occupancy schedules. Special infiltration barriers and heating equipment may be required for a loading dock.

Lighting Loads

The lighting loads usually range between medium and high, depending on the store or the area within the store — make a survey of each area.

Occupancy Loads

The population density and the changes in the population density will vary, depending on the type of store and the area within the store. In most stores, the population density varies from light to medium — make a survey. The occupancy load may or may not have a noticeable affect on the room sensible heat ratio and the heating changeover point, depending on the population density. Refer to the next page for information about pet shop animals.

Equipment Loads

The appliance and equipment loads depend on type of store. Most stores have some small appliances and vending machines. Specialized equipment may be associated with some stores — candy stores, barber shops, beauty shops, dry cleaners, photo studios or shoe stores, for example. TV and appliance stores may have a lot of powered-up equipment on display. Sensible and latent loads are associated with the washers and dryers that are found in a coin laundry. Talk to the manager and make a survey to determine the sensible and latent equipment loads that are associated with a particular store. Use load factors and the usage factors to de-rate "name-plate" inputs. Consider the effects of diversity and exhaust hoods. Refer to other applications for more information.

Cooling Load

Interior areas are not sensitive to solar gains, so the cooling loads and load swings in these areas depend on the size of the internal loads. The perimeter areas are affected by the internal loads and the envelope loads. In these areas, the design loads occur when the internal loads and the envelope loads combine to produce a peak load condition; and large load swings are caused by changes in the population density and the envelope loads (particularly the loads that are associated with glass).

Operating Mode

The interior zones require year-round cooling. The perimeter areas will changeover to heating, but the changeover temperature will depend on the relative size of the internal loads and the envelope loads. (The loads that are associated with the glass areas and the entrance doors are especially important.) Simultaneous heating and cooling is required for incompatible zones.

Humidity control

The humidification loads and the humidification requirements vary, depending on the store or the area within the store. Ask the manager if there are any special humidity control requirements. Changes in the population density will cause the relative humidity to drift, which is acceptable, providing that it stays in the "comfort zone." (Population densities are usually not high enough to cause humidity control problems.) The relative humidity in coin laundry may be higher than for other applications (use 65 percent as an upper limit).

Air Quality

Air quality is not a problem in hard-goods and soft-goods stores, but it could be a problem for some stores — beauty shops, dry cleaners, pet shops, photo studios, pool halls or shoe stores, for example.

Small Stores and Shops — Continued

Ventilation (outdoor air) is always required for any type of store, but some stores may have special exhaust system or ventilation requirements. Special filters are not usually required. Control the odors that are associated with tobacco smoke, food preparation, food service, liquor service and animals. Do not recirculate air from smoking areas, cooking areas and toilets.

Exhaust CFM

The exhaust requirements depend on the activities that are associated with the store. Special exhaust systems may be required for some types of stores. Exhaust all the areas that are associated with activities that produce odors and fumes, and exhaust the toilets (In private toilets, the exhaust systems usually operate intermittently).

Outdoor Air

Large amounts of outdoor air are not required for hard-goods and soft-goods stores. Generous ventilation rates are required for some type of stores — beauty shops, dry cleaners, coin laundries, pet shops, photo studios, pool halls or shoe stores, for example. (Refer to the ventilation requirement table provided in Section 2 of this manual.) Ultimately, the ventilation requirements depend on the population density and on the amount of make up air that is required for the exhaust systems. Refer to table below for information about pet shop animals.

Space Pressure

Most areas can be maintained at a neutral or a slightly positive pressure. Negative pressure is required for areas that contain equipment, animals or materials that produce humidity, odors or airborne contaminants.

Schedules

Ask the manager about the occupancy and equipment operating schedules.

Noise

Normally there are no unusual limitations on noise levels — talk to manager.

Special Problems

Large heating and cooling loads are associated with glass and entrance doors. Wide load swings are caused by changes in the population density and envelope loads. Zones may be required even though the store is relatively small. The design requirements may change with each new tenant.

Laboratory Animal Metabolisim and Outdoor Air Requirements

	Weight (LB)	Sensible	Latent	ACH
Mouse	0.05	1.11	1.65	15
Hamster	0.25	4.02	6.0	15
Rat	0.62	7.77	11.6	15
Guinea Pig	0.90	10.2	15.2	15
Rabbit	5.41	39.2	58.5	10
Cat	6.61	45.6	68.1	10
Primate	12.0	71.3	106.0	15
Dog (Small)	22.7	105.0	161.0	10
Dog (Medium)	50.0	231.0	355.0	10

Sensible and latent heat gains = BTUH; ACH = Air changes per hour (Outdoor air)

Supermarkets

Occupancy
Multipurpose occupancy — perishable and non perishable grocery areas, cold food display case areas, lunch counter, salad bar, deli, bakery, offices, hard and soft goods, flower shop, pharmacy, warehouse and more. Refer to other applications for detailed information about each type of occupancy.

Functions
Heating, cooling, dehumidification and ventilation are required.

Zoning
Zone the office areas, sales areas and storage areas. The food sales area should be divided into perishable food areas, non-perishable food areas and cold food display case areas. General merchandise sales areas (hard and soft goods) can be combined with the non-perishable food sales areas. Zone the food preparation and dining areas, the meat cutting area, the bakery, the florist shop, the storage areas, etc. Some interior spaces may require year-round cooling, but the cold display case area may changeover to heating. The perimeter zones are defined by the exposure and the envelope construction details (the solar and transmission loads that are associated with glass are especially important). Perimeter zones that are not defined by an interior wall can be considered to extend 12 to 18 feet inward from the exterior wall. The area near a checkout counter is usually a separate zone. Large heating loads are associated with the entrance doors (heated vestibules are desirable). All zones must be compatible with the corresponding occupancy schedules. Special infiltration barriers and heating equipment may be required at the entrance and at the loading dock.

Lighting
The lighting loads usually range between medium and high, depending on the area — make a survey.

Occupancy Loads
The population density and the changes in the population density will vary, depending on the time-of-day and day-of-week. A large number of occupants increase the humidity loads, reduce the room sensible heat ratios and reduce the heating changeover points. (Food service causes even larger, occupancy related, humidity loads.) Obtain occupancy information from the manager.

Equipment Loads
The equipment loads vary, depending on the department. The negative sensible and latent loads that are associated with the cold display cases are very important (see the table on the next page.) Also consider the loads that are associated with cooking equipment, food warmers, food service, video games, vending machines and motors — make a thorough survey or each area.

Cooling Load
The interior areas are not sensitive to solar gains, so the cooling load and the load swings depend on the size of the internal loads and negative cooling effect of the cold food display cases. The perimeter areas are affected by the internal loads and the envelope loads. In these areas, the design loads occur when the internal loads and the envelope loads combine to produce a peak load condition; and large load swings are caused by changes in the population density and the envelope loads (particularly the loads that are associated with glass).

Operating Mode
The interior zones require year-round cooling, accept the areas near cold food display cases, which may require heat. The perimeter areas will changeover to heating, but the changeover temperature will depend on the relative size of the internal loads and the envelope loads. (The loads that are associated with the glass areas and the entrance doors are especially important.) Simultaneous heating and cooling is required for incompatible zones.

Humidity Control
The indoor humidity must be controlled year-round, especially in the areas that are associated with cold display cases. (The relative humidity must be held below 55 percent to minimize condensation and frost accumulation on the surfaces of the cold food display cases.) In the produce department, large latent loads are generated by water sprays or by ordinary evaporation. High population densities increase the latent loads. Condensation on cold food display cases reduces the latent load in that area. Reheat or a desiccant dehumidifier may be required to provide the required amount of humidity control.

Air Quality
Ventilation (outdoor air) and standard filters are required for most of the areas (make a survey and talk to the manager to determine if there are any special air quality requirements). Control the odors that are associated with tobacco smoke, food display, food preparation and food service. Do not recirculate air from smoking areas, cooking areas and toilets.

Supermarkets — Continued

Exhaust CFM

Make a survey to determine if there are any special exhaust system requirements. Exhaust all the areas that are associated with activities that produce odors and fumes, and exhaust the toilets.

Outdoor Air

The ventilation requirement that is associated with any area depends on the population density and on the amount of make up air that is required for the exhaust systems — make a survey.

Space Pressure

The space pressure depends on the department. The pressure in the departments that are associated with fumes or odors should be negative with respect to other spaces.

Schedules

Get the occupancy schedules and the equipment operating schedules from the manager.

Noise

Normally, there are no special limitations on noise levels.

Special Problems

Multiple occupancies must be accommodated (some may be incompatible with each other). Large heating and cooling loads are associated with glass and entrance doors. Wide load swings are caused by changes in the population density and envelope loads. Refrigerated cases affect the sensible and latent cooling loads as well as the heating loads. The heat that is absorbed by the display cases can raise the heating changeover point by 10°F or more. (It may be possible for the heat that is absorbed by the display cases to exceed the cooling load.) The cold display cases may cause relatively low room sensible heat ratios in that area, possible below 0.50. Dehumidification with standard HVAC equipment may not be possible. Heating and dehumidification may be required during cold weather. The cold food cases and the HVAC system "fight" each other with a resulting increase in energy consumption. To reduce energy consumption, heat reclaimed from refrigeration systems can be used for HVAC system reheat. Zones are not isolated by walls. Frequent floor plan changes.

Cold Display Case Cooling Effect

	Total Cooling Effect (BTUH / Foot)	Sensible Cooling Effect (BTUH / Foot)	Latent Cooling Effect (BTUH / Foot)	Percent Latent Compared to Total
Frozen Food				
Single deck	245	207	38	15
2 deck	470	400	70	15
3 deck	720	576	144	20
4 or 5 deck	1610	1288	322	20
1 deck - 2 island	2000	1600	400	20
Ice cream				
Single deck	430	366	64	15
1 deck - 2 island	470	400	70	15
Meats				
Single deck	350	298	52	15
Multideck	1095	876	219	20
Dairy				
Multideck	980	784	196	20
Produce				
Single deck	240	204	36	15
Multideck	960	768	192	20

Extracted from 1987 ASHRAE Systems and Applications — page 18.3

Variety Stores

Occupancy

Multipurpose occupancy — offices, hard and soft goods, perishable and non perishable groceries, cold food display case areas, lunch counter, salad bar, deli, bakery, restaurant, garden shop, pet shop, flower shop, offices, beauty shop, pharmacy, auto service, warehouse and more. Refer to other applications for detailed information on each type of occupancy.

Functions

Heating, cooling, dehumidification and ventilation is required. (Winter humidification may be required in some applications.)

Zoning

Each different type of occupancy is a separate zone. Some departments may require two or more zones. Some areas may have to divided into interior zones and perimeter zones. Interior spaces require year-round cooling. The perimeter zones are defined by the exposure and the envelope construction details (the solar and transmission loads that are associated with glass are especially important). Perimeter zones that are not defined by an interior wall can be considered to extend 12 to 18 feet inward from the exterior wall. The area near a checkout counter is usually a separate zone. Large heating loads are associated with the entrance doors (heated vestibules are desirable). Some stores may have areas that have special air quality and exhaust requirements. All zones must be compatible with the corresponding occupancy schedules. Special infiltration barriers and heating equipment may be required for a loading dock.

Lighting Loads

The lighting loads usually range between medium and high, depending on the department — make a survey of each area.

Occupancy Loads

The population density and the changes in the population density will vary, depending on the time-of-day and day-of-week. A large number of occupants increase the humidity loads, reduce the room sensible heat ratios and reduce the heating changeover points. (Food service causes even larger, occupancy related, humidity loads.) Obtain occupancy information from the manager.

Equipment Loads

The appliance and equipment loads vary, depending on the department. Look for office equipment, cooking equipment, food warmers, video games, vending machines and motors. Cold display cases cause negative internal loads. TV and appliance departments may have a lot of powered-up equipment on display. Talk to the manager and make a survey to determine the sensible and latent equipment loads that are associated with each area. Refer to other applications for more information.

Cooling Load

Interior areas are not sensitive to solar gains, so the cooling loads and the load swings in these areas depend on the size of the internal loads (particularly the occupancy loads). The perimeter areas are affected by the internal loads and the envelope loads. In these areas, the design loads occur when the internal loads and the envelope loads combine to produce a peak load condition; and large load swings are caused by changes in the population density and the envelope loads (particularly the occupancy loads and the loads that are that are associated with glass).

Operating Mode

The interior zones require year-round cooling. The perimeter areas will changeover to heating, but the changeover temperature will depend on the relative size of the internal loads and the envelope loads. (The loads that are associated with the occupants, glass areas and the entrance doors are especially important.) Simultaneous heating and cooling is required for incompatible zones.

Humidity Control

Large variations in the population density affect the relative humidity. Close control is not usually required, allow the relative humidity to drift within the comfort zone. Look for unusual sources of moisture. Cold food cases tend to reduce humidity levels. Refer to information about other applications.

Air Quality

The air quality requirements depend on how the area within the store is used. Ventilation (outdoor air) and standard filters are required for most of the areas (make a survey and talk to the manager to determine if there are any special air quality requirements). Control the odors that are associated with tobacco smoke, food preparation and food service. Do not recirculate air from smoking areas, cooking areas and toilets. (Refer to other applications.)

Variety Stores — Continued

Exhaust CFM The exhaust requirements depend on the activities that are associated with each area. Special exhaust systems may be required for some areas. Exhaust all the areas that are associated with activities that produce odors and fumes, and exhaust the toilets

Outdoor Air The ventilation requirement that is associated with any area depends on the population density and on the amount of make up air that is required for the exhaust systems — make a survey.

Space Pressure The space pressure depends on the department. The pressure in the departments that are associated with fumes or odors should be negative with respect to the other spaces.

Schedules Get the occupancy schedules and the equipment operating schedules from the manager.

Noise Normally there are no special limitations on noise levels — talk to the manager.

Special Problems Multiple applications must be accommodated (some may be incompatible with each other). Large heating and cooling loads are associated with glass and entrance doors. Wide load swings are caused by changes in the population density and envelope loads. Some zones are not defined by partitions and walls. Frequent floor plan changes.

Appendix 3
Air-Side Design

A complete air distribution system consists of supply air outlets, returns, air-side devices, duct runs, fans and control dampers. This appendix provides an overview of the work that is associated with designing an air distribution system. This work typically includes the following tasks:

- Select, size and locate the supply air outlets
- Select, size and locate the returns
- Size and locate the trunk and branch ducts
- Select and size the duct fittings
- Select, size and locate the air-side devices
- Calculate the air-side pressure drops
- Establish the fan performance requirements
- Specify the fan RPM and motor horsepower
- Verify that the minimum outdoor air CFM is adequate
- Verify that the desired air balance will be maintained
- Verify that the desired space pressures will be maintained

A3-1 Air Distribution

The air distribution system design procedure consists of selecting, sizing and locating the supply outlets (registers, diffusers or grilles) and the return inlets (registers or grilles). This work is critical. Improper application of air distribution hardware can cause inadequate heating or cooling in some, or all the rooms or zones. Inadequate air distribution also can cause drafts, pockets of stagnant air, poor air quality and noise. A brief overview of air distribution principles is provided below. Refer to ACCA, **Manual T** for comprehensive information about terminology, performance characteristics and design procedures.

Supply Air Outlet Location

Supply air outlets should be selected and portioned so that they can thoroughly mix the supply air with the room air. The supply outlets must not cause drafts, pockets of stagnant air or noise within the occupied zone. (The occupied zone is a volume defined by boundaries that are 2 feet inside the walls and about 6.5 feet above the floor.) The supply air outlet locations also must be compatible with the architectural intent (aesthetics) and the structural details of the building envelope. The supply outlets also must be compatible with the air-side components of the air distribution system (the duct runs and air-side devices).

- Ceiling and high (interior) sidewall terminals can be used when the duct runs are located in a ceiling cavity, an attic or on a roof. Ceiling outlets provide the best cooling performance. Ceiling diffusers (of a special design) must be used for variable volume applications.
- Floor and low sidewall terminals can be used when duct runs are located in a ceiling cavity (that is associated with a downstairs room), in a basement, in a crawl space or underneath a ground slab. In cold climates, perimeter terminals (floor or baseboard) provide the best heating performance when slab or open crawl space construction is required. When possible, vertical discharge outlets should be installed under windows and glass doors (to offset drafts that are caused by cold convection currents).

Supply Air Outlet Sizing

The size of the supply outlet is directly related to the capacity (supply CFM) of the outlet. The size also affects the throw, spread and the pressure drop; and in the case of sidewall and ceiling outlets, it affects the drop. The design objectives for selecting air outlet sizes are:

- No stratification in occupied zone
- Velocities in occupied zone between 20 and 50 FPM
- Supply air thoroughly mixed with the room air
- No collisions between high velocity air streams
- No collisions between high velocity air and surfaces
- No short circuit between supply and return terminals
- No undesirable noise

Supply CFM

The supply CFM for each room or zone can be estimated by using the sensible heat equation (below). In this equation, the load could either be equal to the sensible cooling load or the heating load and the *TD* is the temperature difference between the supply air and the room air.

$$CFM = \frac{Room\ load\ (BTUH)}{1.1 \times TD}$$

Return Location

A return path is required for the air that is supplied to every room or zone. Transfer grilles, transfer ducts or liberal door undercuts can be used to couple two or more rooms to a central return. Since returns have no influence on the air patterns in the room, the location of return opening is not critical. Returns should be sized so the face velocity at the grille is 400 FPM or less. Returns should never be located where a jet of supply air can blow directly into the return opening.

> Returns cannot create velocities that exceed 16 FPM unless the air is within one or two feet of the return. Since stagnate air is defined as air that is moving at less than 16 FPM, air that is gradually drifting toward a return is classified as stagnate.

Models and Configurations
There are many different variations of a given type of supply air outlet. For example, ceiling diffusers may be square or round; some mount flush, but others protrude below the ceiling; square models may have one-, two-, three- or four-way patterns; some models are designed for variable volume applications; some models are designed for constant volume applications — and even if they are similar — some models have more capacity or longer throws than others. Manufacturer's engineering data can be used to compare the performance characteristics of similar devices.

Approach
The air must approach the supply outlet as a smooth, well ordered flow. Any turbulence or distortion of the velocity profile that is immediately upstream from the supply outlet will seriously degrade the performance of the outlet. Boots, collars, necks, extractors, equalizing grids or other accessories should be used to organize the approaching air flow.

Balancing Devices
Supply outlets that are equipped with a damper (registers) should not be used for balancing. All branch supply ducts should be equipped with balancing dampers.

Two Speed Fan Operation
Supply outlets must perform adequately during any operating condition. Check the throw, spread, velocity and noise rating at the air flow rates that corresponds to the highest and lowest blower speeds. (Variable volume applications require outlets that are specifically designed for this type of duty.)

A3-2 Duct Design

Duct runs may be exposed or hidden; depending on the aesthetics of the interior or exterior decor, the available space and the installed cost. There must be no spatial conflicts between the air distribution system and the other building systems — the water, sprinkler, steam, electrical, phone or sound systems, for example. There must be no spatial conflicts between the air distribution system and the structural components of the building.

There are a number of different methods that can be used to size the duct runs, but the equal friction method is usually used for small commercial applications. The equal friction method also is used for large commercial and industrial applications. Occasionally, when the duct system is quite small, an extended plenum design is used. In either case, the duct runs are sized first and then a fan is selected to provide the desired air-side performance. A brief overview of the equal friction duct sizing method is provided below. Refer to ACCA, **Manual Q** for more information about air-side design.

Flow Rate (CFM) Values
Normally, the designer marks the flow rates on a drawing of the duct system. The CFM flowing through any section of a duct system is determined by summing the CFM values that are associated with the trunks and branches that are downstream from the section of interest. These calculations are made by starting at the ends of the duct system and working back to the fan. The accumulated CFM that is associated with each section of duct should be noted on the duct drawing.

Design Friction Rate Value
Low-velocity, low-pressure duct systems are designed for friction rate values that range between 0.08 and 0.20 IN.WG. per 100 feet of duct. One way to select a design friction rate is to base the friction rate on the air velocity requirement. (Air velocity recommendations are provided in **Manual Q**.)

- The design value for the air velocity in the section that is immediately downstream from the fan discharge opening can be used to determine the design friction rate for the supply duct system.
- The design value for the air velocity in the section that is immediately upstream from the fan inlet opening can be used to determine the design friction rate for the return duct system.

Sizing
The size of each section of a low-velocity duct system is determined by the flow through the section (CFM) and the design friction rate (f/100 value). A friction chart or a duct sizing slide rule is required for this work. It is very important to use a friction chart or slide rule that documents the performance of the material that will be used to fabricate the system. For example, do not use a sheet metal chart to size a flexible wire helix duct, a duct that is fabricated from duct board or a duct that is equipped with a fibrous liner. (The ACCA duct sizing slide rule has scales for sheet metal, duct board, duct liner and flexible duct materials.)

A3-3 Duct Fittings

Raw fan power can be used to force air through inefficient fittings, but this wastes energy and produces turbulence that can generate objectionable noise. Aerodynamic fittings — which are characterized by gradual dimensional changes, radius turns, sweep sheets and turning vanes — are preferred. A special effort should be made to make sure that fittings that are installed immediately upstream or downstream from a fan do not create an undesirable "system effect" loss. Detailed information about fitting performance and system effect losses can be found in **Manual Q**.

A3-4 Air-Side Devices

Reheat coils, air volume regulators, VAV boxes and hand dampers provide a few examples of air-side devices that might be installed in a commercial air distribution system. When a specific type of device is required, it must be selected and sized for the intended duty and the pressure loss that is associated with this device must be noted. (The pressure loss that is produced by a device must be added to the other pressure losses that are associated with the duct system.) Refer to **Manual Q** for more information about air-side devices.

A3-5 Pressure Drop Calculations

An estimate of the system resistance must be made before the fan performance can be specified. Calculate the largest pressure loss that is associated with the supply runs and the largest pressure loss that is associated with the return runs. Include the pressure losses that are associated with the straight runs and fittings; all the air handler component pressure losses that are not accounted for in the blower table; all devices that are installed in the duct runs; and the supply outlet and the return inlet. The details of how to make these pressure loss calculations can be found in **Manual Q**.

A3-6 Fan Selection

The manufacturer's blower tables contain the necessary fan performance information. These tables relate CFM and pressure to RPM and motor power. Use the design values for CFM and pressure to determine the required blower RPM and motor horsepower. A complete discussion of fans and fan performance is provided in **Manual Q**.

A3-7 Outdoor Air

During any operating condition, a specific amount of outdoor air is required for ventilation — or outdoor air could be required to replace exhaust air or to pressurize the space. (The outdoor air requirements depend on the application — refer to Section 3, page 3-5.) The air distribution system must be designed to satisfy these outdoor air requirements. This means that the fans, control dampers and the various types of air-side operating controls must be designed to insure that the required amount of outdoor air is provided during every possible operating condition. For some systems, this may only require a properly sized and controlled minimum outdoor air damper. In other cases, a VAV system for example, flow monitoring stations, fan control devices, dampers and interlocking controls might be required to satisfy the outdoor air requirements during any mode of operation.

A3-8 Air Balance

Provisions for testing and balancing the system must be designed into the air distribution system. Balancing dampers are required at every critical point. And, the supply fans, return fans, exhaust fans and operating dampers must be able to provide the desired air balance during any possible operating condition. Refer to **Manual Q**, Section 19 for more information on this subject.

A3-9 Space Pressure

During any operating condition, a specific flow of outdoor air and exhaust air is required to maintain the desired space pressure. (Space pressure requirements depend on the application.) This means that the fans, the control dampers and the air-side operating controls must be designed to maintain the required space pressure. For very simple systems this may only require a properly sized supply fan, an outdoor air damper and a relief (or exhaust) damper. However, a return fan (or exhaust fan), or an interlocked relief damper (or gravity damper) must be used if the system is equipped with an air-side economizer (or if the resistance of the return side of the system exceeds 0.10 Inches Water Gauge). More sophisticated applications may require room pressure sensors, flow monitoring stations, fan control devices, operating dampers and interlocking controls in order to control the pressure in the conditioned spaces. More information about space pressure can be found in **Manual Q**, Section 17.

Appendix 4
Pipe Sizing

The family of HVAC piping systems includes refrigerant piping, water piping, gas piping, oil piping and steam piping. This appendix provides an overview of the work that is associated with sizing the various types of piping systems. This work typically includes some or all the following tasks, depending on the type of piping system.

- Calculate flow rates
- Locate and size the pipe runs
- Select the pipe fittings
- Select and size the speciality hardware
- Calculate pressure drops
- Estimate pump performance requirements
- Specify pump RPM, motor power and impeller size

Other tasks that are associated with pipe system design includes the work that is associated with specifying fabrication methods and selecting materials. In all cases, the system components and the materials must be selected for the intended duty, and they must conform to the applicable code. The design also must allow for expansion that is caused by temperature changes. Hangers, supports and vibration isolators must be installed at appropriate points.

Piping systems may be exposed or hidden, depending on the aesthetics of the interior (or exterior) decor, the available space and the money that is budgeted for the project. There must be no spatial conflicts between a piping system and the other building systems — the duct, sprinkler, steam, electrical, phone, sound or other piping systems, for example. There must be no spatial conflicts between the piping systems and the structural components of the building.

A4-1 Refrigerant Piping

Since the refrigerant changes phase as it circulates through the system, the refrigerant piping can be thought of as three separate piping systems. The pipe between the compressor and the condenser must be designed to carry a hot, high pressure gas; the pipe between the condenser and the metering device must be designed to carry a warm, high pressure liquid and the pipe between the evaporator and the compressor must be designed to carry a cold, low pressure gas. These three pipe runs are called the hot gas line, the liquid line and the suction line, respectively.

When the refrigerant piping is part of a packaged system, there is no design work to do. If a split system is required, the system designer must layout and size the refrigerant lines. Some of the calculations that are required for this work, such as establishing flow rates and calculating pressure drops, are similar to the calculations that would be required for designing any type of piping system. But, the design also must satisfy the following requirements, which are unique to refrigerant piping systems.

- The piping pressure losses must be small. Large losses will seriously degrade the thermal capacity and efficiency of the refrigerant system.
- The piping design must insure that any oil that escapes from the compressor is returned to the compressor at the same rate that it leaves the compressor.
- There must be no significant accumulation of oil in any part of the system, other than the compressor crankcase.
- Liquid refrigerant must not enter the compressor.
- Oil should not drain back and accumulate at the head of an idle compressor.
- Refrigerant should not migrate to the compressor crankcase when the system is shut down.
- The refrigerant should not flash to gas while it is in the liquid line.

Refrigeration Line Pressure Drop Data

Table and charts can be used to estimate the refrigerant line pressure drops. Different sources present this data in different formats, but the format that is the easiest to use relates the pressure drop to refrigeration capacity (tonnage), pipe size (diameter in inches) and pipe length (feet of equivalent length). An example of this type of chart is provided on the next page by Figure A4-1. (Three separate charts are required, one for the hot gas line, a second for the liquid line and a third for suction line. The data also depends on the type of refrigerant. Separate chart-sets have been developed for each type of refrigerant.)

Note that the data presented in Figure A4-1 is based on a 2°F pressure drop. This may seem like a strange way to describe a pressure drop, but it makes sense because there is a one-to-one relationship between the pressure and temperature of a saturated refrigerant. For example, saturated R-22 has a pressure of 68.55 PSIG at 40°F and 61.25 PSIG at 35°F. In this case, a pressure drop of 5 degrees (in temperature units) corresponds to a pressure drop of 7.3 PSIG (in pressure units).

Also note that the data in Figure A4-1 applies to a specific set of conditions, which are a 2°F pressure drop, a 40°F suction temperature and a 105°F condensing temperature. This data must be adjusted if the actual design parameters deviate from these values. This is accomplished by applying correction factors, which are included with the chart.

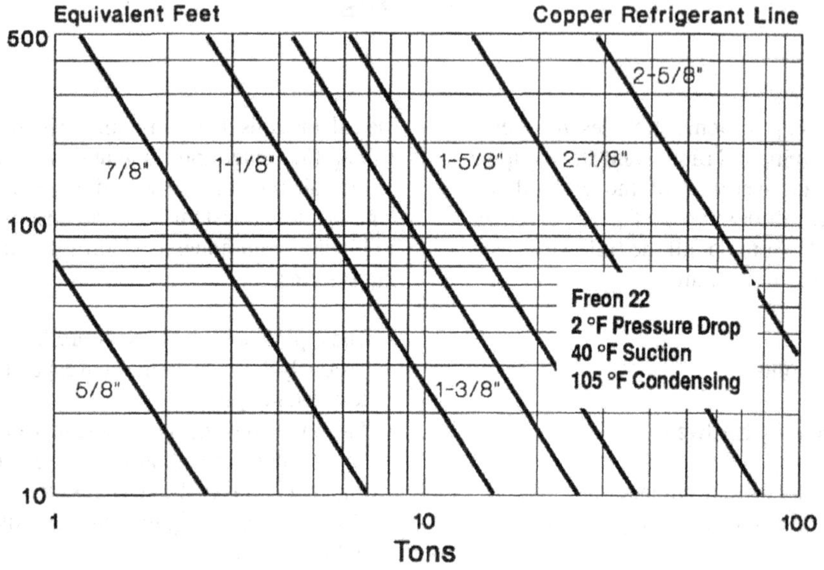

Figure A4-1

Using Refrigeration Line Pressure Drop Data
The pipe sizes can be read directly from the refrigerant line sizing charts. Simply enter the chart at the tonnage value (corrected for suction temperature and condensing temperature, if necessary) and read up to the pipe size and left to the equivalent length values (corrected for pressure drop, if the pressure drop is not 2°F).

The objective is to select a pipe size that provides an equivalent length that is equal to or greater than the actual equivalent length of the pipe run. This requires a trial and error solution. Begin by selecting a pipe size that provides an equivalent length that is equal to or a little larger than 1.5 times the measured length. Use this tentative pipe size to look-up the fitting lengths and to calculate the total equivalent length of the run. If this calculated length is too long or too short for the tentative pipe size, select the next larger or smaller size and repeat the calculation. (Piping manuals provide tables that list the equivalent lengths of copper and steel pipe fittings, valves and other piping hardware.)

Suction Line Sizing
Suction lines are commonly sized for a 2 degree pressure drop. This is a simple procedure, providing there is no riser. However, if there is a riser (because the compressor is above the evaporator or the line loops up and then drops to the compressor), the designer must make sure that the oil will circulate in the direction of the refrigerant flow when the compressor is operating. Information about riser design is provided below.

Hot Gas Line Sizing
Hot gas lines are commonly sized for a 2 degree pressure drop. Again, this is a simple procedure, providing there is no riser. If there is a riser (because the condenser is above the compressor or the line loops up and then drops to the condenser), the designer must make sure that the oil will circulate in the direction of the refrigerant flow when the compressor is operating, and that oil will not drain back to the compressor when it is shut down. Information about riser design is provided below.

Suction and Hot Gas Risers
If a riser is required, the velocity of the gas in the riser must be high enough to pull the oil (which coats the wall of the pipe) up the riser. The required riser size can be determined by referring to refrigerant piping tables that correlate riser size with tonnage. When using these tables, the tonnage "look-up value" is equal to the minimum operating tonnage, which will be established by the capacity control hardware.

Once the riser is sized for the minimum load condition, it is necessary to recheck the pressure drop across the entire run at the full load condition. If the pressure drop associated with the entire suction or hot gas line exceeds 2°F, the horizontal pipe sizes can be increased to compensate for the pressure drop that is associated with the riser.

Note that, a hot gas line should be looped below the compressor before it turns up to the riser. This loop (see Figure

A4-2) provides a reservoir that can hold any oil that drains down the riser when the compressor is off. This arrangement prevents the oil from reaching the head of the compressor.

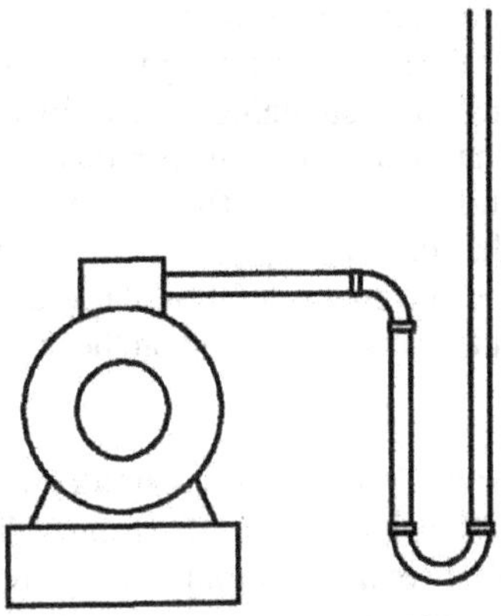

Figure A4-2

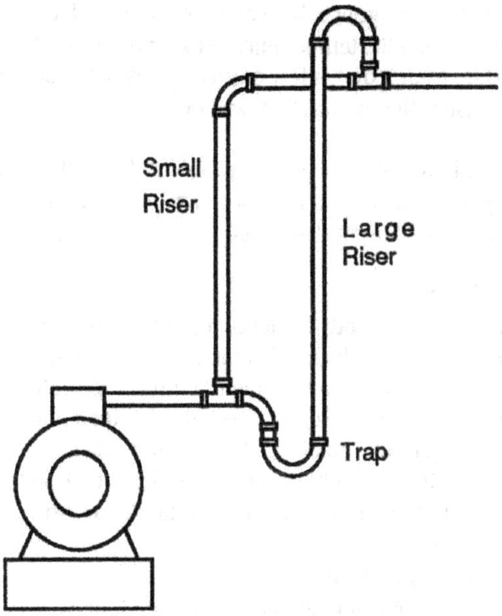

Double Hot-Gas Riser

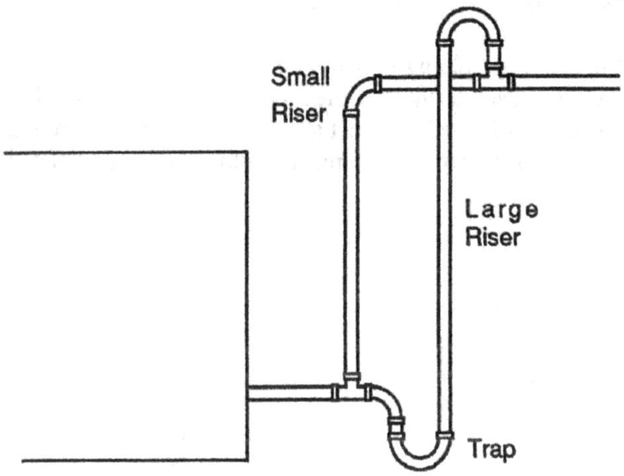

Double Suction Riser

Double Risers

In some cases, particularly if the compressor unloads to a very low capacity, it may be necessary to install a double suction riser or a double hot gas riser. When the compressor operates at low capacity, oil fills the trap and plugs the large riser. When this occurs, the hot gas that flows through the small riser moves fast enough to drag the oil up the riser. In this case, the pressure drop is not excessive because the gas flow rate is low. When the compressor operates at high capacity, the oil is blown out of the trap and the gas flows up both risers. In this case, there is more flow, but since there is more pipe to carry the flow; the oil is dragged up the risers and the pressure drop is not excessive. Refrigerant piping tables can be used to size the pipes that are used for the double riser arrangement.

Liquid Line Sizing

The liquid line also can be sized for a 2°F pressure drop. But, this 2°F pressure drop only accounts for the resistance of the pipe and fittings. Other pressure drops may be associated with refrigerant line accessories, such as valves, filters and driers. These pressure drops must be added to the drop that is associated with the pipe run. And, if there is a liquid line riser, there will be a hydraulic pressure associated with the column of liquid that is contained in the riser. If the direction of flow is up, this hydraulic pressure must be added to the other pressure drops, but if the flow is down, it is subtracted from the other pressure drops.

Refrigerant line accessory manufacturer's catalogs provide information about the pressure drops that are associated with their hardware. Refrigerant piping manuals provide information about the head pressure that is associated with a column of refrigerant. Since this information is normally provided in PSIG units, the liquid line calculations are also made in PSIG units. When required, refrigerant properties tables can be used to correlate a saturated temperature with a saturated pressure.

Once the liquid line pressure drop is calculated, the designer can use this information to determine how much subcooling will be required to prevent flash gas from forming in the liquid line. This is accomplished by subtracting the liquid line pressure drop from the saturated condensing pressure. The result of this calculation will be equal to the pressure at the entrance to the expansion valve. A refrigerant properties table can be used to convert this pressure to a saturated temperature. The smallest amount of subcooling that will prevent flashing is equal to the difference between the saturated condensing temperature and the temperature at the entrance to the expansion valve. In practice, a few extra degrees of subcooling are used to insure that no flashing will occur.

Oil return in liquid lines is not a problem. Since the oil mixes with liquid refrigerant, it is easily transported up risers and it will not accumulate in low spots.

Detail Piping
If a split system is used, the detail piping is usually part of the equipment package. If detail piping is required, get specific piping instructions from the manufacturer, especially if more than one compressor and/or evaporator is piped into a single refrigeration circuit. Incorrect detail piping can cause serious problems, which are characterized by incorrect and unbalanced operating pressures and inadequate oil return.

Codes and Regulations
Sizing procedures and stipulations about methods and materials are extensively codified. Make sure that the design conforms to building codes, fire codes, EPA regulations and insurance underwriter requirements.

A4-2 Water Piping Systems

When a water piping system is designed, the pipe runs are sized first and then the system pressure drop is estimated. This information is used to select a pump(s), which will provide the desired flow and pressure. After this work is completed, the expansion tank and the other hydronic accessories can be selected. A brief overview of this design procedure is provided below.

Flow Rate (GPM) Values
Normally, the designer marks the flow rates on a drawing of the pipe system. The GPM flowing through any section of a pipe system is determined by summing the GPM values that are associated with the mains and branches that are downstream from the section of interest. These calculations are made by starting at the ends of the pipe system and working back to the pump. The accumulated GPM that is associated with each section of pipe should be noted on the drawing.

Sizing Objectives
The size of the water pipe affects installation cost and the pump operating cost. The pipe size also determines the water velocity, which is directly related to problems that are associated with cavitation and noise. Therefore, the water pipe sizes should be based on a design velocity or a design friction rate that will keep these factors within acceptable ranges.

Design Velocity
For hydronic systems, the water velocity design value can range between 1 and 10 FPS, but there is no specific recommended value. Some designers feel that the design velocity should depend on the pipe size. For example, they may reduce the upper limit to 4 FPS when the pipe is two inches in diameter or smaller. There are a number of opinions about the optimum design velocity, but there is not enough documentation to specify a single value.

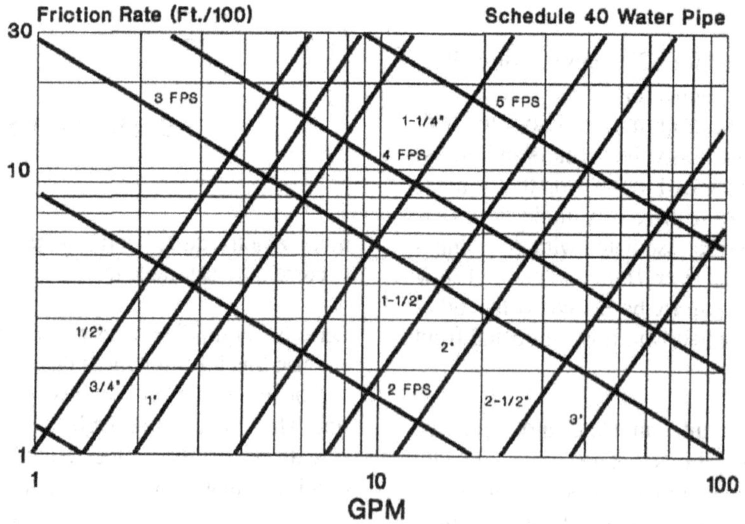

Figure A4-3

Design Friction Rate Value
The friction rate design value also is somewhat arbitrary. Designers typically use friction rate values that range between 1 and 4 feet of pressure loss per 100 feet of pipe. Any value that falls in this range is acceptable, providing that the corresponding velocity is acceptable.

Sizing
The size of each section of a pipe is determined by the flow rate (GPM) through the section and the design friction rate (f/100) value. A pipe sizing friction chart (Figure A4-3 on the previous page) is required for this work. It is very important to use a friction chart that documents the performance of the material that will be used to fabricate the system. Friction charts are published for copper, steel and plastic pipe.

Pressure Drop Calculations
An estimate of the system resistance must be made before the pump can be selected. Calculate the largest pressure loss that is associated with the supply runs and the largest pressure loss that is associated with the return runs. Include the pressure losses associated with the straight runs, fittings, valves and other standard hardware items. Also include the pressure drops that are associated with the water side of the HVAC equipment.

The pressure drop that is associated with the straight runs, fittings and the miscellaneous hardware depends on the friction rate and the total equivalent length of pipe. Refer to piping manuals for tables that list equivalent lengths for the various types of pipe fittings, valves and other incidental hardware. The pressure drops that are associated with water-side components of the HVAC equipment and the pressure drops associated with control valves are documented in the manufacturers' engineering literature.

Pump Selection
The pump manufacturers' performance charts contain the pump performance information. These charts relate GPM and pressure to impeller size and motor power for a specific RPM. Use the design values for GPM and head to determine the required impeller size and motor horsepower. An example of pump performance data is provided by Figure A4-4.

Expansion Tank
Hot water or chilled water system operating pressures must be maintained within certain upper and lower limits during any operating condition (even when the system is idle). This is accomplished by installing an expansion tank. This tank controls system pressure by accommodating the volumetric expansions and contractions that are associated with changing water temperatures. The size of the expansion tank depends on the volume, in gallons, of the hydronic system, the operating temperatures and the operating pressures. Hydronic specialties manufacturers provide comprehensive sizing and installation instructions in their engineering literature. Piping design manuals provide information about the volumetric properties of pipes, and HVAC equipment manufacturers provide information about the holding capacity of the water-side components.

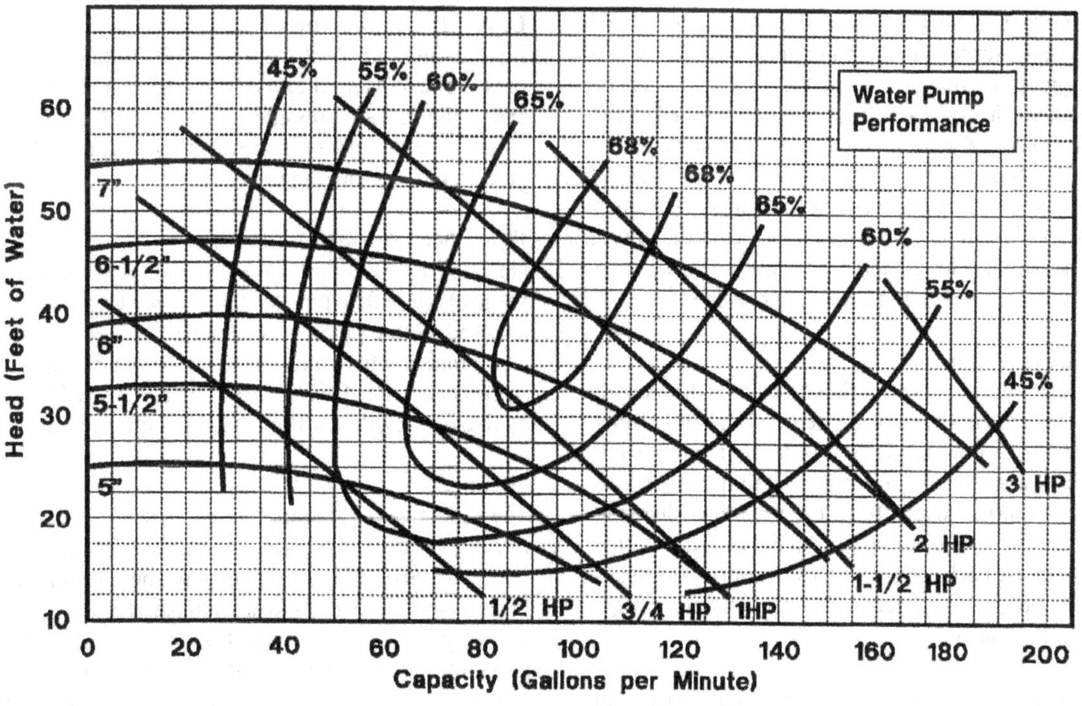

Figure A4-4

Hydronic Specialties and Detail Piping
Pressure reducing valves, air separators and air vents provide a few examples of important hydronic accessories. Hydronic specialties manufacturers provide comprehensive application, sizing and installation information in their engineering literature. Pump, control valve and HVAC equipment manufacturers provide information about the detail piping that is required for a given device.

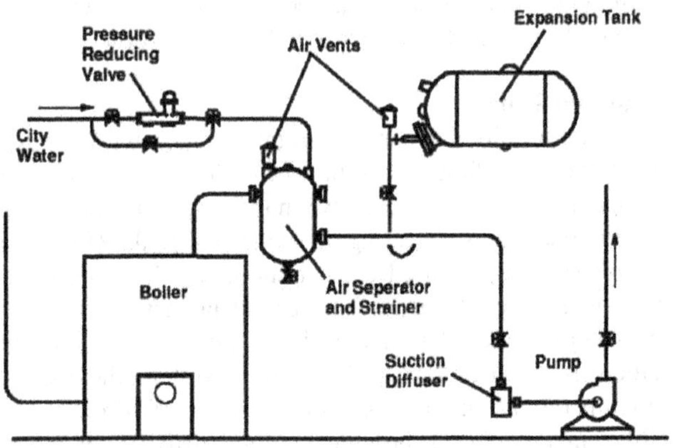

Hydronic Specialties

A4-3 Gas Piping Systems

When a gas piping system is designed, the pipe runs are sized so that the pressure drop from the point of delivery to the gas burning equipment is equal to, or less than, the value for the allowable pressure drop. This sizing procedure is based on the load condition that represents maximum demand. When there are multiple, nonconcurrent loads, the maximum demand value should be reduced by a percentage that reflects the effect of load diversity. And, if new loads will be added in the future, the maximum demand load should be adjusted upward. A brief overview of the design procedure is provided below.

Flow Rate (CFH) Values
Normally, the designer marks the flow rates on a drawing of the pipe system. The cubic feet of gas per hour (CFH) flowing through any section of a pipe system is determined by summing the CFH values that are associated with the mains and branches that are downstream from the section of interest. These calculations are made by starting at the ends of the pipe system and working back to the point of delivery. The accumulated CFH that is associated with each section of pipe should be noted on the drawing.

The input CFH that is required for a piece of equipment can be determined from the manufacturer's specifications. If the input requirements are listed in heating capacity units (BTUH), the CFH input can be calculated by dividing the BTUH value by the heating potential of one cubic foot of gas.

Service Pressure
The pressure at the point of entry can vary from a few ounces to 50 pounds, depending on the operating characteristics of the local gas service. Low-pressure systems (0.50 PSIG and less) are normally used for residential applications and higher pressures are used for commercial and industrial applications.

Pressure Regulators
The pressure at the entrance to the fuel burning equipment must be compatible with the manufacturer's specification. These pressures are usually smaller than the service pressure. One or more pressure regulators can be used to satisfy the manufacturer's requirements. The number of regulators that should be installed is related to the installation cost of the piping system.

- The system pressure can be reduced at the point of delivery and a low-pressure piping system can be used to distribute gas to the equipment. The advantage to this approach is that only one regulator is required. However, if the loads are large or if there are many long runs, large pipes will be required and the installation cost can be high.

- High-pressure gas lines can be used for distribution and the system pressure can be reduced at each piece of equipment. The advantage to this approach is that smaller pipes can be used, but each piece of equipment requires a regulator, which must be purchased and maintained.

Design Pressure Drop
For low-pressure gas systems (0.50 PSIG and less), the design pressure drop is normally 0.30 or 0.50 Inches Water Gauge. When the service pressure is more than 0.50 Inches Water Gauge, the design pressure drop is normally equal to 10 percent of the supply pressure.

Reference Equivalent Length
Gas pipe sizing tables provide sizing information for a wide range of equivalent pipe lengths. However, when these tables are used, all the pipe runs that are associated with the system — no matter what their length — are sized by using a *single reference length*. This *reference length* is equal to the longest equivalent length that is associated with the system.

Sizing
The size of each section of a pipe is determined by the flow rate (CFH) through the section and the reference length. Pipe sizing tables (see Figure A4-5 on the next page) are required for this work. These tables are published in sets. Each table in the set provides sizing data for a particular inlet pressure, pressure drop and piping material. These tables are usually prepared for gas that has a specific gravity of 0.60. A correction might be required if the specific gravity of the gas does

Gas Pipe Sizing Table — Pressure Under 1 Pound 0.50 Inch Water Column Pressure Drop 0.60 Specific Gravity; Schedule-40 Steel Pipe									
Pipe Size	Total Equivalent Length of Pipe in Feet								
(Inches)	50	100	150	200	250	300	400	500	1000
1.00	284	195	157	134	119	108	92	82	56
1.25	583	400	322	275	244	221	189	168	115
1.50	873	600	482	412	366	331	283	251	173
2.00	1681	1156	928	794	704	638	546	484	333
2.50	2680	1842	1479	1266	1122	1017	870	771	530
3.00	4738	3256	2615	2238	1983	1797	1538	1363	937
3.50	6937	4767	3828	3277	2904	2631	2252	1996	1372
4.00	9663	6641	5333	4565	4046	3666	3137	2780	1911

Figure A4-5

not match the specific gravity that was used as the basis for the table. Specific gravity corrections are discussed below.

The pipe size for any section of pipe is determined by entering the table at the reference length and reading down to the desired CFH value and then left to the pipe size. As emphasized above, the same reference length value is used for all the pipe sections that are associated with the system.

Specific Gravity Correction

The specific gravity of natural gas is about 0.60. Gases that have a lower specific gravity occupy a larger volume for a given weight of gas — say, for one pound of gas. In other words, gases with a lower specific gravity are less dense. Therefore, when the specific gravity is less than 0.60, larger pipes must be used to transport a given mass flow (pounds per hour). Conversely, smaller pipes can be used when the specific gravity of the gas is greater than 0.60. A table of specific gravity correction factors is normally provided with the gas pipe sizing tables.

Liquid Petroleum Gas

The gas piping tables can be used for sizing pipe for LPG systems. In this case, the service pressure will be determined by the pressure regulator at the storage tank. Also, note that liquid petroleum gas has a considerably higher specific gravity than natural gas.

Fuel-Air Mixtures

Air can be mixed with gas to control the heat content of the gas. Air-gas mixers are commonly used to make LPG gas behave like natural gas. This allows either fuel to be used without changing the burner setup. Since the LPG fuel-air mixture has a higher specific gravity than natural gas, the distribution piping is designed for natural gas.

Codes and Regulations

Operating pressures, sizing procedures and stipulations about methods and materials are extensively codified. Make sure that the design conforms to building codes, fire codes, utility regulations and insurance underwriter requirements.

A4-4 Oil Piping Systems

Oil piping systems can be classified as pressurized or as nonpressurized. Pressurized systems are used if the fuel oil has to be distributed to many burners (which may be scattered over a large area) or to a few, remotely located burners. Unpressurized systems can be used if one, or a few, burners are located close to the oil tank. A complete oil piping system could consist of one or more burner pumps, a distribution pump, an oil tank, interconnecting piping and various accessories such as filters, hand valves, check valves, relief valves, gauges and air vents.

If the system is pressurized, a distribution pump draws oil from the storage tank and pumps it through the distribution piping. Since this pump has the ability to pump more oil than the burner pumps will use, the excess oil is returned to the tank through a third pipe. In this type of system, the distribution main is the source of supply for the single-stage burner pumps. This distribution main may consist of a single pipe that dead ends after the last burner or it could form a complete loop, which drains back to the oil tank. Examples of these two types of systems are provided on the next page by Figure A4-6.

If the system is not pressurized, the burner pump draws oil directly from the storage tank. The simplest nonpressurized system consists of a single suction pipe that connects a sin-

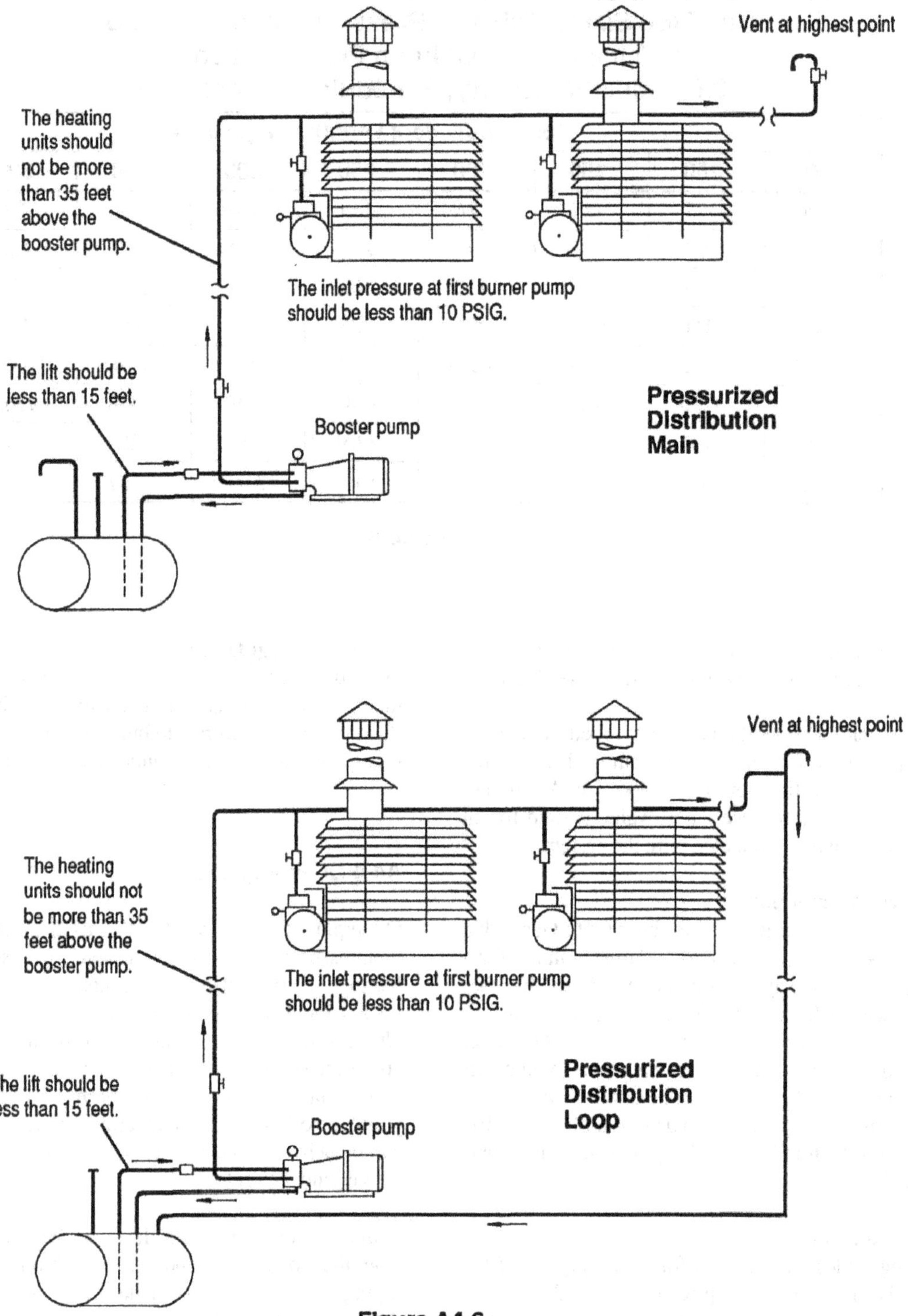

Figure A4-6

gle-stage fuel oil pump to the oil storage tank. This arrangement is normally used when the oil storage tank is above the burner pump, but it can be used (providing that it is air tight) if the lift, that is associated with a tank that is installed below the pump, is not too large. A two-pipe system is normally used when the fuel tank is located below the oil pump. In this case, the burner pump may be a single-stage pump or a two-stage pump, depending on the amount of lift that is required. If there is more than one oil burner, each burner pump will require a separate suction line, but a common pipe can be used for the return line. Examples of these two types of systems are provided on the next page by Figure A4-7.

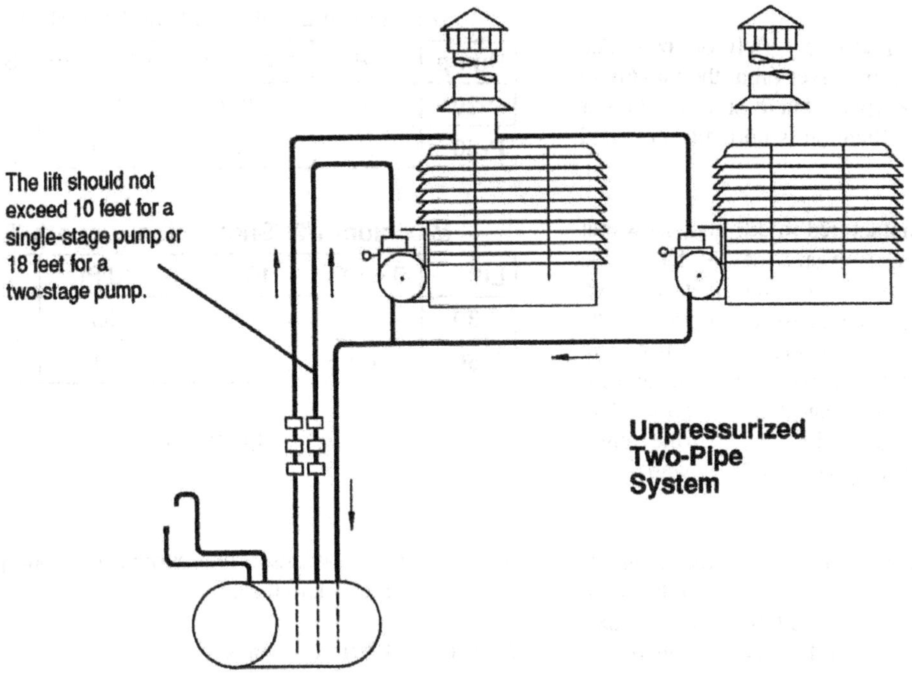

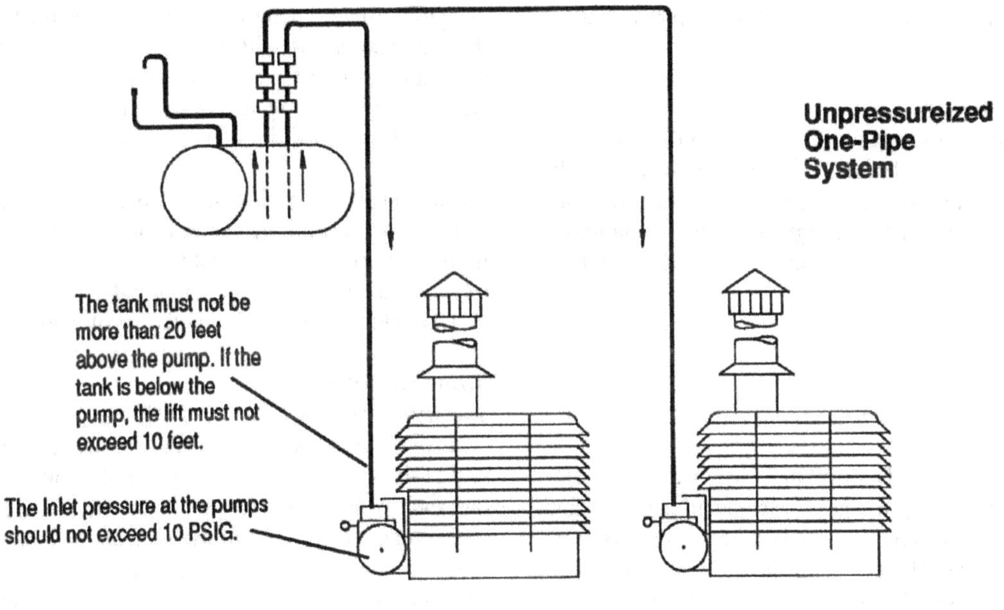

Figure A4-7

Flow Rate (GPH) Values

Normally, the designer marks the flow rates on a sketch or drawing of the pipe system. The gallons of oil per hour (GPH) flowing through any section of a pipe system is determined by summing the GPH values that are associated with the mains and branches that are downstream from the section of interest. These calculations are made by starting at the ends of the pipe system and working back to the point of delivery.

The accumulated GPH that is associated with each oil pump should be noted on the drawing.

The input GPH that is required for a piece of equipment can be determined from the manufacturer's specifications. If the input requirements are listed in heating capacity units (BTUH), the GPH input can be calculated by dividing the BTUH value by the heating potential of gallon of fuel.

Pump Sizing
Burner pumps are manufactured in a variety of sizes. The burner pump should be sized to deliver oil at the maximum oil firing rate, with some capacity held in reserve. The burner pump or a burner pump specification is normally provided with the heating equipment.

Distribution pumps are manufactured in different sizes, but the step from one size to the next size may be large. The distribution pump should be sized to deliver oil at the maximum, combined, oil firing rate that is associated with all the heating units, with some capacity held in reserve. When there are multiple, nonconcurrent loads, the maximum demand value should be reduced by a percentage that reflects the effect of load diversity. And, if new loads will be added in the future, the maximum demand load should be adjusted upward.

Maximum Discharge Line Length (Feet)

Pump	GPH	1/2" Tube	1/2" Pipe	3/4" Pipe
P-30	30	300	800	2500
P-50	50	175	350	1500

Maximum 1/2" Suction Line Length (Feet)

Lift	0 - 7 Ft.	10 Ft.	13 Ft.	15 Ft.
P-30	100	80	63	52
P-50	60	53	41	34

Figure A4-8

Pipe Size
The size of a suction pipe, distribution pipe or return pipe is determined by tables that are (usually) provided by the pump manufacturer (see Figure A4-8). These tables list the maximum allowable pipe length that can be accommodated by a particular pipe size. This allowable length varies, depending on the performance of the pump and on the vertical lift that is associated with the run. (Pipe length calculations should include the equivalent lengths of the fittings, valves and other hardware.) Different tables are required for different grades (weights) of oil.

Codes and Regulations
Sizing procedures and stipulations about methods and materials are usually codified. The details that are associated with locating, anchoring and venting storage tanks also may be in the code. Make sure that the design conforms to building codes, fire codes, utility regulations and insurance underwriter requirements.

A4-5 Steam Piping Systems

Steam piping may consist of a single pipe, which carries steam and condensate or two pipes (one for steam and the other for condensate). The two-pipe system is commonly used for commercial/industrial HVAC applications. Two-pipe systems are further classified by five pressure categories:

- High-pressure — 100 PSIG and above
- Medium-pressure — 15 to 100 PSIG
- Low-pressure — 0 to 15 PSIG
- Vapor — vacuum to 15 PSIG
- Vacuum — vacuum to 15 PSIG

Two-pipe systems also are classified by the condensate return arrangement, which may be identified as a gravity return or a mechanical return. The gravity return can be used if all the heating units are located at a higher level than the boiler or condensate receiver water line. Mechanical return systems use steam traps, condensate pumps or vacuum pumps to force the condensate back to the boiler.

Flow Rate (LB/HR) Values
Normally, the designer marks the flow rates on a sketch or drawing of the pipe system. First, the pounds of steam per hour (LB/HR) that is associated with each heating device is noted on the drawing. Then, the LB/HR flowing through any section of a pipe system is determined by summing the LB/HR values that are associated with the mains and branches that are downstream from the section of interest. These calculations are made by starting at the ends of the pipe system and working back to the point of delivery.

The input LB/HR that is required for a piece of equipment can be determined from the manufacturer's specifications. If the input requirements are listed in heating capacity units (BTUH), the LB/HR input can be calculated by dividing the BTUH value by the heating potential of a pound of steam.

Sizing Objectives
The sizing criterion consists of a design pressure drop for the supply side of the system, a design pressure drop for the return side of the system, a design value for the supply-side friction rate and a design value for the return-side friction rate. Steam piping manuals provide information about sizing criteria that should be applied to low-pressure steam systems and to other types of steam systems. The low-pressure sizing criteria is summarized below.

The steam supply and condensate return lines are sized so that the pressure drop that is associated with either side of the system does not exceed a recommended value. This recommended value depends on the operating pressure of the system. For example, in a low-pressure system, the design value for either side of the system can range from 0.50 to 5.0 PSIG, depending on the supply pressure at the boiler (2 to 20 PSIG).

The steam supply and condensate return lines are also sized so that the friction rate (pressure drop per 100 feet) that is associated with either side of the system does not exceed the maximum allowable value. Again, this allowable value depends on the operating pressure of the system. For example, in a low-pressure system, the supply-side friction rate should not exceed 2 PSIG per 100 feet and the return-side friction rate should not exceed 0.50 PSIG per 100 feet.

Pressure Drop Calculations
Pressure drop calculations are based on the friction rate and the total equivalent length of pipe. Piping design manuals provide tables that document the equivalent lengths of fittings, valves and other piping hardware.

Pipe Size
Steam supply and condensate return pipe sizes can be determined by using charts that are published in steam system design manuals and piping handbooks. These charts can be used to estimate the friction rate (or pressure drop) that is associated with a given pipe size, based on the combination of supply pressure and steam flow rate.

Steam Trap Selection and Sizing
Steam traps are used on mains and risers and they are installed downstream from heat transfer devices. Steam traps are designed to hold the steam in the supply side of the system and allow condensate or air to pass through to the return side of the system. The size of the steam trap depends on the load (LB/HR), the pressure difference across the trap and the desired factor of safety. A number of different steam trap designs are available in a wide range of sizes. Refer to steam trap manufacturer's engineering literature for detailed selection and sizing information.

Condensate Pump Sizing
Condensate pumps can be used to overcome the flow resistance that is associated with the return piping, return risers and the pressure difference between the return side of the system and the boiler. Refer to condensate pump manufacturer's engineering literature for selection and sizing information.

Piping Details
Many of the piping details that are associated with steam mains and runouts are related to condensate drainage and thermal expansion. The piping details that are associated with heat transfer equipment relate to vacuum release and to protecting, isolating and servicing the equipment. Refer to manufacturer's literature and to steam piping handbooks for comprehensive information about piping details.

Codes and Regulations
Operating pressures, sizing procedures, piping details and stipulations about methods and materials are extensively codified. Make sure that the design conforms to all codes.

Appendix 5
Refrigerant-Side Capacity Control

Air conditioning equipment capacity is based on the largest refrigeration load that is expected to occur during any operating condition. However, for most applications, the design load represents an extreme operating condition, which is only experienced for a few hours per year. Most of the time, the refrigeration load is smaller — often much smaller — than the design load. Therefore, some form of capacity control is required. This appendix explains how the various capacity control strategies affect the overall performance of the HVAC system and the refrigeration equipment.

> It is important to make a distinction between system capacity and refrigeration capacity. System capacity is the amount of cooling that is required to maintain the indoor design conditions. Refrigeration capacity is the amount of cooling that must be provided by the refrigeration equipment. These capacities might be equal, or they could be quite different. For example, if a terminal reheat system is used, the required system capacity could vary from 0 to 100 percent of the design load, but the required refrigeration capacity might be fairly steady, depending on the outdoor air load. In this particular situation, the refrigeration capacity is not controlled to maintain the space temperature. However, another type of system might have a one-to-one relationship between the system capacity and the refrigeration capacity, a single-zone constant volume system, for example. In this case, the refrigeration capacity could be controlled to maintain the space temperature.

A5-1 Problems at Part-Load Conditions

A cooling system may have satisfactory performance at the summer design conditions and unsatisfactory performance at part-load conditions. Some of the comfort problems that can be experienced during part-load operation include:

- Poor air circulation
- Inadequate outdoor air for ventilation
- Poor air balance
- Heat required, but not available
- Cooling required, but not available
- Dry-bulb temperature out of control
- Humidity out of control

Some of the mechanical problems that are experienced at part load would include:

- Equipment short cycles
- Unstable operation of control valves or control dampers
- Excessive wear and strain on equipment
- Equipment seriously damaged
- Noise

Some of the efficiency and operating cost problems that are associated with part-load operation would include:

- Poor efficiency at part load
- Extra energy required to compensate for excess capacity
- Cycling penalties
- Electrical demand loads

Many of the problems that are listed above could occur because the overall system design concept does not address the performance requirements that are associated with the application. In this case, the refrigeration cycle control strategy may not be related to the problem. However, if the overall design concept is appropriate for the application, some of these problems could be caused by the refrigeration cycle control strategy. Therefore, it is very important to understand how the various types of refrigeration cycle control strategies affect system performance.

A5-2 DX Coil Capacity

Refrigeration capacity controls are typically associated with compressor capacity adjustments. But, the refrigeration capacity does not simply depend on the compressor capacity— it also depends on how the compressor and the DX coil interact as a system and on how the coil interacts with the load. In order to understand these relationships, it is necessary to understand how the conditions at the refrigerant-side and the air-side of the coil affect the coil performance.

The capacity of a DX coil depends on the CFM that is passing through the coil, the dry-bulb and the wet-bulb temperatures of the entering air, the suction temperature on the refrigerant-side of the coil and the area of the coil. The coil capacity can be increased by increasing the air flow, the entering air temperatures or the area; or by reducing the suction temperature. Conversely, the coil capacity can be reduced by reducing the air flow, the entering air temperatures or the area; or by increasing the suction temperature.

For a given set of entering air temperatures, the coil capacity can be balanced with the load by adjusting the leaving air temperature or the air flow, or by adjusting the temperature and the air flow. (Many small single-zone systems have no way to adjust coil capacity. In this case, the cooling capacity is balanced with the load by starting and stopping the equipment. Note that reheat systems do not adjust the coil capacity, but they do compensate for excess coil capacity by adding heat to the air after it leaves the coil.)

If the coil air flow is regulated by the system controls, the suction temperature (and the leaving air temperature) is usually adjusted by the refrigeration capacity controls. For example, when throttling VAV systems or multizone systems are used, the refrigeration controls maintain a relatively steady coil temperature and the air flow is modulated to balance the load. But, when a constant volume system is used, the system controls must initiate an action that raises the leaving air temperature. For example, a space thermostat could be used to reduce the coil (and compressor) capacity.

Coil air flow can be adjusted by throttling the air flow or by bypassing air around the coil. The throttling action is usually associated with VAV air distribution hardware. The bypass action is usually associated with multizone discharge dampers or a single-zone air handler that is equipped with face-and-bypass dampers.

The temperature of the air leaving the coil is usually adjusted by changing the active coil area. This can be accomplished by reducing the face area or by reducing the number of active rows. When face area is adjusted, the coil is referred to as a face-split coil. When the number of rows are adjusted, the coil is referred to as a row-split coil. In either case, coil capacity is reduced by closing a liquid line solenoid valve.

Face-split coils are characterized by parallel air flows. In this case, the air that flows through the active coil is fully conditioned, but the air that flows through the inactive coil is not conditioned at all. The temperature of the air leaving the coil section is determined by the mixed air temperature that is associated with the two air streams. Figure A5-1 shows how the leaving air temperature is affected when one section of a face-split coil is deactivated. When face-split coils are used, the reduction in sensible and latent coil capacity is proportional to the reduction in face area, regardless of how the coil is circuited.

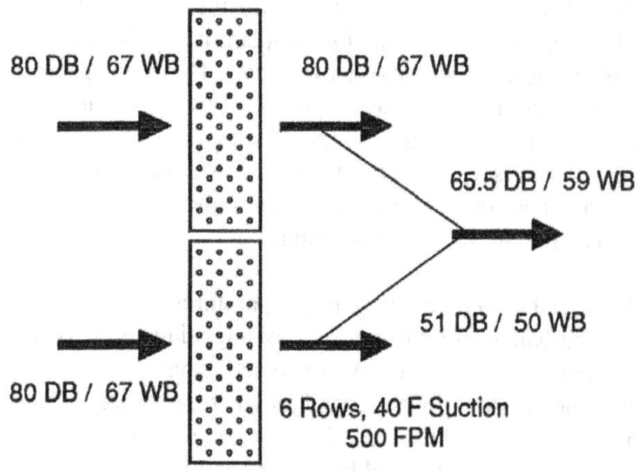

Figure A5-1

Row-split coils are characterized by a series air flow. In this case, the air that flows across the active rows also flows across the inactive rows. But, the temperature of the air leaving the coil is determined by the cooling capacity of the active rows. Figure A5-2 shows how the leaving air temperature is affected when coil rows are deactivated. Note that the reduction in sensible and latent capacity is not proportional to the reduction in the number of rows. In this case, the first few rows do most of the cooling because the air-to-coil temperature difference decreases as the air passes through the coil.

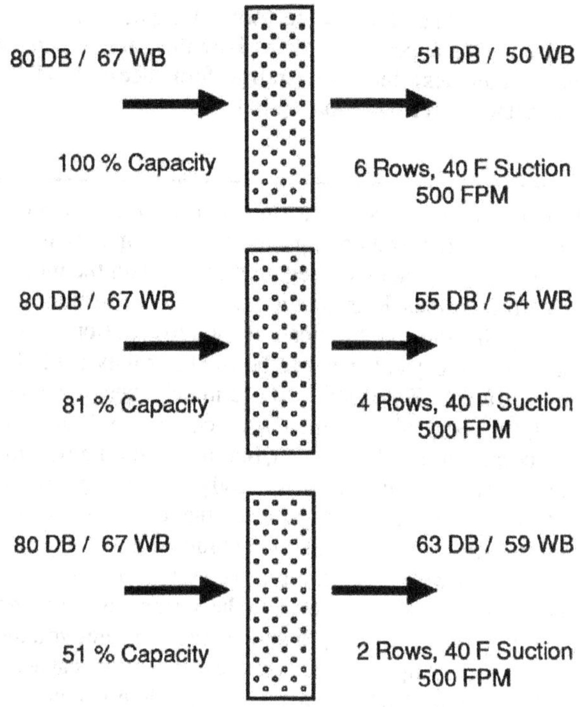

Figure A5-2

When row-split coils are manufactured, the design capacity that is associated with each split (at full-load conditions) will depend on how the coil is circuited. The coil designer can make the capacity of the splits equal or they can be unbalanced. Either way, the expansion valve capacities have to be matched to the capacity of each coil split. Note that the capacity of any particular split will change as the other splits are deactivated. For example, if a coil is split into two equal (50 percent) capacities at full-load conditions, the capacity of one split may increase to 60 or 70 percent of the design load when the other split is deactivated.

A5-3 Low-Side Performance of a Basic DX System

The low-side performance characteristics of a basic DX system are determined by the load. First, consider how a R-22 refrigeration system might operate at a full-load condition,

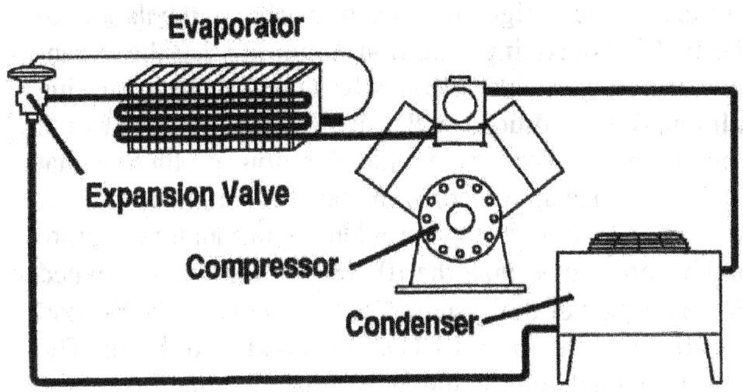

Basic Refrigeration Cycle

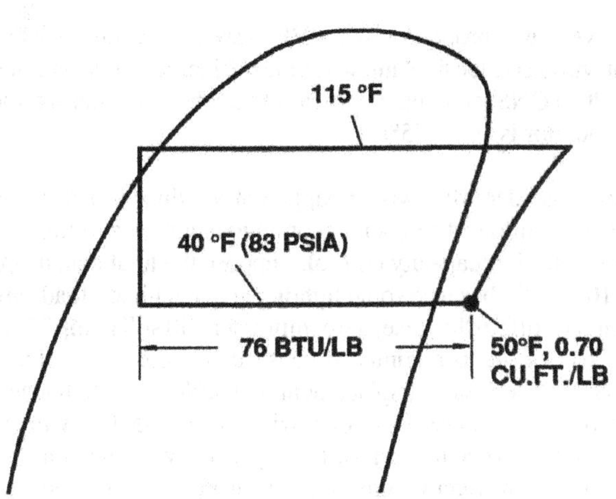

Figure A5-3

which is 20 TONS for this example. A pressure-enthalpy chart (Mollier diagram) can be used for this purpose. Figure A5-3 shows that at full load, the evaporator pressure is equal to 83 PSIA, the evaporator temperature is equal to 40°F, the superheat is equal to 10°F, the cooling capacity of one pound of refrigerant is equal to 76 BTU and the specific volume of the suction gas is about 0.70 CU.FT. per pound. Under these circumstances 52.63 pounds per minute of refrigerant will be circulated and the volume flow through the suction line will be equal to 36.84 CU.FT. per minute.

LB/MIN = 20 x 200 / 76 = 52.63

CU.FT./MIN = 52.63 x 0.70 = 36.84

In order to obtain the desired 20 TON capacity when the suction temperature is 40°F, the equipment manufacturer has to match the evaporator performance characteristics with the condensing unit performance characteristics. Figure A5-4

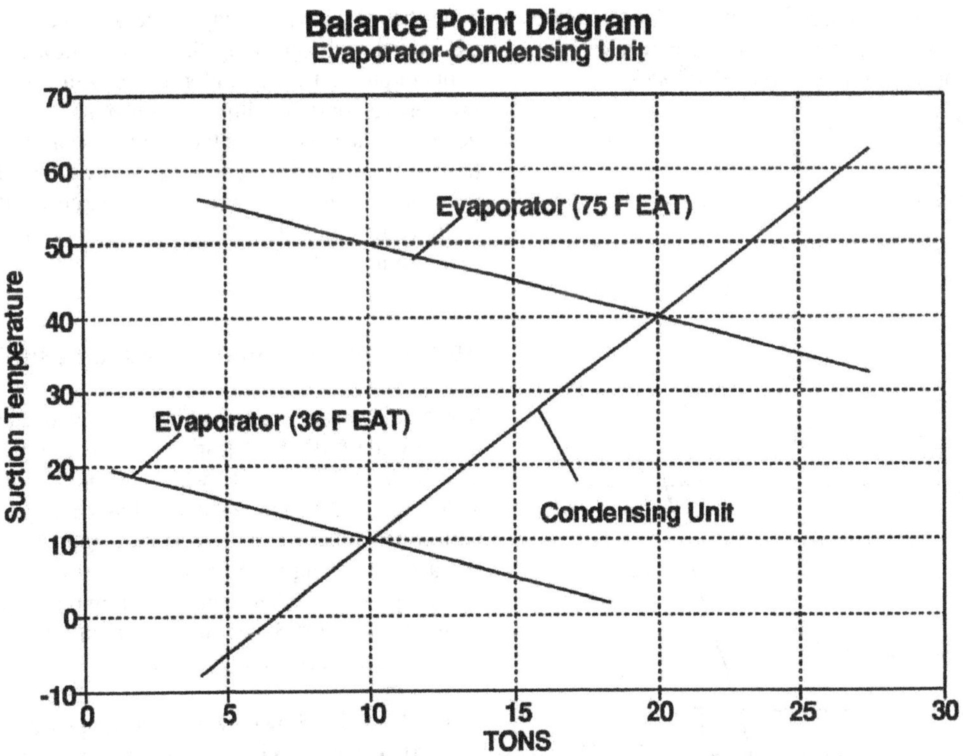

Figure A5-4

shows how a properly sized 40°F evaporator would balance out with a condensing unit when the refrigeration load is equal to 20 TONS and the temperature of the air entering the evaporator is about 75°F.

Now consider what would happen at a reduced load, if the system is allowed to operate continuously with no temperature control or capacity control. Suppose the total load drops to 10 TONS, because some lighting and machinery loads are switched off. In this case, the compressor will still pump 36.84 CU.FT. of gas per minute because it is a constant volume device. (The piston displacement and RPM do not change.) However, the mass flow of refrigerant (LB/MIN) will be reduced. This occurs because the expansion valve will throttle the flow of refrigerant until the system capacity matches the load. Since the volume flow remains constant as the mass flow is reduced, each pound of suction-side gas must occupy a larger volume. In order for this to happen the suction pressure and temperature must drop. (The thermostatic expansion valve only controls the superheat, it does not control the pressure in the evaporator.) This is verified by Figure A5-4, which shows that the compressor capacity will be about 10 TONS when the suction temperature is equal to 10°F.

The Mollier diagram can be used to show how the system operates when the suction temperature is equal to 10°F. Figure A5-5 shows that the evaporator pressure would be equal to 48 PSIA when the evaporator temperature is 10°F. If the superheat is equal to 10°F, the cooling capacity of one pound of refrigerant is about equal to 72 BTU and the specific volume of the suction gas is about 1.3 CU.FT. per pound. Under these circumstances, 28.3 pounds per minute of refrigerant will be circulated and the capacity will be about 10 TONS.

LB/MIN = 36.84 / 1.3 = 28.33

TONS = 28.3 x 72 / 200 = 10.2

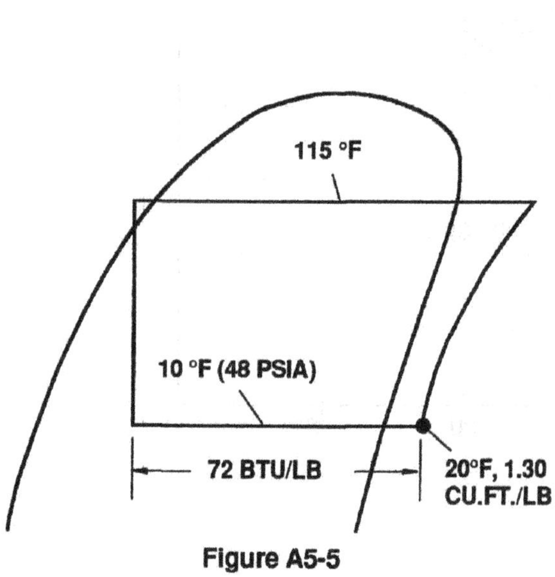

Figure A5-5

Therefore, on the refrigerant-side, the system will balance out at the 10 TON operating condition, assuming that the expansion valve can provide adequate control at this operating condition. But, conditions on the air-side of the system would be out of control. The first and most obvious problem is that the coil would ice up if there is any moisture in the return air. And, even if frosting is not a problem, the room temperature would be too cold because the 10,000 CFM that is associated with the air-side of the system was based on a 20 TON load and a 40°F coil, not on a 10 TON load and a 10°F coil. The new room temperature could be estimated (roughly) by assuming the temperature of the air leaving the coil is about 15°F warmer than the coil — say 25°F. Using the sensible heat equation, it can be shown that the room (return air) temperature would be about 36°F if 10,000 CFM of 25°F air is used to neutralize a 10 TON sensible load.

Room DB = 25 + 120,000 / (1.1 x 10,000) = 35.9 °F

Of course, the 10 TON load would tend to increase as the room temperature drops. This would occur because the transmission and infiltration gains that are associated with the building envelope would increase as the indoor-outdoor temperature difference gets larger. But, even though this example does not account for this effect, it does show what tends to happen to the room temperature if the thermostatic expansion valve is the only device that controls the performance of the refrigeration system.

Note that besides coil icing, there are mechanical problems associated with operating the refrigeration system without temperature and refrigeration cycle controls. For example, problems associated with oil return and hermetic motor overheating could occur when the expansion valve severely throttles the flow of refrigerant. And, at low loads, the expansion valve may "hunt," because it is too large for the load, or it could completely lose control, allowing liquid to flood back to the compressor.

A5-4 High-Side Performance of a Basic DX System

It also is necessary to consider how part-load operation affects the high side of the refrigeration system. In order to obtain the desired 20 TON capacity when the suction temperature is 40°F, the equipment manufacturer has to match the condenser performance characteristics with the compressor performance characteristics. Figure A5-6, on the next page, shows how a condenser would balance out with compressor when the load is equal to 20 TONS and the temperature of the air entering the condenser is equal to 95°F.

Figure A5-6 also shows what would happen at a 10 TON load if the 20 TON system is allowed to operate continuously with no thermostatic control, capacity control or head pressure control. In the example above, the load reduction was caused by a change in the internal load and the outdoor temperature

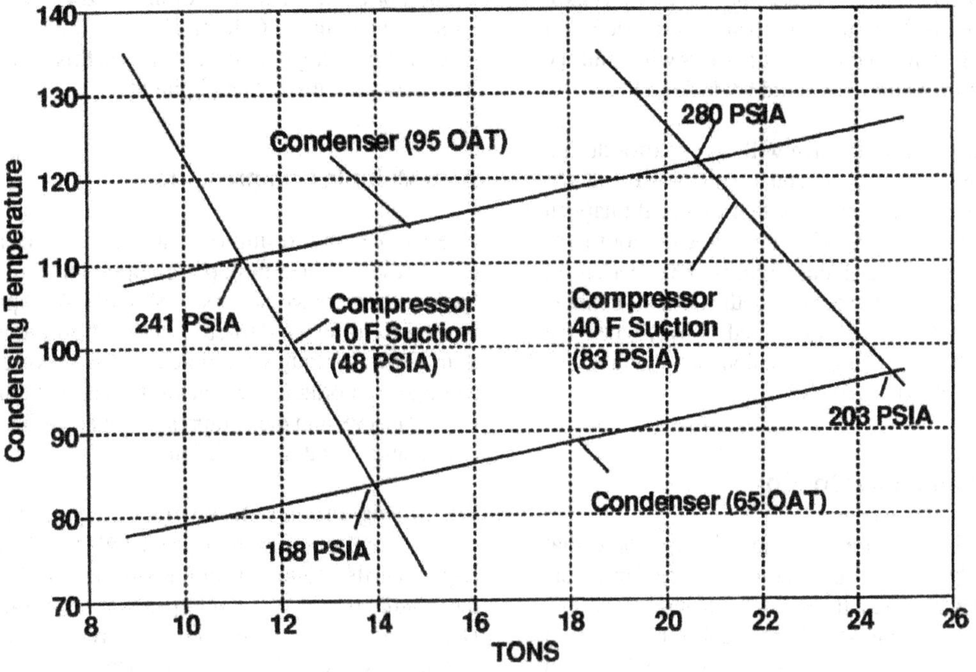

Figure A5-6

did not vary from the 95°F design condition. In this case, the condensing temperature (and pressure) will drop from 122°F to about 110°F (280 PSIA to 241 PSIA). This change could affect the thermostatic expansion valve performance, but in this case, the pressure drop across the valve and the distributor does not change by much (280 - 83 = 197 PSIA vs 241 - 48 = 193 PSIA). And, even if the load dropped to 5 TONS, the pressure drop across the valve and the distributor would still be more than 190 PSIA. (If the pressure drop across the distributor is about 20 PSIA, the pressure drop across the valve would be about 170 PSIA.)

Maintaining a constant pressure drop across the expansion valve is very desirable because the capacity of the valve is related to the pressure drop across the valve. This relationship is demonstrated by Figure A5-7. So, if the outdoor temperature holds steady at 95°F, the pressure drop across the valve will be relatively steady and the part-load performance of the expansion valve will mostly depend on the throttling range of the valve.

A completely different situation occurs when the outdoor temperature is lower, say 65°F. In this case, the pressure drop across the expansion valve will be about 100 PSIA (based on the pressures shown by Figure A5-6 and a 20 PSIA drop across the distributor). Under these circumstances, the flow of refrigerant will be reduced because the pressure drop across the valve is much smaller than the design condition pressure drop. This reduction in flow will cause the superheat to rise, and this will cause the valve to open. The valve will continue

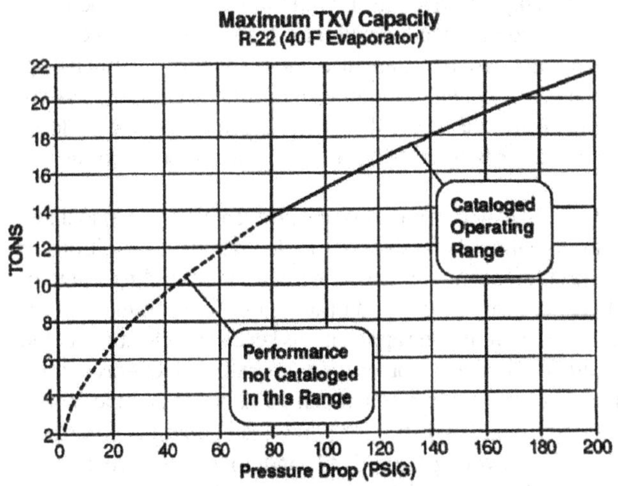

Figure A5-7

to open until the load is satisfied. Figure A5-7 shows that the expansion valve will have about 15 TONS of capacity at 100 PSIA, so this valve will be able to control loads that are 15 TONS or less. Obviously, if the load is greater than 15 TONS, the valve would be driven to the full open position before it can satisfy the load. If this occurs the superheat will continue to rise and the motor/compressor could be damaged.

Whether or not the expansion valve will lose control during low ambient conditions will depend on the load and the outdoor temperature. The probability that this could happen increases if large loads are experienced at low outdoor temperatures. Conversely, this problem might not be experienced if the load falls in step with the decline in the outdoor temperature. In the example that was discussed above, when the outdoor temperature is 65°F, the expansion valve would lose control if the load exceeds 15 TONS.

A5-5 On-Off Temperature Control

Temperature control is required to satisfy the people that occupy the building (or to satisfy the cooling requirements that are associated with production equipment and processes). The simplest way to obtain temperature control is to use a thermostat to start and stop the cooling equipment. This can be done directly, by interlocking the thermostat with the refrigeration equipment motor starters, or indirectly, by closing a liquid line solenoid valve.

If the indirect method is used, the system will pump down before the low-pressure control shuts the system down. This is very desirable because the low-side refrigerant is isolated from the idle compressor. (If the indirect method is used, there is a possibility that the low-side pressure could equalize with the high-side pressure when the compressor is idle. If this happens the compressor could short cycle on the low-pressure switch. Therefore, the controls should be arranged so that the system will not restart unless there is a call for cooling.)

On-off temperature control will eliminate some of the problems that are associated with part-load operation. The room temperature will be maintained, the evaporator will not ice up, efficiency is not substantially degraded and when the system is operating, the mass flow of refrigerant should be large enough to provide adequate oil return and motor cooling.

On-off control does not solve the problems that are associated with operating the system during low ambient conditions or the problems that are associated with the throttling range of the expansion valve. In addition, on-off control may not be suitable for multizone applications because some zones could require cooling when other zones are satisfied. On-off control does not provide precise temperature control and the indoor humidity tends to drift during low-load conditions. Furthermore, short cycling at low-load conditions causes wear on the lubricated surfaces of the compressor and degenerative strain on the motor windings. Short cycling also could cause an overheating condition, due to the low refrigerant mass flow and the high starting currents that are associated with start-up. Short cycling may not harm small tonnage equipment, but this method of control is not recommended for intermediate and large tonnage units. (Ultimately, it is the manufacturers responsibility to approve or disapprove this method of control for a particular piece of equipment.)

A5-6 Multiple Compressors

One simple way to improve capacity control and part-load efficiency is to use multiple compressors. In this case, the cooling capacity would be staged and the compressors would be subject to fewer start-stop cycles during low-load conditions. However, when multiple compressors are connected to a single evaporator, the evaporator capacity, the expansion valve capacity and the compressor capacity must be balanced during any operating condition.

When a single expansion valve (and coil) is used with multiple compressors, the minimum capacity of the valve could be larger than the minimum compressor capacity. Unfortunately, the minimum capacity of an expansion valve is hard to predict. Depending on the equipment performance characteristics, expansion valve performance characteristics, distributor pressure drop, superheat setting and piping details, the minimum capacity of the valve could be as low as 15 percent or as high as 75 percent of its design capacity.

If the expansion valve is oversized for the part-load operating conditions, refrigerant could flood back to the compressor(s). Flood back problems can be avoided by using an evaporator coil that is equipped with two or more expansion valves and refrigerant circuits. This way, the expansion valve capacity can be more closely matched with the part-load compressor capacity. Figure A5-8 shows this type of arrangement.

Note that there are potential lubrication problems associated with connecting two or more compressors to an evaporator

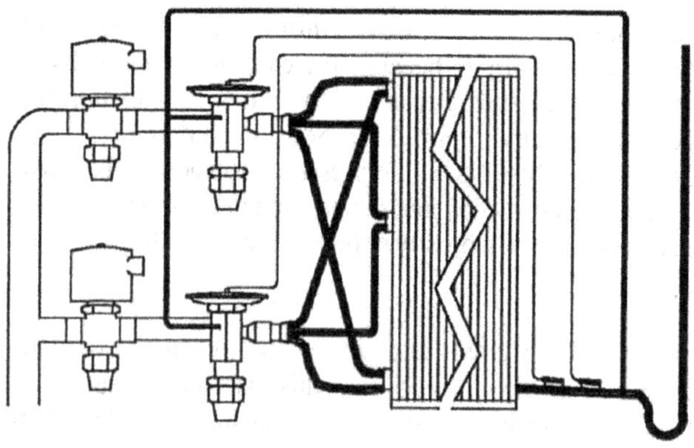

Figure A5-8

that has a single suction-side header (see Figure A5-8). However, the problems that are associated with unbalanced oil return can be avoided, providing that the refrigerant piping details and the equipment layout assure equal crankcase pressures and oil levels during any operating condition.

Another reason for not using a single expansion valve is related to suction pressure control. If the coil capacity is not reduced in step with the compressor capacity, the suction temperature (and pressure) will rise. This behavior is demonstrated by Figure A5-10, which shows that one fully loaded 10 TON compressor and a 20 TON evaporator would operate at a 48°F suction temperature (96 PSIA). However, the part-load suction temperature (and pressure) can be reduced by using two expansion valves and a row-split coil or a face-spilt coil. This behavior also is demonstrated by Figure A5-10, which shows that, one fully loaded 10 TON compressor and a 10 TON coil would operate at a 40°F suction temperature (83 PSIA). (Refer to Section A5-2 for more information about coil configurations.)

Ideally, each compressor should be connected to an independent refrigeration circuit (expansion valve, coil and suction line). This way, the expansion valve and coil capacities are reduced in step with the compressor capacity. With this arrangement, the evaporator operating pressures will be more stable, there will be less variation in refrigerant mass flow through a given circuit and there are no unusual oil return problems. Redundancy is another advantage that is associated with independent circuits. If one system fails, the remaining system(s) will still provide a significant amount of cooling capacity. Figure A5-9 shows a system that has independent circuits.

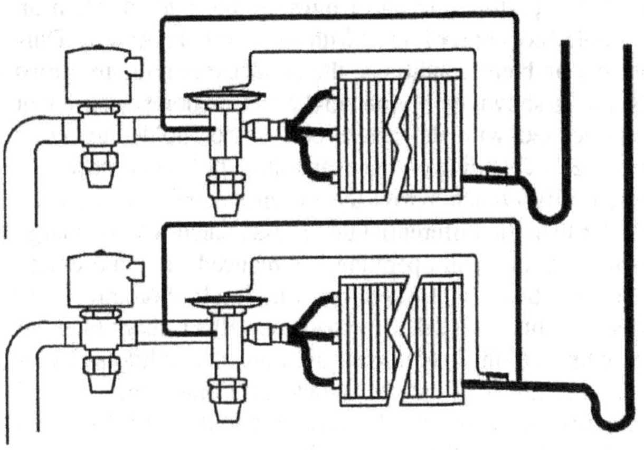

Figure A5-9

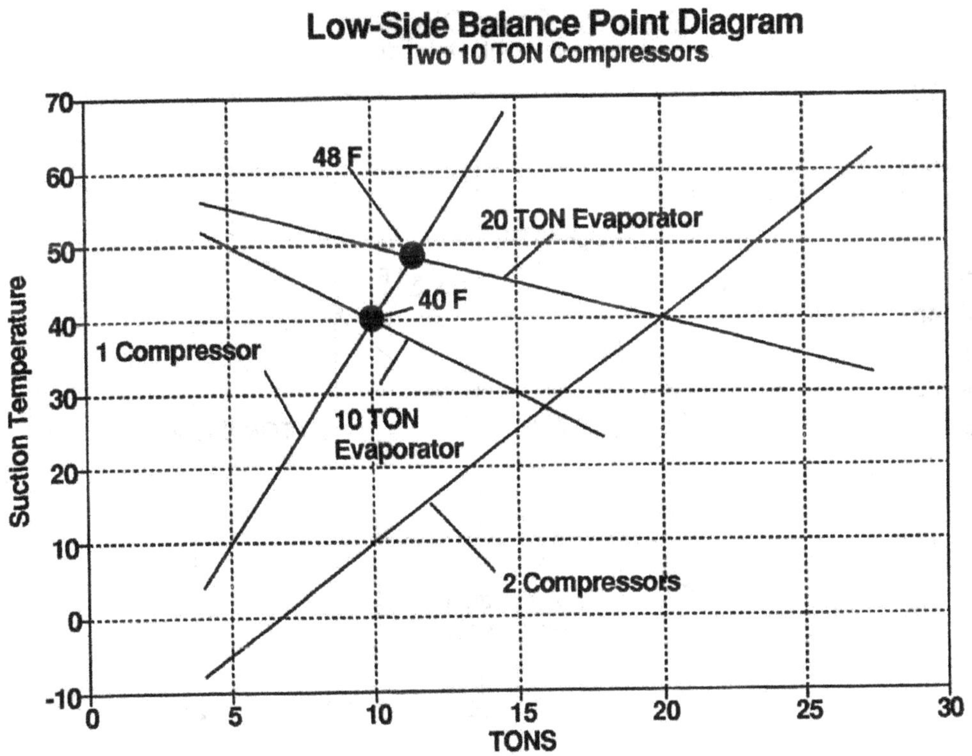

Figure A5-10

High-side performance also must be considered when an air-cooled condenser is used with multiple compressors. During low ambient conditions, the condensing pressure drops off, but as shown by Figure A5-10, the suction pressure that is associated with the active compressor holds steady or increases, depending on the evaporator split. This causes the pressure difference across the expansion valve to be even smaller than the differential that is associated with one large compressor and coil operating at a reduced suction pressure (see subsection A5-4). Therefore, when multiple compressors and coil splits are used, the expansion valve is more likely to lose control of the load during low ambient conditions. However, this problem can be avoided if some type of head pressure control is used. (Head pressure controls are discussed in subsection A5-10.)

A5-7 Cylinder Unloaders

Cylinder unloading provides a more sophisticated way to control piston compressor capacity. Manufacturers have devised a number of different unloading mechanisms, but the net result is that, as the pumping capacity of one or more cylinders is nullified, the compressor capacity is reduced proportionally.

Since the unloading process can be staged, the pumping capacity of a single compressor can be reduced incrementally. The size and number of increments depend on the number of cylinders. Usually, cylinders are unloaded two at a time, so an eight cylinder machine could provide four stages of capacity (100, 75, 50 and 25 percent). However, on hermetic equipment, the mass flow of refrigerant must be large enough to provide adequate oil return and to cool the motor during unloaded operation. Therefore, the lubrication requirement or the motor cooling requirement may establish a lower limit for the unloading process.

When cylinder unloading is used, the step-wise reduction in cooling capacity is equal to, or better than, the reductions that can be obtained by staging compressors. And, if these two capacity control methods are combined, (multiple compressors equipped with unloaders) a large number of capacity steps can be obtained. The obvious benefits of this method of capacity control are that the cooling effect is more continuous, motor cycling is eliminated (or greatly reduced), efficiency is improved and operating cost is reduced.

However, the low-side problems that are associated with compressor unloading are the same as the low-side problems that are associated with compressor staging. In either case, the evaporator capacity, the expansion valve capacity and the

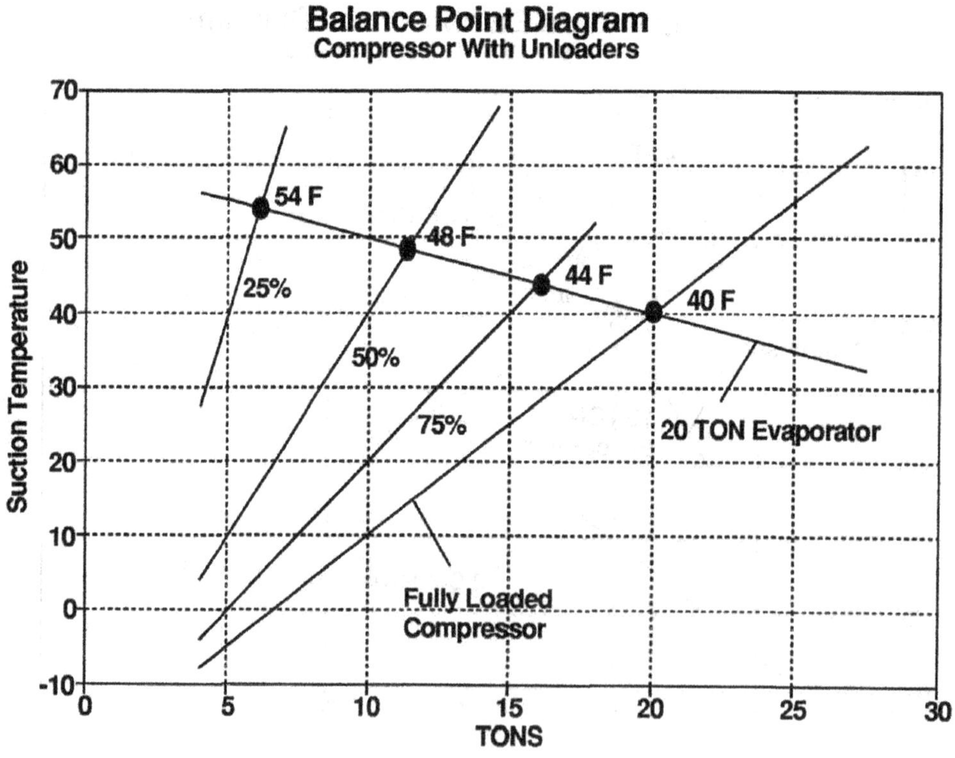

Figure A5-11

compressor capacity must be balanced during any operating condition. As noted above (see subsection A5-6), if a single expansion valve is too large for part-load operation, the valve could "hunt" or liquid could flood back to the compressor. These problems can be avoided if the evaporator is equipped with two or more expansion valves and refrigerant circuits (this arrangement is shown by Figure A5-8).

Another reason for not using a single expansion valve is related to suction temperature and pressure control. If the coil capacity is not reduced in step with the compressor capacity, the suction temperature (pressure) will rise. This behavior is demonstrated on the previous page by Figure A5-11, which shows that the suction temperature increases from 40°F to 54°F (83 PSIA to 105 PSIA) as the compressor unloads from full capacity to 25% capacity. The part-load suction temperature (pressure) can be reduced by using multiple expansion valves and a row-split coil or a face-split coil. (Refer to Section A5-2 for more information on coil configurations.)

Figure A5-12 shows how the system operates if the coil is divided into four, 5-TON, face-split sections. The diagram shows that as the load is gradually reduced, the suction temperature (pressure) will drop until one stage of capacity (coil and compressor) is taken off-line. Then, the suction temperature will quickly increase as the new compressor capacity balances with the new coil capacity. If the load continues to decrease, the suction temperature will move down the new compressor capacity line until the next stage of capacity is taken off-line, and so forth. (This example is only intended to demonstrate the concept. Four stages of unloading probably would not be used with a 20 TON compressor. Also, note that the capacity steps depend on how the coil is split, the coil circuiting and the unloader control setpoints.)

Figure A5-13, on the next page, shows the same information as Figure A5-12, but in a different format. The sawtooth appearance of this suction temperature vs tonnage curve is characteristic of compressors that are equipped with unloaders. Note that in this example, there is a fairly large variation in suction temperature (pressure) as the system cycles through the unloader control points. However, these variations can be reduced.

Figure A5-14, on the next page, shows what would happen if a different set of control points is used. In this case, the fourth stage of capacity (compressor and coil) will be taken off-line when the suction temperature falls to 36 degrees (18 TONS). When this happens, the operating point moves to the 75% capacity line. What happens next depends on the load.

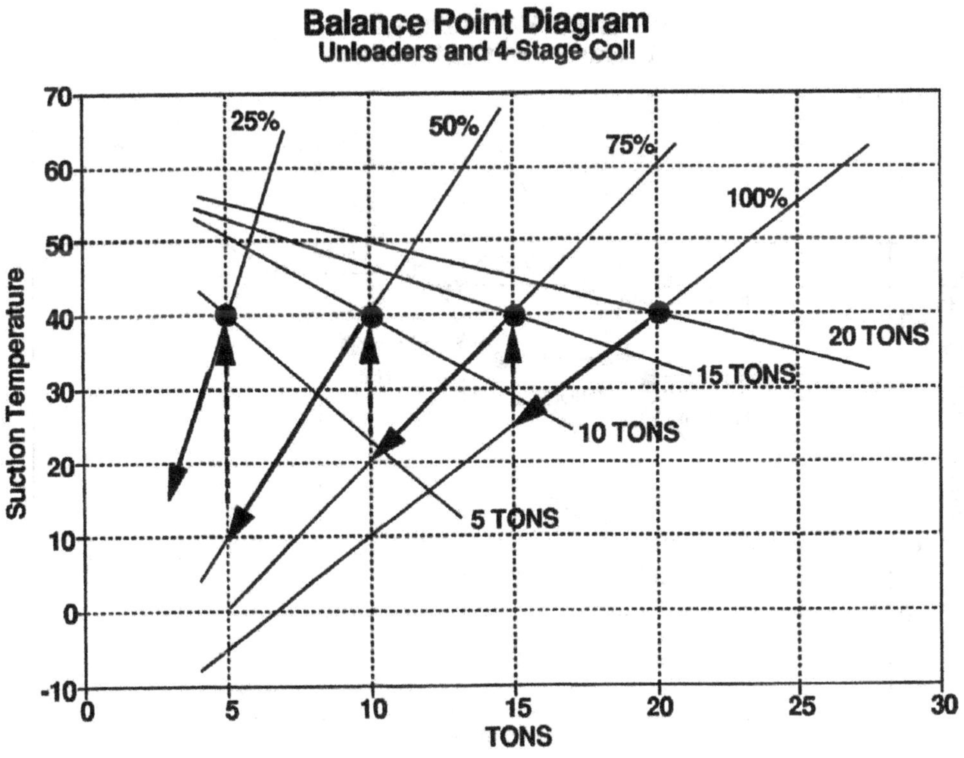

Figure A5-12

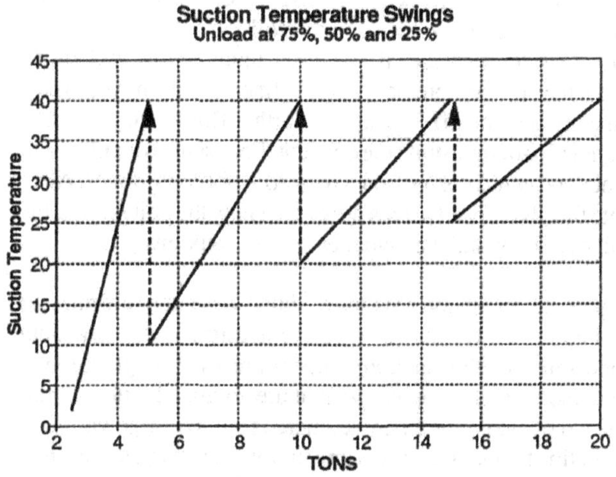

Figure A5-13

If the load is greater than 14 TONS, the suction temperature will rise. If the load is more than 17 TONS, the suction temperature will rise to a 46°F and the fourth stage of capacity will be activated. If the load is less than 14 TONS the suction temperature will fall. In this example, the third stage of capacity (compressor and coil) will be taken off-line when suction temperature falls to 34°F (12 TONS). When this happens, the compressor operating point moves to the 50% capacity line. What happens next will depend on the load.

If the load is greater than 8 TONS, the suction temperature will rise. If the load is more than 11 TONS, the suction temperature will rise to a 44°F and the third stage of capacity will be activated. (If the load is greater than 17 TONS the fourth stage of capacity also will be activated.) If the load is less than 9 TONS the suction temperature will fall. In this example, the first stage of capacity (compressor and coil) will be taken off-line when suction temperature falls 32°F (7 TONS). When this happens, the compressor operating point moves to the 25% capacity line. What happens next will depend on the load.

If the load is greater than 7 TONS, the suction temperature will rise. If the load is more than 5 TONS, the suction temperature will rise to a 42°F and the second stage of capacity will be activated. (If the second stage of capacity cannot accommodate the load, the third and if necessary, the fourth stage will be activated.) If the load is less than 4 TONS the suction temperature will fall and if the load is low enough, the system will shut down on the low-pressure control.

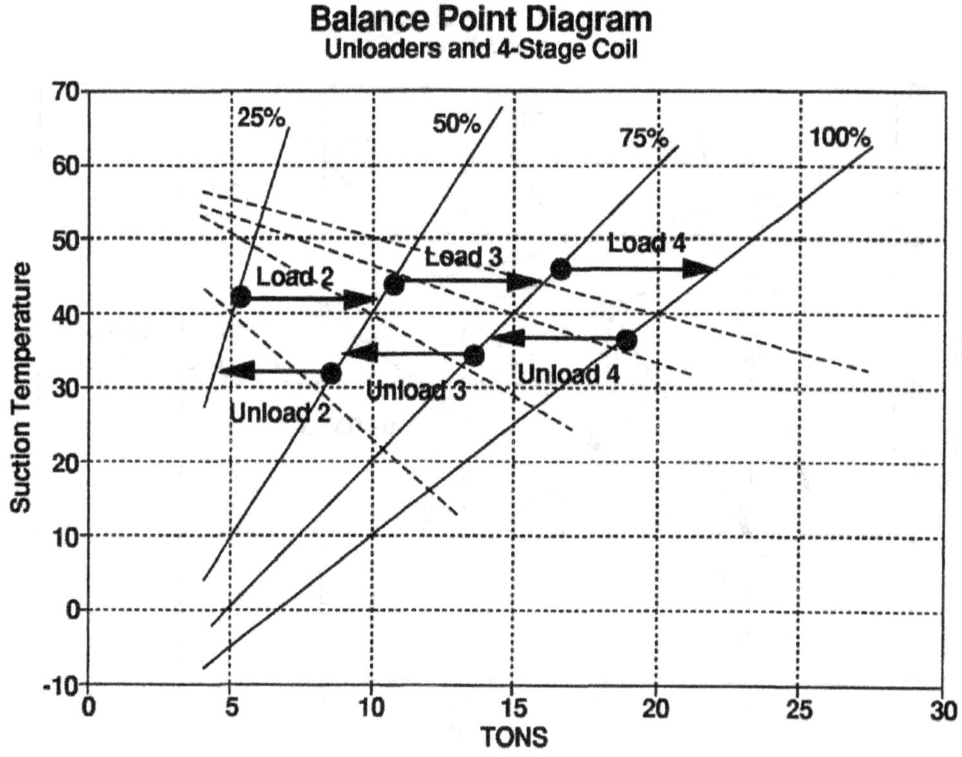

Figure A5-14

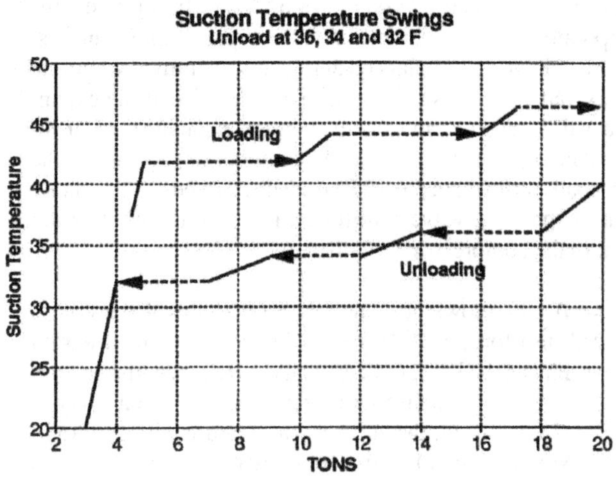

Figure A5-15

Figure A5-15 shows the same information as Figure A5-14, but in a different format. The step-wise changes in the suction temperature vs tonnage curve are similar to those found in Figure A5-13, but the size of the temperature swings (and the corresponding suction pressure swings) have been reduced. It is the system designers responsibility to select control points that are compatible with the requirements of the application.

High-side performance also must be considered when an air-cooled condenser is used with unloaders. During low ambient conditions, there will be a considerable drop in the condensing pressure, but as shown by Figure A5-15, the suction temperature (and pressure) that is associated with a fully unloaded compressor does not drop too much (usually 8 to 12 PSIA). If the compressor reloads, the suction pressure will rise above this minimum value. This behavior causes the pressure difference across the expansion valve to be smaller than the differential that would be associated with one large compressor operating at a reduced suction pressure (see subsection A5-4). When compressor unloaders and coil splits are used, the expansion valve is more likely to lose control of the load during low ambient conditions. However, the problems that are associated with low ambient operation can be avoided if some type of head pressure control is used. (Head pressure controls are discussed in subsection A5-10.)

A5-8 Speed Control

As the rotational speed (RPM) of the compressor decreases, the compressor capacity is reduced proportionally. Figure

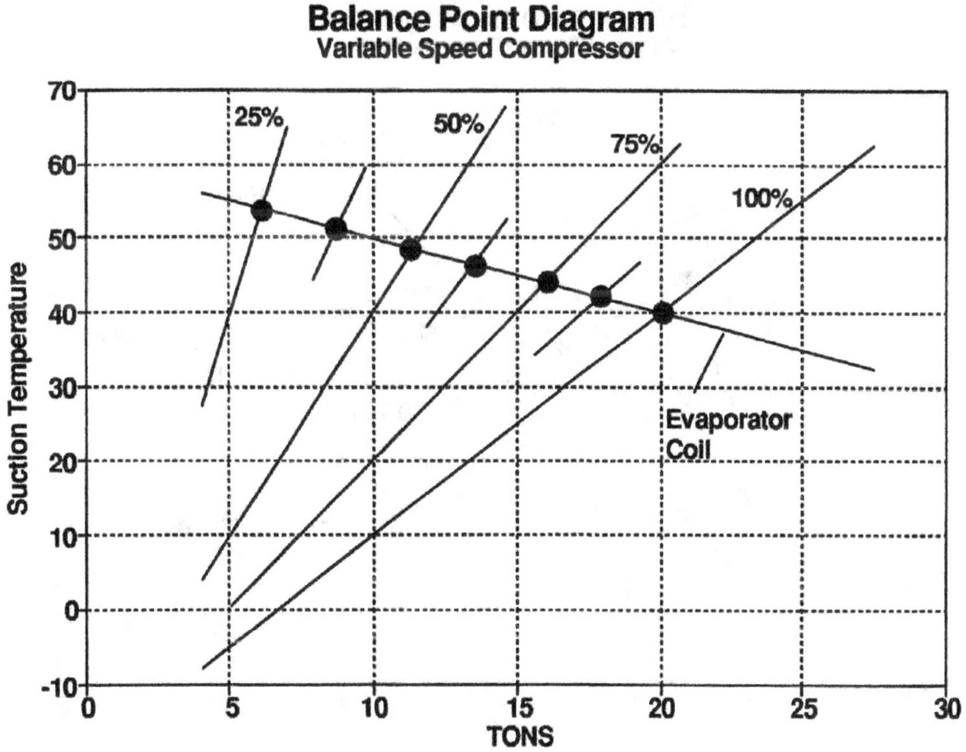

Figure A5-16

A5-16, on the previous page, shows how a variable speed compressor would balance out with a single evaporator as the RPM is reduced. In this case, the operating point slides along the evaporator curve and the suction temperature (pressure) continuously increases.

As with any other type of capacity control, the mass flow of refrigerant must be large enough to provide adequate oil return and to cool the motor during unloaded operation. Therefore, the lubrication requirement or the motor cooling requirement, may establish a lower limit for the unloading process. The minimum capacity also may be determined by the limitations of the electronic devices that are used to control the motor speed.

The benefits of variable speed capacity control are that the reduction in cooling capacity is continuous and precise, motor cycling is eliminated and the mechanical complexity of the unloading hardware is reduced. And, under certain circumstances, variable speed systems can be more efficient than constant speed systems. However, the overall system efficiency is affected by the efficiency of the motor speed controller, which may decrease as the motor speed is reduced.

The low-side problems that are associated with speed control are the same as the problems that are associated with compressor staging and compressor unloading. In all cases, the evaporator capacity, expansion valve capacity and compressor capacity must be balanced during any operating condition. As explained in subsections A5-6 and A5-7 — if the expansion valve capacity and the evaporator capacity are not reduced in step with the decrease in compressor capacity — the part-load performance could be characterized by an inadequate expansion valve throttling range, or liquid flooding back to the compressor.

Figure A5-17 shows how the system operates if the coil is divided into four, 5-TON, face-split sections. The diagram shows that, as the RPM (and capacity) is reduced, the suction temperature will increase until one stage of coil capacity is taken off-line. Then, the suction temperature will drop (and the RPM will change) as the compressor capacity balances with the new evaporator capacity. The net effect is that the suction pressure stays fairly steady as the compressor capacity is reduced. (This example is only intended to demonstrate the concept. Four stages of coil capacity probably would not be used with a 20 TON compressor. Also, note that other tonnage splits are possible, depending on whether the coil is face-split or row-split and on the coil circuiting. And, if the application is compatible with evaporator fan control — a VAV system for example — a wide range of evaporator capacities can be obtained.)

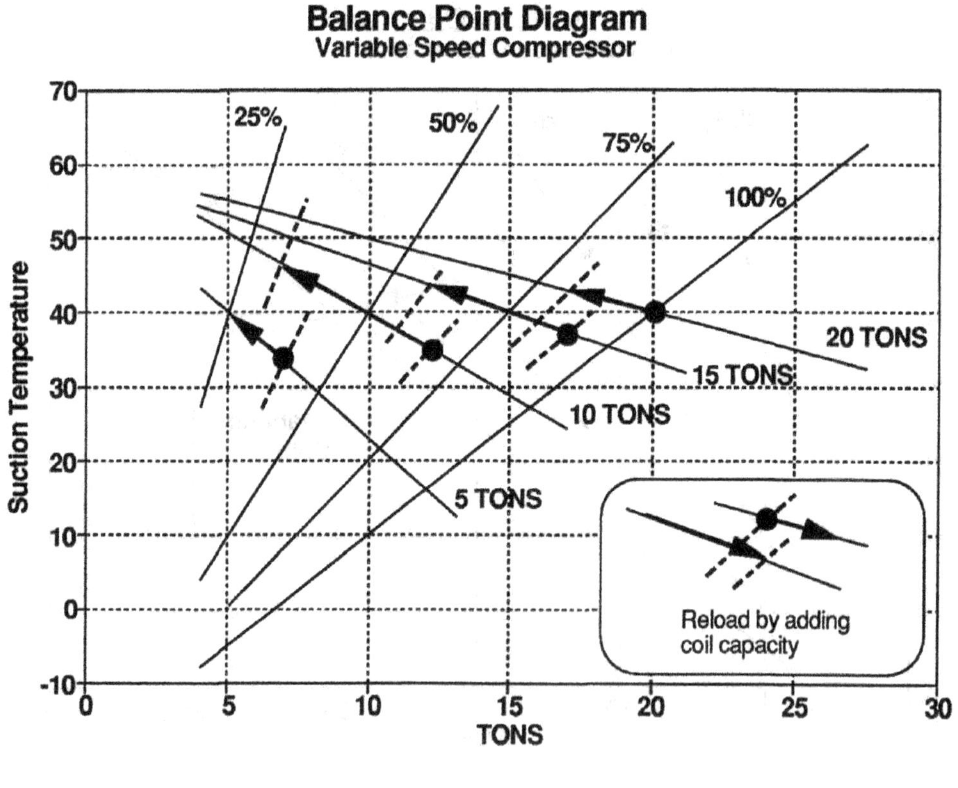

Figure A5-17

High-side performance also must be considered when an air-cooled condenser is used with speed control. During low ambient conditions, there will be a considerable drop in the condensing pressure, but as shown by Figure A5-17, the suction temperature (and pressure) that is associated with part-load operation stays fairly steady. This results in a relatively small pressure difference across the expansion valve. Under these conditions the expansion valve can lose control of the load (as explained in subsection A5-4). However, the problems that are associated with low ambient operation can be avoided if some type of head pressure control is used. (Head pressure controls are discussed in subsection A5-10.)

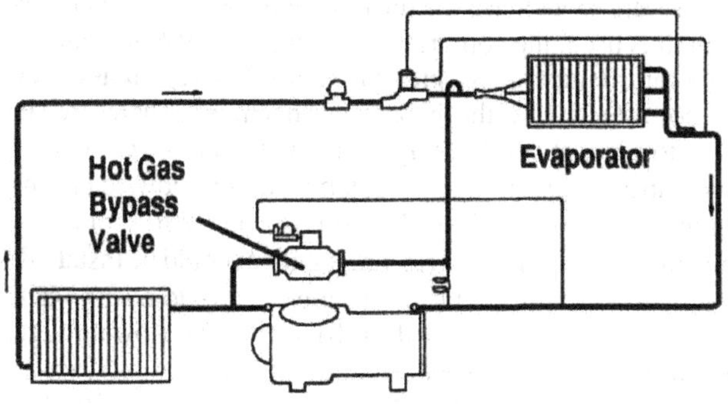

Figure A5-18

A5-9 Hot Gas Bypass

Constant temperature systems, such as VAV systems or multizone systems, control the cooling capacity that is delivered to the conditioned space by adjusting the cooling coil air flow. In this case, the load on the coil and the load on the compressor diminish as the coil air flow is reduced. Normally, when the load falls past a certain point, the compressor — even if it is equipped with unloaders — would be stopped by the low-pressure control. But, this is unacceptable for these types of systems because a continuous supply of cold air is required, regardless of the size of the load. In this case, hot gas bypass can be used to keep the compressor running. (Since hot gas bypass, by itself, does not control capacity, it would not be used to control the capacity of a constant volume system.)

When hot gas bypass is installed, some of the hot gas that is discharged from compressor is routed directly to the low-side of the refrigeration system. This creates an artificial compressor load that compensates for a reduction in the actual coil load. Hot gas bypass can be used on any type or size compressor. It can be used as the only method of capacity control or it can be combined with another method of capacity control. For example, small compressors are not usually equipped with unloaders, so hot gas bypass could be used to control the compressor capacity throughout the whole range of part-load operating conditions. But, hot gas bypass also can be used to keep a larger, more sophisticated, compressor running after it has been completely unloaded.

The benefits of the hot gas bypass method of capacity control are that the cooling coil capacity can be modulated all the way down to a no-load condition and that compressor cycling is prevented. However, the part-load system efficiency is not improved because the compressor must still pump the refrigerant that is associated with creating and neutralizing the artificial load.

Two basic methods are used to introduce hot gas into the low-side of the system. The first method, which is the preferred method, is shown by Figure A5-18. This method introduces the hot gas between the expansion valve and the evaporator. The second method, which is shown by Figure A5-19, introduces the hot gas between the evaporator and the compressor. Both of these methods are discussed below.

When hot gas is introduced upstream of the evaporator, (Figure A5-18) it increases the rate of evaporation inside the coil, and there is a corresponding rise in the superheat temperature at the coil outlet. When the expansion valve superheat bulb senses this rise, it will cause the expansion valve to open, and more refrigerant will be fed to the coil. The net effect is that, the evaporator pressure is stabilized, the refrigerant mass flow increases and the compressor continues to operate. And, since refrigerant mass flow and superheat are maintained, there are no problems associated with oil return, motor cooling or liquid slugging when the actual cooling load is small. (Since the minimum evaporator temperature can be controlled, this method can be used on water chillers.)

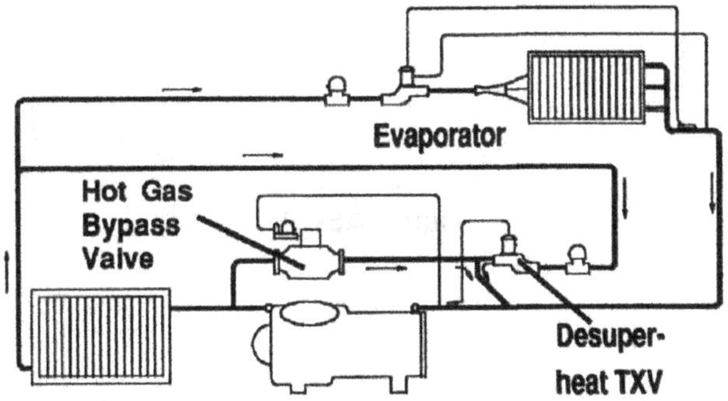

Figure A5-19

If the evaporator is too far away from the compressor, (35 to 50 feet) or if there is a large pressure drop associated with the distributor, the hot gas should be introduced between the evaporator and the compressor (Figure A5-19). In this case, an additional expansion valve is required to desuperheat the hot gas and to increase the refrigerant mass flow. When this method is used, the compressor continues to operate, but the evaporator temperature, pressure and refrigerant mass flow are not affected by the bypass arrangement. Therefore, in order to prevent coil frosting, some method must be used to maintain a minimum evaporator temperature at low-load conditions. This could be done by using solenoids to shut down coil circuits, or a suction pressure regulator could be installed. Note that oil return problems are more likely to occur when this method is used because the refrigerant mass flow through the evaporator and the suction header continue to decrease as the cooling load diminishes. And, if the components of the hot gas bypass system are improperly sized or positioned, the compressor could be damaged by high operating temperatures or liquid floodback. (Since evaporator temperature is not controlled when hot gas is bypassed, this method should not be used on water chillers.)

Hot gas bypass valves can be set to crack open when the suction temperature (pressure) falls to some selected value. Once it begins to open, it takes about 6 additional degrees of temperature drop to drive the valve to the full open position. However, when the valve is completely open, the coil must still be warm enough to prevent icing.

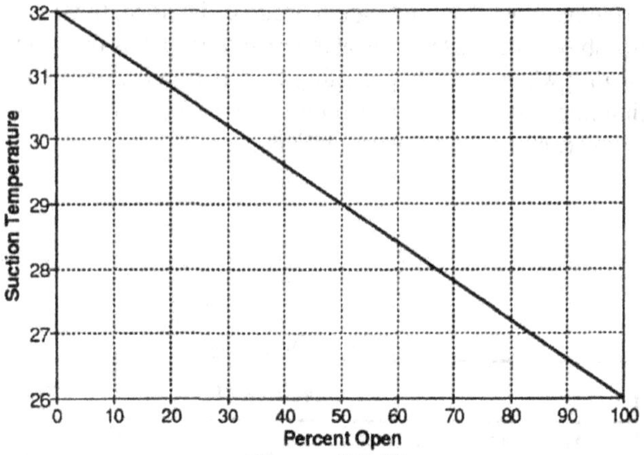

Figure A5-20

The minimum allowable coil temperature varies between 20°F and 28°F depending on the air flow. At reduced air flow, the coil temperature should not be allowed to fall below 28°F. Therefore, as shown by Figure A5-20, the bypass valve must begin to open at (approximately) 32°F. Note that, this 32°F temperature is a minimum value. The valve could be set to open at a temperature that is higher than 32°F, but if unloaders are used, the setpoint would have to be a few degrees below the temperature that is associated with the last stage of unloading. Figure A5-21 shows how a hot-gas valve might be sequenced with the compressor unloaders.

High-side performance also must be considered when an air-cooled condenser is used with hot gas bypass. When the bypass valve opens, it will reduce the load on the condenser and cause a drop in the condensing pressure. This drop is additive to the usual reduction in condensing pressure that is associated with low ambient operation. However, the suction temperature (and pressure) that is associated with hot gas bypass operation stays fairly steady (see Figure A5-21). This results in a relatively small pressure difference across the expansion valve. Under these conditions the expansion valve can lose control of the load (as explained in subsection A5-4). However, the problems that are associated with low ambient operation can be avoided if some type of head pressure control is used. (Head pressure controls are discussed in subsection A5-10.)

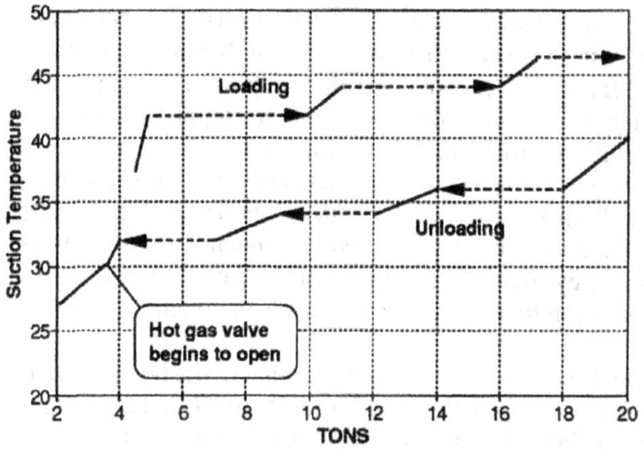

Figure A5-21

A5-10 Head Pressure Control

Condenser capacity and operating pressure (condensing temperature) are based on the heat sink temperature that is associated with hot weather operation. If a water-cooled condenser is used, the heat sink temperature may be relatively constant during any season. But, if an air-cooled or evaporative condenser is used, the heat sink temperature can range from more than 100°F to below 0°F.

Low heat sink temperatures create a problem because the condensing pressure decreases as the heat sink temperature drops. This causes the pressure at the high side of the expansion valve to decrease. Since the pressure on the low side of the expansion valve (evaporator pressure) is under control and reasonably stable, the pressure drop across the expansion valve decreases by about the same amount. If the pressure drop across the expansion valve gets too small, the expansion valve may not be able to feed enough refrigerant to the evaporator. If this happens, the system will lose control over the load, and in some cases, the compressor could overheat (refer to subsection A5-4). Therefore, the capacity of the condenser must be reduced as the heat sink temperature declines.

Air-Cooled Condensers
The capacity of an air-cooled condenser can be controlled by adjusting fan capacity. If fan control is used, the fans can be cycled on and off, or the fan speed can be adjusted. However, if on-off fan control is used, there must be two or more fans, because cycling a single fan causes abrupt swings in the condenser pressure and short on-off fan cycles. If two or more fans are used, or if speed control is employed, the condenser capacity can be reduced more gradually.

The capacity of an air-cooled condenser also can be controlled by using dampers to throttle the air flow. This method provides continuous modulation all the way down to the no flow condition. Damper control also can be combined with fan control. In this case, fan control is used until the outdoor temperature drops to a certain level and then the dampers begin to close as it gets colder.

Fan control or damper control may not be satisfactory if the system has to operate in very cold weather. In this case, the condenser capacity can be controlled by flooding the condenser with refrigerant. This method is effective when the outdoor temperature is as low as 0°F.

Figure A5-22 shows how a three-way flood-back valve can be used to control the condensing pressure. If the condensing pressure is too low, the hot-gas port cracks open and the liquid outlet port starts to close. This causes liquid to backup in the condenser. The valve will continue to modulate until the condensing pressure is under control. The action is reversed if conditions cause the condensing pressure to rise.

Evaporative Condensers
The capacity of evaporative condensers can be controlled by adjusting fan capacity or by modulating dampers. During cool to cold weather, the water could be shut off. In this case, the dry capacity will be about half the design (wet) capacity. (Indoor installation or some form of freeze protection is required if the wet capacity is required during very cold weather.) It is not usually required, but it is also possible to use flood-back control.

Water-Cooled Condensers
The water that is supplied to a water-cooled condenser could come from a city water main (if laws permit), a well or a water tower. If the water comes from a water main or a well, the temperature is fairly constant. But, if a water tower is used, the temperature of the water leaving the tower decreases as the outdoor temperature drops. Therefore, the tower fans, the piping circuits, the pumps and the associated controls must be designed to provide a supply of reasonably constant temperature water (to the condenser) during any type of weather. (Freeze protection is required if the water tower has to operate in very cold weather.)

Providing that the entering temperature of the condenser water is under control, the problems that are associated with controlling the condenser capacity are minimal. In this case,

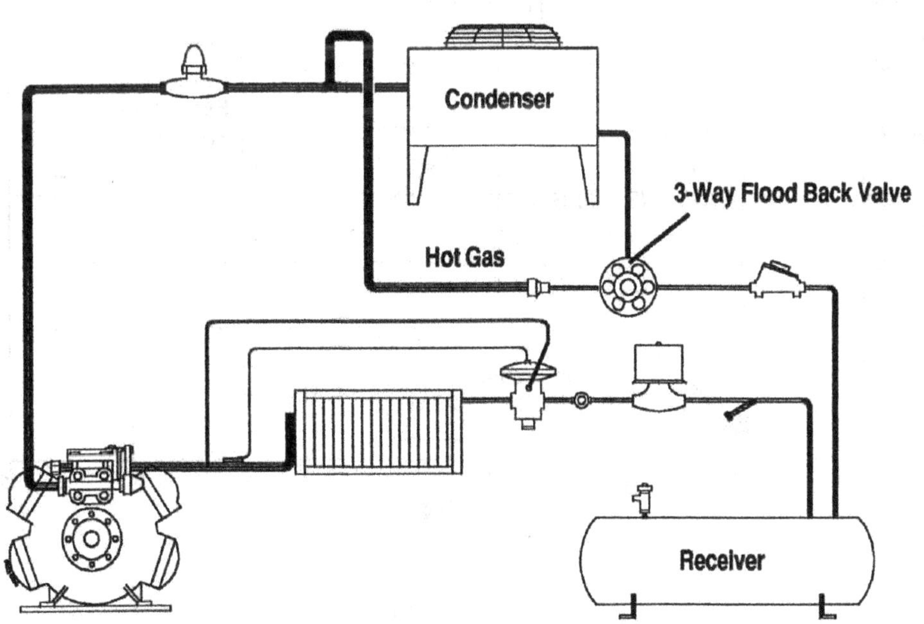

Figure A5-22

the condenser capacity can be controlled by using a water regulating valve to maintain the desired condensing pressure.

Air-side economizers save energy and they reduce the problems that are associated with operating an air-cooled condenser or a water tower during cool or cold weather. When an air-side economizer is used, the refrigeration system can be shut down when the outdoor temperature drops below the changeover temperature (50 to 60°F).

A5-11 Evaporator Pressure Regulators

If a valve (restriction) is installed between the evaporator and the compressor, the evaporator pressure will be higher than the suction pressure at the compressor. The difference between the two pressures will be equal to the pressure drop across the valve. In this case, the evaporator temperature will be determined by the evaporator pressure and the system capacity (tonnage) will be determined by the suction pressure at the compressor. (Since the compressor is a constant volume device, the compressor capacity is directly related to the specific volume of the suction gas, which depends on the suction pressure.)

If the restriction is suction pressure regulator, the evaporator pressure and temperature can be held constant as the cooling capacity of evaporator decreases. This is exactly what is needed to maintain a constant coil temperature when the coil air flow is adjusted to satisfy the load (constant temperature, variable volume application). Therefore, if the compressor is not equipped with capacity control hardware, a suction pressure regulator could be used as a capacity control device. However, the minimum compressor capacity is limited by the oil return or the motor cooling requirements. Therefore, this device should not be used if severe throttling is required.

Figure A5-23 shows how a pressure regulator maintains a steady 40°F coil temperature (83 PSIA) when the load is equal to 50 percent of the design load. In this case, the compressor operates at 10°F (48 PSIA), so there must be a 35 PSIA drop across the regulator. The suction pressure at the compressor determines the refrigerant mass flow and the compressor capacity. Since the refrigerant flow through the coil is the same as the flow through the compressor, the coil capacity is equal to the compressor capacity (10 TONS).

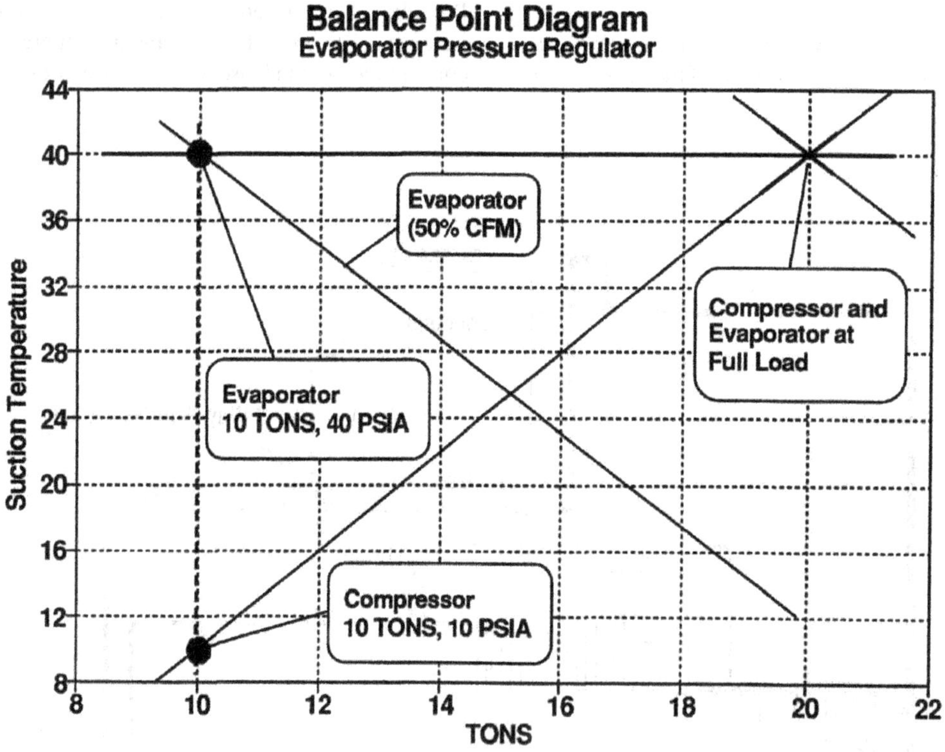

Figure A5-23

A5-12 Screw, Centrifugal and Scroll Compressors

The capacity of screw compressors and centrifugal compressors can be continuously modulated by hardware that is built into the machine. A slide valve is used to adjust screw compressor capacity and inlet vanes are used to adjust centrifugal compressor capacity. The capacity of a scroll compressor can be modulated by adjusting the speed.

The benefits associated with these types of compressors are that the reduction in cooling capacity is continuous and precise, and that motor cycling is eliminated. The part-load efficiency of screw and centrifugal compressors is quite good. The part-load efficiency of a variable speed scroll compressor also can be good, but it depends on the efficiency of the motor speed controller.

As with any other type of capacity control, the mass flow of refrigerant must be large enough to provide adequate oil return and to cool the motor during unloaded operation. Therefore, the lubrication requirement or the motor cooling requirement, may establish a lower limit for the unloading process. The minimum capacity also may be determined by the limitations of the devices that are used to control the capacity.

Figure A5-16 on page A5-11, shows how a modulated compressor would balance out with a single evaporator as the capacity is reduced. In this case, the operating point slides along the evaporator curve and the suction temperature (pressure) continuously increases. However, the low-side problems that are associated with modulating control are the same as the problems that are associated with compressor staging and compressor unloading (see subsections A5-6 and A5-7). As indicated by Figure A5-17 on page A5-12, the expansion valve capacity and the evaporator capacity can be balanced with the part-load compressor capacity by using two or more coil splits.

High-side performance also must be considered when an air-cooled condenser is used with modulating control. During low ambient conditions, there will be a considerable drop in the condensing pressure, but as shown by Figure A5-17, the suction temperature (and pressure) that is associated with part-load operation stays fairly steady. This results in a relatively small pressure difference across the expansion valve. Under these conditions the expansion valve can lose control of the load (as explained in subsection A5-4). However, the problems that are associated with low ambient operation can be avoided if some type of head pressure control is used. (Head pressure controls are discussed in subsection A5-10.)

A5-13 Water Chillers

Two types of shell and tube evaporators are used to produce chilled water for hydronic cooling systems—the flooded type and the direct expansion type. In a flooded evaporator, the water flows through the tubes and the refrigerant fills the lower part of the shell. In a direct expansion (dry) evaporator, the refrigerant flows through the tubes and the water flows through the shell. Most manufacturers use the direct expansion type for small and medium tonnage equipment.

The capacity of a direct expansion evaporator depends on the water flow (GPM) that is passing through the tubes, the temperature of the entering water, the temperature of the leaving water, the refrigerant-side temperature, the heat transfer area of the tubes and the compressor capacity. But, the water flow is usually regulated by the system controls. (The water flow will be relatively constant if the cooling coils are controlled by three-way valves or it will vary if the cooling coils are controlled by two-way valves.) Therefore, the temperature of the return water or the leaving water temperature is used to control the capacity of the refrigeration equipment. At part-load, the evaporator capacity can be reduced by decreasing the compressor capacity and the number of active tubes. Therefore, the tube bundle is usually divided into two or more refrigerant circuits and the compressor (or compressors) is (are) normally equipped with unloaders or another type of capacity control hardware. And, if the minimum load is very small, hot gas bypass can be used to keep the compressor running.

The high-side problems that are associated with chilled water equipment are the same as the high-side problems that apply to all-air equipment. Both types of equipment require head pressure controls if low ambient operation is required.

A5-14 Controls

Refrigeration cycle capacity controls may have to start and stop compressors, unload compressors, modulate compressor capacity, activate a hot gas bypass valve and operate liquid line solenoid valves. These controls may act directly or indirectly, they may be electric, electronic, pneumatic, or they may respond to suction pressure.

Usually, the refrigeration capacity is controlled by a thermostat. Depending on the type of system, the thermostat may control the temperature in the conditioned space, or it may maintain a constant (or scheduled) supply air temperature, return air temperature, supply water temperature or return water temperature.

The simplest systems use a space thermostat to start and stop a compressor. This can be done directly, by controlling a relay in the motor starter, or indirectly, by closing a liquid line solenoid (which initiates a pump-down cycle). More sophisticated controls are required when the refrigeration capacity is modulated or reduced in stages.

Two-stage thermostats can be used to cycle two compressors on and off, or to activate one stage of unloading before the compressor is cycled off. In either case, the compressor can be controlled directly or pumped down. If unloading is controlled directly, the unloaders will be activated by an electric or pneumatic signal. If unloading is controlled indirectly, a liquid line solenoid will be closed and the unloaders will be activated by a change in the suction pressure.

If more than two stages of capacity control are required, the equipment can be controlled by a step controller or by a microprocessor. Either device can be used to control the compressor or the unloaders directly, or indirectly. Microprocessors, which offer more control strategy options than step controllers, are commonly provided with new equipment.

Hot gas bypass valves are normally controlled by suction pressure. Hot gas bypass can be used as the only method of capacity control or it can be combined with unloaders or with some type of modulating control. If the compressor is equipped with unloaders, the hot gas bypass will be activated after the suction pressure has dropped below the control point that activates the last stage of unloading.

The stability of a control system is affected by the response time of a controlled variable and the response time of the control. When the supply air temperature or the supply water temperature is the controlled variable, the response time is usually very short. If direct control is used it will have a shorter response than indirect control. Historically, direct control — based on a supply temperature — was not recommended for some types of systems, a water chiller, for example. However, microprocessor controls are now being successfully used in these situations.

Appendix 6
Venting Gas and Oil Heating Equipment

A vent is required for natural gas, LP gas and oil-fired heating equipment. This vent must be designed so that all the combustion gases are expelled to the outdoors. Usually, these combustion gases must be discharged before condensation occurs in the vent system. This appendix summarizes the concepts and principles that are associated with venting gas- and oil-fired equipment.

A6-1 Terms and Definitions

There is a tendency to use the terms flue, stack and vent interchangeably, but this is not consistent with industry practice. In this appendix, these terms are used in accordance with the following definitions.

- The term "vent" applies to any piping or chimney system that is used to expel the products of combustion to the outdoors. A vent could be under positive pressure or it could be under negative pressure.
- The term "flue" applies to equipment that has a draft hood, draft diverter or barometric. The flue pipe is located between the flue outlet and the draft hood, diverter or barometric. Undiluted combustion gases flow through the flue pipe. If the equipment has a built-in draft diverter there is no external flue pipe.
- The term "stack" applies to equipment that uses dilution air (equipment that has a draft hood, draft diverter or a barometric). Diluted gases flow through the stack. Therefore, the stack is downstream from the device that is used to introduce the dilution air.
- The term "lateral" refers to the horizontal (or slightly inclined) portion of a single appliance vent system.
- The term "vent connector" refers to a pipe — which usually has a rise and an offset — that connects an appliance to a common vent.

A6-2 Factors That Effect Vent Performance

The vent design must consider the pressure inside the vent, the temperature of the vent gases, the ambient temperature surrounding the vent, the rate at which the combustion gases are generated, the conductivity of the vent walls, the thermal mass of the vent walls, the height of the vent, the length of the lateral(s), the vent gas dew point and the condensate drain requirements.

In addition, a vent must terminate at a location that insures that the venting performance will not be affected by the structural components of the building. And, the vent must terminate at a location that insures that the vented gases will not reenter the building through the fresh air intakes, or through the windows or doors. Finally, adequate clearances must be provided between the venting system and the building components (or any other materials that are stored in the building).

Pressure in the Vent

The first thing to consider is whether the vent will operate at a positive or a negative pressure. If the vent is under positive pressure, it must be air tight; or dangerous gases will be discharged into the building. The vent does not have to be air tight if it operates under a negative pressure. In fact, the potential for moisture condensation is slightly reduced if indoor air is drawn into the vent.

The pressure in the vent will depend on whether or not the combustion system is fan assisted. But, even if a fan is used, the vent may not be under positive pressure. Some fan-assisted combustion systems have powerful fans that pressurize everything that is downstream from the fan (mechanical draft); others, which only have enough fan power to move the combustion air through the burner and the heat exchanger, do not pressurize the vent.

> If a fan forces air into the combustion chamber, it is referred to as a forced draft, power burner, power combustion or pressure-fired system. If the fan exhausts the combustion chamber, it is referred to as an induced draft, induced vent or power vent system. If a fan forces outside air into a sealed combustion chamber and forces the combustion gases out of the vent, it is referred to as a direct vent system.

If fan-assisted heating equipment pressurizes the vent, the equipment manufacturer will provide proprietary vent sizing information. (In this case, the vent sizing problem is similar to a duct design problem — for a given rate of gas flow, the resistance of the vent must be matched to the external pressure that is produced by the fan.) The manufacturer also should provide information about the methods and materials that are required to install the vent. Depending on the temperature of the vent gas, the vent material could be plastic or metal. Of course, all the joints must be air tight when the vent operates under a positive pressure.

If fan-assisted heating equipment does not pressurize the vent (natural draft), or if no fan is used (atmospheric burner), the vent will operate under negative pressure and the venting power will come from the stack effect. In this case, the manufacturer is still required to provide the necessary vent sizing information, particularly if the product has a fan-assisted combustion system. However, it is possible that a manufacturer may not provide vent sizing information. In this case, the National Fuel Gas Code (NFGC), the American Gas Association (AGA), the Gas Appliance Manufacturers Association (GAMA), the Hydronics Institute and prefabricated vent and chimney manufacturers provide generic tables, which can be used to design the vent. More information about these tables is provided in Sections A6-7 and A6-8.

Some fan-assisted heating equipment may partially pressurize the vent. In this case, the manufacturer should provide specific instructions that explain how to design and install the venting system.

Vent Gas Temperature

The vent gas temperature is important. It affects the venting power of a (natural draft) vent, stack or chimney and it determines the potential for condensation to form in the vent. Venting power increases the potential for condensation decreases as the vent gas temperature increases. The vent gas temperature also affects the type of material that can be used to build or fabricate the vent (see Section A6-3).

The temperature of the vent gas that is produced by gas-fired equipment is not as hot as the vent gas that is produced by oil-fired equipment (or solid fuel equipment). However, some newly developed oil-fired condensing furnaces and condensing boilers have relatively low vent gas temperatures.

Approximate Vent-Gas Temperatures	
Gas-fired, condensing	100 to 130 °F
Gas-fired, mid-efficiency, no draft hood	275 to 350 °F
Natural gas, low efficiency, draft hood	360 °F
LP gas, low efficiency, draft hood	360 °F
Gas-fired, low efficiency, no draft hood	460 °F
Oil-fired, low efficiency	560 °F
Oil-fired, mid-efficiency	400 °F

Table A6-1

Besides the type of fuel, the vent gas temperature depends on the efficiency of the equipment, the amount of excess air and the amount of dilution air (if the equipment has a draft hood, diverter or barometric) that is mixed with the flue gas. Table A6-1 provides a few examples of the vent gas temperatures that are associated with various types of equipment, but this table should not be used to design a vent system. Always refer to the manufacturer's installation literature for information about the vent gas temperature that is associated with a particular piece of equipment.

Ambient Temperature

The ambient temperature surrounding the vent is important because it affects the venting power of a natural draft vent, stack or chimney, and because it determines the potential for condensation to form in the vent. Venting power increases and the potential for condensation increases as the ambient temperature decreases.

Generation Rate

The rate at which the vent gases are generated depends on the input capacity (BTUH) of the burner. Therefore, instead of using vent gas flow rate data (CFM or CFH), a vent, stack or chimney is sized according to the input BTUH rating that is associated with a particular piece of equipment.

Conductivity of the Vent Wall

Wall losses, due to conduction, will cause the temperature of the vent gas to decrease as it moves through the vent. This temperature drop reduces venting power of the vent and increases the possibility that water vapor (or acid, in the case of oil-fired equipment) will condense in the vent. Therefore, double-wall vents and chimneys, which are designed to reduce the wall losses, are recommended for negative vent pressure, non-condensing gas- and oil-fired equipment. (Manufactured double-wall chimneys, or masonry chimneys, are required if the vent gas temperature exceeds the temperature limits that are associated with a B- or L-vent.)

Metal, single-wall vents are not recommended because they have relatively large conduction losses. They also require more clearance between the vent wall and any combustible material, including the structural components of the building. However, if codes permit, they could be used if the designer is sure that condensation will not occur. But, this would depend on the initial temperature of the vent gas, the length of the vent and the ambient temperature.

Thermal Mass of the Vent

The thermal mass of the vent is important because it affects the temperature of the vent gas during startup. If the walls have a large mass (masonry chimney), a considerable amount of heat will be extracted from the vent gas during the beginning of each operating cycle. This could cause moisture or acidic condensation to form in the vent and it degrades venting power. Eventually, if the burner fires long enough, the vent might be heated to a suitable operating temperature and the performance of the vent would be restored. However, it is possible — even probable — that a massive vent will not be heated to a suitable operating temperature when the burner cycles to satisfy an off-peak load. (Part-load venting problems are exacerbated if the heating equipment is oversized).

The vent walls will be cold during startup, so some condensation can be expected to occur in any type of vent system right after the burner begins to fire. This transient buildup of condensation is referred to as wet time. After a few minutes, the vent heats up and this condensation usually disappears. The amount of condensation that is produced during the wet time and the power of the drying action depends on the type of equipment. Since mid-efficiency, gas- and oil-fired equipment have a longer wet time and less drying action, this equipment should not be vented into a chimney that has a lot of thermal mass or a passageway that is surrounded by cold temperatures.

Metal vents and manufactured metal chimneys have an advantage over masonry chimneys because there is only a small amount of thermal mass associated with the metal walls. If a masonry chimney is used, it should have an air space between the tile liner and the masonry work. This way, the thermal mass that is associated with the tile liner is much smaller than the thermal mass of the masonry walls.

Even if a masonry chimney has an air space between the tile liner and the chimney walls, it may not be suitable for venting some types of equipment. More specifically, condensation problems could be experienced if the temperature of the vent gas is too low. For example, moisture condensation could occur when a mid-efficiency, gas-fired, induced draft unit is vented through a masonry chimney; or, acid condensation could occur if a mid-efficiency oil-fired unit is vented through a masonry chimney. In these situations the chimney should be fitted with a double-wall liner.

Geometry of the Vent and the Lateral(s)

The height of the vent is not a factor when the vent is pressurized because the gases are forced through the vent by a fan. However, the flow resistance that is associated with the straight runs and the fittings is important. If there is too much resistance, the fan will not provide the required flow rate. Therefore, when the equipment pressurizes the vent, the equipment manufacturer must supply the information that is required to design a vent that is compatible with the fan.

The height of the vent is important if fan-assisted heating equipment does not pressurize the vent, or if there is no fan. In this case, the venting power depends on the height of the vent, stack or chimney. (The venting power increases with the height.) Most natural draft installations require a vent that is at least five feet higher than the top of the equipment (always refer to manufacturer's instructions and to codes). Vented recessed wall heaters or wall furnaces require a vent that terminates at least 12 feet above the bottom of the heater (always refer to manufacturer's instructions and to codes).

The height of the vent is a primary factor that affects venting power of a single appliance vent system. However, if the gases from two or more pieces of natural draft equipment are vented into a common vent, the height of the connector risers are as important, or even more important, than the height of the common vent. (Combined venting situations are described in section A6-6)

The venting power of unpressurized vents also depends on the flow resistance that is associated with the straight runs and the fittings. If there is too much resistance, the vent will not produce the required venting power. Tables, which may be generic (NFGC, AGA, GAMA, Hydronics Institute or prefabricated vent and chimney manufacturers) or product specific (published by the equipment manufacturer) must be used to design a negative pressure vent. More information about these tables is provided in sections A6-7 and A6-8.

The resistance to the flow of the vented gas depends on the gas flow rate, the diameter of the vent and the total equivalent length (*TEL*) that is associated with the straight runs and the fittings. Venting power of a vent, whether it is pressurized or not, can be increased by increasing the diameter or by reducing the TEL of the vent. (But, at some point, the vent size becomes too large to establish a draft.)

A lateral or a vent connector is required when a vertical vent cannot be connected directly to the fitting that is located on the heating equipment. When a lateral or vent connector is required, it increases the resistance to the gas flow and reduces the temperature of the vent gas. The flow resistance of a lateral or vent connector is accounted for by the tables that are used to size the vent system. The heat loss that is associated with the lateral or the vent connector walls can be minimized by using a double-wall pipe. (A double-wall pipe is especially important if a mid-efficiency, gas-fired, induced draft furnace is vented through a lateral or vent connector.)

In general, the horizontal run that is associated with a lateral or vent connector should be kept as short as possible, but the rise that is associated with a vent connector is important. (The equipment will not vent properly if the vent connector rise is too short, even if the common vent is very tall.)

Vent Gas Dew Point (Water Vapor)

The vent gas dew point (water vapor) depends on the type of fuel, the amount of excess air and the amount of dilution air that is mixed with the flue gas. (Dilution air enters the vent through a hood, diverter or barometric.) This temperature could range from less than 100°F, when natural gas is burned with a lot of excess and dilution air, to about 135°F; when natural gas is burned with no excess air and no dilution air. (The dew points that are associated with LP gas are lower than the dew points that are associated with natural gas.)

A low dew point temperature is desirable because it reduces the possibility of condensation. Therefore, the worst situation (limiting case) occurs when natural gas is burned with no surplus air (135°F dew point).

Vent Gas Dew Point (Acid)
With oil-fired equipment, there are two dew points — one is associated with the water vapor in the vent gas, the other is associated with the formation of acid. The acid dew point is the more important dew point, because acidic condensation occurs at a higher temperature than water vapor condensation; and because acid is more destructive than water.

The acid condensation dew point depends on the sulfur content of the oil. The acid dew point may fall between 250°F and 300°F if the oil has a high sulfur content, or it could fall between 225°F and 240°F if the oil has a low sulfur content. In either case, the acid dew point is not greatly affected by dilution air. Therefore, with oil-fired equipment, dilution air is not desirable because it lowers the stack gas temperature, but does not lower the acid dew point temperature.

Condensate Drain Requirements
Heat exchangers can be classified as non-condensing or as condensing. A condensate drain will be required if condensation occurs in the heat exchanger. In this case, the drain is connected to the heating equipment. If no condensation occurs in the heat exchanger, a condensate drain may still be required, depending on the situation (see Table A6-2). In this case, the drain is connected to a trap that is built into the vent system. Refer to manufacturer's installation instructions for information about condensate drain requirements.

A6-3 Vent and Chimney Operating Temperatures

Plastic, single-wall vents can be used when a gas-tight, condensation resistant vent is required. PVC (Polyvinyl Chloride) can be used if the vent gas temperature is below 140°F, CPVC (Chlorinated Polyvinyl Chloride) can be used if the vent gas temperature is below 180°F and PEI (Polyetherimid) can be used when the vent gas temperature is as high as 400°F. (Some venting system and equipment manufacturers offer proprietary vents that can withstand temperatures that approach 500 °F.) *Always use the vent pipe material that is recommended by the heating equipment manufacturer.*

Metal vents are used when the vent gas temperature is in the 300°F to 500°F range. In this case, the vent could be a single-wall pipe or a double-wall pipe. Double-wall vents consist of two concentric metal pipes, which are separated by an air space. Type-B vents, which have a steel outer wall and an aluminum inner wall, are only used to vent gas-fired equipment. Type-L vents, which have a steel outer wall and a stainless steel inner wall, are designed for oil-fired equipment, but they also can be used for gas-fired equipment.

Manufactured, metal chimneys are designed to vent medium to high temperature gases (500 °F to more than 2000 °F). These chimneys also have a double wall, but in this case, an insulating material is sandwiched between the walls. These chimneys can be used to vent oil-fired equipment and some codes may require a chimney for large gas-fired equipment. (They could be used in place of a B-vent or an L-vent, but the extra expense would not be justified.)

Masonry chimneys can withstand high vent gas temperatures, but they can cause problems when the vent gas temperature is moderate (say less than 300°F), because they are more likely to cause the vent gas to condense. Masonry chimneys also should have a double-wall construction, with an insulating air space between the masonry walls and the tile liner.

Flue Materials

	Temperature	Application
Various Types of Gas-Tight, Plastic and Metal Venting Materials	100 to 480 °F	PVC, CPVC, PEI and proprietary products used to vent gas-fired equipment (1)
Single-Wall Metal Pipe	Refer to codes and to manufacturer's recommendations	Aluminum, galvanized steel or stainless steel; depending on the application (2)
B- & BW-Vent	To 480 °F	Gas equipment
L-Vent	To 500 °F	Oil equipment (2)
Factory Built Chimney	500 to 2200 °F	Oil or gas equipment (2)
Masonry Chimney (tile liner and air space)	360 to 1800 °F	Gas or oil equipment (double-wall liner may be required for mid-efficiency equipment) (3)

1) Some high temperature plastic materials could be used to vent mid-efficiency (80 to 85 percent) oil-fired equipment if acidic condensation is expected or unavoidable.

2) Stainless steel vents can resist the heat that is associated with oil-fired vent-gases, but when the inner wall temperature exceeds 250°F, it loses it's ability to resist acidic damage.

3) Masonry chimneys also are susceptible to acidic damage.

A6-4 Venting Requirements — Gas-Fired Equipment

Four venting categories have been assigned to gas-fired equipment — which could be a furnace or a boiler. The two factors that determine what category is assigned to a particular piece of equipment are the vent pressure and the efficiency of the equipment. Another important consideration is whether the combustion system is designed to operate wet or dry. This information and the vent system performance requirements that are associated with each category are summarized by Table A6-2.

As indicated by Table A6-2, there are some venting problems associated with induced draft, mid-efficiency equipment that qualifies for Category 1 and some forced draft, mid-efficiency equipment that qualifies for Category III — based on the efficiency rating. (This equipment is characterized by a marginal difference between the vent gas temperature and the condensing temperature of the vent gas.) This type of mid-efficiency equipment is likely to produce condensation in the venting systems that have traditionally been used for higher temperature Category I and III equipment. (It seems that two more categories, say I-A and III-A, could be created to cover

Gas-Fired Equipment Venting Categories

		Category I	Category II	Category III	Category IV
Operating Charac teristics (1992)	Pressure in the vent	Negative	Negative	Positive	Positive
	Annual efficiency	Below 83 Percent	Above 83 Percent	Below 83 Percent	Above 83 Percent
	Condensation	Not acceptable (1)	Possible (in vent)	Possible (3)	In heat exchanger
Design Require- ments	Gas-tight vent	No	No	Yes	Yes
	Corrosion resistant vent	No (1)	Yes	Possible (3)	Yes
	Vent into masonry chimney	Permitted (1 & 2)	No	No	No
	Combined venting	Permitted	No	No	No
	Condensate drain	Not required (1)	Ask Manufacturer	Possible (3)	At equipment
	Source of information regarding design and Installation	Fuel gas code, heating equipment and vent system manufacturers	Manufacturer's literature	Manufacturer's literature	Manufacturer's literature
Old Stan- dard (4)	Vent-gas dew point temp.	More than 140 °F	Less than 140 °F		
	Temperature of vent-gas (5)	Above 275 °F	Below 275 °F		

Note 1 — Usually, there is no problem when high-vent-gas-temperature equipment is vented into a double-wall vent or into a lined masonry chimney; but condensation could occur if mid-efficency, (78 - 83 %) mechanical draft equipment is vented into a vent that has highly conductive walls, cold walls or massive walls. In this case, design a vent system that minimizes the wall losses (use double-wall pipe for the whole run and avoid long runs through cold spaces).

Note 2 — Install a B-Vent inside of the masonry chimney and use a double-wall connector when venting mid-efficency, (78 - 83 %) mechanical draft equipment.

Note 3 — Condensation in the vent is possible with some types mid-efficiency (78 - 83 %), direct-vent equipment — depending on the ambient temperature and the conductivity of the vent walls. In this case, design a vent system that minimizes the wall losses (use insulated pipe and avoid long runs through cold spaces). A corrosion resistant flue and a drain may be required if condensation cannot be prevented (refer to the manufacturer's recommendations).

Note 4 — This criteria is obsolete. Currently (1992), the criterion that is used to categorize the equipment is based on whether its efficiency rating is more or less than 83 %.

Note 5 — The dew point of the vent gas depends on the fuel (natural or LP gas), the amount of excess air and the amount of dilution air. The limiting case occurs when the dew point of the vent gas is at a maximum, which is about 135 °F. This maximum is produced when natural gas is burned with no excess air or dilution air. Therefore 275 °F = 135 °F dew point + 140 °F rise.

Table A6-2

some types of condensation prone, mid-efficency, gas-fired equipment.)

Also note that the four gas equipment venting categories are usually associated with smaller, residential type equipment (225,000 BTUH or less). This is because a wide variety of high efficiency equipment is available in this capacity range. Most of the (larger) commercial equipment falls into Category I, which is covered by the National Fuel Gas Code. (Always refer to the manufacturer's instructions when installing Category II, III and IV equipment, regardless of the heating capacity.)

A6-5 Venting Requirements — Oil-Fired Equipment

Depending on the burner design, oil-fired furnaces and boilers may require a natural draft vent or a vent that operates under positive pressure. If a natural draft vent is required, the manufacturer's installation instructions will provide a recommended connector size, vent size and vent height. If a positive pressure vent is required, the manufacturer's installation instructions will provide the necessary design and installation information. (There also is some natural draft and mechanical draft vent sizing information published in the sixth edition of the Hydronics Institute's boiler test and rating standard.)

The four venting categories that are assigned to gas-fired equipment do not apply to oil-fired equipment. (Oil-fired equipment may be categorized in the near future.) But, as noted above, oil-fired equipment could be classified by the pressure in the vent, which could be negative or positive.

Low efficiency oil-fired equipment produces relatively hot vent gas temperatures — around 500°F — so the water vapor condensation problems that are associated with gas-fired equipment do not (usually) apply to this type of oil-fired equipment. However, if the vent wall temperature falls below the acid dew point, the vent will still be vulnerable to acid damage.

Mid-efficency, oil-fired equipment (80 to 85 percent) produces cooler vent gas temperatures — around 400°F — so the water vapor condensation problems that are associated with gas-fired equipment could apply to this type of oil-fired equipment. And, with this type of equipment, the possibility of vent damage — due to acidic condensation — is much more likely to occur.

Natural draft vents might be built out of single-wall steel pipe, Type-L vent pipe, a masonry material — or they could be a steel, factory-built chimney (refer to the equipment manufacturer's instructions, local and national codes). Positive pressure vents, which must be gas-tight, should be built in accordance with manufacturer's installation instructions and local codes. In either case, the vent must be designed to prevent acidic condensation or it must be made of materials that will not be damaged by acid.

Acidic condensation may be prevented by using well insulated, double-wall chimneys. (L-vents, which only have an air space between the walls, may not be adequate.) Also, try to avoid outdoor installations and long runs through cold spaces. If the vent has to be routed through a cold space, it should have enough insulation to prevent acid condensation. If acid condensation is possible, the vent can be built out of an acid resistant material. (If the vent gas temperature is below 480°F a plastic vent material could be used, subject to UL approval.)

A few manufacturers offer high efficiency (90 percent or more) oil-fired equipment, which produces even lower vent gas temperatures. In this case, the equipment should be vented according to the manufacturer's installation instructions. (Plastic pipe is suitable for venting this type of equipment.)

A6-6 Combined Venting

A gas water heater and Category I gas-fired heating equipment can be vented into a common vent. In fact, when mid-efficiency heating equipment is used, combined venting is quite desirable. With this arrangement, the water heater will make the vent warmer and it will provide some dilution air. (Dilution air lowers the dew point of the vent gas). Therefore, when a water heater is vented with a mid-efficency heating unit, the potential for condensation problems is reduced.

It also is acceptable to vent multiple Category I gas-fired heating units into the same vent. If all the units are in the same room, they can be vented through a manifold. If the units are located on different floors, they can be vented into a common vent, stack or chimney. (In multistory venting situations, the various pieces of equipment shall be installed in rooms that are separate from the occupied areas. And, these equipment rooms shall be supplied with a source of outdoor air.) Information about designing these types of vent systems is published in the **National Fuel Gas Code** and in booklets that are provided by vent and chimney system manufacturers.

The vent that is used with Category II, III and IV gas-fired equipment should not be used by any other equipment (no combined venting). However, it may be permissible to route this type of vent through a chimney that is used for another appliance, but it must be so noted by the equipment manufacturer and permitted by the applicable codes.

Two or more oil-fired units can be vented into the same stack or chimney. Refer to manufacturer's installation instructions for the information that is required to design manifolds or breeching.

A6-7 Vent Sizing — Gas Equipment

Single appliance vent sizes depend on the material and construction techniques that are used to build the vent, the BTUH capacity of the equipment, the height of the vent and the length of the lateral. Multiple appliance vent sizes (including manifold sizes and multistory vent sizes) depend on the type of material and construction techniques that are used to build the vent, the BTUH capacity of the equipment, the height of the vent and the height of the vent connector rise.

Tables that can be used to size vents for Category I gas-fired (atmospheric or fan-assisted natural draft) equipment can be found in the **National Fuel Gas Code**, the American Gas Association (AGA) venting tables, the Gas Appliance Manufacturers Association (GAMA) venting tables; in booklets published by factory-built, vent (and chimney) system manufacturers; and (possibly) in the installation instructions that are published by the heating equipment manufacturer. These sources contain one set of tables that can be used to size single appliance vents and a second set of tables that can be used to size vents that serve two or more appliances (combined venting). Information about manifold venting and multistory venting also may be included with the basic vent sizing tables.

Category I Gas Vent Sizing (Atmospheric and Fan-Assisted)

	NFGC AGA GAMA Tables	Vent System Manufacturers	ASHRAE Handbook
Single-Wall Metal Vent	SA CV		Theoretical equations and charts that can be used to design masonry or metal chimneys, stacks and vents.
Double-Wall Metal Vent	SA CV	SA CV	
Masonry Chimney	SA CV		
Factory-Built Chimney		Yes	
Manifold Venting		Yes	
Multistory Vent Design	Yes	Yes	
Combustion Air Supply	Yes	Yes	Yes
Materials and Clearances	Yes	Clearances	Yes

1) SA = Single appliance; CV = Combined venting
2) Do not use for Category II, III or IV
3) Do not use these tables if vent is under positive pressure
4) These vent sizing tables can be used to find the maximum and the minimum size. Do not oversize vents, (especially when venting mid-efficiency equipment, which typically has lower vent-gas temperatures and lower vent-gas flow rates than older, less efficent equipment.

Information about sizing vents for Category II, III and IV gas-fired equipment is proprietary. Refer to the manufacturer's installation instructions for this information.

The four, gas-fired, equipment venting categories are usually associated with residential size equipment. However, the same principles apply to all sizes of gas-fired equipment, even if the category is not included on the manufacturer's nameplate label. In this case, the generic tables can be used to size vents for natural draft equipment; but proprietary tables, which should be supplied by the manufacturer, should be used to size positive pressure vents.

A6-8 Vent Sizing — Oil-Fired Equipment

Oil-fired equipment may operate with negative or positive pressure in the vent. Information about sizing vents for oil-fired equipment can usually be found in the installation instructions that are published by the heating equipment manufacturer. Also refer to the **June 1989 Test and Rating Standard for Heating Boilers, Sixth Edition**, which is published by the Hydronics Institute. And, for equipment that requires a natural draft, sizing and design information also can be found in the manuals that are published by L-vent and factory-built, chimney system manufacturers. (There is an obvious need for more information about venting oil-fired equipment. Currently, useful information is being generated by researchers at the Brookheaven National Laboratory. Some of this information may be published in 1993.)

A6-9 Methods, Materials and Clearances

Generic information about installation methods, materials and clearances can be found in the **National Fuel Gas Code**, the **Uniform Mechanical Code**, in Handbooks published by the American Gas Association and the Gas Appliance Manufacturers Association, the **ASHRAE Equipment Handbook** and in vent-system manufacturers' manuals. Product specific information is provided in the installation instructions that are published by the heating equipment manufacturer.

A6-10 Combustion Air

An adequate source of combustion air is required. If indoor air is used as the source of combustion air, building codes usually specify the combustion air opening requirements. If there is no codified guidance, refer to the sources that are mentioned in the preceding paragraph.

A6-11 Draft Pressure Controls

Some types of heating equipment require a device that can be used to control the draft pressure. Refer to the manufacturer's installation instructions for information about this subject.

A6-12 Codes and Regulations

Local codes, regulations and insurance underwriters' specifications establish the vent system design and installation requirements. These documents may simply defer to national codes and standards, or they may adopt the national codes with some revisions, or they may be proprietary codes that were developed by the local authority. In any case, the vent system should be installed in accordance with the local codes. If there is a conflict between the local code and the manufacturer's installation instructions, it should be brought to the attention of the local authority.

Index

C (Continued)

C (Continued)

D

E

E (Continued)

F

F (Continued)

G

H

H (Continued)

I

L

M

M (Continued)

N

O

P

P (Continued)

R

S

S (Continued)

T

U

U (Continued)

W

Z

Notes

Notes

Notes

Notes